IUTAM Symposium on Nonlinear Stochastic Dynamics

SOLID MECHANICS AND ITS APPLICATIONS
Volume 110

Series Editor: G.M.L. GLADWELL
Department of Civil Engineering
University of Waterloo
Waterloo, Ontario, Canada N2L 3GI

Aims and Scope of the Series

The fundamental questions arising in mechanics are: *Why?*, *How?*, and *How much?* The aim of this series is to provide lucid accounts written by authoritative researchers giving vision and insight in answering these questions on the subject of mechanics as it relates to solids.

The scope of the series covers the entire spectrum of solid mechanics. Thus it includes the foundation of mechanics; variational formulations; computational mechanics; statics, kinematics and dynamics of rigid and elastic bodies: vibrations of solids and structures; dynamical systems and chaos; the theories of elasticity, plasticity and viscoelasticity; composite materials; rods, beams, shells and membranes; structural control and stability; soils, rocks and geomechanics; fracture; tribology; experimental mechanics; biomechanics and machine design.

The median level of presentation is the first year graduate student. Some texts are monographs defining the current state of the field; others are accessible to final year undergraduates; but essentially the emphasis is on readability and clarity.

IUTAM Symposium on

Nonlinear Stochastic Dynamics

Proceedings of the IUTAM Symposium held in Monticello, Illinois, U.S.A., 26–30 August 2002

Edited by

N. SRI NAMACHCHIVAYA

University of Illinois at Urbana-Champaign, U.S.A.

and

Y.K. LIN

Florida Atlantic University, U.S.A.

SPRINGER SCIENCE+BUSINESS MEDIA, B.V.

A C.I.P. Catalogue record for this book is available from the Library of Congress.

ISBN 978-94-010-3985-7 ISBN 978-94-010-0179-3 (eBook)
DOI 10.1007/978-94-010-0179-3

Printed on acid-free paper

Contents

Preface

Advances in nonlinear stochastic dynamics have been rapid in the past decade, with applications to almost all fields of engineering. The overall goal of these investigations has been to formulate and develop methods of analysis for complex interactions between noise and inherent nonlinearities in real mechanical and structural systems, and to suppress undesirable vibrations which can lead to failure. It has brought about an enormous increase in the number of scientists working in this area, along with the discovery of many important new phenomena.

Progress in theoretical research (stochastic flow, Lyapunov exponent, multiplicative ergodic theory, stochastic bifurcation theory, large deviations, chaos theory, etc.) has also led to the development of reliable numerical algorithms for computing standard statistical measures of response and stability in random dynamical systems, system identification and stochastic control, and to a better understanding of nonlinear phenomena. Practical applications of these new fundamental results are beginning to appear across the entire spectrum of mechanics; for example, stability of aircraft at high angles of attack, wave propagation in random media, mixing and transport phenomenon in fluid mechanics, and queuing networks. Some of the results are even contrary to intuition, such as stabilization by noise and stochastic resonance.

The range of subjects in which nonlinearity and noise play a significant role is enormous, too large for a symposium of modest size to cover adequately at any one time. For the purpose of promoting common interest in the emerging areas of nonlinear and stochastic dynamics, the Symposium brought together forty-six (46) speakers from the engineering, physics and applied mathematics communities, from Canada, China, Czech Republic, France, Germany, Italy, Norway, Russia, Slovenia, South Africa, United Kingdom, and United States. The five *lead lectures* were given by Professor S. T. Ariaratnam (Waterloo, Canada), Professor Ludwig Arnold (Breman, Germany), Professor P. R. Kumar (Urbana, USA), Professor Walter Wedig (Karlsruhe, Germany), and Professor Kurt Wiesenfeld (Atlanta, USA). In addition to the invited speakers, there were thirty (30) participants mainly from the United States and

Canada, from the areas of Engineering, Mechanics, Physics and Applied Mathematics. The Symposium provided a forum for the discussion of developments in nonlinear and stochastic dynamics and reviewed both fundamental theory and its applications in structural and mechanical engineering, physics, signal-processing, and other areas of science and technology. Subject areas included, but were not limited to:

1 Multi-scale phenomena in stochastic dynamics

2 Stability and bifurcations of stochastic systems

3 Control estimation of nonlinear stochastic systems

4 Modeling and random vibrations

5 Computational methods in stochastic and chaotic dynamics.

For the engineering and physics community, the Symposium provided the opportunity to obtain first-hand information on recent developments in related mathematical areas, and to point out directions for further research. The applied mathematics community benefited primarily from the opportunity to obtain information on various possible applications.

The Symposium was held at the Robert Allerton Park and Conference Center, a secluded retreat surrounded by a 1,500 acre park located just 26 miles from the University of Illinois at Urbana-Champaign. The park offered individuals a unique opportunity to experience nature at its best and provided the perfect setting for this Conference. The Symposium was co-chaired by Professor N. Sri Namachchivaya of University of Illinois at Urbana-Champaign, and Professor Y. K. Lin of Florida Atlantic University. The scientific (organizing) committee consisted of Professors S. T. Ariaratnam (*Canada*), Pierre Bernard (*France*), R. Narayana Iyengar (*India*), Peter V. E. McClintock (*United Kingdom*), Arvid Naess (*Norway*), and Werner Schiehlen (*Germany*). The opening remarks were given by Professor Sri Namachchivaya, Provost Richard Herman (University of Illinois at Urbana-Champaign) and Professor Schiehlen (Vice President, IUTAM). The inaugural address was given by Professor Lin, in which he recounted the past history of IUTAM symposia on stochastic and nonlinear systems. During the Symposium, we also took the opportunity to recognize the contributions of three world class scientists.

Professor Ludwig Arnold is one of the pioneers in the field of random dynamical systems, and has been conducting research at the forefront in the areas of stochastic dynamics, stochastic stability, multiplicative ergodic theory, stochastic bifurcations, and stochastic flows since the 1970s. He has been a mentor for many participants in the Symposium. We celebrated his 65th birthday during the symposium banquet.

Professor Harry H. Hilton whose contribution to visco-elasticity are considered to be landmarks in the evolution of the field. His professional career at Illinois spans over 50 years, including an 11-year appointment as Head of the Aeronautical and Astronautical Engineering Department from 1974-1985. During his tenure, he played an influential role in attracting many of the great minds who have contributed to this field. We celebrated his 75th birthday during the symposium banquet.

Professor Werner Schiehlen is a world authority on multi-body-dynamics and one of the most influential mechanists in Europe. His works in dynamical systems have served as cornerstones for many recent developments in multi-body-dynamics. He has served in the General Assembly and the Bureau of IUTAM from 1978 to the present in various capacities. We recognized Professor Werner Schiehlen's contribution to IUTAM during the symposium reception.

The Symposium was made possible through the support of the IUTAM and the National Science Foundation under Grant No. 0129587. The encouragement and support of Professor Dick H. van Campen, Secretary-General of IUTAM and Dr. Alison Flatau (Dynamic Systems and Control Program) and Dr. Henry A. Warchall (Applied Mathematics Program) of the National Science Foundation are greatly appreciated. Any opinions, findings, and conclusions or recommendations expressed in the proceedings are those of the author(s) and do not necessarily reflect the views of the National Science Foundation. We like to thank the Symposium Program Coordinators, Ms. Diane Jeffers and Ms. Linda Conway, for their hard work, which went far beyond the call of duty. We also like to thank Ms. Frances Bridges for her efficient work on the preparation of the proceedings. Finally it was a pleasure to work with the professional staff of Kluwer Academic Publishers, especially Dr. Arno Schouwenburg and Ms. Nathalie Jacobs.

N. Sri Namachchivaya, Urbana, Illinois, USA
Y. K. Lin, Boca Raton, Florida, USA
March 31, 2003

Figure 1. Group picture of some of the participants.

Figure 2. 1. E. Dowell, 2. W. Schiehlen, 3. Y.K. Lin, 4. H. Hilton, 5. N. Sri Namachchivaya, 6. P.V.E. McClintock, 7. L. Arnold, 8. S. Ariaratnam, 9. W. Wedig, 10. T. Whalen, 11. B. Balachandran, 12. S. Masri, 13. A. Kovaleva, 14. R. Khasminskii, 15. P. Baxendale, 16. P. Bernard, 17. J. Naprstek, 18. R. Ibrahim, 19. H. Davies, 20. G. Cai, 21. W. Zhu, 22. P. Imkeller, 23. M. Freidlin, 24. A. Ams, 25. L. Schimansky-Geier, 26. Z. Yang, 27. B. Eckhardt, 28. W. Langford 29. C. Bucher, 30. F. Colonius, 31. M. Vasta, 32. M. Dimentberg 33. W.C. Xie, 34. C. Schiehlen, 35. A. Stefanovska, 36. Y. Wihstutz, 37. L. Vedula, 38. A. Mallik, 39. C. Pahlajani, 40. H. Schurz, 41. E. Brown, 42. R. Song, 43. R. Sowers, 44. H. Van Roessel, 45. G. Lythe, 46. I. Poloskov, 47. J. Acebrón, 48. M. Evstigneev, 49. M. Fofana, 50. F. Ma, 51. A. Bulsara, 52. A. Naess, 53. M. Grigoriu, 54. J. Park, 55. D. Kok, 56. L. Bergman, 57. D. Jeffers, 58. M. Doyle, 59. A. Langford.

Figure 3. Ludwig Arnold (left) and Harry H. Hilton (right) cutting their Birthday Cakes as Rafail Khasminskii looks on.

Figure 4. Werner Schiehlen at the reception given in his honor.

I

PART 1: MULTI-SCALE PHENOMENA

Part-I: MULTI-SCALE PHENOMENA

Nonlinear stochastic systems are at the center of many engineering disciplines, and the multi-scale phenomena that arise in their study are the focus of the *first* part of the Proceedings. The first *Lead Lecture* of the Symposium given by Ludwig Arnold of the University of Bremen, Germany, is the lead paper of this part. The rest of the papers presented during the Symposium, on the general topic of *multi-scale phenomena in stochastic dynamics* are compiled alphabetically, based on the last name of the first author.

Stochastic averaging has become an important technique to replace, often in some limiting regime, a complicated system by a simpler, or lower-dimensional one. Stochastic averaging is at the heart a problem of weak convergence of stochastic processes, or more exactly, weak convergence of the laws of Markov processes. The underpinning of classical averaging is a separation of time scales; there is a coordinate which varies slowly and a coordinate which varies quickly. As the difference in the speeds of the two coordinates increases, it becomes possible to fix , in a sense, the slowly varying coordinate, and carry out a long-term average in the quickly varying component. The following three papers dealt with the theoretical side of stochastic averaging.

For a dynamical system with slow and fast variables, Ludwig Arnold presented an approximation of the slow motion by the solution of a nonlinear stochastic differential equation (in meteorology known as Hasselmann's equation) which was recently proved by Yuri Kifer. He presented several examples which support the usefulness of the nonlinear diffusion approximation, including the Lorenz-Maas model from climatology. Both Pierre Bernard and Richard Sowers used the martingale problem, one of the modern frameworks for considering convergence of the laws of Markov processes. Pierre Bernard and Richard Sowers identified the generators of the limiting laws of the reduced Markov processes, for single-well (Khasminskii's result) and multi-well (Freidlin and Wentzell's result) potentials, respectively.

Stochastic Resonance is a phenomenon in which a periodic signal in a nonlinear system is enhanced by adding external noise of optimal intensity to the input. Over the past decade, stochastic resonances or associated phenomena have been observed in a wide variety of contexts, including chemical reactions, semiconductor devices, magneto-elastic strip and sensory neurons. Thus, the Symposium included papers and discussions on stochastic resonance and related topics which include large deviations and metastability.

Mark Freidlin presented results on the action functional when the Wiener process is replaced by the random walk. They are useful in the solutions of various problems (exit problem, wavefront propagation, and stochastic resonance) related to the large deviations for the Wiener process. Peter Imkeller gave a first glimpse of the stochastic resonance phenomena in a rigorous setting with two-state Markov chain with discrete time. He also pointed out that the reduced dynamics were robust, only if small scale fluctuations were carefully eliminated inside the potential wells. Sean Meyn developed model reduction based on the assumption of V-uniform ergodicity and gave a novel definition of metastability via exit rates, which will be helpful in the identification of metastable sets even for large scale systems. Peter McClintock presented theoretical and experimental results on noise-induced fluctuations in non-linear systems for finite noise intensity. The non-equilibrium distribution for the whole phase space was determined numerically for different noise intensities. Modeling issues related to stochastic resonance can also be found in Part-IV of this proceedings.

LINEAR AND NONLINEAR DIFFUSION APPROXIMATION OF THE SLOW MOTION IN SYSTEMS WITH TWO TIME SCALES

Ludwig Arnold
Institut für Dynamische Systeme
Fachbereich 3
Universität Bremen, Postfach 33 04 40
D-28334 Bremen, Germany
arnold@math.uni-bremen.de
http://www.math.uni-bremen.de/~arnold

Abstract For a dynamical system with slow and fast variables the popular method of averaging enables one to derive an equation for the slow variables alone whose solution approximates the original slow motion on a finite time interval.

If the fast variables are sufficiently "random" the error term in the averaging procedure is described by a central limit theorem, i.e., the scaled error is Gaussian in the weak limit and satisfies a linear SDE (linear diffusion, or Gaussian approximation).

We will present an approximation of the slow motion by the solution of a *nonlinear* SDE (in meteorology known as *Hasselmann's equation*) which was recently proved by Yuri Kifer. Although this nonlinear diffusion approximation holds in general, as the previous one's, only on a finite time interval, it is at least in principle capable of correctly describing important long-term qualitative features of the slow motion.

We present several examples which support the usefulness of the nonlinear diffusion approximation, including the Lorenz-Maas model from climatology.

Keywords: Averaging principle, multiscale system, central limit theorem, Gaussian approximation, Hasselmann's equation, linear and nonlinear diffusion approximation, stochastic climate models

1. Introduction

There has been a recent burst of activity in the area of multiscale systems aiming at reducing the original system to one for the slow variables alone by replacing the fast degrees of freedom by noise.

N. Sri Namachchivaya and Y.K. Lin (eds.),
IUTAM Symposium on Nonlinear Stochastic Dynamics, 5-18.

This is of course part of a now classical program. It started with the method of averaging centuries ago, although the rigorous mathematical foundation was provided only some 40 years ago by Bogolyubov and Mitropolskii [5]. It continued with Khasminskii's 1966 paper [17] in which he considered the case of a random differential equation where the fast motion is a stochastic process. He proved not only the averaging principle, but also that the scaled error in the averaging principle is Gaussian in the weak limit and satisfies a linear stochastic differential equation (SDE), provided that the fast stochastic process is sufficiently mixing.

Khasminskii [16] also considered the case where the averaged vector field vanishes. In this case one needs time of order $1/\varepsilon^2$ to observe non-trivial effects. More precisely, Khasminskii obtains a nonlinear SDE as the weak limit of the slow motion.

Khasminskii's 1968 paper [18] on the averaging principle for SDE has been of extreme importance in the physics and engineering literature up to date, with hundreds (maybe even more) of applications. This was further developed by Papanicolaou, Stroock and Varadhan [30] using the powerful martingale method.

Freidlin in his monumental 1978 paper [8] developed the theory of large deviations in averaging by studying the phenomena observable only on an exponential time scale $e^{c/\varepsilon}$ (like exit of the slow variable from the neighborhood of a stable steady state of the averaged equation, invariant measures, hopping between local attractors etc.). An account of Khasminskii's and Freidlin's theories is given in Chapter 7 of Freidlin and Wentzell's book [9].

On the physics side Hasselmann in his visionary 1976 paper [10] proposed to improve deterministic climate models by incorporating the influence of atmospheric weather in form of random noise. We have recast *Hasselmann's program* in modern mathematical language in [2].

While Hasselmann's paper mainly deals with the method of averaging and the corresponding central limit theorem, he also proposed to approximate climate by a *nonlinear* SDE in order to model long-term phenomena. This SDE has defied rigorous mathematical analysis until the recent work by Kifer [23].

The papers by Just, Kantz, Rödenbeck and Helm [14] and Kantz, Just and Baba [15] formally derive Hasselmann's equation through perturbation and averaging techniques on the level of the Fokker-Planck equation.

The nonlinear diffusion approximation proved to be of superior quality in a recent extensive numerical case study of a coupled ocean-atmosphere model by Arnold, Imkeller and Wu [3]. (See Section 6).

In climatology, the interest in reducing multiscale models to stochastic equations is presently particularly vivid. See the work of Majda, Timofeyev and Vanden-Eijnden [28, 29] about general strategies for replacing the unresolved degrees of freedom in atmosphere-ocean models by noise and for the treatment of several rather complex illustrative examples. See also the Proceedings volume on *Stochastic Climate Models* [13] edited by P. Imkeller and J.-S. von Storch, and the Special Issue of the journal *Stochastics and Dynamics*, also entitled *Stochastic Climate Models*, edited by P. Imkeller and A. Monahan [12].

As we will almost exclusively (except in section 6) deal with the simpler case where the fast motion is a stationary stochastic process being sufficiently mixing, we refer to the work of Kifer [19, 20, 21, 22] for the case where the fast motion is a sufficiently chaotic dynamical system.

2. Averaging

Let

$$\dot{z} = h(z)$$

be an ODE in some Euclidean space. In many applications we can naturally separate the variables as

$$z = (x, y)$$

with strongly differing "response times" $\tau_y \approx \varepsilon \ll \tau_x \approx 1$, so that y is the vector of *fast* variables, while x is the vector of *slow* variables. Introducing the small scaling parameter ε into the equation $\dot{z} = h(z)$ yields the equivalent coupled system of the two ODE

$$\dot{x}_t^\varepsilon = f(x_t^\varepsilon, y_t^\varepsilon), \quad x_0^\varepsilon = x \in \mathbb{R}^d, \tag{1}$$

$$\dot{y}_t^\varepsilon = \frac{1}{\varepsilon} g(x_t^\varepsilon, y_t^\varepsilon), \quad y_0^\varepsilon = y \in \mathbb{R}^m. \tag{2}$$

The general problem of model reduction consists of approximating the slow part $x_t^\varepsilon(x, y)$ of the solution of (1,2) as accurately as possible (and on a time interval as long as possible) *not* by solving (1,2) explicitly, but by using only some long-term "statistical" properties of $y_t^\varepsilon(x, y)$ in order to obtain an equation for the slow variables alone.

This program has been only partly implemented. For the sake of focussing presentation we restrict ourselves to the case where y is not coupled to x, in fact, to the case

$$\dot{x}_t^\varepsilon = f(x_t^\varepsilon, \xi_{t/\varepsilon}), \quad x_0^\varepsilon = x \in \mathbb{R}^d, \quad t \in [0, T]. \tag{3}$$

Here ξ_t is an ergodic stationary stochastic process on some probability space $(\Omega, \mathbb{F}, \mathbb{P})$ with values in $\mathbb{R}^m$, cadlag (i.e., right-continuous with left-hand limits) trajectories and invariant probability distribution $\mathbb{P}\{\xi_t \in \cdot\} = \rho(\cdot)$. The vector field $f : \mathbb{R}^d \times \mathbb{R}^m \to \mathbb{R}^d$ is assumed to be continuous in (x, ξ), with

$f(\cdot,\xi)$ globally Lipschitz-continuous uniformly with respect to ξ, and $f(0,\cdot)$ bounded (instead of the uniform bounds it suffices for Theorem 1 that the Lipschitz constant $L(\xi)$ of $f(\cdot,\xi)$ and $f(0,\xi)$ are in $L^2(\rho)$).

We can average out the fast process $\xi_{t/\varepsilon}$ from (3) keeping the slow variable x "frozen" and arrive at the *averaged vector field*

$$\bar{f}(x) := \mathbb{E}\, f(x,\xi_0) = \int_{\mathbb{R}^m} f(x,\xi)\rho(d\xi) = \lim_{T\to\infty} \frac{1}{T}\int_0^T f(x,\xi_t)\,dt \quad \mathbb{P}\text{-a. s.}\,, \tag{4}$$

which is again globally Lipschitz continuous and of at most linear growth in x.

Hence the ODE (3) as well as the *averaged equation*

$$\dot{\bar{x}}_t = \bar{f}(\bar{x}_t), \qquad \bar{x}_0 = x \in \mathbb{R}^d, \tag{5}$$

both have global solutions $x_t^\varepsilon(x)$ resp., $\bar{x}_t(x)$ for any initial value $x \in \mathbb{R}^d$ and $t \in [0,T]$ which we want to compare.

Theorem 1 (Averaging Principle) *Let the above assumptions on f and ξ_t be satisfied, and let $x_t^\varepsilon(x)$ be the solution of (3) and $\bar{x}_t(x)$ the solution of (5). Assume that the $d \times d$ covariance (or correlation) matrix*

$$R_x(t) := \operatorname{cov}(f(x,\xi_t), f(x,\xi_0)) = \mathbb{E}(f(x,\xi_t) - \bar{f}(x))(f(x,\xi_0) - \bar{f}(x))' \tag{6}$$

satisfies

$$\lim_{t\to\infty} R_x(t) = 0 \quad \textit{for all } x \in \mathbb{R}^d. \tag{7}$$

Then for each $T > 0$ there exists a constant C_T independent of x and ε such that for all $x \in \mathbb{R}^d$ and $\varepsilon > 0$

$$\mathbb{E} \max_{0\le t\le T} \|x_t^\varepsilon(x) - \bar{x}_t(x)\|^2 \le C_T\, \varepsilon. \tag{8}$$

The fact that the decay of correlation to zero expressed by (7) is sufficient for the statement (8) is discussed by Freidlin and Wentzell [9, p. 218] and can also be easily deduced from Khasminskii's original paper [17].

Our program in the remainder of this paper is to replace $\bar{x}_t(x)$ in (8) by a "better" approximation of $x_t^\varepsilon(x)$ in the sense that

– the mean square error is smaller than in (8),

– the approximation is capable of catching long-term features of $x_t^\varepsilon(x)$ which go beyond a fixed time interval.

3. Linear Diffusion Approximation

We now look at the error $x_t^\varepsilon - \bar{x}_t$ made in the Averaging Principle. If ξ_t is a periodic function, this error can be asymptotically expanded in powers of ε (see Sanders and Verhulst [33] for a precise statement).

If, however, ξ_t is a sufficiently mixing stochastic process, the error is of a completely different character: It has order $\sqrt{\varepsilon}$, and no further term of the asymptotic expansion can be written down.

Let ξ_t be a stationary stochastic process with values in $\mathbb{R}^m$ with cadlag trajectories. We define the sub-σ-algebras of $\mathbb{F}$ generated by ξ_t as

$$\mathbb{F}_s^t := \sigma(\xi_u : s \le u \le t), \quad -\infty \le s \le t \le \infty,$$

and the *mixing coefficient* of ξ_t by

$$\alpha(\tau) := \sup_{A \in \mathbb{F}_{-\infty}^0, B \in \mathbb{F}_\tau^\infty} |\mathbb{P}(A)\mathbb{P}(B) - \mathbb{P}(A \cap B)|, \quad \tau \in \mathbb{R}^+.$$

The stochastic process ξ_t is called *strongly mixing*, if the (non-increasing) mixing coefficient satisfies

$$\lim_{\tau \to \infty} \alpha(\tau) = 0. \tag{9}$$

It follows that a strongly mixing stationary stochastic process is ergodic and, plugged into a function f satisfying our assumptions above, satisfies property (7). Good sources for all this are Rozanov [32, Chap. IV] and Ibragimov and Linnik [11, Chapter 17].

We are now in a position to formulate Khasminskii's *Central Limit Theorem* [17], also known as *Gaussian* or *linear diffusion approximation* (as opposed to the *nonlinear* diffusion approximation introduced in the next section).

Theorem 2 (Weak Linear Diffusion Approximation) *Let $f : \mathbb{R}^d \times \mathbb{R}^m$, $(x, \xi) \mapsto f(x, \xi)$, be continuous, and let $f(\cdot, \xi)$ together with its first and second derivatives with respect to x be bounded uniformly with respect to ξ.*

Let ξ_t be a stationary stochastic process with values in $\mathbb{R}^m$ and cadlag trajectories which is strongly mixing and satisfies, moreover, the condition

$$\int_0^\infty \tau\, \alpha(\tau)\, d\tau < \infty. \tag{10}$$

Denote by $x_t^\varepsilon(x)$ and $\bar{x}_t(x)$ the solution of (3) and (5), resp. Then the Averaging Principle (Theorem 1) holds, and, moreover, for all $x \in \mathbb{R}^d$

$$\zeta_t^{\varepsilon,x} := \frac{1}{\sqrt{\varepsilon}}(x_t^\varepsilon(x) - \bar{x}_t(x)) \Rightarrow \zeta_t^x \quad \text{as } \varepsilon \to 0 \quad \text{on } [0, T]. \tag{11}$$

Here $\Rightarrow$ means weak convergence (i.e., the probability distribution of the left-hand side converges to the probability distribution of the right-hand side as $\varepsilon \to 0$), and ζ_t^x is a Gauss-Markov (or Ornstein-Uhlenbeck) process on $[0, T]$ given by the solution of the linear SDE

$$d\zeta_t^x = D\bar{f}(\bar{x}_t(x))\zeta_t^x dt + \sqrt{\sigma(\bar{x}_t(x))}\, dW_t, \quad \zeta_0^x = 0, \tag{12}$$

driven by the solution $\bar{x}_t(x)$ of the averaged ODE (5). In (12) the $d \times d$ *matrix*

$$D\bar{f}(x) := \left(\frac{\partial^i \bar{f}}{\partial x_j}\right)(x)$$

is the Jacobian of the averaged vector field $\bar{f}$, W_t *is a standard Brownian motion in* $\mathbb{R}^d$, *and* $\sqrt{\sigma(x)}$ *is the square root of the nonnegative-definite* $d \times d$ *matrix* $\sigma(x)$ *defined as the "nonlinear average"*

$$\sigma(x) := \lim_{T\to\infty} \frac{1}{T} \int_0^T \int_0^T R_x(t-s)dtds = \int_{-\infty}^{\infty} R_x(t)dt = 2\pi S_x(0), \quad (13)$$

where $R_x(t)$ *is the correlation matrix (6) of* $\varphi_t := f(x, \xi_t)$ *for frozen* x, *and* $S_x(0)$ *is its spectral density matrix at frequency* $\lambda = 0$.

For a proof see the original paper by Khasminskii [17] or Freidlin and Wentzell [9, Chapter 7]. By our assumptions, the limits in (13) exist and define a C^2 function, and $\sigma(\cdot)$ and its first and second derivatives are bounded. (10) means that the decay of the mixing coefficient $\alpha(\tau)$ is at least of polynomial order

$$\alpha(\tau) \sim \frac{1}{\tau^{2+\gamma}} \quad \text{as } \tau \to \infty \text{ for some } \gamma > 0.$$

The result of Theorem 2 is that we can replace the averaged equation (5) by the more informative pair of equations

$$\dot{\bar{x}}_t = \bar{f}(\bar{x}_t), \quad \bar{x}_0 = x, \quad (14)$$

$$d\zeta_t^x = D\bar{f}(\bar{x}_t(x))\zeta_t^x dt + \sqrt{\sigma(\bar{x}_t(x))}\, dW_t, \quad \zeta_0^x = 0, \quad (15)$$

which has skew product structure. We can approximate the original process $x_t^\varepsilon(x) = \bar{x}_t(x) + \sqrt{\varepsilon}\,\zeta_t^{\varepsilon,x}$ with $\zeta_t^{\varepsilon,x} \Rightarrow \zeta_t^x$ by the Gaussian process (also known in physics as *Van Kampen's approximation*)

$$\eta_t^{\varepsilon,x} := \bar{x}_t(x) + \sqrt{\varepsilon}\,\zeta_t^x. \quad (16)$$

As a by-product of his work [23] on the nonlinear diffusion approximation Kifer proved the following strong approximation of $x_t^\varepsilon(x)$ by $\eta_t^{\varepsilon,x}$.

Theorem 3 (Strong Linear Diffusion Approximation) *Let the assumptions of Theorem 2 hold true. Then for each* $\varepsilon > 0$ *there is a possibly richer probability space which carries a d-dimensional Brownian motion* W_t *and on which we can redefine* ξ_t *without changing its distribution such that for these* ξ_t *and* W_t *the solutions of (3), (5) and (12) satisfy for each* $x \in \mathbb{R}^d$ *and* $\varepsilon > 0$

$$\mathbb{E} \max_{0\le t\le T} \|x_t^\varepsilon(x) - (\bar{x}_t(x) + \sqrt{\varepsilon}\,\zeta_t^x)\|^2 \le C_{\delta,T}\, \varepsilon^{1+2\delta} \quad (17)$$

for any $\delta < 1/(90d + 177)$, *where* $C_{\delta,T}$ *is independent of* x *and* ε.

"Changing the probability space" in applications should not create any problem as such a space is a measure-theoretic underpinning and typically neither fixed nor unique anyway.

A comparison of Theorems 1 and 3 reveals that the linear diffusion approximation, i.e., the solution of the averaged equation (5) *corrected* by the solution of the SDE (12) is a *better* approximation in the L^2 sense on a finite time interval than is the averaged motion.

However, the crucial drawback of the linear diffusion approximation (16) is that it will typically not be able to model correctly the long-term behaviour of the original slow motion $x_t^\varepsilon(x)$ (see our examples in sections 5 and 6).

4. Nonlinear Diffusion Approximation

The aforementioned drawback is partly remedied by another approximation which was proposed by Hasselmann [10, Sect. 4] in the context of stochastic climate models to which deterministic general circulation models should be reduced.

Unlike the linear diffusion approximation which is a (Gaussian) correction of the *solution* of the averaged equation (5), the *nonlinear diffusion approximation* is a state-dependent correction of the averaged *equation* itself, namely

$$dy_t^\varepsilon = \bar{f}(y_t^\varepsilon)\,dt + \sqrt{\varepsilon}\,\sqrt{\sigma(y_t^\varepsilon)}\,dW_t, \quad y_0^\varepsilon = x, \tag{18}$$

where $\bar{f}$ and σ are defined above. Using Doob's and the Gronwall-Bellman inequality and the fact that $\bar{f}$ and σ are Lipschitz and bounded it follows that for each $x \in \mathbb{R}^d$ and $\varepsilon > 0$

$$\mathbb{E} \max_{0\le t\le T} \|y_t^\varepsilon(x) - \bar{x}_t(x)\|^2 \le C_T\,\varepsilon, \tag{19}$$

where C_T is independent of x and ε. Hence $\bar{x}_t(x)$ is as close to $y_t^\varepsilon(x)$ as it is to $x_t^\varepsilon(x)$, namely $O(\varepsilon)$. The question now arises whether $x_t^\varepsilon(x)$ and $y_t^\varepsilon(x)$ are closer to each other than $O(\varepsilon)$.

This turns out to be indeed the case. Here is Kifer's result [23].

Theorem 4 (Nonlinear Diffusion Approximation) *Let the assumptions of Theorem 2 hold true. Then for each* $\varepsilon > 0$ *there is a possibly richer probability space which carries a d-dimensional Brownian motion* W_t *and on which we can redefine* ξ_t *without changing its distribution such that for these* ξ_t *and* W_t *the solution* $x_t^\varepsilon(x)$ *of (3) and the solution* $y_t^\varepsilon(x)$ *of (18) satisfy for each* $x \in \mathbb{R}^d$ *and* $\varepsilon > 0$

$$\mathbb{E} \max_{0\le t\le T} \|x_t^\varepsilon(x) - y_t^\varepsilon(x)\|^2 \le C_{\delta,T}\,\varepsilon^{1+2\delta} \tag{20}$$

for any $\delta < 1/(90d + 177)$, where $C_{\delta,T}$ is independent of x and ε.

A sketch of the proof of this theorem can be found in the Appendix.

As a result, the linear diffusion approximation $\bar{x}_t(x)+\sqrt{\varepsilon}\,\zeta_t^x$ and the nonlinear diffusion approximation $y_t^\varepsilon(x)$ approximate the original slow motion $x_t^\varepsilon(x)$ on a finite time interval $[0,T]$ equally well, namely with an L^2 error of order $O(\varepsilon^{1/2+\delta})$. However, the nonlinear SDE (18) is more likely to catch the correct long-term and qualitative behavior of the slow motion.

5. Toy Examples

Consider the scalar ODE

$$\dot{x}_t = f(x_t) + g(x_t)\xi_{t/\varepsilon}, \quad x_0 = x \in \mathbb{R}^1, \tag{21}$$

where ξ_t is a mean zero ergodic stationary process with all the properties assuring the validity of the linear and nonlinear diffusion approximation. Then

$$\bar{f}(x) = f(x), \quad \sigma(x) = c^2 g(x)^2, \quad c^2 := 2\int_0^\infty \mathbb{E}\,\xi_t\xi_0\,dt = 2\pi S(0), \tag{22}$$

where $S(\lambda)$ is the spectral density of ξ_t. Assume $c > 0$. The approximations introduced above are

$$\begin{aligned}
\dot{\bar{x}}_t &= f(\bar{x}_t), \quad \bar{x}_0 = x, &(23)\\
d\zeta_t^x &= Df(\bar{x}_t)\zeta_t^x\,dt + cg(\bar{x}_t)\,dW_t, \quad \zeta_0^x = 0, &(24)\\
\eta_t^{\varepsilon,x} &= \bar{x}_t(x) + \sqrt{\varepsilon}\,\zeta_t^x, \quad \eta_0^{\varepsilon,x} = x, &(25)\\
dy_t^\varepsilon &= f(y_t^\varepsilon)\,dt + \sqrt{\varepsilon}\,cg(y_t^\varepsilon)\,dW_t, \quad y_0^\varepsilon = x. &(26)
\end{aligned}$$

We consider two particular cases.

Additive noise on pitchfork bifurcation

Consider the family of ODE parametrized by $\alpha \in \mathbb{R}^1$ given by

$$\dot{x}_t = \alpha x_t - x_t^3 + \xi_{t/\varepsilon}, \quad x_0 = x \in \mathbb{R}^1. \tag{27}$$

In this case

$$\begin{aligned}
\dot{\bar{x}}_t &= \alpha\bar{x}_t - \bar{x}_t^3, \quad \bar{x}_0 = x, &(28)\\
d\zeta_t^x &= (\alpha - 3(\bar{x}_t(x))^2)\zeta_t^x\,dt + c\,dW_t, \quad \zeta_0^x = 0, &(29)\\
\eta_t^{\varepsilon,x} &= \bar{x}_t(x) + \sqrt{\varepsilon}\,\zeta_t^x, \quad \eta_0^{\varepsilon,x} = x, &(30)\\
dy_t^\varepsilon &= (\alpha y_t^\varepsilon - (y_t^\varepsilon)^3)\,dt + \sqrt{\varepsilon}\,c\,dW_t, \quad y_0^\varepsilon = x. &(31)
\end{aligned}$$

The averaged equation (28) undergoes a pitchfork bifurcation at $\alpha = 0$ and has the two stable steady states at $\pm\sqrt{\alpha}$ for $\alpha > 0$ to which all solutions with

$x \neq 0$ converge. Hence the linear diffusion approximation (29,30) is for $x > 0$ and large times

$$\eta_t^{\varepsilon,\pm x} = \pm\sqrt{\alpha} + \sqrt{\varepsilon}\,\zeta_t, \tag{32}$$

where ζ_t is an Ornstein-Uhlenbeck process with stationary distribution $\mathbb{N}\,(0, c^2/4\alpha)$. Hence, in the long run, we observe a Gaussian process centered either at $\sqrt{\alpha}$ or at $-\sqrt{\alpha}$, depending on the sign of the initial value x of the averaged equation.

In contrast, the nonlinear diffusion approximation (31) has for all α a unique stationary measure with density

$$p_\alpha(x) = N_\alpha \exp\left(\frac{1}{\varepsilon c^2}(\alpha x^2 - \frac{x^4}{2})\right) \tag{33}$$

which is unimodal for $\alpha < 0$ and bimodal for $\alpha > 0$, so that the family (31) undergoes a P-bifurcation at $\alpha = 0$. Moreover, the above measure is also unique in the sense of the theory of random dynamical systems (see Arnold [1, Sect. 9.2.2]), its disintegration is a random Dirac measure at $\tilde{y}(\omega)$, and its Lyapunov exponent is strictly negative for all α, implying that there is no D-bifurcation. One can thus say that "additive noise destroys a pitchfork bifurcation" – which is the title of the paper by Crauel and Flandoli [6]. However, a typical solution of the nonlinear diffusion approximation converges exponentially fast to the stationary stochastic process $\tilde{y}(\theta_t\omega)$ which has mean zero and fluctuates between the peaks $\pm\sqrt{\alpha}$ of (33).

Evidently, the qualitative behavior of the nonlinear diffusion approximation is closer to the behavior of the original equation as is the linear diffusion approximation. There are, however, differences between x_t^ε and y_t^ε as far as large deviations from $\bar{x}_t$ are concerned.

Multiplicative noise on pitchfork bifurcation
Now consider

$$\dot{x}_t = (\alpha + \xi_{t/\varepsilon})x_t - x_t^3, \quad x_0 = x \in \mathbb{R}^1, \tag{34}$$

which can be explicitly solved and undergoes a stochastic pitchfork bifurcation at $\alpha = 0$ (for details see [1, Sect. 9.3.5]), with a stable random Dirac measure on the positive as well as negative half-axis for parameter values $\alpha > 0$.

In this case the approximations are

$$\begin{aligned}
\dot{\bar{x}}_t &= \alpha\bar{x}_t - \bar{x}_t^3, \quad \bar{x}_0 = x, && (35)\\
d\zeta_t^x &= (\alpha - 3(\bar{x}_t(x))^2)\zeta_t^x\,dt + c\bar{x}_t(x)\,dW_t, \quad \zeta_0^x = 0, && (36)\\
\eta_t^{\varepsilon,x} &= \bar{x}_t(x) + \sqrt{\varepsilon}\,\zeta_t^x, \quad \eta_0^{\varepsilon,x} = x, && (37)\\
dy_t^\varepsilon &= (\alpha y_t^\varepsilon - (y_t^\varepsilon)^3)\,dt + \sqrt{\varepsilon}\,c y_t^\varepsilon\,dW_t, \quad y_0^\varepsilon = x,\\
&= ((\alpha - \frac{1}{2}\varepsilon c^2)y_t^\varepsilon - (y_t^\varepsilon)^3)\,dt + \sqrt{\varepsilon}\,c y_t^\varepsilon \circ dW_t, && (38)
\end{aligned}$$

where (38) is an SDE in the Stratonovich sense.

The linear diffusion approximation is the same as for the additive noise case, the only difference being that the Ornstein-Uhlenbeck process ζ_t in (32) has stationary distribution $\mathbb{N}(0, c^2/4)$ independent of α.

The SDE (38) for the nonlinear diffusion approximation can also be explicitly solved (see [1, Sect. 9.3.2]) and again models the long-term behavior of (34) better than (37). It undergoes a stochastic pitchfork bifurcation, but, compared with the original family (34), the bifurcation point is delayed from $\alpha = 0$ to $\alpha = \frac{1}{2}\varepsilon c^2 > 0$.

We consider the latter phenomenon as a deficiency of the nonlinear diffusion approximation which leads back to the work of Khasminskii [16] and a discussion in the paper of Kifer [23] (cf., the discussion around (6) and Remark 3.4): One can try to improve the nonlinear diffusion approximation (18) by adding a small drift term,

$$dz_t^\varepsilon = (\bar{f}(z_t^\varepsilon) + \varepsilon D(z_t^\varepsilon))\, dt + \sqrt{\varepsilon}\, \sqrt{\sigma(z_t^\varepsilon)}\, dW_t, \quad z_0^\varepsilon = x, \tag{39}$$

where

$$D(x) := \int_0^\infty \mathbb{E}\,(Df(x,\xi_t) - D\bar{f}(x))(f(x,\xi_0) - \bar{f}(x))\, dt. \tag{40}$$

Kifer proved that if D is bounded Theorem 4 remains true if $y_t^\varepsilon(x)$ is replaced with $z_t^\varepsilon(x)$. He also conjectured that, after the usual redefinition of ξ_t,

$$\lim_{\varepsilon \to 0} \mathbb{E} \max_{0 \le t \le T/\varepsilon} \|x_t^\varepsilon(x) - z_t^\varepsilon(x)\|^2 = 0. \tag{41}$$

The extension of the validity of L^2 convergence from a fixed time interval $[0, T]$ to an interval of the form $[0, T/\varepsilon]$ would be a substantial improvement.

For the additive noise case above we obtain $D(x) \equiv 0$, while for the multiplicative noise case $D(x) = \frac{1}{2}c^2 x$ which cancels the Stratonovich correction term, so that the stochastic pitchfork bifurcation point of z_t^ε moves back to its original position $\alpha = 0$.

6. The Lorenz-Maas Atmosphere-Ocean Model

We briefly mention the extensive numerical case study of the following coupled ocean-atmosphere model recently undertaken by Arnold, Imkeller and Wu [3]. The model was proposed by van der Schrier and van Veen [35, 36] and couples the ocean model by Maas [27] with the Lorenz-84 model of the atmosphere [26]. This 6-dimensional model (3-dimensional slow ocean $x = (x_1, x_2, x_3)$ and 3-dimensional fast atmosphere $y = (y_1, y_2, y_3)$) is considered the simplest known atmosphere-ocean model which is derived in a controlled manner from first principles.

The model is of the general form of fully coupled equations (1,2) (hence is not covered by the theorems presented) and reads

$$\dot{x}_1 = x_1x_3 + a_1x_2x_3 - x_1, \tag{42}$$
$$\dot{x}_2 = x_2x_3 - a_1x_1x_3 - a_2x_2 + a_3 + a_4\mathbf{y}_1, \tag{43}$$
$$\dot{x}_3 = -a_5(x_1^2 + x_2^2) - a_6x_3 + a_7, \tag{44}$$
$$\dot{y}_1 = \frac{1}{\varepsilon}(-y_2^2 - y_3^2 - b_1y_1 + b_2 + b_3\mathbf{x}_2), \tag{45}$$
$$\dot{y}_2 = \frac{1}{\varepsilon}(y_1y_2 - b_4y_1y_3 - y_2 + b_5), \tag{46}$$
$$\dot{y}_3 = \frac{1}{\varepsilon}(b_6y_1y_2 + y_1y_3 - y_3), \tag{47}$$

where representative values of the constants $(a_i)_{i=1,\dots,7}$ and $(b_i)_{i=1,\dots,6}$ are given by van Veen [35] and in [3]. The coupling term a_4y_1 in (43) describes the heat flux transfered by the atmosphere at the ocean surface, while the coupling term b_3x_2 in (45) represents the modulation of the geostrophic wind due to north-south temperature differences.

Our numerical study [3] consists of the calculation of the (typically several) invariant measures $\mu_x(dy)$ of the fast atmosphere (45,46,47) with frozen x_2, of averaging the slow ocean (42,43,44) with respect to these invariant measures (i.e., evaluating $\varphi(x) := \int y_1\, \mu_x(dy)$) and of determining the coefficients (in particular $\sigma(x)$) of the linear as well as nonlinear diffusion approximation of the ocean. In these simulations the nonlinear diffusion approximation proved to be a qualitatively more accurate approximation of the ocean and in most situations outperformed the other two models.

Appendix: Sketch of the Proof of Theorem 4.3

We basically follow Kifer [23].

Suppose we have found a common probability space for ξ_t and W_t. We start with

$$x_t^\varepsilon(x) - y_t^\varepsilon(x) = \int_0^t (f(x_s^\varepsilon(x), \xi_{s/\varepsilon}) - \bar{f}(y_s^\varepsilon(x)))\, ds - \sqrt{\varepsilon}\int_0^t \sqrt{\sigma(y_s^\varepsilon(x))}\, dW_s.$$

By using the Taylor expansion of f, σ and $\bar{f}$ at $\bar{x}_t(x)$ and introducing the centered quantity $F(x,\xi_t) := f(x,\xi_t) - \bar{f}(x)$ it is not hard to see that the task can be reduced to proving that

$$\mathbb{E}\max_{0\le t\le T}\left\| \sqrt{\varepsilon}\int_0^{t/\varepsilon} F(\bar{x}_{s\varepsilon}(x), \xi_s)ds - \int_0^t \sqrt{\sigma(\bar{x}_s(x))}\, dW_s \right\|^2 \le C_T\varepsilon^{2\delta}. \tag{1}$$

Let us first look at the simple case where $F(x,\xi)$ is independent of x. The *invariance principle* says that

$$\sqrt{\varepsilon}\int_0^{t/\varepsilon} F(\xi_s)ds \Rightarrow \sqrt{\sigma}\, W_t \quad \text{as } \varepsilon \to 0, \tag{2}$$

where

$$\sigma := \int_{-\infty}^{\infty} \mathbb{E}\, F(\xi_t)F(\xi_0)'dt.$$

It can be proved via the representation

$$S_t := \int_0^t F(\xi_s)ds = M_t + X_t - X_0,$$

where M_t is a martingale with stationary increments and

$$X_t := -\int_{s=t}^{\infty} \mathbb{E}\left(F(\xi_s)|\mathbb{F}_{-\infty}^t\right) ds$$

is a stationary process, using standard martingale limit theorems for showing that $\sqrt{\varepsilon}M_{t/\varepsilon} \Rightarrow \sqrt{\sigma}W_t$ and noting that $\sqrt{\varepsilon}X_{t/\varepsilon} \to 0$ in probability (see e.g., Liptser and Shiryayev [25, Chap. 9]). This reasoning can be extended to our case, so that we have

$$\sqrt{\varepsilon}\int_0^{t/\varepsilon} F(\bar{x}_{s\varepsilon}(x), \xi_s)ds \Rightarrow \int_0^t \sqrt{\sigma(\bar{x}_s(x))}\, dW_s \quad \text{as } \varepsilon \to 0. \tag{3}$$

As we want to estimate the error between the two random variables in (2) or (3) in L^2 we have to redefine ξ_t on a new probability space on which there is also a Brownian motion which makes the error of order $\varepsilon^{2\delta}$.

In dimension $d = 1$ one could use the above martingale approximation of S_t and Skorokhod's embedding theorem to obtain an appropriate Brownian motion (as done for the *almost sure invariance principles*, see Philipp and Stout [31]).

To overcome the dimension restriction due to Skorokhod's theorem one can utilize ideas developed by Berkes and Philipp [4] and Kuelbs and Philipp [24] which work in any number of dimensions and are based on the following key result of Strassen [34] and Dudley [7].

Theorem A1 *Let (S, d) be a separable metric space, and let $\mathbb{P}_1$ and $\mathbb{P}_2$ be probability measures on the Borel sets $\mathbb{B}$ of S for which*

$$\rho(\mathbb{P}_1, \mathbb{P}_2) < \alpha, \quad \rho \text{ the Prokhorov distance.}$$

Then there exists a probability space $(\Omega, \mathbb{F}, \mathbb{P})$ and S-valued random variables X_1 and X_2 with distributions $\mathbb{P}_1$ and $\mathbb{P}_2$, resp., such that

$$\mathbb{P}\{d(X_1, X_2) > \alpha\} \leq \alpha.$$

The "message" of Theorem A1 is that at the cost of a new probability space, closeness in distribution can be converted into closeness of random variables.

To utilize this idea, we decompose our integral into blocks:

$$\sqrt{\varepsilon}\int_0^{t/\varepsilon} = \sum_k Y_k + \sum_k Z_k, \quad Y_k = \sqrt{\varepsilon}\int_{S_k}^{T_k}, \quad Z_k = \sqrt{\varepsilon}\int_{T_k}^{S_{k+1}},$$

where the Y_k are "large" blocks which are asymptotically independent due to the mixing assumption since $S_{k+1} - T_k \to \infty$, and the Z_k are "small" blocks which are asymptotically of negligible size.

We now redefine ξ_t and find W_t via the Strassen-Dudley Theorem A1 such that

$$\mathbb{E}\max_k \|Y_k - \int_{\varepsilon S_k}^{\varepsilon T_k} \sqrt{\sigma(\bar{x}_s(x))}\, dW_s\|^2 \leq \beta,$$

where β is small. We skip further details.

It is believed that the optimal δ in Theorems 3 and 4 is close to 1.

References

[1] L. Arnold, *Random dynamical systems*. Springer-Verlag, Berlin Heidelberg New York, 1998.

[2] L. Arnold, *Hasselmann's program revisited: the analysis of stochasticity in deterministic climate models*. In P. Imkeller and J.-S. von Storch, editors, *Stochastic Climate Models*, volume 49 of *Progress in Probability*, pp. 141–157, Birkhäuser, Boston Basel Stuttgart, 2001.

[3] L. Arnold, P. Imkeller, and Y. Wu, *Reduction of deterministic coupled atmosphere-ocean models to stochastic ocean models: A numerical case study of the Lorenz-Maas system*, Report, Institut für Dynamische Systeme, Universität Bremen, 2002.

[4] I. Berkes and W. Philipp, "Approximation theorems for independent and weakly dependent random vectors," *Ann. Probab.*, vol. 7, pp. 29–54, 1979.

[5] N. Bogolyubov and Y. Mitropolskii, *Asymptotic methods in the theory of nonlinear oscillations*, Gordon and Breach, NY, 1961.

[6] H. Crauel and F. Flandoli, "Additive noise destroys a pitchfork bifurcation," *Journal of Dynamics and Differential Equations*, vol. 10, pp. 259–274, 1998.

[7] R. Dudley, "Distances of probability measure and random variables," *Ann. Math. Statist.*, vol. 39, pp. 1563–1572, 1968.

[8] M. Freidlin, "The averaging principle and theorems on large deviations," *Russian Math. Surveys*, vol. 33, pp. 117–176, 1978.

[9] M. Freidlin and A. Wentzell, *Random perturbations of dynamical systems*. Springer-Verlag, Berlin, Heidelberg, New York, 1998. Second edition.

[10] K. Hasselmann, "Stochastic climate models: Part I. Theory," *Tellus*, vol. 28, pp. 473–485, 1976.

[11] I. A. Ibragimov and Y. V. Linnik, *Independent and stationary sequences of random variables*. Wolters-Noordhoff, Groningen, 1971.

[12] P. Imkeller and A. Monahan, editors. *Stochastic Climate Models*, volume 2 of *Stochastics and Dynamics*, World Scientific, Singapore, 2002. Special Issue.

[13] P. Imkeller and J.-S. von Storch, editors. *Stochastic Climate Models*, volume 49 of *Progress in Probability*, Birkhäuser, Boston Basel Stuttgart, 2001.

[14] W. Just, H. Kantz, C. Rödenbeck, and M. Helm, "Stochastic modelling: replacing fast degrees of freedom by noise," *J. Phys. A: Math. Gen.*, vol. 34, pp. 3199–3213, 2001.

[15] H. Kantz, W. Just, and N. Baba, "Replacing fast chaotic degrees of freedom by noise: a formally exact result," Preprint, 2002.

[16] R. Z. Khasminskii, "A limit theorem for solutions of differential equations with random right–hand side," *Theory of Probability and its Applications*, vol. 11, pp. 390–406, 1966.

[17] R. Z. Khasminskii, "On stochastic processes defined by differential equations with a small parameter," *Theory of Probability and its Applications*, vol. 11, pp. 211–228, 1966.

[18] R. Z. Khasminskii, "On the principle of averaging for Itô stochastic differential equations," (in Russian). *Kybernetika*, vol. 4, pp. 260–279, 1968.

[19] Y. Kifer, "Large deviations in dynamical systems and stochastic processes," *Trans. Amer. Math. Soc.*, vol. 321, pp. 505–524, 1990.

[20] Y. Kifer, "Averaging in dynamical systems and large deviations," *Invent. math.*, vol. 110, pp. 337–370, 1992.

[21] Y. Kifer, "Limit theorems in averaging for dynamical systems," *Ergodic Theory and Dynamical Systems*, vol. 15, pp. 1143–1172, 1995.

[22] Y. Kifer, *Averaging and climate models*. In P. Imkeller and J.-S. von Storch, editors, *Stochastic Climate Models*, volume 49 of *Progress in Probability*, pages 171–188, Birkhäuser, Boston Basel Stuttgart, 2001.

[23] Y. Kifer, L^2 diffusion approximation for slow motion in averaging. Preprint, Institute of Mathematics, Hebrew University, 2002.

[24] J. Kuelbs and W. Philipp, "Almost sure invariance principles for partial sums of mixing B-valued random variables," *Ann. Probab.*, vol. 8, pp. 1003–1036, 1980.

[25] R. Liptser and A. Shiryayev, *Theory of martingales*, Kluwer, Dordrecht, 1989.

[26] E. Lorenz, "Irregularity: a fundamental property of the atmosphere," *Tellus*, vol. 36A, pp. 98–110, 1984.

[27] L. Maas, "A simple model for the three-dimensional, thermally and wind-driven ocean circulation," *Tellus*, vol. 46A, pp. 671–680, 1994.

[28] A. Majda, I. Timofeyev, and E. Vanden-Eijnden, "A mathematical framework for stochastic climate models," *Communications on Pure and Applied Mathematics*, vol. 54, pp. 891–974, 2001.

[29] A. Majda, I. Timofeyev, and E. Vanden-Eijnden, A priori tests of a stochastic mode reduction strategy, preprint, Courant Institute of Mathematical Sciences, New York University, 2002.

[30] G. Papanicolaou, D. Stroock, and S. R. S. Varadhan. *Martingale approach to some limit theorems*. Duke University Mathematics Series III, 1977.

[31] W. Philipp and W. Stout, *Almost sure invariance principles for partial sums of weakly dependent random variables*, Mem. Amer. Math. Soc. 161, 1975.

[32] Y. A. Rozanov, *Stationary random processes*, Holden-Day, San Francisco, 1967.

[33] J. A. Sanders and F. Verhulst, *Averaging methods in nonlinear dynamical systems*, Springer-Verlag, Berlin, Heidelberg, New York, 1985.

[34] V. Strassen, "The existence of probability measures with given marginals," *Ann. Math. Statist.*, vol. 36, pp. 423–439, 1965.

[35] L. van Veen, *The Maas-Lorenz model*. Manuscript, Department of Mathematics, Universiteit Utrecht, 2001.

[36] L. van Veen, *Time scale interaction in low-order climate models*, PhD thesis, Department of Mathematics, Universiteit Utrecht, 2002.

EXPERIMENTS ON LARGE FLUCTUATIONS

A. Bandrivskyy, S. Beri, D. G. Luchinsky, and P. V. E. McClintock
Department of Physics, Lancaster University,
Lancaster LA1 4YB, UK
p.v.e.mcclintock@lancaster.ac.uk

Abstract Noise-induced fluctuations in non-linear systems are studied theoretically and experimentally for finite noise intensity. After reviewing briefly the basis of the Hamiltonian theory of fluctuations, the non-equilibrium distribution is calculated for the whole phase space, and determined numerically for different values of noise intensity. It is found that there are singularities which are shifted with respect to those seen in the zero-noise limit. A corresponding shift in the optimal escape path is observed and discussed. A new technique facilitating ultra-fast Monte Carlo simulation at extremely weak noise intensity is introduced and demonstrated.

1. Introduction

A system subject to noise typically spends most of its time fluctuating close to an attractor of some kind, corresponding, for example, to the minimum of a potential well. Just occasionally, however, the system may make a fluctuation that is much larger than average, taking it far from the attractor. It is these rare *large fluctuations* that are responsible for most of the interesting and important events that occur in stochastic nonlinear systems, such as transitions between coexisting attractors. Remarkably, when they happen, and despite their origin in random processes, such events occur in an almost deterministic way. In the limit of zero noise intensity ($D \rightarrow 0$), enormous theoretical progress has been made over many years, particularly through the development of the Hamiltonian theory of fluctuations [10, 24, 6, 7] and the equivalent path-integral approach to the problem [28, 27] and through the introduction of important concepts such as the *prehistory probability density* [5], *optimal paths* [21, 6, 30] and *non-equilibrium potential* [11, 12]. Many aspects of the theory have been tested through analogue electronic experiments [19]. In real experiments, however, the noise intensity is necessarily finite, and numerical experiments are limited by the exponentially growing compu-

N. Sri Namachchivaya and Y.K. Lin (eds.),
IUTAM Symposium on Nonlinear Stochastic Dynamics, 19-28.

tation time required as $D \to 0$. Only in the last decade have experimental techniques been developed that enable optimal paths to be observed and measured directly [5]. The experiments have illuminated aspects of the theory that were not always properly understood, and they have helped to stimulate further theoretical progress, especially in relation to nonequilibrium systems.

In this paper we review briefly the conceptual basis of the theory and then we extend it to include finite noise corrections. The non-equilibrium distribution is calculated, and the changes (from the $D \to 0$ case) in the topology of its singularities are discussed. We then describe how analogue electronic and numerical experiments have led to recent advances. We also propose a new technique capable of decreasing exponentially the computation time in Monte Carlo simulations.

2. Basis of the Theory

The basis of the theory is well established [21, 10, 24, 7, 25, 26, 30, 29] and here we just summarise the central results. We consider a noise-driven non-linear system described by the Langevin equations

$$\begin{aligned} \dot{x}_i &= f_i(x,t) + \sigma_{ij}\xi_j(t), \quad i = 1,\ldots,n, \\ \langle \xi_i(t)\rangle &= 0, \quad \langle \xi_i(t)\xi_j(0)\rangle = DQ_{ij}\delta(t), \end{aligned} \tag{1}$$

where x_i are n state variables of the system, $f_i(x,t)$ are components of the deterministic force, $\xi_i(t)$ are n white Gaussian noise terms of intensity D mixed by a noise matrix σ_{ij}; $Q_{ij} \equiv \sigma_{ik}\sigma_{jk}$ is the diffusion matrix. The Fokker-Planck equation corresponding to (1) is

$$\frac{\partial \rho}{\partial t} = -\frac{\partial}{\partial x_i}(f_i\rho) + \frac{D}{2}Q_{ij}\frac{\partial^2 \rho}{\partial x_i \partial x_j}, \tag{2}$$

On substituting the asymptotic form for the probability distribution

$$\rho(x,t) = z(x,t)\exp\left(-S(x,t)/D\right), \quad D \to 0 \tag{3}$$

into (2) and collecting the terms $\propto D^{-1}$ we have

$$\begin{aligned} \tfrac{\partial S}{\partial t} + \tfrac{1}{2}p_iQ_{ij}p_j + p_if_i = 0, \\ p_i = \tfrac{\partial S}{\partial x_i}, H = \tfrac{1}{2}p_iQ_{ij}p_j + p_if_i, \end{aligned} \tag{4}$$

which can be viewed as a Hamilton-Jacobi equation of an auxiliary classical action S, where H is the Wentzel-Freidlin Hamiltonian [10]. Equation (4) is solved by covering the state-space with a family of trajectories obtained by integrating the Hamiltonian equations of motion

$$\begin{cases} \dot{x}_i = \frac{\partial H}{\partial p_i} = f_i + Q_{ij} p_j \\ \dot{p}_i = -\frac{\partial H}{\partial x_i} = -\frac{\partial f_j}{\partial x_i} p_j \\ \dot{S} = S_t + \frac{\partial S}{\partial x_i} \dot{x}_i = \frac{1}{2} p_i Q_{ij} p_j \end{cases} \tag{5}$$

with given initial conditions. It is important to note that the diffusion matrix is non-negative definite and the action is non-decreasing function along the trajectories. Although in general (5) is non-integrable and S is not differentiable, it is nonetheless differentiable almost everywhere [10, 14], thus justifying its use as a generalized non-equilibrium potential [11, 14]. The more general case of coloured noise can be treated in the same way: it can be shown [3, 8, 9] that a system driven by coloured noise is equivalent to a system of higher dimensionality driven by white noise, so the extension of the theory is straightforward.

3. Basis of the Experiments

The theory predicts that, for limit $D \to 0$, large fluctuations follow optimal paths corresponding to deterministic trajectories of the Hamiltonian auxiliary system. Some scepticism about the applicability of the theory of optimal paths to the real ($D \neq 0$)) system persisted until the first experiments [5] were performed. Both analogue and digital simulations were carried out using essentially the same procedure: several dynamical variables, and the random force where accessible, are recorded simultaneously from the system. The statistics of all actual trajectories along which the system moves to a particular subspace of the coordinate space are then analysed. In line with theoretical expectations, the experiments [5, 16, 15, 19, 20, 13], demonstrated that the so-called *prehistory probability distributions* of such trajectories, moving the system from the equilibrium state to the remote state, are sharply peaked about the optimal fluctuational path which minimizes the action S. Information about stochastic processes obtained in this way is far more detailed than that from measurements of stationary probability distributions. Not only do we count the rare events, but we also learn how, on average, each of them comes about.

Such experiments have verified Onsager's proposition that there is symmetry between the average growth and average decay of fluctuations for a system in thermal equilibrium. Both the fluctuational and relaxational parts of trajectories were collected from an analogue electronic circuit [20, 17, 18] modelling the overdamped motion of a Brownian particle

$$\dot{x} = f(x) + \xi(t), \quad \langle \xi(t) \rangle = 0, \quad \langle \xi(t)\xi(0) \rangle = D\delta(t), \tag{6}$$

in a Duffing potential $U(x) = -x^2/2 + x^4/4, \quad f(x) = -U'(x)$. It was found [19] that the fluctuational and relaxational optimal paths, identified as the maxima of the prehistory density, were symmetrical in time. This was in accord with theory: the Hamilton-Jacobi equation in this case takes the form $H = p^2/2 + f(x)p = 0$; the Hamiltonian equations of motion are $\dot{x} = p - U'(x), \quad \dot{p} = pU''(x)$; and the two solutions of the Hamilton-Jacobi equation, for $p = 0$ and $p = -2f(x)$, correspond to the the relaxational ($\dot{x} = -U'(x)$) and fluctuational ($\dot{x} = +U'(x)$) deterministic paths respectively. It is now understood [15] that the effective momenta p_i that appear in (5) are ultimately related to the optimal fluctuational force corresponding to the optimal path. In the particular case of analogue or numerical simulations, where the noise is external and therefore accessible to measurement, the momentum can be identified as the average value of the force driving the fluctuation.

4. Finite Noise Intensity

The results in section 2 concern fluctuational dynamics for $D \to 0$. In this section the theory is extended to include finite noise corrections for $D \neq 0$ considering the next to the leading terms in the WKB expansion (3). As a definite example of a 2–D system, we consider the periodically-driven, bistable, overdamped Duffing oscillator.

$$\begin{cases} \dot{x}_1 = x_1 - x_1^3 + A\sin(x_2) + \xi(t), \\ \dot{x}_2 = \Omega. \end{cases} \tag{7}$$

A and Ω are parameters of the system. The periodic driving breaks the detailed balance applicable in (6): the two symmetrical equilibrium points turn into two stable limit cycles separated by a saddle cycle. The parameters A and Ω are chosen to be outside the perturbative and adiabatic regimes, viz the driving amplitude is larger than the characteristic diffusion length $\sim \sqrt{D}$, and the period Ω^{-1} is comparable with the escape time. Equations (5) describe a set of trajectories emanating from the stationary state of the system (7) and covering its state space. Analysis of (5) shows that, in the general case of a non-equilibrium system such as (7), $S(x)$ is a multi-valued function of position x in the state space [11, 14], i.e. more than one trajectory may arrive at the same point x, and the probability distribution takes the form [6]

$$\rho(x) = \sum_i z_i(x) e^{-\frac{S_i(x)}{D}}, \tag{8}$$

where index i runs over the subset of trajectories arriving at x. However, in the limit of $D \to 0$ the only trajectories contributing to the sum are those that provide the global minimum of the action, i.e. the optimal paths. Escape from the basin of attraction of a stationary state of the system (7) occurs via the so-

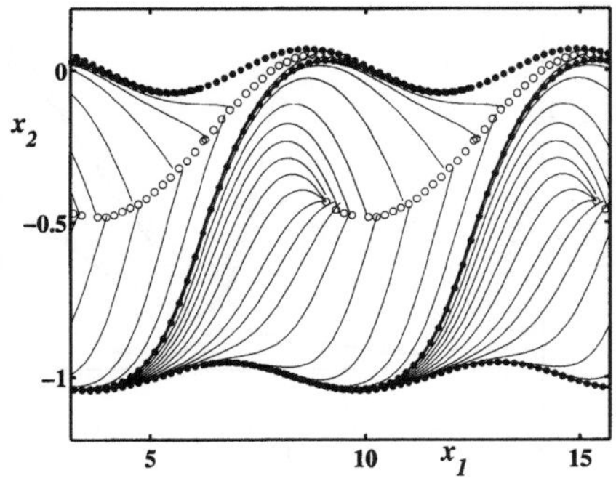

Figure 1. The limit $D \to 0$: escape paths and switching line for the periodically-driven overdamped Duffing oscillator with $A = 0.1$. Optimal paths are shown as thin full lines and the switching line by open circles. The position of the MPEP and of the stable and unstable states are shown by dotted lines.

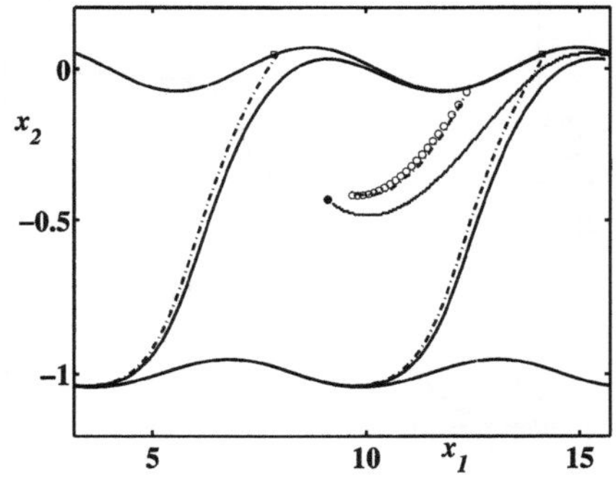

Figure 2. Finite noise intensity, $D = 0.01$. The calculated switching line (dash-dotted) is compared with Monte Carlo simulations (open circles). Both it and the escape path (dash-dotted) are seen to differ from their $D \to$ equivalents (full curves). The stable and unstable solutions are shown by full lines.

called *most probable escape path* [4, 23] (MPEP) that emanates (at $t \to -\infty$) from the stationary state and arrives (at $t \to +\infty$) on the saddle cycle of (7).

As a result of the global minimization, the state space of a two dimensional non-equilibrium system is separated by so-called *switching lines* into areas to which the system arrives along different optimal paths [6]. The switching line and the most probable escape paths for the Duffing system (7) are shown in Fig. 1. It can be seen, in particular, that in the limit $D \to 0$ the MPEP is the only path along which the auxiliary Hamiltonian system for (7) reaches the boundary of the basin of attraction of the stationary state (cf., however, a discussion of special cases [22]).

In the regime $0 \neq D \ll 1$, we consider the next-to-leading order of approximation in D in the solution of the Fokker-Plank equation. Keeping terms proportional to D^0 in the WKB expansion, one obtains an equation for the prefactor in (3) [21]

$$\frac{\partial z}{\partial t} + Q_{ij}\frac{\partial S}{\partial x_j}\frac{\partial z}{\partial x_i} + \frac{\partial}{\partial x_i}(f_i z) + \frac{1}{2} z Q_{ij} W_{ij} = 0. \tag{9}$$

Solution of (9) requires simultaneous integration of the equation for the second derivative of the action $W_{ij} = \partial_i \partial_j S$ along the characteristics of (5). It can be found by differentiating the Hamilton-Jacobi equation (4) twice [26] with respect to the position on the unstable Lagrangian manifold formed by trajectories emanating from the stationary state of the system. The resulting set of

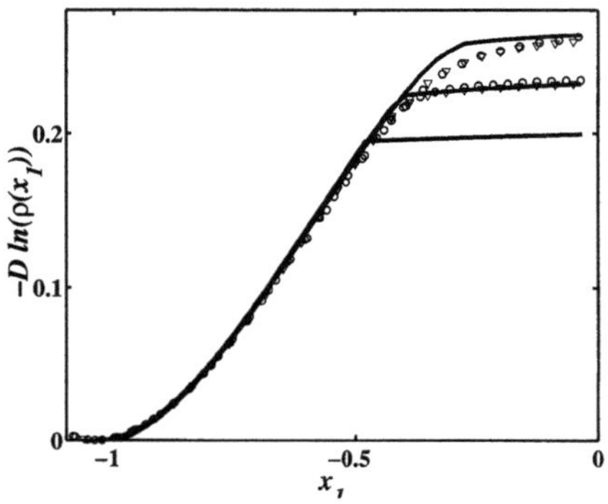

Figure 3. Cross section of the logarithm of the calculated probability density distribution for system (7) with noise intensities D = 0.02, 0.01 and 0 compared with the results of fast Monte Carlo simulations.

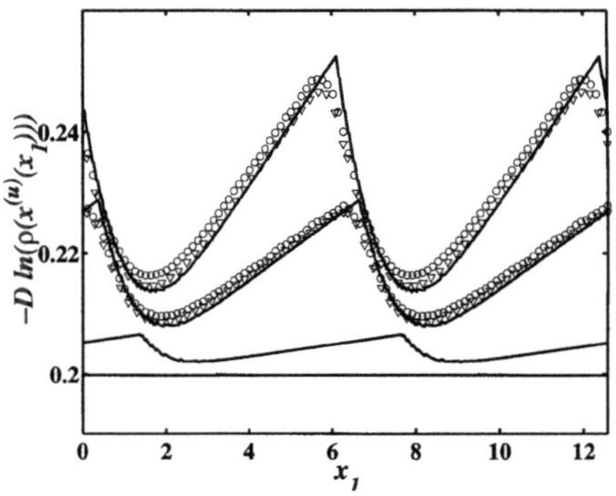

Figure 4. Logarithm of the probability density calculated along the saddle boundary of system (7), for D = 0.01, 0.005 and 0.001, compared to numerical simulations (circles) and direct integration of the Fokker-Planck equation (triangles).

equations that have to be added to the system (5) is of form

$$\begin{cases} \dot{z} = -z\left(\frac{\partial f_i}{\partial x_i} + \frac{1}{2}Q_{ij}W_{ij}\right), \\ \dot{W}_{ij} = -p_m\frac{\partial^2 f_m}{\partial x_i \partial x_j} - W_{im}\frac{\partial f_m}{\partial x_j} \\ -W_{jm}\frac{\partial f_m}{\partial x_i} - W_{jm}W_{ik}Q_{km}. \end{cases} \tag{10}$$

Analysis of the extended system (5),(10) leads to the results shown in Fig. 2. Note that the position of the switching line is now defined by the condition $z_1 \times \exp(-S_1/D) = z_2 \times \exp(-S_2/D)$ where z_i and S_i are respectively the prefactor and action calculated along different trajectories arriving to the state x. The switching line is therefore shifted and crosses the saddle cycle – which is important because infinitely many optimal paths can now cross the saddle cycle. Moreover, the $D \to 0$ MPEP limit ceases to be the most probable escape path and its role is taken by a different path that crosses the saddle cycle.

In Fig 3 the logarithm of the probability distribution, calculated by asymptotic analysis of the Fokker-Planck equation, is compared with the results of the Monte-Carlo simulations described below. In Fig. 4 the logarithm of the probability distribution at the boundary is shown for different noise intensities. The effect of the finite noise intensity is to introduce a modulation in the probability at the boundary as confirmed by the simulations and by numerical integration of the Fokker-Planck equation. The theory discussed above has also been applied to the inverted Van der Pol oscillator, yielding similar results.

5. Fast Monte Carlo Technique

In this section we introduce a modified Monte Carlo algorithm to improve exponentially the efficiency of the numerical integration of Langevin equation (6). Standard Monte Carlo-based methods allow one to measure the discrete probability distribution $\rho(x)$ for (6); the potential can then be deduced as $U(x) \propto -D \ln \rho(x)$ within the basin of attraction. For very small noise intensity, the large fluctuations required to build the distribution far from the equilibrium point become exponentially rare. The conventional technique therefore becomes unapplicable in practice, either for studying the probability distribution far from the equilibrium state, or for investigating the dynamics of the optimal escape path.

In the modified Monte Carlo scheme [1], the simulation is performed in steps, starting from the vicinity of a stable state and gradually moving away from it. Two pieces of the potential as given by two successive steps of the simulation are shown in Fig. 5. We start by building a discrete probability distribution using the conventional Monte Carlo technique. Then we fix two probability levels corresponding respectively to the two potential levels U_i and U_f and the two coordinates x_i and x_f. We start the simulation from the level U_i, and then reinject the system each time it crosses the boundary x_i. Application of the constraint $x \geq x_i$ means that the system cannot relax towards the stable state, and therefore that a new part of the probability distribution must be built. The two pieces of the inferred potential can then be merged at the point x_f. Continuing the procedure the probability distribution and the corresponding potential can be built up step-by-step for the whole region of interest. As shown in Fig. 6, the potential constructed (from 13 pieces) by our algorithm coincides closely with the Duffing potential. The efficiency of the method is easily estimated: using N steps to build the distribution, if T is the amount of time required by conventional Monte Carlo, the modified method requires only $T_{\text{mod}} \approx NT \exp[-(N-1)U_i/D]$. The same efficiency can be reached in a 2-dimensional non-equilibrium system with the same procedure, but in this case a boundary line of constant probability must be found instead of the boundary point. When the system relaxes across the boundary towards the equilibrium state, it is reinjected at the boundary according to the a reinjection location probability (RLP). The reinjection location probability must be calculated at each step, together with the constant probability line. In Figs. 3 and 4 the experimental points obtained with the modified Monte Carlo method are compared with the theoretical results, showing very good agreement.

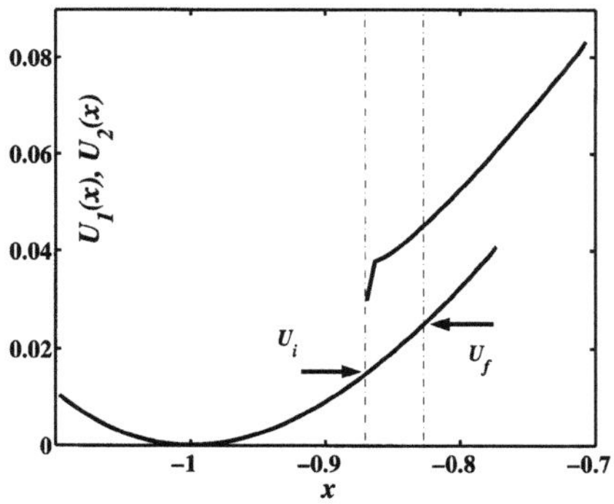

Figure 5. The first and the second pieces of the inferred potential for the system (6) ($D = 0.005$).

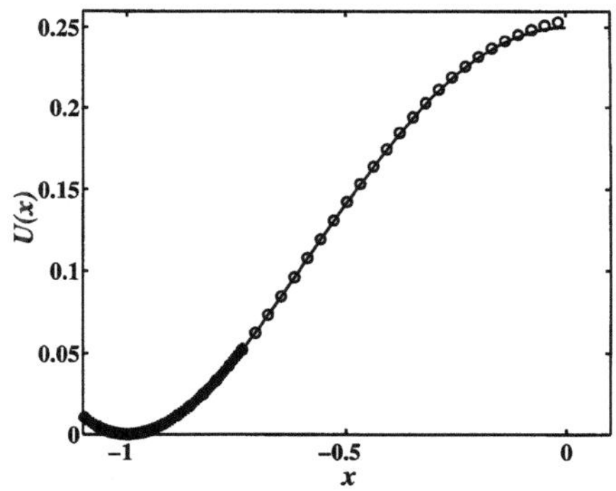

Figure 6. The Duffing potential (thin line) is compared with the potential as inferred by standard Monte Carlo (bold line) or by the modified technique (circles) using $D = 0.005$.

6. Conclusion

In this paper, some non-trivial properties of stochastic non-equilibrium systems have been revealed. In particular, the non-equilibrium potential has been calculated for the whole state-space by computation of the prefactor in (3). Its singularities are shown to be shifted with respect to those for $D \to 0$; the MPEP for the escape problem is also shifted. The experimental investigations were greatly facilitated by a new fast Monte Carlo technique. By exponentially shortening the computation time, it has led to the first direct comparison of the theory and the experiments when $D \to 0$ yielding good agreement.

Acknowledgments

We are very grateful for valuable discussions of these matters with many co-workers including, especially, Mark Dykman, Riccardo Mannella, Vadim Smelyanskiy and Slava Soskin. The work was supported by the Engineering and Physical Sciences Research Council (UK).

References

[1] A. Bandrivskyy, *PhD thesis*, in preparation. Lancaster University, Lancaster, 2003.

[2] . Bandrivskyy, S. Beri, D. G. Luchinsky, R. Mannella, and P. V. E. McClintock, "Fast Monte Carlo simulations and singularities in the probability distributions of non-equilibrium system," Los Alamos archive http://arxiv.org/abs/nlin.AO/0212038.

[3] M. I. Dykman, "Large fluctuations and fluctuational transitions in systems driven by colored Gaussian noise – a high frequency noise," *Phys. Rev. A*, vol. 42, pp. 2020–2029,

1990.

[4] M. I. Dykman and M. A. Krivoglaz, "Theory of the fluctuational transitions between the stable states of a nonlinear oscillator," *Sov. Phys. – JETP*, vol. 50, pp. 30–37, 1979.

[5] M. I. Dykman, P. V. E. McClintock, V. N. Smelyanskiy, N. D. Stein, and N. G. Stocks, "Optimal paths and the prehistory problem for large fluctuations in noise driven systems," *Phys. Rev. Let.*, vol. 68, no. 18, pp. 2718–2721, 1992.

[6] M. I. Dykman, M. M. Millonas, and V. N. Smelyanskiy, "Observable and hidden features of large fluctuations in nonequilibrium systems," *Phys. Lett. A*, vol. 195, no. 1, pp. 53–58, 1994.

[7] M. I. Dykman, E. Mori, J. Ross, and P. Hunt, "Large fluctuations and optimal paths in chemical kinetics," *J. Chem. Phys.*, vol. 100, no. 8, pp. 5735–5750, 1994.

[8] S. J. B. Einchcomb and A. J. McKane, "Escape rates in bistable systems induced by quasi-monochromatic noise," *Phys. Rev. E*, vol. 49, pp. 259–266, 1994.

[9] S. J. B. Einchcomb and A. J. McKane, "Use of Hamiltonian mechanics in systems driven by colored noise," *Phys. Rev. E*, vol. 51, no. 4, pp. 2974–2981, 1995.

[10] M. Freidlin and A. D. Wentzel, *Random Perturbations in Dynamical Systems*. Springer, New-York, 1984.

[11] R. Graham and T. Tel, "Existence of a potential for dissipative dynamical systems," *Phys. Rev. Lett.*, vol. 52, pp. 9–12, 1984.

[12] R. Graham and T. Tel, "Weak noise limit of Fokker-Plank models and nondifferentiable potentials for dissipative dynamical systems," *Phys. Rev. A*, vol. 31, no. 2, pp. 1109–1122, 1985.

[13] J. Hales, A. Zhukov, R. Roy, and M. I. Dykman, "Dynamics of activated escape and its observation in a semiconductor laser," *Phys. Rev. Lett.*, vol. 85, no. 1, pp. 78–81, 2000.

[14] H. R. Jauslin, "Nondifferentiable potentials for nonequilibrium steady states," *Physica A*, vol. 144, pp. 179–191, 1987.

[15] D. G. Luchinsky, "On the nature of large fluctuations in equilibrium systems: observation of an optimal force," *J. Phys. A*, vol. 30, no. 16, pp. L577–L583, 1997.

[16] D. G. Luchinsky, R. S. Maier, R. Mannella, P. V. E. McClintock, and D. L. Stein, "Experiments on critical phenomena in a noisy exit problem," *Phys. Rev. Lett.*, vol. 79, no. 17, no. 3117–3120, 1997.

[17] D. G. Luchinsky, R. Mannella, P. V. E. McClintock, and N. G. Stocks, "Stochastic resonance in electrical circuits– i: Conventional stochastic resonance," *IEEE Trans. on Circuits and Systems II: Analog and Digital Signal Processing*, vol. 46, pp. 1205–1214, 1999.

[18] D. G. Luchinsky, R. Mannella, P. V. E. McClintock, and N. G. Stocks, "Stochastic resonance in electrical circuits– ii: Nonconventional stochastic resonance," *IEEE Trans. on Circuits and Systems II: Analog and Digital Signal Processing*, vol. 46, pp. 1215–1224, 1999.

[19] D. G. Luchinsky and P. V. E. McClintock, "Irreversibility of classical fluctuations studied in analogue electrical circuits," *Nature*, vol. 389, no. 6650, pp. 463–466, 1997.

[20] D. G. Luchinsky, P. V. E. McClintock, and M. I. Dykman, "Analogue studies of nonlinear systems," *Reports on Progress in Physics*, vol. 61, no. 8, pp. 889–997, 1998.

[21] D. Ludwig, "Persistence of dynamical systems under random perturbations," *SIAM Rev.*, vol. 17, no. 4, pp. 605–640, 1975.

[22] R. S. Maier and D. L. Stein, "Transition-rate theory for non-gradient drift fields," *Phys. Rev. Lett.*, vol. 69, pp. 3691–3695, 1992.

[23] R. S. Maier and D. L. Stein, "Effect of focusing and caustics on exit phenomena in systems lacking detailed balance," *Phys. Rev. Lett.*, vol. 71, pp. 1783–1786, 1993.

[24] R. S. Maier and D. L. Stein, "Escape problem for irreversible systems," *Phys. Rev. E*, vol. 48, no. 2, pp. 931–938, 1993.

[25] R. S. Maier and D. L. Stein, "A scaling theory of bifurcations in the symmetrical weak-noise escape problem," *J. Stat. Phys.*, vol. 83, pp. 291–357, 1996.

[26] R. S. Maier and D. L. Stein, "Limiting exit location distributions in the stochastic exit problem," *SIAM J. Appl. Math.*, vol. 57, no. 3, pp. 752–790, 1997.

[27] A. J. McKane, "Noise-induced escape rate over a potential barrier: Results for a general noise," *Phys. Rev. A*, vol. 40, no. 7, pp. 4050–4053, 1989.

[28] L. Onsager and S. Machlup, "Fluctuations and irreversible processes," *Phys. Rev.*, vol. 91, no. 6, pp. 1505–1512, 1953.

[29] V. N. Smelyanskiy, M. I. Dykman, and B. Golding, "Time oscillations of escape rates in periodically driven systems," *Phys. Rev. Lett.*, vol. 82, no. 16, pp. 3193–3197, 1999.

[30] V. N. Smelyanskiy, M. I. Dykman, and R. S. Maier, "Topological features of large fluctuations to the interior of a limit cycle," *Phys. Rev. E*, vol. 55, no. 3, pp. 2369–2391, 1997.

STOCHASTIC AVERAGING: SOME METHODS AND APPLICATIONS

P. Bernard
Laboratoire de Mathématiques Appliquées
Université Blaise Pascal/CNRS
63 177 Aubière Cedex
FRANCE
Pierre.Bernard@math.univ-bpclermont.fr

Abstract When studying the dynamics of a structure submitted to some actions of stochastic type, if the excitation is not of too high level, the dynamical geometric properties of the structure will be determinant in the response analysis.

In this contribution, we only deal with dimension one. Amplitude-phase variables, which are inspired by action-angle variables for Hamiltonian systems, are introduced. The main interest is that they are computable in an explicit form, and are well adapted to give results on the maximum of the mechanical variable under consideration, with inequalities which are conservative for reliability analysis. A slow and a fast process then appear. The idea is the same as Lagrange's variation of constants approach: the slow process, when considered in the unperturbed system, is a first integral of motion. In the stochastic context, the martingale formulation is used, and diffusion approximation results obtained concerning the slow process (which usually is not a diffusion process). These results extend Khasminskii's results. This is the stochastic averaging method.

The same method was used by N. Sri Namachchivaya and R. B. Sowers in [20] to obtain a diffusion approximation of the Hamiltonian itself.

Using this limit averaged diffusion process as an approximation of the slow process, and some results concerning the asymptotic behavior (with respect to level) of level crossings by a one dimensional diffusion process, approximations of the probability distribution of the maximum of the absolute value of the displacement of a strongly nonlinear oscillator on a given time interval are obtained. These formulae, compared with other formulae in the literature, proved to be much better.

Keywords: Stochastic dynamical systems, stochastic averaging, maximum of a process

N. Sri Namachchivaya and Y.K. Lin (eds.),
IUTAM Symposium on Nonlinear Stochastic Dynamics, 29-41.

1. Introduction

The idea of averaging goes back to J.L.Lagrange, in his "Mécanique Analytique", published in 1788. It is in fact the well known "variation des constantes" method.

Let Lagrange explain the main idea: *In the problems of mechanics which we can only solve by approximation, one usually finds the first solution by taking into account only the main forces which act on the body, to extend this solution to other forces which one can call perturbations. The simplest course is to conserve the form of the first solution while making variable the arbitrary constants which it contains; the reason for this is that if the quantities which we have neglected and which we want to take into account are very small, the new variables will be almost constant and we can apply to them the usual methods of approximation. So we have reduced the problem to finding the equations between these variables.*

After their introduction in Astronomy (Clairaut, Lagrange, Laplace), these methods were extended to nonlinear problems in dynamical systems (Jacobi, Poincaré, van der Pol). Stratonovich gave the first stochastic version at the level of rigor of physics, and Khasminskii gave the first proof of a Law of Large Numbers, a Central Limit Theorem and a Diffusion Approximation result. He also gave the first averaging result for Itô stochastic differential equations in [10].

The results presented here are obtained through very weak conditions, which can be verified in practical applications. Noise is White Noise, but the results are available in the case of parametric stochastic excitation. As the method rests on the martingale formulation of the problem, the assumption is that the martingale problem is well posed. In the case of diffusion processes, this is a non-explosion of solutions condition. One can not hope for weaker assumptions.

2. Averaging Method for ODE's

The main reference here is V. Arnold's book [1]. One can also consult [19].

Let's consider a differential system on the manifold $(0,\infty)\times S^1$, such that the first component I is a first integral:

$$\left\{\begin{array}{rcl} \dot{I} &=& 0, \\ \dot{\phi} &=& \omega(I,\phi). \end{array}\right. \tag{1}$$

Let this system be perturbed as follows:

$$\left\{\begin{array}{rcl} \dot{I} &=& \varepsilon g(I,\phi), \\ \dot{\phi} &=& \omega(I,\phi)+\varepsilon f(I,\phi). \end{array}\right. \tag{2}$$

Now, I is a slowly varying function, while ϕ has fast variation, as long as $\omega(I, \phi)$ stays away from 0. It is easy to prove that, under some natural assumptions, on any time interval $[O, T]$ bounded uniformly in ε, the solutions of (2) stay very close to the solutions of (1). But if one is concerned with middle long time intervals, of order $O(1/\varepsilon)$ $(1 \ll T \ll 1/\varepsilon)$, then the solutions of (2) will significantly turn away from the solutions of (1). To understand what happens, assume I is a constant on $[0, T]$, upto a term of order ε. Then, upto a term of order two in ε, assuming some regularity of g,

$$I(T) = I_0 + \int_0^T \varepsilon g(I_0, \phi) dt.$$

As T is large, if ϕ satisfies ergodic properties, the time average $1/T \int_0^T g(I_0, \phi)\, dt$ is the same as the space average $\bar{g}(I) = \int_{S^1} g(I, \phi) d\phi$.

Hence, at first order: $I(T) = I_0 + \varepsilon T \bar{g}(I_0)$. The system describing this evolution is the so-called averaged system:

$$\dot{J} = \varepsilon \bar{g}(J); J(0) = I_0. \tag{3}$$

2.1 Averaging result

Consider the system on $(0, \infty) \times S^1$. *Assume that there is a domain* $U \subset I\!R^n$ *and a constant* c *such that* $0 < c \leq \omega(I, \phi)$ *for all* $(I, \phi) \in U \times S^1$. *(This means that there is really a rotation, so that averaging is possible). Consider an initial value* I_0 *in* U, *such that the solution* $I(t)$ *remains in* U *for* $0 \leq t \leq T/\varepsilon$.

Then, if J *denotes the solution of the averaged equation (3),* $|I(t) - J(t)| \leq C\varepsilon$ *for all* $t \leq T/\varepsilon$.

A proof can be found in [1].

2.2 Example

An important example is the case of a Hamiltonian System, in Poincaré-Arnold's action-angle variables (I, θ). This was used as a model for our general framework. As is known, using these variables, Hamilton's equations can be written:

$$\begin{cases} \dot{I} &= -\frac{\partial H}{\partial \theta} = 0, \\ \dot{\theta} &= \frac{\partial H}{\partial I} = \omega(I). \end{cases}$$

In such coordinates, H does not depend on θ, hence $H(I, \theta) = H(I)$ and $\omega(I) = \frac{dH}{dI}$. So action-angle variables are ideal for our purpose. Unfortunately, such variables generally are not computable. That is why some other variables, thought as generalized action-angle variables, will be introduced. These variables will be called amplitude-angle variables. Thanks to their global character, these variables will be used also in the stochastic case. Of course, we lose the symplectic property, which implies that the new equations are not of Hamilton's form.

2.3 Amplitude-angle variables

The inspiration here is again action-angle variables than Van der Pol or Krylov-Bogoliubov-Mitropolski (KBM) variables. Let us consider the nonlinear conservative oscillator with restoring forces deriving from a potential G. The dynamical equation is:

$$\ddot{x} + \nabla G(x) = 0.$$

In the following, we assume that G is strictly convex, even, and that

$$lim_{x \to \infty} G(x) = +\infty.$$

These assumptions will be satisfied in the case of the Duffing oscillator.

Now, let us define the amplitude-angle variables (A, Ψ) on the phase space of this oscillator, using Lagrange's idea. First, A should be a first integral of the unperturbed system. Good candidates are functions of the Hamiltonian, which, in current variables, is the energy $H(x, y) = \frac{1}{2}y^2 + G(x)$. We define A as being the positive solution of $G(A) = H(x, y)$. Now, extending Van der Pol or KBM approaches, we want that the change of variables $P : (x, y) \mapsto (A, \Psi)$ be such that

$$\left\{ \begin{array}{lcl} x & = & Acos\Psi, \\ y & = & A\omega(A, \Psi)sin\Psi; \end{array} \right.$$

where $\omega(A, \Psi)$ is to be defined. It is easily checked that

$$\omega(A, \Psi) = (2\frac{G(A) - G(Acos\Psi)}{A^2 sin^2\Psi})^{1/2}. \quad (4)$$

If G is regular enough, P defines a nice diffeomorphism [15].

Remarks.

- There is generally no global regular change of variables of the form (4) such that ω is a function of A only.

- Several authors, see for example [22], [16], have introduced similar but different change of variables. The amplitude variable introduced in the present contribution is very convenient for one important application we had in mind: the law of the maximum of the displacement of a stochastically excited nonlinear oscillator.

2.3.1 Example. The Duffing oscillator: $G(x) = \frac{x^2}{2} + \lambda\frac{x^4}{4}$. Then,

$$\omega(A, \Psi) = (1 + \frac{\lambda}{2}A^2(1 + cos^2\psi))^{1/2}.$$

Note that the mean value $\frac{1}{2\pi}\int_0^{2\pi}\omega^2(A,\Psi)d\Psi$ is $\bar{\omega}^2(A) = 1 + \frac{3}{4}\lambda A^2$, which is exactly the expression used by R. Bouc in [6].

3. Stochastic Averaging

Let us consider the dynamics of an oscillator with rigidity potential G such that $\nabla G = g$, with small damping and small external and parametric stochastic excitation driven by a white noise $\dot{w}$. That is to say consider the equation:

$$\ddot{x} + \varepsilon f(x, \dot{x}) + \nabla G(x) = \varepsilon^n h(x, \dot{x})\dot{w}, \tag{5}$$

In this equation, $\varepsilon \geq 0$ is a small parameter, $n > 0$ a real constant, f, h are C^∞ functions, and G satisfies the same assumptions as in paragraph 2.3. In the position-velocity space, with variables (x, y), (5) can be written as a Stratonovich stochastic differential equation (SSDE):

$$\begin{cases} \delta X_t &= Y_t dt, \\ \delta Y_t &= -(g(X_t) + \varepsilon f(X_t, Y_t))dt + \varepsilon^n h(X_t, Y_t)\delta W_t; \end{cases} \tag{6}$$

where $(W_t, t > 0)$ is a scalar standard Brownian Motion and the notation δ means that the SDE is in Stratonovich sense.

We proceed as in the deterministic case.

First step. Time is rescaled (fastened) by introducing the new processes $(X^\varepsilon, Y^\varepsilon)$ as follows:

$$(X^\varepsilon, Y^\varepsilon)(t) = (X, Y)(t/\varepsilon).$$

Hence, the probability distribution of the process $(X^\varepsilon, Y^\varepsilon)$ is a weak solution of:

$$\begin{cases} \delta X_t^\varepsilon &= \frac{1}{\varepsilon}Y_t^\varepsilon dt, \\ \delta Y_t^\varepsilon &= -(\frac{1}{\varepsilon}g(X_t^\varepsilon) + f(X_t^\varepsilon, Y_t^\varepsilon))dt + \varepsilon^{n-1/2}h(X_t^\varepsilon, Y_t^\varepsilon)\delta W_t^\varepsilon; \end{cases} \tag{7}$$

where $W^\varepsilon(t) = \sqrt{\varepsilon}W(t/\varepsilon)$ is also a standard Brownian Motion.

By a weak solution, we mean here that the probability distribution of the process $(X^\varepsilon, Y^\varepsilon)$ is the same as the probability distribution of the solution of this SDE, if same initial conditions are used.

Second step. The diffeomorphism P defined in paragraph 2.3 is applied, so that we get Amplitude-Angle processes $(A^\varepsilon, \Psi^\varepsilon)$. The process A^ε is a slow process, while Ψ^ε is a fastly rotating process. The quantity $\omega(A, \Psi)$ being defined by (4), the following system of SSDE's is obtained:

$$\delta(A^\varepsilon, \Psi^\varepsilon) = B^\varepsilon(A^\varepsilon, \Psi^\varepsilon)dt + \Gamma^\varepsilon(A^\varepsilon, \Psi^\varepsilon)\delta W_t^\varepsilon, \tag{8}$$

where

$$\begin{aligned} B^\varepsilon(A, \Psi) &= -\frac{1}{\varepsilon}\begin{bmatrix} 0 \\ \omega(A, \Psi) \end{bmatrix} \\ &\quad -\frac{\omega(A, \Psi)}{g(A)}\begin{bmatrix} Af(A\cos\Psi, A\omega(A, \Psi)sin\Psi)sin\Psi \\ f(A\cos\Psi, A\omega(A, \Psi)sin\Psi)cos\Psi \end{bmatrix}, \end{aligned} \tag{9}$$

and

$$\Gamma(A, \Psi) = \varepsilon^{n-1/2}\frac{\omega(A, \Psi)}{g(A)}\begin{bmatrix} Ah(A\cos\Psi, A\omega(A, \Psi)sin\Psi)sin\Psi \\ h(A\cos\Psi, A\omega(A, \Psi)sin\Psi)cos\Psi \end{bmatrix}. \tag{10}$$

The asymptotic behavior of (8) as $\varepsilon \to 0$ is now considered. It will result in the application of a general theorem, proved by A. Rachad [15], using the martingale formulation of the problem. The reader interested in these methods is referred to this contribution, and also to [20].

3.1 Limit theorems

Some results of weak convergence about the SDE (8) can be stated. Let us make the following assumptions, available on $(0, \infty)xS^1$:

(H0) The vector field B^ε is of class C^0 and Γ_k^ε is C^1, for $k = 1..d$,

(H1) There exists a unique weak solution of the SDE (8),

(H2) There exists a constant $c > 0$ such that $\omega(A, \Psi) \geq c$.

Theorem 3.1 (Diffusion Approximation) *Assume $n = 1/2$. Under the hypothesis (H0), (H1), and (H2), for any $(A_0, \Psi_0) \in (0, \infty) \times S^1$ the family of processes $A^\varepsilon_{A_0,\Psi_0}$ converges weakly in $C([0,T], \mathbb{R}_+^*)$, as $\varepsilon \to 0$, to a Markov process solution of the martingale problem associated witho $\bar{L}_2$, defined by formula (11), with initial condition A_0, if the last problem is well posed.*

$$\bar{L}_2 f = \int_{S^1}[b_1\frac{df}{dA} + \sum_{k=1}^{d} L_{k,1}f]m(A, d\Phi). \tag{11}$$

A law of large numbers and a central limit theorem can also be obtained [15, 4].

4. Averaging a Stochastic Nonlinear Oscillator

The following dynamical equation, obtained after rescaling in time and space the dynamical equation of a nonlinear oscillator with external white noise, is considered in this section:

$$\ddot{x} + \varepsilon \dot{x} + g(x) = \sqrt{2\varepsilon}\dot{w}. \tag{12}$$

One important example is the Duffing oscillator. The case of Lienard Oscillator (including van der Pol oscillator):

$$\ddot{x} + \varepsilon f(x)\dot{x} + g(x) = \sqrt{2\varepsilon}\dot{w}. \tag{13}$$

is also in the scope of our method, and its analysis can be found in [15].

Equation (12) is written under the form of a Stratonovitch SDE, and then the method is applied. We obtain a SDE for amplitude-angle variables, of the form (8) with $k = 1$, and coefficients:

$$\begin{aligned} b_1(A,\Psi) &= -\frac{A^2\omega(A,\Psi)^2\sin^2\Psi}{G(A)}, b_2(A,\Psi) = -\frac{A\omega(A,\Psi)^2\sin\psi\cos\Psi}{G(A)}, \\ \gamma_1(A,\Psi) &= \sqrt{2}\frac{\omega(A,\Psi)A\sin\Psi}{G(A)}, \gamma_2(A,\Psi) = \sqrt{2}\frac{\omega(A,\Psi)\cos\Psi}{G(A)}; \end{aligned}$$

where G is a positive primitive of g. Of course, G is assumed to satisfy all properties as in paragraph (2.3).

Then, according to Theorem 3.1, for any $T > 0$, the process A^ε converges weakly in $C([0,T],(0,\infty))$ to the diffusion process $\bar{A}$ with infinitesimal generator $\bar{L}$:

$$\bar{L} = b(a)\frac{d}{da} + \frac{\Sigma^2(a)}{2}\frac{d^2}{da^2}, \tag{14}$$

where

$$\begin{aligned} b(a) &= \int_{S_1}\left[b_1(a,\psi) + \frac{1}{2}\left[\gamma_1\frac{\partial\gamma_1}{\partial a} + \gamma_2\frac{\partial\gamma_1}{\partial\psi}\right]\right] m(a,d\psi), \\ \Sigma^2(a) &= \int_{-\pi}^{\pi}\gamma_1^2 m(a,d\psi), \end{aligned}$$

5. Application to Approximations of the Law of the Maximum of the Displacement of a Nonlinear Oscillator

One of the nice consequences of the stochastic averaging method is that the amplitude process is approximated by a diffusion process. And there are a large number of results concerning approximations of the probability distribution of the maximum of a one-dimensional diffusion process. A quite complete study of this aspect can be found in [7].

5.1 Some results about one-dimensional diffusions

5.1.1 Definitions. Let $A = (A_t, t \geq 0)$ be a one dimensional diffusion process with state space $]0, +\infty[$, satisfying the following SDE:

$$dA_t = b(A_t)dt + \Sigma(A_t)dW_t, \quad A_0 = a_0, \tag{15}$$

where $(W_t, t \geq 0)$ is a standard Wiener process, b and Σ are smooth functions on $]0, +\infty[$ such that $\Sigma(a) \neq 0$ for all $a > 0$.

Let S be the **scale function** of the process A [7]. It is characteristic of the spatial behavior of process A.

Concerning the process A solution of (15), its infinitesimal generator is

$$L = b(a)\frac{d}{da} + \frac{\Sigma^2(a)}{2}\frac{d^2}{da^2}. \tag{16}$$

The function S satisfies $LS = 0$. For any constant c in $]0, +\infty[$ and for all $a \in]0, +\infty[$ let us define:

$$S(a) = \int_c^a w(x)dx, w(a) = \exp\left(\int_c^a u(x)dx\right), u(a) = -\frac{2b(a)}{\Sigma^2(a)}. \tag{17}$$

Define the function $G_{a,b}$ on $]0, +\infty[\times]0, +\infty[$ by:

$$G_{a,b}(x,y) = \begin{cases} \frac{(x-a)(b-y)}{b-a} & \text{for} \quad a \leq x \leq y \leq b, \\ \frac{(y-a)(b-x)}{b-a} & \text{for} \quad a \leq y \leq x \leq b \\ 0 & \text{otherwise.} \end{cases} \tag{18}$$

Let μ be the **speed measure** of the diffusion A which is characteristic of the time flow along trajectories of the process A, [7].

Concerning the diffusion A with drift b and diffusion coefficient Σ, the speed measure is absolutely continuous with respect to Lebesgue measure on $]0, +\infty[$ and

$$\mu(da) = \frac{2da}{\Sigma^2(a)w^2(a)}, \tag{19}$$

where $w(a)$ was defined in (17).

5.1.2 Properties. Let A be a diffusion process solution of the SDE (15) and let S be its scale function.

An important result concerning the behavior of A is the following [11]:

if $\lim_{a\to 0} S(a) = -\infty$ and $\lim_{a\to+\infty} S(a) = +\infty$, the non-explosion of A is ensured.

Moreover, if $\frac{au(a)}{\ln a} \to +\infty$ when $a \to +\infty$ then $\lim_{a\to+\infty} S(a) = +\infty$.

Besides, according to S. M. Berman [2], if in addition to previous assumptions on the scale function, the speed measure μ is finite, there is a stationary distribution for the process A with density p of the form:

$$p(a) = \frac{K}{\Sigma^2(a)w(a)} \text{ for } a \in I, \tag{20}$$

where w is defined by (17) and K is a normalizing constant.

5.2 Approximation of the probability distribution of the maximum of a one-dimensional diffusion

Some useful notations Let A be the stationary solution of (15) and let T be a strictly positive real number. The maximum of A under a fixed interval $[0,T]$ is $M_A^T = \sup_{[0,T]} A_t$.

The random variable M_A^T is linked to another random variable τ_a, the first passage time of A at level a, which is defined by $\tau_a = \inf(t \geq 0; A_t \geq a)$.

The probability distribution function of τ_a is $F_{\tau_a}(T) = P(\tau_a < T)$ for all $T \geq 0$ and all real $a > 0$.

Let us denote by

$$K(a, t/a_0) \quad = \quad P(\tau_a \leq t/A_0 = a_0), \tag{21}$$

the repartition of the first passage time τ_a conditioned by $A_0 = a_0$.

By the definition of τ_a, for all real $T > 0$ and all real $a > 0$, we have:

$$P(\tau_a > T) = P(M_A^T < a).$$

Our aim is to give asymptotic expressions of the repartition function of M_A^T when $a \to +\infty$. Many results are available. P. Mandl's [13] result was retained for this presentation. For a complete bibliography, see [7].

P. Mandl's result If A does not explode and if the speed measure μ of A is finite, the following equality holds [13]:

$$\lim_{a \to +\infty} K(a, t\gamma(a)/a_0) = 1 - \exp(-t) \text{ for } a_0 \in (0, \infty), \tag{22}$$

where

$$\gamma(a) = \left(\int_{(0,\infty)} \mu(dx) \right) S(a), \tag{23}$$

and S is the scale function.

Then, for all initial probability measure, the following approximation is deduced from (22):

$$P(\tau_a \leq t) \simeq 1 - \exp\left(- \frac{t}{(\int_{(0,\infty)} \mu(dx))S(a)} \right) \ (a \to +\infty). \tag{24}$$

Hence, the following asymptotic approximation is available:

$$P(M_A^T \leq a) \simeq \exp\left(-\frac{T}{(\int_{(0,\infty)} \mu(dx))S(a)}\right) \quad (a \to +\infty), \tag{25}$$

where μ and S are respectively given by (19) and (17).

Remark. If we consider $A_\varepsilon(t) = A(\varepsilon t)$, the speed measure of A_ε is ε times the speed measure of A, so that the repartition functions of $M_{A_\varepsilon}^{\varepsilon T}$ and M_A^T have the same asymptotics.

5.3 Approximating the law of the maximum of the stationary displacement of a stochastic nonlinear oscillator

Let X^ε be the stationary displacement of the nonlinear oscillator with small damping and small stochastic excitation (5), that is the first component of the solution of (6), and T a strictly positive real number. There are many approximate formulas in the literature for the pdf of the maximum M_X^T of $|X|$ over $[0, T]$. Simulations show that most proposed approximations for $F_{M_X^T}(\alpha) = P(M_X^T < \alpha)$ are accurate when ε and the threshold α are high enough [7]. But the case we are interested in is precisely when ε is small. Then, it is necessary to develop a technique to obtain approximations for $F_{M_X^T}$ which will be available for small values of ε. In this section, a new procedure is proposed to answer this question, based on the stochastic averaging method.

The first step of this method is to setup the dynamics using the Amplitude-Angle processes defined in paragraph (2.3). Then, using the diffusion approximation result, the amplitude A^ε is approximated by the averaged amplitude $\bar{A}$, which is a one-dimensional diffusion process whose infinitesimal generator is (14). The pdf of M_X^T, the maximum of X over the fixed time interval $[0, T]$ is approximated by the pdf of the maximum of the amplitude. With used definition of the amplitude, following inequality is always true:

$$|X_t| \leq A_t, \text{ for all } t \geq 0.$$

Hence the pdf of the maximum of $|X|$ and the maximum of its amplitude defined before are linked by:

$$P(\sup_{[0,T]} |X_t| > u) \leq P(\sup_{[0,T]} A_t > u).$$

Moreover, when the velocity is null, $|X| = A$, which happens precisely when the displacement $|X|$ reaches its maximum. Hence replacing the pdf of the maximum of $|X|$ by the maximum of the amplitude A over the same fixed interval seems to be a good approximation, and anyway there is always an inequality which is in the sense of reliability. Now, we use the results on the law of the maximum of a one-dimensional diffusion presented in the preceding paragraph.

Then, the process A^ε converges weakly in $C(\mathbb{R}_+, \mathbb{R}_+^*)$ to the diffusion process $\bar{A}$ with infinitesimal generator $\bar{L}$ defined by (15) where:

$$\begin{aligned} b(a) &= \int_{S_1} \left[c_1(a,\psi) + \frac{1}{2}\left(\gamma_1 \frac{\partial \gamma_1}{\partial a} + \gamma_2 \frac{\partial \gamma_1}{\partial \psi} \right) \right] m(a, d\psi), \\ \frac{\Sigma^2(a)}{2} &= \int_{S_1} \frac{1}{2} \gamma_1^2 m(a, d\psi). \end{aligned} \tag{26}$$

Applying this result to the Duffing oscillator Example 2.3.1, A^ε converges weakly, as $\varepsilon \to 0$, to the one-dimensional diffusion $\bar{A}$ with infinitesimal generator $\bar{L}$ characterized by:

$$\begin{aligned} \omega(a,\psi) &= \sqrt{1 + \frac{\lambda}{2} a^2 (1 + \cos^2 \psi)}, \\ b(a) &= \frac{1}{a + \lambda a^3} + \frac{\int_0^{2\pi} \sin^2 \psi I(a,\psi) d\psi}{\int_0^{2\pi} I(a,\psi) d\psi} \left[-\frac{1 + 3\lambda a^2}{a(1+\lambda a^2)^2} \right], \\ \Sigma^2(a) &= \frac{4}{1+\lambda a^2} \frac{\int_0^{2\pi} \sin^2 \psi I(a,\psi) d\psi}{\int_0^{2\pi} I(a,\psi) d\psi}, \\ I(a,\psi) &= \sqrt{1 - \frac{\lambda a^2}{2(1+\lambda a^2)} \sin^2 \psi}. \end{aligned}$$

6. Conclusion

A stochastic averaging method, resting on some Amplitude-Angle variables, was introduced. The diffusion approximation result gives good approximation for the pdf of the Amplitude process associated with a one dimensional stochastically excited highly nonlinear oscillator, as long as damping and stochastic terms are not too high. Using the asymptotics obtained this way, approximation formulas were given for the pdf of the maximum of the displacement of such an oscillator on a fixed time interval. This approach was applied to the case of Duffing oscillator. Comparison with other approximation formulas available in the literature, and with Monte-Carlo simulation, showed that this method is very performant. The next step is to extend it to multi-degrees-of-freedom nonlinear oscillators.

Acknowledgments

We thank EDF (Electricity of France) and the CNRS for support.

References

[1] V. Arnold, Chapitres supplémentaires de la théorie des équations différentielles ordinaires, *Editions MIR (1980), Librairie du Gobe*, 1996.

[2] S. M. Berman, "Sojourns and extremes of stationary processes," *The Annals of Proba*, vol. 10, pp. 1–46, 1982.

[3] P. Bernard and A. Rachad, "Averaging a nonlinear stochastic oscillator," *Fourth International Conference on Stochastic Structural Dynamics*, University Notre Dame, Indiana, USA, Balkema editor, pp. 3–10, 1998.

[4] P. Bernard and A. Rachad, "Moyennisation d'un oscillateur stochastique quasi-conservatif," *C.R.A.S.*, t.331, série 1, pp. 1–4, 2000.

[5] N. Bogoliubov and Y. Mitropolsky, "Asymptotic Methods in the Theory of Nonlinear Oscillations," *Gordon & Breach*, 1961.

[6] R. Bouc, "The power spectral density of response for a strongly nonlinear random oscillator," *Journal of Sound and Vibrations*, vol. 175, no. 3, pp. 317–331, 1994.

[7] S. Espinouze. *Manuscrit de Thèse de l'Université Blaise Pascal*, 2002.

[8] W. Feller, "Diffusion processes in one dimension," *Trans. of Amer. Math. Soc.*, vol. 77, pp. 1–31, 1954.

[9] M. I. Freidlin and A. D. Wentzell, *Random Perturbations of Dynamical Systems*. Springer Verlag, 1984.

[10] R. Z. Khasminskii, "On the principles of averaging Itô stochastic differential equations," *Kybernetica*, vol. 4, pp. 260–279, 1968.

[11] N. Ikeda and S. Watanabe, *Stochastic differential equations and diffusion processes*, North-Holland, 1981.

[12] J. L. Lagrange, *Mécanique Analytique*, A. Blanchard, Paris, 1788.

[13] P. Mandl, *Analytic Treatment of One Dimensional Markov Processes*, Springer Verlag, Prague, 1968.

[14] G. C. Papanicolaou, D. Stroock, and S. R. S. Varadhan, "Martingale Approach to Some Limit Theorems," *Duke Turbulence Conference*, 1976.

[15] A. Rachad. *Manuscrit de Thèse de l'Université Blaise Pascal*, 1999.

[16] J. R. Red Horse and P. D. Spanos, "A generalization to stochastic averaging in random vibrations," *International Journal of Nonlinear Mechanics*, vol. 27, no. 1, pp. 85-101, 1992.

[17] J. B. Roberts, "First passage time for oscillators with nonlinear restoring force," *Jal of Sound and Vibration*, vol. 56, no. 1, pp. 71–86, 1978.

[18] J. B. Roberts and P. D. Spanos, "Stochastic Averaging: an approximate method of solving random vibration problems," *International Journal of Nonlinear Mechanics*, vol. 21, no. 2, pp. 111-134, 1986.

[19] J. A. Sanders and F. Verhulst, *Averaging Methods in Nonlinear Dynamical System*. Springer Verlag, 1985.

[20] N. Sri Namachchivaya and R. B. Sowers, "Rigorous Stochastic Averaging at a Center with Additive Noise," *Meccanica*, vol. 37, pp. 85-114, 2002.

[21] R. L. Stratonovitch, *Topics in the theory of random noise*, vols. 1 and 2, Gordon & Breach, New York, 1963.

[22] W. M. Tien, N. Sri Namachchivaya, and V. T. Coppola, "Stochastic Averaging using Elliptic Functions to Study Nonlinear Stochastic Systems," *International Journal of Nonlinear Dynamics*, vol. 4, pp. 373–387, 1993.

[23] W. Q. Zhu, Z. L. Huang, and Y. Q. Yang, "Stochastic Averaging of Quasi-Integrable Hamiltonian Systems," *Journal of Applied Mechanics*, vol. 64, pp. 985–984, 1997.

NOISE SENSITIVITY OF STOCHASTIC RESONANCE AND OTHER PROBLEMS RELATED TO LARGE DEVIATIONS

Mark Freidlin
Department of Mathematics
University of Maryland, College Park
Maryland, USA
mif@math.umd.edu

Abstract The random walk (approximating the Wiener process in the area of normal deviations) and the Wiener process have, in general, different asymptotics of the large deviation probabilities. Therefore, special arrangements in the numerical procedures should be made to calculate solutions of various problems related to the large deviations for the Wiener process and for the corresponding PDE's with a small parameter. Exit problem, wavefront propagation, stochastic resonance are among these problems. We calculate the action functional when the Wiener process is replaced by the random walk, find the relation between the small parameters which provide convergence to the solution of the problem with the Gaussian white noise as the perturbation, and calculate the correction term.

Keywords: Stochastic resonance, computer simulation, numerical methods for PDE's with a small parameter, large deviations.

1. Introduction

Problems related to large deviations for stochastic processes attracted attention of many physicists and engineers in the recent years. Exit problem and stochastic resonance are examples of such problems. Another example is given by wavefront propagation as a result of an interaction of a stochastic transport of particles and a chemical reaction.

If the stochastic component of the system (noise) is of the white noise type, such problems, as a rule, can be reformulated in the terms of corresponding parabolic (or elliptic, in the stationary case) differential equations. The noise, usually, is assumed to be small, so that the equations have a small parameter in the higher derivatives. For instance, the wavefront propagation problem, in an

N. Sri Namachchivaya and Y.K. Lin (eds.),
IUTAM Symposium on Nonlinear Stochastic Dynamics, 43-55.

appropriate time scale, can be written as follows:

$$\frac{\partial u^{\varepsilon}(t,x)}{\partial t} = \frac{\varepsilon}{2}\Delta u^{\varepsilon} + \frac{1}{\varepsilon} f(x,u^{\varepsilon}), t > 0, x \in R^n,$$

$$u^{\varepsilon}(0,x) = g(x) \geq 0, \quad 0 < \varepsilon \ll 1 \tag{1}$$

Of course, the Laplacian Δ in (1) can be replaced by a general second order elliptic operator, the equation can include a drift term of order 1 as $\varepsilon \downarrow 0$, reaction-diffusion systems can be considered [2], [3]. But, as a rule, we restrict ourselves here with the simplest situation.

Let W_t be the Wiener process in R^n, and assume, for brevity, that $f(x,u) = c(x,u)u$. Then the solution of (1) satisfies the following integral equation in the space of trajectories [2]:

$$u^{\varepsilon}(t,x) = Eg(x+\sqrt{\varepsilon}W_t) exp\left\{\frac{1}{\varepsilon}\int_0^t c(x+\sqrt{\varepsilon}W_s,\, u^{\varepsilon}(t-s,x+\sqrt{\varepsilon}W_s))ds\right\}. \tag{2}$$

Here "E" means the expectation with respect the Wiener measure in the space $C_{0T}(R^n)$ of the continuous functions on $[0,T]$ with values in R^n. It is clear from the form of the right hand side of (2), that the asymptotic behavior of $u^{\varepsilon}(t,x)$ as $\varepsilon \downarrow 0$ is related to the large deviations for the process $\sqrt{\varepsilon}W_t$.

In the stochastic resonance problem for a system perturbed by a small white noise, we deal with the equation

$$\frac{\partial u^{\varepsilon}(t,x)}{\partial t} = \frac{\varepsilon}{2}\Delta u^{\varepsilon}(t,x) + b(x)\cdot\nabla u^{\varepsilon}(t,x),\ t > 0,\ x \in R^n,\ u^{\varepsilon}(0,x) = g(x) \tag{3}$$

Here $b(x)$ is a smooth vector field in R^n which has two or more asymptotically stable equilibriums $O_1, \ldots, O_m$. Transitions between these equilibriums are defined by the large deviations of the process X_t^{ε},

$$\dot{X}_t^{\varepsilon} = b(X_t^{\varepsilon}) + \sqrt{\varepsilon}\dot{W}_t, \qquad X_0^{\varepsilon} = x \in R^n, \tag{4}$$

from the dynamical system $\dot{X}_t = b(X_t)$, $X_0 = x \in R^n$.

Roughly speaking, for each time scale $T = T(\varepsilon) = e^{\frac{\lambda}{\varepsilon}}$, $\lambda > 0$, and each initial point $x \in R^n$, there exists an equilibrium O_k, $k = k(x,\lambda)$ (metastable state for given x and λ), such that the perturbed system spends in a neighborhood of O_k most of the time if $\varepsilon \ll 1$. If now the vector field slowly changes with time, say, in a periodic way, then the metastable state may jump form one equilibrium to another. This leads to oscillations with amplitude of order 1 as $\varepsilon \downarrow 0$. These oscillations are called stochastic resonance (see details in [5]). Thus to study stochastic resonance, we should estimate the order of transitions (hierarchy of cycles [1], [5], [6]), metastable states, and transition times.

Let us replace one-dimensional Wiener process W_t by the symmetric random walk $\xi_t^{\delta,1}$, with time steps δ and space step $\sqrt{\delta}$, so that $\xi_{t+\delta} - \xi_t = \pm\sqrt{\delta}$ with equal probabilities. It is well known that $\xi_t^{\delta,1}$, $0 \le t \le T$, converges to the Wiener process. The random walk $\xi_t^{\delta} = (\xi_t^{\delta,1}, \ldots, \xi_t^{\delta,n})$, $t \in N_\delta = \{0, \delta, \ldots, k\delta, \ldots\}$, on the n-dimensional lattice $Z^n_{\sqrt{\delta}}$ ($\sqrt{\delta}$ is the step of the lattice), where $\xi_t^{\delta,k}$ are independent and identically distributed, converges as $\delta \downarrow 0$ to the n-dimensional Wiener process. Therefore, such an approximation which is equivalent to replacement of the differential equation by an appropriate difference equation, works in many problems for second orders parabolic and elliptic equations.

One could expect that the sequence of transitions between the equilibriums O_k and transition times for X_t^ε are close to such characteristics for the process $X_t^{\delta,\varepsilon}$,

$$X_{t+\delta}^{\delta,\varepsilon} - X_t^{\varepsilon,\delta} = b(X_t^{\delta,\varepsilon})\delta + \sqrt{\varepsilon}(\xi_{t+\delta}^{\delta} - \xi_t^{\delta}), \;\; X_0^{\delta,\varepsilon} = x, \tag{5}$$

where the Wiener process is replaced by a random walk with $\delta \ll 1$. In a similar way, equation (2) could be approximated by the equation

$$u^{\varepsilon,\delta}(t,x) = E_x^{\delta} g(X_t^{\delta,\varepsilon}) exp\{\frac{\delta}{\varepsilon} \sum_{k=0}^{\delta^{-1}t-1} (c(X_{k\delta}^{\delta,\varepsilon}, u^{\varepsilon,\delta}(t-k\delta, X_{k\delta}^{\delta,\varepsilon}))\} \tag{6}$$

where $X_{k\delta}^{\delta,\varepsilon}$ is defined by (5) with $b(x) \equiv 0$.

But it turns out that the large deviation asymptotics for process $X_t^{\delta,\varepsilon}$, $0 \le t \le T$, as ε, $\delta \downarrow 0$, is, in general, different from such an asymptotics for process X_t^ε. Therefore, if say, we take $\delta = \varepsilon \downarrow 0$, the result obtained by computer simulation are different from what we, actually, want to calculate.

Actually, we face here a deeper problem: The main reason why the noises in many physical models are considered Gaussian is the Central Limit Theorem (CLT). But the CLT guarantees closeness to Gaussian distributions just for the normal deviations. Large deviations for a sum of independent random variables and for the limiting Gaussian distribution are, in general, different. The same is true for stochastic process. For instance, in this paper we consider the question how one should choose a random walk which approximates the Wiener process and has the same large deviation asymptotics. Therefore, in the problems related to large deviations a special attention should be paid to the assumptions concerning the noise.

2. Action Functional for the Process $X_t^{\delta,\varepsilon}$

Let X_t^θ, $0 \le t \le T$, be a family of continuous n-dimensional stochastic processes depending on a parameter $\theta > 0$. Let $\lambda(\theta) \to \infty$ as $\theta \downarrow 0$. Recall [6] that, if $\lambda(\theta)S_{0T}(\varphi)$, $\varphi \in C_{0T}(R^n)$, is the action functional for the family

X_t^θ as $\theta \downarrow 0$ in the space $C_{0T}(R^n)$, then, for any "regular enough" set $A \subset C_{0T}(R^n)$

$$\lim_{\theta \downarrow 0} \lambda^{-1}(\theta) ln P\{X^\theta_\cdot \in A\} = -inf\,\{S_{0T}(\varphi) : \varphi \in A\}$$

If $F[\varphi]$ is a bounded continuous functional on $C_{0T}(R^n)$, then

$$\lim_{\theta \downarrow 0} \lambda^{-1}(\theta) ln E exp\{\lambda(\theta) F([X^\theta_\cdot])\} = \inf_{\varphi \in C_{0T}(R^n)} \{F[\varphi] - S_{0T}(\varphi)\}$$

These equalities, actually, can be taken as the definition of the action functional.

It is convenient for us to deal with continuous processes on $[0, T]$. So that we will consider the process $X_t^{\delta,\varepsilon}$, defined by (5) just for $t = k\delta$, k is an integer, for all $t \in [0, T]$: for $t \in [k\delta, (k+1)\delta]$ we define $X_t^{\delta,\varepsilon}$ as the linear function connecting the points $(k\delta, X_{k\delta}^{\delta,\varepsilon})$ and $((k+1)\delta, X_{(k+1)\delta}^{\delta,\varepsilon})$.

Let $\varepsilon, \delta \downarrow 0$ so that $\frac{\delta}{\varepsilon} = \mu^2$, where $\mu > 0$ is a fixed constant. Consider, first, the one-dimensional case and assume that $b(x) \equiv 0$. Then $X_{m\delta}^{\delta,\varepsilon} - x$ is the sum of m independent random variables $\eta_i^\delta = \sqrt{\varepsilon}(\xi^\delta_{(i+1)\delta} - \xi^\delta_{i\delta})$; $\eta_i^\delta = \pm\sqrt{\delta\varepsilon}$ with probabilities $\frac{1}{2}$. To calculate the action functional for $X_t^{\delta,\varepsilon}$ as $\varepsilon, \delta \downarrow 0$, $\frac{\delta}{\varepsilon} = \mu^2$, according to [8], §§4.2-4.3, one should, first, calculate $G_0^{\mu,\varepsilon}(z) = \frac{1}{\delta} ln E exp\{z\eta_i^\delta\}$. Then one should find $G_0^\mu(z) = \lim_{\varepsilon \downarrow 0} \varepsilon G_0^{\mu,\varepsilon}(z\varepsilon^{-1})$. In our case,

$$G_0^{\mu,\varepsilon}(z) = \frac{1}{\delta} ln \frac{e^{z\sqrt{\delta\varepsilon}} + e^{-z\sqrt{\delta\varepsilon}}}{2}, \quad G_0^\mu(z) = \frac{1}{\mu^2} ln \frac{e^{\mu z} + e^{-\mu z}}{2}$$

The function $G_0^\mu(z)$ is convex. Let $L_0^\mu(u)$ be the Legendre transform of $G_0^\mu(z)$: $L_0^\mu(u) = \sup_Z [uz - G_0^\mu(z)]$. Finding this supremum, we have:

$$L_0^\mu(u) = \begin{cases} \frac{1}{2\mu^2}[(1+\mu u) ln(1+\mu u) + (1-\mu u) ln(1-\mu u)], & u \le \frac{1}{\mu} \\ +\infty, u > \frac{1}{\mu} \end{cases} \tag{7}$$

Then the action functional for $X_t^{\delta,\varepsilon}$ as $\varepsilon, \delta \downarrow 0$, $\frac{\delta}{\varepsilon} = \mu^2$ in $C_{0T}(R^1)$ is equal $\lambda(\varepsilon) S_{0,T}^{\mu,0}(\varphi)$, where $\lambda(\varepsilon) = \varepsilon^{-1}$, $S_{0,T}^{\mu,0}(\varphi) = \int_0^T L_0^\mu(\dot{\varphi}_s) ds$, if $\varphi \in C_{0T}(R^1)$ is absolutely continuous and $\varphi_0 = x$ and $S_{0,T}^{\mu,0}(\varphi) = +\infty$ for the rest of $C_{0T}(R^1)$.

Consider now n-dimensional random walk $X_t^{\delta,\varepsilon}$ defined by (5) with $b(x) \equiv 0$. Since the components of this process are independent, we conclude that the action functional for $X_t^{\delta,\varepsilon}$ as $\varepsilon, \delta \downarrow 0$, $\delta\varepsilon^{-1} = \mu^2$, in $C_{0T}(R^n)$ is equal

$$\frac{1}{\varepsilon} \tilde{S}_{0T}^\mu(\varphi) = \frac{1}{\varepsilon} \sum_{i=1}^n S_{0T}^\mu(\varphi^i), \ \varphi_t = (\varphi_t^1, \cdots, \varphi_t^n) \in C_{0T}(R^n)$$

To calculate the action functional in the case $b(x) \neq 0$, note that the map $R : \varphi \to X$ in $C_{0T}(R^n)$, defined by the equation $X_t = x + \int_0^t b(X_s)ds + (\varphi_t - \varphi_0)$ is continuous, and $R^{-1}(X) = \varphi$ is defined in the unique way if we assume that $\varphi_0 = 0$. Thus, because of contraction principle ([6], Th. 3.3.1), we conclude that the action functional for the family $X_t^{\delta,\varepsilon}$ defined by (5) is equal $\frac{1}{\varepsilon}\tilde{S}^{\mu}_{0,T}(R^{-1}(\varphi))$. We sum up this as the following theorem.

Theorem 1 *The action functional for the family $X_t^{\delta,\varepsilon}$ in the space C_{0T} as ε, $\delta \downarrow 0$ so that $\delta\varepsilon^{-1} = \mu^2 = const$ is equal $\frac{1}{\varepsilon}S^{\mu}_{0T}(\varphi)$, where*

$$S^{\mu}_{0T}(\varphi) = \begin{cases} \int_0^T \sum_{i=1}^{n} L_0^{\mu}(\dot{\varphi}_s^i - b^i(\varphi_s))ds, \text{ if } \varphi_t = (\varphi_t^1, \cdots, \varphi_t^n) \in C_{0T}(R^n) \text{ is} \\ \text{absolutely continuous and} \varphi_0 = x, \\ \quad +\infty, \text{ for the rest of } C_{0T}(R^n), \end{cases}$$

and $L_0^{\mu}(u)$ is defined by (7).

Lemma 2 *Let $L_0^{\mu}(u)$ be defined by (7). Then*

(i) *$L_0^{\mu}(u)$in the interval $|u| < \frac{1}{\mu}$has the expansion: $L_0^{\mu}(u) = \frac{1}{2\mu^2}\sum_{k=1}^{\infty}\frac{\mu^{2k}u^{2k}}{k(2k-1)}$;*

(ii) *$L_0^{\mu}(u) \geq \frac{u^2}{2}$; $L_0^{\mu}(u) = \frac{u^2}{2} + \frac{\mu^2 u^4}{12} + o(\mu^4), \mu \downarrow 0$; $L_0^{\mu}(\frac{1}{\mu}) \geq \frac{ln2}{\mu^2}$.*

The statement (i) follows from the expansion of $ln(1 + \mu u)$. The second statement follows from the first since all the terms in the expansion for$L_0^{\mu}(u)$are nonnegative. The third statement follows from the first. The last one follows from the definition of $L_0^{\mu}(u)$.

So far we considered the case $\varepsilon, \delta \downarrow 0$, $\delta\varepsilon^{-1} = \mu^2 = const > 0$. Consider now the case $\delta\varepsilon^{-1} \downarrow 0$. Using the fact that $G_0^{\mu}(u) = \frac{u^2}{2} + o(u^2), u \to 0$, and Theorem 4.4.1 from [8], we can derive the following result.

Theorem 3 *The action functional for the family$X_t^{\delta,\varepsilon}$as $\varepsilon, \delta \downarrow 0$, $\delta\varepsilon^{-1} \downarrow 0$ in $C_{0T}(R^n)$is equal $\varepsilon^{-1}S_{0T}(\varphi)$, where*

$$S_{0T}(\varphi) = \begin{cases} \frac{1}{2}\int_0^T |\dot{\varphi}_s - b(\varphi_s)|^2 ds, \text{ if } \varphi \in C_{0T}(R^n) \\ \text{is absolutely continuous and } \varphi_0 = x; \\ +\infty, \text{for the rest of } C_{0T}(R^n). \end{cases}$$

Consider briefly the case of variable diffusion matrix $a(x) = \sigma^2(x)$, $\det a(x) > 0$. The process X_t^{ε} in this case is defined by the equation

$$\dot{X}_t^{\varepsilon} = b(X_t^{\varepsilon}) + \sqrt{\varepsilon}\sigma(X_t^{\varepsilon})\dot{W}_t, \;\; X_t^{\varepsilon} = x \in R^n \tag{8}$$

This process can be approximated by the random walk

$$X_{t+\delta}^{\delta,\varepsilon} - X_t^{\varepsilon,\delta} = b(X_t^{\delta,\varepsilon})\delta + \sqrt{\varepsilon}\sigma(X_t^{\varepsilon})(\xi_{t+\delta}^{\delta} - \xi_t^{\delta}), \quad X_0^{\delta,\varepsilon} = x \in R^n \tag{9}$$

This is a generalization of (5) and $\xi_t^\delta$is the same as in (5).

The action functional $\varepsilon^{-1}\hat{S}_{0T}^{\mu}(\varphi)$ for the family of process $X_t^{\delta,\varepsilon}$ defined by (9) as $\varepsilon, \delta \downarrow 0$, $\delta\varepsilon^{-1} = \mu^2 > 0$ is given by the formula $\varepsilon^{-1}\hat{S}_{0T}^{\mu}(\varphi) = \varepsilon^{-1}\tilde{S}_{0T}^{\mu}(\int_0^t \sigma^{-1}(\varphi_s)[\dot{\varphi}_s - b(\varphi_s)]ds)$.

If $\varepsilon, \delta \downarrow 0$, $\delta\varepsilon^{-1} \downarrow 0$, then the action functional

$$\hat{\hat{S}}_{0T}(\varphi) = \begin{cases} \int_0^T |\sigma^{-1}(\varphi_s)(\dot{\varphi}_s - b(\varphi_s))|^2 ds \\ \qquad\qquad \text{for } \varphi \in C_{OT}(R^n) \\ \qquad\qquad \text{absolutely continuous}, \varphi_0 = x; \\ \infty, \text{ for the rest of } C_{OT}(R^n) \end{cases}$$

3. Wavefront Propagation Problem

Equation (1) is considered in this section. Assume that the nonlinear term $f(x,u) = c(x,u)u$ is of KPP type [2]: $c(x,u)$ is Lipschitz continuous, $c(x,u) > 0$ for $u < 1$ and $x \in R^n$, $c(x,u) < 0$ for $u > 1$, $x \in R^n$, $c(x) = c(x,0) = \max_{u \geq 0} c(x,u)$.

One can prove (see [2], §6.2), that, except a thin set, $\lim_{\varepsilon \downarrow 0} u^\varepsilon(t,x)$ is equal to 0 or to 1. The set $G_t \subset R^n$, where $\lim_{\varepsilon \downarrow 0} u^\varepsilon(t,x) = 1$, $t > 0$, is non-decreasing as t grows. So that to estimate $u^\varepsilon(t,x)$ for $\varepsilon \ll 1$, we need to describe the evolution of the sets G_t. The boundary of this set (the interface between the domains where $u^\varepsilon(t,x)$ is close to 0 or to 1) is the position of the wavefront at time $t > 0$. The support of the initial function we denote by G_0: $G_0 = \{x \in R^n : g(x) > 0\}$.

Consider the case when $c(x,0) = c(x) = c = const$. Then the motion of the front can be described by the Huygens principle [2], [3]:

$$G_t = \{x \in R^n : \rho(x, G_0) < t\sqrt{2c}\}, \tag{10}$$

where $\rho(\,.\,,\,.\,)$ is the Euclidian distance in R^n. If $\rho(x,G_0) > t\sqrt{2c}$, $\lim_{\varepsilon \downarrow 0} u^\varepsilon(t,x) = 0$. In particular, if the initial function is positive just in a small neighborhood of a point x_0, say, in a ball of radius $\sqrt{\varepsilon}$ with the center at x_0, then G_t is the ball of radius $t\sqrt{2c}$ with the center at x_0.

Consider now the approximating problem on the lattice $N_\delta \times Z^n_{\sqrt{\delta}}$:

$$u^{\delta,\varepsilon}(t,x) = E_x^\delta g(X_t^{\delta,\varepsilon}) exp\{\tfrac{\delta}{\varepsilon} \sum_{k=0}^{\delta^{-1}t-1} (c(X_{k\delta}^{\delta,\varepsilon}, u^{\delta,\varepsilon}(t-k\delta, X_{k\delta}^{\delta,\varepsilon}))\}, \quad t \in N_\delta \backslash \{0\},\ x \in Z^n_{\sqrt{\delta}} \tag{11}$$

Problem (11) has a unique solution for any $\varepsilon, \delta > 0$ (we assume that $c(x, u)$ is Lipschitz continuous). This solution can be constructed by successive approximations.

To study the limiting behavior of $u^{\delta,\varepsilon}(t, x)$ as $\varepsilon, \delta \downarrow 0$, $\delta\varepsilon^{-1} = \mu^2 > 0$, consider the equation for $v = v(e)$:

$$L^\mu(v(e), e) = c, \ e \in R^n, \ \ |e| = 1,$$

where

$$\begin{aligned} L^\mu(u) &= \sum_{k=1}^{n} L_0^\mu(u_k) \\ &= \tfrac{1}{2\mu^2} \sum_{k=1}^{n} [(1+\mu u_k)ln(1+\mu u_k) + (1-\mu u_k)ln(1-\mu u_k)], \\ u &= (u_1, \cdots, u_n), \in R^n \end{aligned}$$

For any unit vector $e \in R^n$, there exists a unique positive solution of this equation if

$$c \leq \tfrac{1}{2\mu^2} \sum_{k=1}^{n} [(1+\tfrac{e_k}{||e||})ln(1+\tfrac{e_k}{||e||}) + (1-\tfrac{e_k}{||e||})ln(1-\tfrac{e_k}{||e||})] \tag{12}$$

where $||e|| = \max\limits_{1 \leq i \leq n} |e_i|$, $e = (e_1, \ldots, e_n)$. Define

$$v^*(e) = v^*_{c,\mu}(e) = \begin{cases} v(e), & \text{if (12) is satisfied,} \\ \frac{1}{\mu||e||}, & \text{if (12) is not satisfied.} \end{cases}$$

If μ is small enough, condition (12) is satisfied for any unit vector $e \in R^n$.

Let $\prod = \{x \in R^n : \ |x| \leq v^*(\frac{x}{|x|})\}$. The set $\prod$ can be characterized also as the set where $L^\mu(x) \leq c$: $\prod = \{x \in R^n : \ L^\mu(x) \leq c\}$. Thus $\prod$ is a bounded convex set in R^n, since $L^\mu(x)$ is convex. The set $\prod$ is symmetric with respect to the origin.

Let us introduce a new metric $\rho^\mu(\,.\,,\,.\,)$ as the Finsler metric in R^n corresponding to the unit sphere $\prod$.

Theorem 4 *Let $f(x, u)$ is of KPP type and $c(x, 0) = c = const$, let $\varepsilon, \delta \downarrow 0$ so that $\delta\varepsilon^{-1} = \mu^2 = const > 0$. Then the motion of the front is described by the Huygens principle, which in the Finsler metric $\rho^\mu(\,.\,,\,.\,)$corresponds to the homogeneous and isotropic velocity field and the absolute value of the speed is equal to 1:* $\lim\limits_{\varepsilon,\delta\downarrow 0;\delta\varepsilon^{-1}=\mu^2} u^{\delta,\varepsilon}(t, x) = 1$ *if* $x \in G_t = \{x \in R^n : \rho^\mu(G_0, x) < t\}$, $\lim\limits_{\varepsilon,\delta\downarrow 0;\delta\varepsilon^{-1}=\mu^2} u^{\delta,\varepsilon}(t, x) = 0$ *if* $\rho^\mu(G_0, x) > t$.

The proof of this theorem is similar to the proof of Theorem 6.2.2 from [2], so that we give just a sketch of the proof.

First, we show that the solution of $u^{\delta,\varepsilon}(t,x)$ of equation (11) have the properties:

$$0 \le u^{\delta,\varepsilon}(t,x) \le \inf_{x\in R^n} gx(x), \tag{13}$$

$$\overline{lim}_{\varepsilon,\delta\downarrow 0;\delta\varepsilon^{-1}=\mu^2} u^{\delta,\varepsilon}(t,x) = 1, \tag{14}$$

To prove (13) and (14), we can use the strong Markov property of $X_t^{\delta,\varepsilon}$. For example, to prove (13), assume that a point $(t_0, x_0) \in N_\delta \times Z^n_{\sqrt{\delta}}$ exists such that $A = u^{\delta,\varepsilon}(t_0,x_0) > 1 \vee (\max_x g(x)) = B$. Consider the set $D = \{t \in N_\delta, x \in Z^n_{\sqrt{\delta}} : \quad u^{\delta,\varepsilon}(t,x) > B + \frac{A-B}{2}\}$, and let τ_D be the first time when $(t_0 - s,\, X_s^{\delta,\varepsilon})$ leaves D: $\tau = min\{s : (t_0 - s,\, X_s^{\delta,\varepsilon} \notin D\}$. It follows from the strong Markov property and (11), that

$$\begin{aligned} u^{\delta,\varepsilon}(t_0,x_0) &= E_{x_0} u^{\delta,\varepsilon}(t-\tau, X_\tau^{\delta,\varepsilon}) exp\{\tfrac{\delta}{\varepsilon}\sum_{k=0}^{\delta^{-1}\tau-1} c(X_{k\tau}^{\delta,\varepsilon}, u^{\delta,\varepsilon}(t_0 - k\delta, X_{k\delta}^{\delta,\varepsilon}))\} \\ &\le (B + \tfrac{A-B}{2}) = \tfrac{A+B}{2} < A \end{aligned} \tag{15}$$

We use here that $c(x,u) < 0$ for $u > 1$. Inequality (15) contradicts our assumption that $u(t_0,x_0) = A$. Equality (14) can be proved in a similar way.

Now, using the properties of the action functional and the fact that $c(x,0) \ge c(x,u)$ for $u \ge 0$, we have

$$\begin{aligned} u^{\delta,\varepsilon}(t,x) &= E_x g(X_t^{\delta,\varepsilon}) exp\{\tfrac{\delta}{\varepsilon}\sum_0^{t/\delta-1} c(X_s^{\delta,\varepsilon}, u^{\delta,\varepsilon}(t-s, X_s^{\delta,\varepsilon}))\} \\ &\le E_x g(X_t^{\delta,\varepsilon}) e^{\frac{ct}{\varepsilon}} = e^{\frac{ct}{\varepsilon}} ||g|| P\{X_t^{\delta,\varepsilon} \in G_0\} \\ &= ||g|| e^{\frac{ct}{\varepsilon}} exp\{\tfrac{1}{\varepsilon} inf\{\textstyle\int_0^t L^\mu(\dot\varphi_s)ds : \varphi_0 = x, \varphi_t \in G_0\}\}, \\ \|g\| &= \sup_{x\in R^n} g(x). \end{aligned} \tag{16}$$

Taking into account that the integrand in the variational problem depends on $\dot\varphi_s$ only, we conclude that the infimum in the right hand side of (16) is equal to $\inf_{y\in G_0} tL^\mu(\frac{y-x}{t})$.

The inequality $\inf_{y\in G_0} L^\mu(\frac{x-y}{t}) > c$ is equivalent to the inequality $\frac{|x-y|}{t} > v^*(\frac{x-y}{|x-y|})$. Thus $\inf_{y\in G_0} tL^\mu(\frac{y-x}{t}) > ct$ if and only if $\rho^\mu(G_0,x) > t$. Taking this into account, we conclude from (16): $0 \le u^{\delta,\varepsilon}(t,x) \le ||g|| \exp\{\frac{1}{\varepsilon}[ct - \inf_{y\in G_0} tL^\mu(\frac{y-x}{t})]\} \to 0$, as $\varepsilon, \delta \downarrow 0$, $\delta\varepsilon^{-1} = \mu$, if $\rho^\mu(G_0,x) > t$.

To prove that $\lim_{\varepsilon,\delta\downarrow 0;\delta\varepsilon^{-1}=\mu^2} u^{\delta,\varepsilon}(t,x) = 1$ if $\rho^\mu(G_0,x) < t$, $\varepsilon,\delta \downarrow 0$, $\delta\varepsilon^{-1} = \mu^2$, it is sufficient, because of (14), to check that in this domain,

$$\underline{lim}_{\varepsilon,\delta\downarrow 0;\delta\varepsilon^{-1}=\mu^2} \quad u^{\delta,\varepsilon}(t,x) \geq 1 \tag{17}$$

The last inequality can be proved in two steps. First, we prove that for any $h_1 > 0$ there exists $h_2 > 0$ such that

$$u^{\delta,\varepsilon}(t,x) > exp\{-\frac{h_1}{\varepsilon}\} \tag{18}$$

for (t,x) such that $\rho^\mu(G_0,x) = (1+h_2)t$, $0 < t \leq T$, and $\varepsilon < 0$, $\delta = \mu^2\varepsilon$ are small enough. This can be proved similarly to Lemma 3.1 from [3], if one takes into account that the action functional for $X_t^{\delta,\varepsilon}$ in $C_{0T}(R^n)$ as $\varepsilon,\delta \downarrow 0$, $\delta\varepsilon^{-1} = \mu^2 > 0$ is given by Theorem 2.1. Using bound (18), the proof of (17) is carried out in the way similar to the proof of (13). (Compare with Theorem 3.1 [3]).

I will mention now some differences in the front motion for (2) and (11) as $\varepsilon,\delta \downarrow 0$, $\delta\varepsilon^{-1} = \mu^2 = const > 0$:

1 The asymptotic wavefront velocity, in the case of (2), is a homogeneous and isotropic field. In the case of (11), the speed of the front $v_\mu^*(e)$ in the direction defined by the unit vector $e \in R^n$ is invariant with respect to a smaller group, namely with respect to the subgroup of the orthogonal group, which keeps unchanged the lattice $Z^n_{\sqrt{\delta}}$.

2 The front speed $v^*(e) = v^*_{c,\mu}(e)$ in the case of (11) is always less than the speed in (2): $v^*(e) < \sqrt{2c}$. Moreover, the speed in the case of (2) tends to infinity as $c = c(x,0)$ grows. But $v^*_{c,\mu}(e)$ always satisfies the inequality $v^*_{c,\mu}(e) \leq \frac{1}{\mu\|e\|} \leq \frac{\sqrt{n}}{\mu}$. This follows form Lemma 2.1.

3 One can see form Lemma 2.1(i), that $\lim_{\mu\downarrow 0} v^*_{c,\mu}(e) = \sqrt{2c}$, so that as $\mu \downarrow 0$, $v^*_{c,\mu}(e)$ approaches the asymptotic wavefront speed in (2). Moreover, using Lemma 2.1(i), we can find the correction term when $\mu \ll 1$:

$$v^*_{c,\mu}(e) = \sqrt{2c} - \mu^2 \frac{(2c)^{3/2} \sum_1^n e_i^4}{3} + O(\mu^4), \quad \mu \downarrow 0 \tag{19}$$

Thus if μ is not small the difference in the front motion for (2) and (11) can be essential. On the other hand, if μ is small the speeds of the fronts in both cases are close and the formula (19) gives the correction term.

In the same way as Theorem 3.1, the following result can be proved if one uses Theorem 2.2 instead of Theorem 2.1.

Theorem 5 *Let* $\varepsilon, \delta \downarrow 0$ *and* $\delta\varepsilon^{-1} \downarrow 0$. *Let* $c(x, 0) = c = const.$ *Then* $\lim_{\varepsilon,\delta\downarrow 0;\delta\varepsilon^{-1}\downarrow 0} u^{\delta,\varepsilon}(t,x) = 1$ *if* $\rho(G_0, x) < t\sqrt{2c}$, *and* $\lim_{\varepsilon,\delta\downarrow 0;\delta\varepsilon^{-1}\downarrow 0} u^{\delta,\varepsilon}(t,x) = 0$ *if* $\rho(G_0,x) > t\sqrt{2c}$ *uniformly in any compact domain in the space* $\{t > 0, x \in R^n\}$ *which has empty intersection with the set* $\{(t,x) : \rho(G_0,x) = t\sqrt{2c}\}$. *Here* $\rho(.,.)$ *is the Euclidian metric in* R^n.

4. Exit Problem and Stochastic Resonance

Let the vector field $b(x)$, $x \in R^n$, have an asymptotically stable equilibrium at a point $O \in R^n$. Let a bounded domain $G \subset R^n$ be attracted to O : $\lim_{t\to\infty} X_t = O$, for each trajectory $\dot{X}_t = b(X_t)$, $X_0 = x \in G$. Assume that G has a smooth boundary ∂G and $b(x)\cdot n(x)|_{x\in\partial G} < 0$, where $n(x)$ is the exterior normal to ∂G. Denote by $\tau = \tau^\varepsilon(\tau^\delta = \tau^{\delta,\varepsilon})$ the first exit time from G for the process $X_t^\varepsilon(X_t^{\delta,\varepsilon})$: $\tau^\varepsilon = \min\{t : X_t^\varepsilon \notin G\}$, $\tau^{\delta,\varepsilon} = \min\{t \in N_\delta : X_t^{\delta,\varepsilon} \notin G\}$. To describe the asymptotic behavior of τ^ε and $\tau^{\delta,\varepsilon}$, introduce the functions $V(x)$, $V^\mu(x)$:

$$
\begin{aligned}
V(x) &= \inf\left\{\tfrac{1}{2}\textstyle\int_{T_1}^{T_2} |\dot\varphi_s - b(\varphi_s)|^2 ds : \varphi_{T_1} = 0,\right.\\
&\qquad\qquad \left.\varphi_{T_2} = x, -\infty \le T_1 < T_2 \le \infty\right\},\\
V^\mu(x) &= \inf\left\{\tfrac{1}{2}\textstyle\int_{T_1}^{T_2} L^\mu(\dot\varphi_s - b(\varphi_s))ds : \varphi_{T_1} = 0,\right.\\
&\qquad\qquad \left.\varphi_{T_2} = x, -\infty \le T_1 < T_2 \le \infty\right\}.
\end{aligned}
$$

The functions $V(x)$ and $V^\mu(x)$ are non-negative. They can be found as the solutions of corresponding Jacobi-Hamilton Equations:

$$
\begin{aligned}
&\tfrac{1}{2}|\nabla V(x)|^2 + b(x)\cdot\nabla V(x) = 0,\ \ V(0) = 0,\ \ V(x) > 0 \text{ if } x \ne 0;\\
&\frac{1}{\mu^2}\sum_{K=1}^{n} ln\frac{e^{\mu\frac{\partial V^\mu(x)}{\partial x^k}} + e^{-\mu\frac{\partial V^\mu(x)}{\partial x^k}}}{2} + b(x)\cdot\nabla V^\mu(x) = 0,\ V^\mu(0) = 0,\\
&V^\mu(x) > 0 \text{ if } x \ne 0.
\end{aligned}
\tag{20}
$$

Theorem 6 *Let* G *be a bounded domain in* R^n *with a smooth boundary* ∂G *and* $O \in G$. *Assume that* $b(O) = 0$ *and the domain* G *is attracted to* O. *Let* $b(x)\cdot n(x)|_{x\in\partial G} < 0$. *Put* $V_0 = \min_{x\in\partial G} V(x)$, $V_0^\mu = \min_{x\in\partial G} V^\mu(x)$.

Then for any $x \in G$ and $h > 0$,

$$
\lim_{\varepsilon,\delta\downarrow 0;\delta\varepsilon^{-1}=\mu^2} \varepsilon ln E_x\tau^{\delta,\varepsilon} = V_0^\mu,\ \ \lim_{\varepsilon,\delta\varepsilon^{-1}\downarrow 0} \varepsilon ln E_x\tau^{\delta,\varepsilon} = \lim_{\varepsilon\downarrow 0} \varepsilon ln E_x\tau^\varepsilon = V_0,
$$

$$\lim_{\varepsilon,\delta\downarrow 0;\delta\varepsilon^{-1}=\mu^2} P_x\{e^{\frac{V_0^\mu-h}{\varepsilon}} < \tau^{\delta,\varepsilon} < e^{\frac{V_0^\mu+h}{\varepsilon}}\} = 1,$$

$$\lim_{\varepsilon,\delta\varepsilon^{-1}\downarrow 0} P_x\{e^{\frac{V_0-h}{\varepsilon}} < \tau^{\delta,\varepsilon} < e^{\frac{V_0+h}{\varepsilon}}\} = \lim_{\varepsilon\downarrow 0} P_x\{e^{\frac{V_0-h}{\varepsilon}} < \tau^{\varepsilon} < e^{\frac{V_0+h}{\varepsilon}}\} = 1.$$

If $\min_{x\in\partial G} V^\mu(x)(\min_{x\in\partial G} V(x))$ is achieved just at one point $x_*^\mu \in \partial G(x_* \in \partial G)$, then

$$\lim_{\varepsilon,\delta\downarrow 0;\delta\varepsilon^{-1}=\mu^2} P_x\{|X_{\tau^{\delta,\varepsilon}}^{\delta,\varepsilon} - x_*^\mu| > h\} = 0,$$

$$\lim_{\varepsilon,\delta\varepsilon^{-1}\downarrow 0} P_x\{|X_{\tau^{\delta,\varepsilon}}^{\delta,\varepsilon} - x_*| > h\} = \lim_{\varepsilon\downarrow 0} P_x\{|X_{\tau^\varepsilon}^{\varepsilon} - x_*| > h\} = 0$$

Taking into account Theorem 2.1 and 2.2, the proof of this theorem is a modification of the proof of the results in Ch. 4 of [6] and we omit it.

I will mention some peculiarities of the exit problem for the processes $X_t^{\delta,\varepsilon}$:

1 For each $\mu > 0$, $V_0^\mu > V_0$, since $L^\mu(z) > \frac{|z|^2}{2}$ for $z \in R^n$, $z \neq 0$.

2 Since $L^\mu(z) = \infty$, if $||z|| = \max_{1\le i\le n} |z_i| > \frac{1}{\mu}$, V_0^μ can be equal to $+\infty$.

3 Consider $V^\mu(x)$ for $\mu \ll 1$. Assume that $b(x) = -\nabla U(x) + l(x)$ where the function $U(x)$ and vector field $l(x)$ are smooth enough and $\nabla U(x)\cdot l(x) = 0$, $x \in R^n$. The equation (20) for $V^\mu(x)$ has the form

$$\nabla U(x)\cdot\nabla V^\mu(x) - l(x)\cdot V^\mu(x) = \frac{1}{\mu^2}\sum_{K=1}^n ln\frac{e^{\mu\frac{\partial V^\mu(x)}{\partial x^k}} + e^{-\mu\frac{\partial V^\mu(x)}{\partial x^k}}}{2}$$
$$= \frac{|\nabla V^\mu(x)|^2}{2} - \frac{\mu^2}{12}\sum_{K=1}^n (\frac{\partial V^\mu(x)}{\partial x^K})^4 + o(\mu^2), \mu\downarrow 0 \quad (21)$$

We will look for the solution of (21) with the condition $V^\mu(0) = 0$ in the form: $V^\mu(x) = V_0^\mu(x) + \mu^2 V_1^\mu(x) + \cdots$. Substituting this series in (21), we find that $V_0^\mu(x) = 2U(x)$. The function $V_1^\mu(x)$ satisfies the equation

$$\nabla V_1^\mu(x)\cdot(\nabla U(x) + l(x)) = \frac{4}{3}\sum_{k=1}^n (\frac{\partial U}{\partial x^k})^4, \quad V_1^\mu(0) = 0 \quad (22)$$

Let z_t be the solution of the equation

$$\dot{z}_t = \nabla U(z_t) + l(z_t), \quad z_0 = x \in G\cap\{U(x) < V_0\}, \quad t < 0.$$

There exists a unique solution of this problem and $\lim_{t\to-\infty} z_t = 0$. The solution of problem (22) can be written in the form $V_1^\mu(x) = \frac{4}{3}\int_{-\infty}^0 \sum_{k=1}^n [\frac{\partial U}{\partial x^k}(z_t)]^4 dt$.

Consider two Lipschitz continuous vector fields $b_1(x)$ and $b_2(x)$, $x \in R^n$. Assume that each of them has two asymptotically stable equilibriums O_i^j: $b_i(O_i^j) = 0, i, j \in \{1,2\}$. Denote by D_i^j the basin of O_i^j for the field $b_i(x)$. Assume, that $R^n = D_1^1 \cup D_1^2 \cup \Gamma_1$, $R^n = D_2^1 \cup D_2^2 \cup \Gamma_2$, where Γ_1, Γ_2 are smooth $(n-1)$ dimensional manifolds separating the basins D_1^1 and D_1^2 and D_2^1 and D_2^2, respectively. Let, for brevity, $O_2^1 \in D_1^1$, $O_2^2 \in D_1^2$, $O_1^1 \in D_2^1$, $O_1^2 \in D_2^2$.

Consider stochastic processes

$$\dot{X}_1^\varepsilon(t) = b_1(X_1^\varepsilon(t)) + \sqrt{\varepsilon}\dot{W}_t, \quad X_1^\varepsilon(0) = x_1,$$

$$\dot{X}_2^\varepsilon(t) = b_2(X_2^\varepsilon(t)) + \sqrt{\varepsilon}\dot{W}_t, \quad X_2^\varepsilon(0) = x_2.$$

Let $\rho > 0$ be so small that ρ-neighborhood E_i^j of O_i^j belongs to D_i^j

$$\begin{aligned} \tau_i^j &= min\{t : X_i^\varepsilon(t) \in E_i^j\};\ i, j \in \{1,2\}; \\ V_i^{j,k} &= inf\{\textstyle\int_{T_1}^{T_2} |\dot{\varphi}_s - b_i(\varphi_s)|^2 ds : \varphi_{T_1} = O_i^j, \varphi_{T_2} \in \Gamma_i\}; \\ & i, j, k \in \{1,2\}, j \neq k \end{aligned} \tag{23}$$

It follows from Chapter 4 of [6], that $\lim_{\varepsilon\downarrow 0} \varepsilon ln E_x \tau_i^k = V_i^{j,k}$ if $x \in D_i^j$ and $j \neq k$, $\lim_{\varepsilon\downarrow 0} P_x\{e^{\frac{V_i^{j,k}-h}{\varepsilon}} < \tau_i^k < e^{\frac{V_i^{j,k}+h}{\varepsilon}}\} = 1$ for any $h > 0$, $x \in D_i^j$, $j \neq k$. This implies, that if $V_i^{1,2} < V_i^{2,1}$, then the process $X_i^\varepsilon(t)$ spends most of the time as $\varepsilon \downarrow 0$ in a neighborhood of O_i^2. If $V_i^{1,2} > V_i^{2,1}$, then $X_i^\varepsilon(t)$ spends most of the time near O_i^1.

Suppose $V_1^{1,2} > V_1^{2,1}$ and $V_2^{1,2} > V_2^{2,1}$, so that the first process $X_1^\varepsilon(t)$ spends most of the time near O_1^1 and $X_2^\varepsilon(t)$ spends most of the time near O_2^2. Let $v(t)$ be a 1-periodic function such that $v(t) = 1$ for $t \in [0, \bar{t}]$, $v(t) = 2$ for $t \in [\bar{t}, 1]$, $\bar{t} \in (0,1)$. Consider the process $X_t^{\varepsilon,T}$:

$$\dot{X}_t^{\varepsilon,T} = b_{v(\frac{t}{T})}(X_t^{\varepsilon,T}) + \sqrt{\varepsilon}\dot{W}_t, \quad X_0^{\varepsilon,T} = x, \quad T > 0. \tag{24}$$

Let $T = e^{\frac{\lambda}{\varepsilon}}$, $\lambda > (V_1^{2,1} \vee V_2^{1,2})$. One can check that then $X_t^{\varepsilon,T}$ will be, in a sense, close as $\varepsilon \downarrow 0$ to a periodic function with the period $T = T(\varepsilon)$. The function has two values: O_1^1 on the interval $[\varphi, \varphi + \bar{t}T(\varepsilon))$, and O_2^2 on the interval $[\varphi + \bar{t}T(\varepsilon), \varphi + T(\varepsilon))$; the phase φ is defined by the initial point. The transitions from O_1^1 to O_2^2 and back occur due to the small noise and their frequency is defined by the periodic oscillations of $v(\frac{t}{T})$. This effect is called stochastic resonance. The general mathematical treatment one can find in [5].

Suppose that we replaced the Wiener process in (24) by the random walk ξ_t^δ:

$$X_{t+\delta}^{\delta,\varepsilon,T} - X_t^{\delta,\varepsilon,T} = \delta b_{v(\frac{t}{T})}(X_t^{\delta,\varepsilon,T}) + \sqrt{\varepsilon}(\xi_{t+\delta}^\delta - \xi_t^\delta), \;\; t \in N_\delta. \qquad (25)$$

Then logarithmic asymptotics of the transition times for $X_t^{\delta,\varepsilon,T}$ as $\varepsilon, \delta \downarrow 0$, $\delta\varepsilon^{-1} = \mu^2$ is defined by the numbers $V_i^{j,k,\mu}$, which are defined by (23) but with $\int_{T_1}^{T_2} |\dot{\varphi}_t - b_i(\varphi_s)|^2 ds$ replaced by $\int_{T_1}^{T_2} L^\mu(\dot{\varphi}_t - b_i(\varphi_s))ds$. This, actually, follows from Theorem 4.1. Therefore $V_i^{j,k,\mu} > V_i^{j,k}$ for any $\mu > 0$. But one can check that $\lim_{\mu\downarrow 0} V_i^{j,k,\mu} = V_i^{j,k}$, so that if $\varepsilon, \delta, \mu \ll 1$ we can expect that the results of computer simulation of $X_t^{\delta,\varepsilon,T}$ are close to real behavior of $X_t^{\varepsilon,T}$. One should keep in mind that, if μ is not small enough, $V_i^{j,k,\mu}$ can be equal to infinity and the transitions can be impossible.

If the vector fields $b_i(x)$ have more than 2 stable equilibriums, the white noise perturbations and random walk perturbations can lead to different hierarchies of cycles (see [1]) and different conditions for stochastic resonance.

References

[1] M. I. Freidlin, "Sublimiting distributions and stabilization of solutions of parabolic equations with a small parameters," *Soviet Math, Dokl.*, vol. 18, no. 4, pp. 1114–1118, 1977.

[2] M. I. Freidlin, *Functional Integration and Partial Differential Equations*, Princeton Univ. Press, 1985.

[3] M. I. Freidlin, "Semi-linear PDE's and limit theorems for large deviations," *Lectures Notes in Mathematics*, 1527, Springer Verlag, 1991.

[4] M. I. Freidlin, *Markov Processes and Differential Equations: Asymptotic Problems*, Birkhäuer, 1996.

[5] M. I. Freidlin, "Quasi-deterministic approximation, metastability and stochastic resonance," *Physica D*, vol. 137, pp. 333–352, 2000.

[6] M. I. Freidlin and A. D. Wentzell, *Random Perturbations of Dynamical Systems*, Second edition, Springer Verlag, 1998.

[7] L. Gammaitoni, P. Hånggi, P. Jung, and F. Marchesooni, "Stochastic resonance," *Reviews in Modern Physics*, vol. 70, no. 1, pp. 224–287, 2000

[8] A. D. Wentzell, *Limit Theorems on Large deviations for Markov Stochastic Processes*, Kluwer, 1991.

THE REDUCTION OF POTENTIAL DIFFUSIONS TO FINITE STATE MARKOV CHAINS AND STOCHASTIC RESONANCE

Peter Imkeller
Institut für Mathematik, Humboldt-Universität zu Berlin
Unter den Linden 6, 10099 Berlin, Germany
imkeller@mathematik.hu-berlin.de

Ilya Pavlyukevich
Institut für Mathematik, Technische Universität Berlin
Straße des 17. Juni 136, 10623 Berlin, Germany
pavljuke@math.tu-berlin.de

Abstract Consider a dynamical system describing the motion of a particle in a double well potential with a periodic perturbation of very small frequency, and a white noise perturbation of intensity ε. If its trajectories amplify the small periodic perturbation in a 'best possible way', it is said to be in stochastic resonance. A lower bound for the ratio of amplitude and logarithm of the period above which quasi-deterministic periodic behavior can be observed is obtained via large deviations theory. However, to obtain optimality, periodicity of trajectories has to be studied by means of a measure of quality of tuning such as spectral power amplification. In the particular setting where the potential alternates every half period between two spatially antisymmetric double well states we encounter a surprise. The stochastic resonance pattern is not correctly described by the reduced dynamics associated with a two state Markov chain whose periodic hopping rates between the potential minima mimic the large (spatial) scale motion of the diffusion. Only if small scale fluctuations inside the potential wells where the diffusion spends most of its time are carefully eliminated, the reduced dynamics is robust.

Keywords: Stochastic resonance, spectral gap, stochastic differential equation, energy balance model, potential diffusion, effective dynamics.

N. Sri Namachchivaya and Y.K. Lin (eds.),
IUTAM Symposium on Nonlinear Stochastic Dynamics, 57-69.

1. Background and Paradigm

The physical effect we shall discuss is related to the dynamics of an overdamped Brownian particle in a periodically changing double well potential landscape described by a potential function $U(x,t)$, $x \in \mathbb{R}$, $t \geq 0$. To catch its essentials and at the same time to simplify the system we will work with a function U switching discontinuously every half period between two antisymmetric time independent potential states U_1 and U_2. In the strip $(x,t) \in \mathbb{R} \times [0,1)$

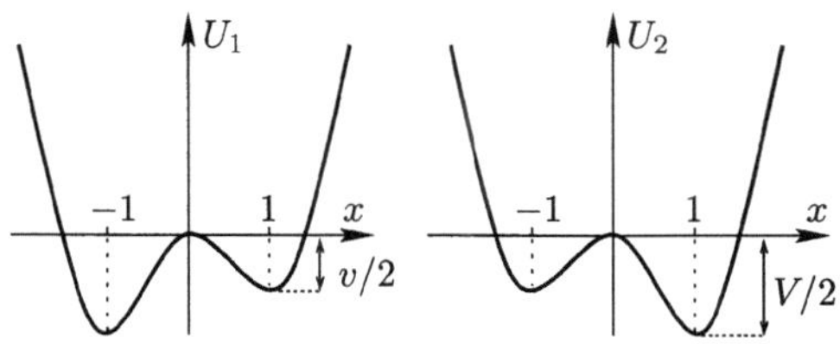

Figure 1. Time-periodic potential U.

it is defined by the formula

$$U(x,t) = \begin{cases} U_1(x), & t \in [0, \frac{1}{2}), \\ U_2(x) = U_1(-x), & t \in [\frac{1}{2}, 1). \end{cases} \tag{1}$$

It is periodically extended for all times t by the relation $U(\cdot,t) = U(\cdot,t+1)$, see Figure 1. We assume that the potential has two local minima at ± 1 and a local maximum at 0, that $U_1(-1) = -\frac{V}{2}$, $U_1(1) = -\frac{v}{2}$, $\frac{2}{3} < \frac{v}{V} < 1$, and $U_1(0) = 0$. We also suppose that the extrema of U are not degenerate, i.e. the curvatures at these points do not vanish.

A trajectory of a Brownian particle in this potential is described by the SDE

$$dX_t^{\varepsilon,T} = -U'(X_t^{\varepsilon,T}, \frac{t}{T})\, dt + \sqrt{\varepsilon}\, dW_t, \quad X_0^{\varepsilon,T} = x \in \mathbb{R}, \tag{2}$$

where $\varepsilon > 0$ is the noise intensity, and $T > 0$ the period.

The system is said to be in stochastic resonance if the trajectories of the diffusion given by (2) amplify the small periodic forcing in the potential in a 'best possible way'. To find the stochastic resonance point, we have to determine an *optimal tuning* $\varepsilon = \varepsilon(T)$, i.e., to find the noise intensity for which the trajectories $X^{\varepsilon,T}$ look 'as periodic as possible'. Of course, in these terms stochastic resonance is a rather unprecise concept. To make it precise requires at least that we are able to *measure* periodicity in diffusion trajectories.

Stochastic resonance as a tuning effect of periodically perturbed stochastic systems was first discussed in the context of a simple qualitative explanation of global glacial cycles in papers by C. Nicolis [15] and Benzi et al. [2, 3,

4]. In the context discussed there, the periodically varying potential function interprets X as the globally averaged annual temperature T of the earth, and describes the balance of incoming and outgoing radiative energy in the earth's atmosphere, normalized by the heat capacity factor c:

$$-\frac{\partial}{\partial T}U(T,t) = \frac{1}{c}[Q(t)(1-a(T(t))) - \sigma T(t)^4]. \tag{3}$$

In the simplest case considered here the second term on the right hand side, the total energy flux emitted by the earth, is assumed to be given by the Stefan–Boltzmann law of a black body radiator.

The first term on the right hand side, describing the absorbed energy flux depends on two factors. The *global solar* function $Q(t)$ describes the flux of the solar energy which reaches the earth at time t. It is supposed to fluctuate with a very long period of about 10^5 years and a small amplitude around an average value, and periodicity is attributed to eccentricities in the earth's orbit around the sun caused by the gravitational influence of exterior planets of the solar system. To obtain the solar radiation absorbed by the atmosphere, one has to multiply the solar function by the proportion of the radiation reaching the atmosphere which is not reflected to space, where the reflected ratio is measured by a purely temperature dependent albedo function $a(T)$. In the simple Budyko-Sellers setting (see Budyko [5] and Sellers [18]) the graph of the albedo function takes the shape of a ramp, and therefore U is a continuously varying potential function with a very large period, and in general time dependent local extrema. The potential function in (1) of our outset is just a simple caricature of the one encountered in the glacial cycles' example, still retaining its essential features.

2. Freidlin's Approach

Using large deviations theory, M. Freidlin [9] explains periodicity as a quasi-deterministic property of diffusion trajectories for very large period lengths.

For instance, let us study a Brownian particle in the time homogeneous double well potential U_1. Its motion is determined by the SDE

$$dX_t^\varepsilon = -U_1'(X_t^\varepsilon)\,dt + \sqrt{\varepsilon}\,dW_t. \tag{4}$$

For small ε, this stochastic system can be considered as a deterministic dynamical system $\dot{x} = -U_1'(x)$ perturbed by white noise of small intensity. The Freidlin-Wentzell theory of large deviations [10] allows to study asymptotic properties of (4) in the small noise limit in terms of the geometry of the potential U_1. It is intuitively clear that for ε small the trajectories of (4) spend most of the time very close to the *meta-stable states* ± 1. Jumps between the wells occur. The dynamics of these very rare transitions can be investigated by means of the *quasipotential*. It describes the work to be done by the diffusion in order to move between points in the potential landscape. Let, for instance, $X_0^\varepsilon = -1$ and x belong to the left well. Then the quasipotential $V(-1,x)$ equals $2(U_1(x) - U_1(-1))$ and thus corresponds to twice the height of the potential barrier between -1 and x. If x belongs to the right well and $0 < x \leq 1$, then $V(-1,x) = 2(U_1(0) - U_1(-1))$. Only the way 'up' in the potential landscape contributes to the quasipotential; the way 'down' requires no work and is free. Quasipotentials are defined for rather general classes of stochastic systems by means of *action functionals*, for details see [10].

Let us define the first hitting time of the point $y \in \mathbb{R}$ by

$$\tau_y^\varepsilon = \inf\{t \geq 0 \, : \, X_t^\varepsilon = y\}.$$

Then the quasipotential at x and y determines the exponential order of τ_y^ε if the diffusion starts in x (under the law $\mathbf{P}_x$) in the limit of small noise.

Theorem 1 ('transition law') *For all $\delta > 0$ the following holds:*

$$\lim_{\varepsilon \downarrow 0} \mathbf{P}_x\big(e^{\frac{1}{\varepsilon}(V(x,y)-\delta)} < \tau_y^\varepsilon < e^{\frac{1}{\varepsilon}(V(x,y)+\delta)}\big) = 1.$$

The most important statement of the theorem is that the system (4) has two different intrinsic time scales of exponential order: the exit time from the left well is of the order $e^{V/\varepsilon}$ whereas the exit time from the right well of the order $e^{v/\varepsilon}$. This has the following consequence. If we consider the diffusion (4) on the exponentially long time intervals $T_\varepsilon \propto e^{\lambda/\varepsilon}$, then for $0 < \lambda < v$ the trajectories typically cannot leave the initial well, whereas for $\lambda > v$ they spend most of their time near the deep valley (in probability). In other words, on different exponential time scales the system (4) has different meta-stable states.

For the time inhomogeneous system (2) this description of meta-stable behaviour has the following consequences. Let the period $T = T_\varepsilon$ be such that $\lim_{\varepsilon \downarrow 0} \varepsilon \log T_\varepsilon = \lambda > 0$. Then for $\lambda < v$ the diffusion does not have enough energy to leave even the shallow well during one half period, and therefore, as in the time homogeneous case, is stuck in its initial well. However, if $\lambda > v$, the following effect of quasi-periodicity appears.

Theorem 2 ([9]) *Let the process $X^{\varepsilon,T}$ satisfy (2), and*

$$\lim_{\varepsilon \downarrow 0} \varepsilon \log T_\varepsilon > v. \tag{5}$$

Then for all $A > 0$ and $\delta > 0$ the following holds true:

$$\Lambda\{t \in [0, A] \, : \, |X_{T_\varepsilon t}^{\varepsilon,T} - \phi(t)| > \delta\} \to 0$$

in $\mathbf{P}_x$*-probability as* $\varepsilon \to 0$*, where* $\Lambda\{\cdot\}$ *denotes Lebesgue measure on* $\mathbb{R}$*, and*

$$\phi(t) = \begin{cases} -1, & t(\text{mod } 1) \in [0, \frac{1}{2}), \\ 1, & t(\text{mod } 1) \in [\frac{1}{2}, 1), \end{cases}$$

is the coordinate of the global minimum of $U(\cdot, t)$*, see Figure 2.*

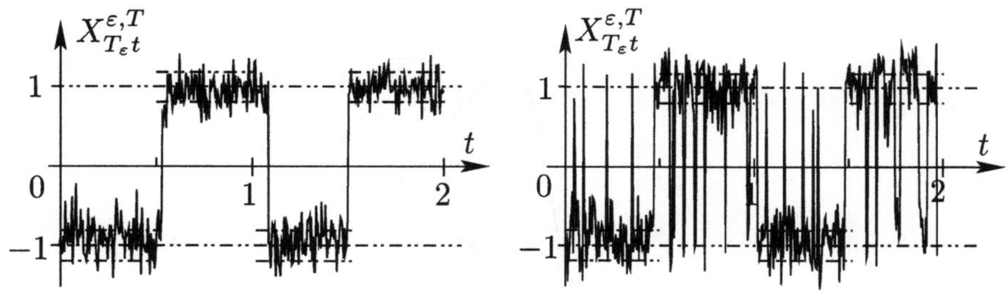

Figure 2. On time intervals satisfying condition (5) the diffusion $X^{\varepsilon,T}$ is close to the deterministic periodic function ϕ.

Theorem 2 suggests a measure of periodicity of diffusion trajectories: the Lebesgue measure of those times the trajectories spend outside of a δ-tube around the deterministic discontinuous periodic function ϕ. Condition (5) on period T_ε and noise intensity ε provides a family of tunings. Now a resonance point in the sense explained above would require to choose ε such that the above Lebesgue measure is minimal for given large period. Such an extreme value is not suggested by Freidlin's approach.

3. Spectral Power Amplification

The coefficient of *spectral power amplification* (SPA) is another measure of periodic tuning of random trajectories. In fact it appears to be the physicists' favorite measure, see e.g. [4, 14, 11, 1, 6]. For the diffusion (2) it is given by

$$\eta^X(\varepsilon, T) = \left| \int_0^1 \mathbf{E}_\mu(X^{\varepsilon,T}_{Ts}) \cdot e^{2\pi i s}\, ds \right|^2. \tag{6}$$

The function η^X depending on noise intensity and the period of time variation of the potential has a clear physical meaning. It describes the amount of energy carried by the average path of the diffusion with noise amplitude ε on the frequency $\frac{2\pi}{T}$. The expectation $\mathbf{E}_\mu$ indicates that averages are taken with respect to the invariant law of $X^{\varepsilon,T}$. This will be explained in detail later.

Figure 3 borrowed from [1] where Ω corresponds to our $\frac{2\pi}{T}$ and D to the diffusion intensity ε shows that physicists expect a local maximum of the function $\varepsilon \mapsto \eta^X(\varepsilon, \cdot)$. It is taken for a value of ε for which the random sample paths have their strongest periodic component. In fact, Figure 3 depicts not the SPA coefficient of the diffusion itself, but of its so-called 'effective dynamics'. The 'effective dynamics' of a diffusion in a double well potential is described by a two-state Markov chain living on the meta-stable states ± 1. It reflects only the *interwell* dynamics of the diffusion and neglects small fluctuations inside the wells. It is generally believed in the physical literature that the 'effective dynamics' adequately describes the asymptotic properties of the diffusion in the limit of small noise.

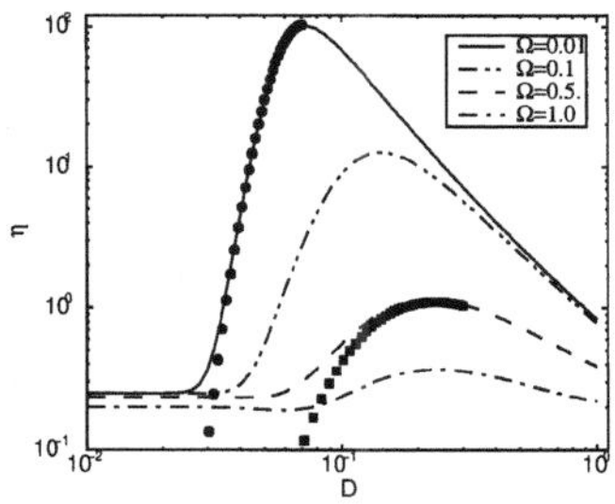

Figure 3. SPA coefficient as a function of noise amplitude has a well pronounced maximum depending at the frequency of periodic perturbation [1].

If periodic tuning is measured by SPA, the 'optimal tuning' or *stochastic resonance point* is determined by finding the argument $\varepsilon = \varepsilon(T)$ of a local maximum of $\varepsilon \mapsto \eta^X(\varepsilon, \cdot)$.

The key step to solve this problem consists in determining the density μ of the invariant law of $(X_{Tt}^{\varepsilon,T})_{t\geq 0}$. We now follow [17] and [13]. By enlarging the state space we can eliminate the inhomogeneity due to periodicity of U and consider a two-dimensional time homogeneous Markov process $(X_{Tt}^{\varepsilon,T}, t \pmod 1)$ which possesses an invariant law in the usual sense. The invariant density is a positive solution of the forward Kolmogorov (Fokker-Planck) equation $A_{\varepsilon,T}^* \mu = 0$, where

$$A_{\varepsilon,T}^* \cdot = -\frac{1}{T}\frac{\partial}{\partial t} \cdot + \frac{\varepsilon}{2}\frac{\partial^2}{\partial x^2} \cdot + \frac{\partial}{\partial x}\left(\cdot \frac{\partial}{\partial x} U\right)$$

is the formal adjoint of the infinitesimal generator of the two-dimensional diffusion. From the time periodicity and time-space antisymmetry of the potential U (1) the following constraints arise: $\mu(x,t) = \mu(-x, t+\frac{1}{2})$ and $\mu(x,t) = \mu(x, t+1)$, $(x,t) \in \mathbb{R} \times \mathbb{R}_+$.

These equations may be summarized in the following boundary-value problem used to determine μ. It is easily seen to reduce to the Fokker–Planck equation $A_{\varepsilon,T}^* \mu = 0$ in the strip $(x,t) \in \mathbb{R} \times [0, \frac{1}{2}]$ with boundary condition $\mu(x, 0) = \mu(-x, \frac{1}{2})$, $x \in \mathbb{R}$.

4. The Spectral Gap

In (1) the time dependent potential U is supposed to be a step function of the time variable. In the half period region $(x,t) \in \mathbb{R} \times (0, \frac{1}{2})$ it therefore is identical to a time independent double well potential U_1, and therefore the Fokker–Planck equation turns into a one-dimensional parabolic PDE

$$\frac{1}{T}\frac{\partial}{\partial t}\mu(x,t) = \frac{\varepsilon}{2}\frac{\partial^2}{\partial x^2}\mu(x,t) + \frac{\partial}{\partial x}\left(\mu(x,t)\frac{\partial}{\partial x}U_1(x)\right). \tag{7}$$

Let L_ε^* denote the second order differential operator appearing on the right hand side of (7).

To determine μ we shall expand the solution of (7) into a Fourier series with respect to the system of eigenfunctions of the operator L_ε^*. If U_1 is smooth and increases 'fast enough' at infinity (for example, as fast as x^4), the operator L_ε^* is essentially self-adjoint in $\mathcal{L}^2(\mathbb{R}, e^{\frac{2U_1}{\varepsilon}}\,dx)$, its spectrum is discrete and non-positive, and the corresponding eigenspaces are one-dimensional. Denoting by $\|\cdot\|$ and $\langle\cdot,\cdot\rangle$ the norm and the inner product in $\mathcal{L}^2(\mathbb{R}, e^{\frac{2U_1}{\varepsilon}}\,dx)$ we consider the following formal Floquet type expansion

$$\mu(x,t) = \sum_{k=0}^{\infty} a_k \frac{\Psi_k(x)}{\|\Psi_k\|} e^{-T\lambda_k t}, \quad (x,t) \in \mathbb{R} \times [0, \tfrac{1}{2}], \tag{8}$$

where $\{-\lambda_k, \frac{\Psi_k}{\|\Psi_k\|}\}_{k\geq 0}$ is the orthonormal basis corresponding associated with the spectral decomposition of L_ε^*. The eigenvalues are numbered in increasing order and the Fourier coefficients a_k are obtained from the boundary condition $\mu(x,0) = \mu(-x, \frac{1}{2})$, $x \in \mathbb{R}$.

Here is the key observation opening the route towards finding local maxima of the SPA coefficient. The terms in the sum (8) decay in time exponentially fast with rates λ_k, and therefore the terms corresponding to larger eigenvalues contribute less than the ones belonging to the low lying eigenvalues. This underlines their key importance. Fortunately, in the case of a double well potential there is a gap in the spectrum between λ_1 and λ_2.

Theorem 3 ('spectral gap') *In the limit of small noise, the following asymptotics holds:*

$$\lambda_0 = \lambda_0(\varepsilon) = 0, \textit{ and } \Psi_0 = e^{-\frac{2U_1}{\varepsilon}},$$

$$\lambda_1 = \lambda_1(\varepsilon) = \frac{1}{2\pi}\sqrt{U_1''(1)|U_1''(0)|}\cdot e^{-v/\varepsilon}(1 + \mathcal{O}(\varepsilon)),$$

$$\lambda_2 = \lambda_2(\varepsilon) \geq C > 0 \textit{ uniformly in } \varepsilon.$$

The result of Theorem 3 plays a crucial role in our analysis. Due to the *spectral gap* only the first two terms of (8) can contribute essentially as indicated by Theorem to the SPA coefficient η^X.

5. Asymptotics of the SPA Coefficient

As indicated by Theorem 3 we now have to investigate the asymptotics of the first two Fourier coefficients a_0 and a_1.

Theorem 4

$$a_0 = \|\Psi_0\|,$$
$$a_1 = \frac{\|\Psi_1\|}{\|\Psi_0\|^2} \cdot \frac{\langle \Psi_0(-\cdot), \Psi_1 \rangle}{\|\Psi_1\|^2 - e^{-\frac{1}{2}T\lambda_1}\langle \Psi_1(-\cdot), \Psi_1 \rangle} + r$$

where r tends to 0 in the limit of small noise and for $T \geq \exp\{(v+\delta)/\varepsilon\}$, and where δ is positive and sufficiently small.

Recalling the definition (6) of the SPA coefficient we set

$$S^X(\varepsilon, T) = \int_0^{\frac{1}{2}} \mathbf{E}_\mu X_{Ts}^{\varepsilon,T} \cdot e^{2\pi i s}\, ds, \tag{9}$$

to identify $\eta^X = 4|S^X|^2$.

Theorem 5 *Consider $T \geq \exp\{(v+\delta)/\varepsilon\}$ for δ positive and sufficiently small. Then in the small noise limit $\varepsilon \to 0$ the following asymptotic expansion holds true*

$$S^X = \frac{1}{\pi i} b_0 + \frac{1}{\pi i - \frac{1}{2}\lambda_1 T} b_1 + r_1$$

where the rest term r_1 tends to zero and the coefficients are given by

$$b_0 = \frac{\int y\, e^{-\frac{2U_1(y)}{\varepsilon}}\, dy}{\int e^{-\frac{2U_1(y)}{\varepsilon}}\, dy},$$
$$b_1 = -\frac{1 + e^{-\frac{1}{2}T\lambda_1}}{2} \cdot \frac{\int y\, \Psi_1(y)\, dy}{\int e^{-\frac{2U_1(y)}{\varepsilon}}\, dy} \cdot \frac{\langle \Psi_0(-\cdot), \Psi_1 \rangle}{\|\Psi_1\|^2 - e^{-\frac{1}{2}T\lambda_1}\langle \Psi_1(-\cdot), \Psi_1 \rangle}.$$

Finally,

$$\eta^X = b_0^2 \frac{4}{\pi^2} \frac{(\lambda_1 T)^2}{4\pi^2 + (\lambda_1 T)^2} + R, \tag{10}$$

where R tends to zero with ε.

Equipped with our knowledge of the first two Fourier coefficients let us now come to the main point of this investigation, the *resonance* behaviour of the SPA coefficient η^X. We have to find out whether it has a local maximum in ε. The following Lemma is obtained by application of Laplace's method of asymptotic expansions of singular integrals, see [8, 16], see also [17, 13]. It contains precise asymptotic estimates of the main contributions of our Fourier expansion to η^X.

Lemma 6 ('Laplace's method') *In the small noise limit, the following holds true:*

$$b_0 = -1 - \frac{1}{4}\frac{U_1^{(3)}(-1)}{U_1''(-1)^2}\varepsilon + \mathcal{O}(\varepsilon^2),$$
$$b_1 = -1 + \mathcal{O}(\varepsilon),$$

and consequently

$$b_0^2 = 1 + \frac{1}{2}\frac{U_1^{(3)}(-1)}{U_1''(-1)^2}\varepsilon + \mathcal{O}(\varepsilon^2), \tag{11}$$
$$(b_0 - b_1)^2 = \mathcal{O}(\varepsilon^2).$$

Now we can formulate our main theorem.

Theorem 7 *Let us fix δ positive and sufficiently small and $\Delta > v + \delta$. Let also $U_1(x) - 2U_1(-x) < v + V$ for all $x \in \mathbb{R}$ (the potential is not strongly asymmetric!). Then for $T \to \infty$ and ε from the domain*

$$\frac{v+\delta}{\log T} \leq \varepsilon \leq \frac{\Delta}{\log T} \tag{12}$$

the following asymptotic expansion for the SPA coefficient holds:

$$\eta^X(\varepsilon, T) = \frac{4}{\pi^2}\left(1 + \frac{1}{2}\frac{U_1^{(3)}(-1)}{U_1''(-1)^2}\varepsilon\right) + \mathcal{O}\left(\frac{1}{\log^2 T}\right).$$

This result has the following surprising consequences.

Corollary 8 *For $T \to \infty$ and $\varepsilon \in [\frac{v+\delta}{\log T}, \frac{\Delta}{\log T}]$ the SPA coefficient is a decreasing function of ε if $U_1^{(3)}(-1) < 0$ and an increasing function of ε if $U_1^{(3)}(-1) > 0$.*

Thus, the SPA coefficient as quality measure for tuning shows *no resonance* in a domain above Freidlin's threshold for quasi-deterministic periodicity (Theorem 2). This contradicts the physical intuition for the 'effective dynamics'. The reason for this surprising phenomenon can only be hidden in the *intrawell* behaviour of the diffusion which was neglected when passing to the reduced Markov chain. We return to this question later. Let us next study mathematically the 'effective dynamics' of the diffusion (2).

6. The 'Effective Dynamics': Two-state Markov Chains

The idea of approximation of diffusions in potential landscapes by appropriate finite state Markov chains in the context of stochastic resonance is due to Eckmann and Thomas [7], and C. Nicolis [15]. It was further developed by McNamara and Wiesenfeld [14]. In this section we follow [17, 13]. The discrete time case was studied in [12].

In order to catch the main features of the *interwell* hoppings of the diffusion (2) we consider the time inhomogeneous Markov chain $Y^{\varepsilon,T}$ living on the diffusion's meta-stable states ± 1. The infinitesimal generator of $Y^{\varepsilon,T}$ is periodic in time and is given by

$$Q_{\varepsilon,T}(t) = \begin{cases} \begin{pmatrix} -\varphi & \varphi \\ \psi & -\psi \end{pmatrix}, & \frac{t}{T} \ (\mathrm{mod}\ 1) \in [0, \frac{1}{2}), \\ \begin{pmatrix} -\psi & \psi \\ \varphi & -\varphi \end{pmatrix}, & \frac{t}{T} \ (\mathrm{mod}\ 1) \in [\frac{1}{2}, 1). \end{cases}$$

The transition rates φ and ψ have to be chosen appropriately, so that the Markov chain mimics the interwell motion of the diffusion as closely as possible. To exponential order they should correspond to the inverses of the Kramers' transition times (see Theorem 1). In their definition we even take care of the first sub-exponential expansion term:

$$\varphi = \frac{1}{2\pi}\sqrt{U_1''(-1)|U_1''(0)|} \cdot e^{-V/\varepsilon} \text{ and } \psi = \frac{1}{2\pi}\sqrt{U_1''(1)|U_1''(0)|} \cdot e^{-v/\varepsilon}.$$

In this simplified setting, it is no surprise that quantities important for our analysis can be explicitly calculated. For instance, the invariant measure of $Y_{Tt}^{\varepsilon,T}$ can be obtained as a solution of a forward Kolmogorov equation and is given by

$$\begin{aligned} \nu^-(t) &= \frac{\psi}{\varphi+\psi} + \frac{\varphi-\psi}{\varphi+\psi}\frac{e^{-(\varphi+\psi)Tt}}{1+e^{-\frac{1}{2}(\varphi+\psi)Tt}}, \\ \nu^+(t) &= \frac{\varphi}{\varphi+\psi} - \frac{\varphi-\psi}{\varphi+\psi}\frac{e^{-(\varphi+\psi)Tt}}{1+e^{-\frac{1}{2}(\varphi+\psi)Tt}}, \qquad t \in [0, \tfrac{1}{2}], \end{aligned}$$

and $\nu^{\pm}(t) = \nu^{\mp}(t + \frac{1}{2})$ for $t \geq 0$.

The SPA coefficient η^Y for the Markov chain $Y^{\varepsilon,T}$ is defined analogously to (6). It also can be explicitly described.

Theorem 9 *For all $\varepsilon > 0$ and $T > 0$ the following holds:*

$$\eta^Y(\varepsilon, T) = \frac{4}{\pi^2}\frac{T^2(\varphi-\psi)^2}{4\pi^2 + T^2(\varphi+\psi)^2}. \tag{13}$$

Compare (13) with (10). Since $(\varphi \pm \psi)^2 \approx \lambda_1^2$ in the limit of small ε, the expressions for η^X and η^Y differ only in the 'geometric' pre-factor b_0^2 and the asymptotically negligible rest term R.

The exact formula (13) allows to give a complete account of the local maxima of η^Y as a function of noise intensity for large periods T.

Theorem 10 *In the limit* $T \to \infty$ *the function* $\varepsilon \mapsto \eta^Y(\varepsilon, T)$ *has a local maximum at*

$$\varepsilon(T) \approx \frac{v+V}{2} \frac{1}{\log T}.$$

The 'resonance' behaviors of η^X and η^Y are quite different. Whereas the diffusion's SPA has no extremum for small ε, the Markov chain's *always* has. What can be responsible for this discrepancy? Note that the Markov chain mimics only the *interwell* dynamics of the diffusion. Thus, the SPA coefficient η^Y measures only the spectral energy contributed by interwell jumps. On the other hand, η^X also counts the numerous *intrawell* fluctuations of the diffusion. These fluctuations have small energy. But since the diffusion spends most of its time near ± 1 very subtle local asymmetries of the potential at these points, expressed for example by the signs of the third derivative of U_1 in the wells' bottoms, become important and destroy optimal tuning.

To underpin this heuristics mathematically, let us now make the idea of neglecting the diffusion's intrawell fluctuations precise. For example, we *cut off* those excursions which have not enough energy to reach half the height of the potential barrier between the wells. Consider the cut-off function g defined by

$$g(x) = \begin{cases} -1, & x \in [x_1, x_2], \\ 1, & x \in [y_1, y_2], \\ x, & \text{otherwise}, \end{cases}$$

where $x_1 < -1 < x_2 < 0$ and $0 < y_1 < 1 < y_2$ are such that $U_1(x_1) = U_1(x_2) = -\frac{V}{4}$ and $U_1(y_1) = U_1(y_2) = -\frac{v}{4}$. After cutting off small fluctuations, it is clear that we have to study a modified SPA coefficient of the form

$$\widetilde{\eta}^X(\varepsilon, T) = \left| \int_0^1 \mathbf{E}_\mu \left[g(X_{Ts}^{\varepsilon,T}) \right] e^{2\pi i s} \, ds \right|^2 .$$

Following the steps of Section 5 we obtain a formula for $\widetilde{\eta}^X$ which is quite similar to (10) and (13):

$$\widetilde{\eta}^X(\varepsilon, T) = \widetilde{b}_0^2 \frac{4}{\pi^2} \frac{(\lambda_1 T)^2}{4\pi^2 + (\lambda_1 T)^2} + \widetilde{R},$$

where $\widetilde{R}$ is a small rest term, and

$$\widetilde{b}_0^2 = \left(\frac{\int g(y) e^{-\frac{2U_1(y)}{\varepsilon}} dy}{\int e^{-\frac{2U_1(y)}{\varepsilon}} dy} \right)^2 = 1 - 4\sqrt{\frac{U_1''(-1)}{U_1''(1)}} e^{-\frac{V-v}{\varepsilon}} (1 + \mathcal{O}(\varepsilon))$$

compare to (11).

The modified geometric pre-factor $\widetilde{b}_0^2$ is essentially smaller than its counterpart b_0^2. This affects the SPA coefficient $\widetilde{\eta}^X$ crucially: in the limit of large period and small noise its behaviour now approaches the one of the reduced Markov chain.

Theorem 11 *Let the assumptions of Theorem 7 hold. Then for any $\gamma > 1$ in the limit $T \to \infty$ the function $\varepsilon \mapsto \widetilde{\eta}^X(\varepsilon, T)$ has a local maximum on*

$$\left[\frac{1}{\gamma} \frac{v+V}{2} \frac{1}{\log T}, \gamma \frac{v+V}{2} \frac{1}{\log T} \right].$$

In other words, the optimal tuning for the measure of goodness $\widetilde{\eta}^X$ exists and is given approximately by

$$\varepsilon(T) \approx \frac{v+V}{2} \frac{1}{\log T}.$$

References

[1] V. S. Anishchenko, A. B. Neiman, F. Moss, and L. Schimansky-Geier, "Stochastic resonance: noise-enhanced order," *Physics–Uspekhi*, vol. 42, no. 1, pp. 7–36, 1999.

[2] R. Benzi, G. Parisi, A. Sutera, and A. Vulpiani, "The mechanism of stochastic resonance," *J. Phys. A*, vol. 14, pp. 453–457, 1981.

[3] R. Benzi, G. Parisi, A. Sutera, and A. Vulpiani, "Stochastic resonance in climatic changes," *Tellus*, vol. 34, pp. 10–16, 1982.

[4] R. Benzi, G. Parisi, A. Sutera, and A. Vulpiani, "A theory of stochastic resonance in climatic change," *SIAM J. Appl. Math.*, vol. 43, pp. 563–578, 1983.

[5] M. I. Budyko, "The effect of solar radiation variations on the climate of the earth," *Tellus*, vol. 21, pp. 611–619, 1969.

[6] M. I. Dykman, D. G. Luchinskii, R. Mannella, P. V. E. McClintock, N. D. Stein, and N. G. Stocks, "Stochastic resonance in perspective," *Nuovo Cimento D*, vol. 17, pp. 661–683, 1995.

[7] J.-P. Eckmann and L. E. Thomas, "Remarks on stochastic resonance," *J. Phys. A*, vol. 15, pp. 261–266, 1982.

[8] A. Erdélyi, *Asymptotic expansions.* Dover Publications, Inc., New York, 1956.

[9] M. I. Freidlin. "Quasi-deterministic approximation, metastability and stochastic resonance," *Physica D*, vol. 137, no. 3-4, pp. 333–352, 2000.

[10] M. I. Freidlin and A. D. Wentzell. "Random perturbations of dynamical systems," volume 260 of *Grundlehren der Mathematischen Wissenschaften.* Springer, New York, NY, second edition, 1998.

[11] L. Gammaitoni, P. Hänggi, P. Jung, and F. Marchesoni, "Stochastic resonance," *Reviews of Modern Physics*, vol. 70, pp. 223–287, January 1998.

[12] P. Imkeller and I. Pavlyukevich. "Stochastic resonance in two-state Markov chains," *Arch. Math.*, vol. 77, no. 1, pp. 107–115, 2001.

[13] P. Imkeller and I. Pavlyukevich, "Model reduction and stochastic resonance," *Stochastics and Dynamics*, vol. 2, no. 4, pp. 463–506, 2002.

[14] B. McNamara and K. Wiesenfeld. "Theory of stochastic resonance," *Physical Review A (General Physics)*, vol. 39, pp. 4854–4869, May 1989.

[15] C. Nicolis. "Stochastic aspects of climatic transitions — responses to periodic forcing," *Tellus*, vol. 34, pp. 1–9, 1982.

[16] F. W. J. Olver, *Asymptotics and special functions.* Computer Science and Applied Mathematics. Academic Press, a subsidiary of Harcourt Brace Jovanovich, Publishers, New York - London, 1974.

[17] I. E. Pavlyukevich, *Stochastic Resonance*. PhD thesis, Humboldt-Universität, Berlin, 2002. Logos–Verlag, ISBN 3-89722-960-9.

[18] W. B. Sellers, "A global climate model based on the energy balance of the earth-atmosphere system," *J. Appl. Meteor.*, vol. 8, pp. 301–320, 1969.

METASTABILITY OF DIFFUSION PROCESSES

C. Schütte,* W. Huisinga
Department of Mathematics and Computer Science
Free University
Berlin, Germany.
schuette,huisinga@math.fu-berlin.de

S. Meyn†
Department of Electrical and Computer Engineering and
the Coordinated Science Laboratory
University of Illinois at Urbana-Champaign
Urbana, IL 61801, USA
meyn@nichols.csl.uiuc.edu

Abstract Diffusion models arising in analysis of real world systems are typically far too complex for exact solution, or even meaningful simulation. The purpose of this paper is to develop foundations for model reduction, and new modeling techniques for diffusion models. Based on the main assumption of *V-uniform ergodicity* of the diffusion process it is shown that real eigenfunctions provide a decomposition of the state space into so-called metastable sets. We give a novel definition of metastability via exit rates which seems to be promising for a algorithmic identification of metastable sets even for large scale systems.

1. Introduction

Diffusion models are a popular alternative to the classical description of complex processes in terms of large sets of ordinary differential equations that give rise to chaotic dynamics. Examples are varied and come from the such diverse fields as molecular dynamics [4], materials science [18], or internet security [3]. However, even diffusion models in practice are far too com-

*Partially supported by the DFG Research Center "Mathematics for key technologies" (FZT 86) in Berlin, and within DFG priority program "Analysis, Modeling and Simulation of Multiscale Problems" (SPP 1095).
†Supported in part by NSF Grant ECS 99 72957

N. Sri Namachchivaya and Y.K. Lin (eds.),
IUTAM Symposium on Nonlinear Stochastic Dynamics, 71-81.

plex for exact solution, or even long-term simulation. Recently, various approaches to the essential features of such diffusion models have been proposed. They range from novel algorithmic methods based on transfer operator theory that decompose the state space into 'metastable' sets [14, 15] and transition path computation between these sets [19, 4], to more theoretical approaches to model reduction techniques based on variants of the classical Wentzell–Freidlin theory (see e.g., [2, 5, 13]), or the analysis of transition times via the theory of quasi-stationary distributions of Markov process as introduced in [16, 17, 6]. The approaches closest to this article are [2, 5]. However, while the present article defines and investigates the essential features of diffusion processes via resent results for spectral theory, [5] mainly exploits large deviation theory and [2] is mainly concerned with the limit of vanishing noise intensity (which is not the case in the present article).

The approach via quasi-stationary distributions and the mentioned algorithmic transfer operator approach, both, rely on the analysis of eigenfunctions of the full or some restricted semigroup of the diffusion process. This article corresponds to this discussion, combined with recent results concerning large deviations and spectral theory for ψ-irreducible Markov processes [1, 8, 9]. It summarizes the article [7] of the present authors. There, the theory is worked out in detail, while we here aim to provide a rapidly accessible reference to the results in [7] combined with a short nontrivial illustration.

We consider general multivariate diffusion processes $\boldsymbol{X}$ which are governed by the stochastic differential equation

$$\mathrm{d}X \quad = \quad u(X)\,\mathrm{d}t + A(X)\,\mathrm{d}W,, \tag{1}$$

where $u : \mathbb{R}^d \to \mathbb{R}^d$ with measurable components, W denotes d-dimensional Brownian motion, and $A(x)$ a $d \times d$ matrix with measurable entries. A special case is the so-called Smoluchowski or high friction Langevin equation on $\mathbb{R}^d$, where the function u is given by a potential U, $u = -\nabla U$, and the matrix A is state-independent and a multiple of the identity matrix, $A = \sigma I$, such that the process obeys $\mathrm{d}X = -\nabla U(X)\,\mathrm{d}t + \sigma\,\mathrm{d}W$. Choosing $d = 1$ and U to be the perturbed three well potential shown in Figure 1, we will use this case for illustration.

2. V-uniform Ergodicity and Spectral Gaps

Here we assume that the reader is acquainted with the basics of the general theory of Markov processes, especially with the definitions of ψ-irreducibility, (Harris) recurrence, and aperiodicity. The state space $\mathbb{X}$ considered herein is assumed to be an open, connected subset of $\mathbb{R}^d$. The ergodic theory and spectral theory described here is based upon the vector space setting developed in [10, Chapter 16]. Let $V : \mathbb{X} \to [1, \infty)$ be a given function, and denote by

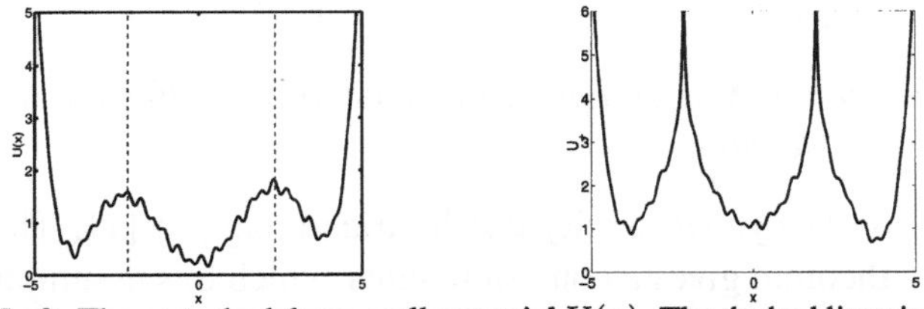

Figure 1. Left: The perturbed three–well potential $U(x)$. The dashed lines indicate the wells separated by the two saddle points with maximal barriers. Right: Corresponding potential of the so-called twisted process due to Proposition 4 below based on the eigenfunction h_3 of our illustrative example (see Figure 2, middle).

L_∞^V the vector space of measurable functions $h : \mathbb{X} \to \mathbb{C}$ satisfying

$$\|h\|_V := \sup_{x \in \mathbb{X}} \frac{|h(x)|}{V(x)} < \infty.$$

The vector space $\mathcal{M}_1^V$ is the set of complex-valued measures ν on the Borel sigma-field $\mathcal{B} = \mathcal{B}(\mathbb{X})$ such that

$$\|\nu\|_V := \int_{\mathbb{X}} V(x)|\nu(dx)| < \infty.$$

The induced operator norm is defined by

$$|\!|\!|P|\!|\!|_V := \sup \frac{\|Ph\|_V}{\|h\|_V},$$

where the supremum is over $h \in L_\infty^V$, $\|h\|_V \neq 0$. For later reference, we define the function space

$$C^V = \{g \in L_\infty^V : \|P^t g - g\|_V \to 0,\ t \downarrow 0\}.$$

2.1 Diffusion semigroups

Based on the previous definitions this section will present some spectral theory for the semigroup associated with a hypoelliptic diffusion process: We assume that time is continuous, $\mathbb{T} := \mathbb{R}_+$, and that the diffusion $\boldsymbol{X} = \{\mathbb{X}(t) : t \in \mathbb{T}\}$ evolves on $\mathbb{X}$ due to (1). Its differential generator (the generator of the associated Fokker-Planck equation) is given by

$$\mathcal{D}h = \sum_i u_i(x) \frac{\partial}{\partial x_i} h(x) + \tfrac{1}{2} \sum_{ij} \Sigma_{ij}(x) \frac{\partial^2}{\partial x_i \partial x_j} h(x) \qquad (2)$$

with $\Sigma_{ij}(x)$ being the entries of the $d \times d$ matrix $\Sigma(X) = A(x)A(x)^T$. In more compact notation, $\mathcal{D} = -u \cdot \nabla + \frac{1}{2}\text{trace}(\Sigma \nabla^2)$. In the following we assume that

the Markov process $\boldsymbol{X}$ is an aperiodic, hypoelliptic diffusion, with continuous sample paths. (3)

The definition of hypoellipticity can be found, e.g., in [11, Theorem 3.3]. The following theorem gives a condition under which this assumption holds:

Theorem 1 *Suppose that $\boldsymbol{X}$ is a diffusion with generator given in (2), and suppose that the generator is hypoelliptic. Suppose moreover that there is a state $x_0 \in \mathbb{X}$ that is 'reachable' in the following sense: For any $x \in \mathbb{X}$, and any open set O whose closure contains x_0, we have that $P^t(x, O) > 0$, for all $t \in \mathbb{T}$ sufficiently large. Then, the Markov process is ψ-irreducible and aperiodic with*

$$\psi(\cdot) \quad = \quad \int_0^\infty e^{-t} P^t(x_0, \cdot)\, \mathrm{d}t. \tag{4}$$

The RHS of (4) is known as the resolvent kernel corresponding to the transition function P^t of the diffusion defined in (1). Formulations and characterizations of the *spectral gap* of our diffusion $\boldsymbol{X}$ are facilitated by three different **generators**:

1 The *extended generator* $\mathcal{A}$: We write $g = \mathcal{A}f$ if the adapted stochastic process $(M_f(t), \mathcal{F}_t)$ is a local martingale, where $\mathcal{F}_t = \sigma(\mathbb{X}(s); 0 \leq s \leq t)$, and

$$M_f(t) := f(\mathbb{X}(t)) - f(\mathbb{X}(0)) - \int_0^t g(\mathbb{X}(s))\, \mathrm{d}s\,. \tag{5}$$

2 The *differential generator* $\mathcal{D}$: Defined on $C^2(\mathbb{X})$ via $\mathcal{D}f = -u \cdot \nabla f + \frac{1}{2}\mathrm{trace}\,(\Sigma \nabla^2 f)$.

3 The *strong generator* $\mathcal{D}_V$: For a given $V \colon \mathbb{X} \to [1, \infty]$, finite a.e., we write $g = \mathcal{D}_V f$ if $f, g \in C^V$, and $|||\frac{P^t f - f}{t} - g|||_V \to 0$ for $t \downarrow 0$.

The extended generator $\mathcal{A}$ is a true extension of $\mathcal{D}$ in the sense that $\mathcal{A}f = \mathcal{D}f$ a.e. $[\psi]$ when $f \in C^2(\mathbb{X})$. The extended generator and differential generator are used in criteria for stability and to obtain bounds on the 'essential spectrum' of the associated semigroup. The strong generator is used to define a **spectral gap**: For a given $V \colon \mathbb{X} \to [1, \infty]$, finite a.e.,

1 The *spectrum* $\mathcal{D}_V$ is the set of $\Lambda \in \mathbb{C}$ such that the inverse $[I\Lambda - \mathcal{D}_V]^{-1}$ does not exist as a bounded linear operator on L_∞^V;

2 The generator admits a *spectral gap* if the set $\mathcal{D}_V \cap \{z \in \mathbb{C} : \mathrm{Re}(z) \geq -\varepsilon\}$ is finite for sufficiently small $\varepsilon > 0$;

3 The Markov process is called *V-uniformly ergodic* if there is a spectral gap; $\{0\} = \mathcal{D}_V \cap \{z \in \mathbb{C} : \mathrm{Re}(z) = 0\}$; and the eigenvalue $\Lambda = 0$ is simple.

The following 'drift condition' characterizes V-uniform ergodicity and is central to this paper. It is useful that we may use the extended generator, and not the strong generator in (V4). A function $s\colon \mathbb{X} \to [0, \infty)$ is called *small* if

$$P^T(x, A) \geq \varepsilon s(x)\nu(A), \qquad x \in \mathbb{X}.$$

for some probability distribution ν on $\mathcal{B}(\mathbb{X})$, $\varepsilon > 0$, and $T < \infty$.

(V4) *For constants $b < \infty$, $0 < \bar{\Gamma} < \infty$, a small function $s\colon \mathbb{X} \to [0, \infty)$, and a function $V : \mathbb{X} \to [1, \infty)$, one has $\mathcal{A}V \leq -\bar{\Gamma}V + bs$.*

Condition (V4) is equivalent only to a spectral gap.

Theorem 2 *Suppose that $\boldsymbol{X}$ is ψ-irreducible and aperiodic, and suppose that (V4) holds for some $V\colon \mathbb{X} \to [1, \infty)$. Then $\boldsymbol{X}$ is V-uniformly ergodic. Conversely, if the Markov process $\boldsymbol{X}$ is V_0-uniformly ergodic then there exists a solution to (V4) with $V \in L_\infty^{V_0}$.*

2.2 Non–probabilistic semigroups

For a given function $F\colon \mathbb{X} \to \mathbb{R} \cup \{\infty\}$ we consider the following *positive semigroup*, for $A \in \mathcal{B}, x \in \mathbb{X}, t \in \mathbb{T}$,

$$P_F^t(x, A) = \mathsf{E}_x\Big[\mathbf{1}(\mathbb{X}(t) \in A) \exp\Big(-\int_0^t F(\mathbb{X}(s))\,\mathrm{d}s\Big)\Big].$$

A strong generator can be defined in analogy with the probabilistic semigroup.

For an arbitrary positive semigroup $\{\widehat{P}^t\}$ the definitions of irreducibility, small sets and measures, and other set classifications remain the same in this non-probabilistic setting. For a given function $V\colon \mathbb{X} \to [1, \infty]$, finite a.e., the *$V$-spectral radius* of $\{\widehat{P}^t\}$ is given by

$$\mathrm{sr}_V(\{\widehat{P}^t\}) := \lim_{T\to\infty} \Big(|\!|\!|\widehat{P}^T|\!|\!|_V\Big)^{1/T}.$$

Closely related is the *Perron-Frobenius eigenvalue*, defined for any small pair (s, ν) with $s \in \mathcal{B}^+$, $\nu \in \mathcal{M}^+$, via

$$\mathrm{pfe}(\{\widehat{P}^t\}) \quad := \quad \lim_{T\to\infty} \Big(\nu\widehat{P}^T s\Big)^{1/T} \tag{6}$$

A straightforward generalization of [12, Proposition 3.4] shows that these definition are independent of the particular small pair chosen when the process is ψ-irreducible.

In analogy with V-uniform ergodicity, the semigroup $\{\widehat{P}^t : t \in \mathbb{R}_+\}$ with generator $\widehat{\mathcal{D}}_V$ is called *V-uniform* if the following conditions are satisfied:

1 The constant $\Lambda := -\log(\mathrm{sr}_V(F))$ is a *simple eigenvalue*, i.e., the associated eigenspace is a one-dimensional subspace of L_∞^V.

2 The generator admits spectral gap: for sufficiently small $\varepsilon > 0$,

$$\{\Lambda\} = \widehat{\mathcal{D}}_V \cap \{z \in \mathbb{C} : \mathrm{Re}(z) \geq \Lambda - \varepsilon\}.$$

In [7], the reader may find an analog of Theorem 2 which states the V-uniformity of the semigroup $\{\widehat{P}_F^t\}$ under the assumption of a drift condition similar to (V4).

3. Metastability and Exit Rates

For a given set $A \in \mathcal{B}$ we define the stopping times,

$$\tau_A := \inf\Big\{t > 0 : X(t) \in A\Big\},\ \varrho_A := \inf\Big\{t > 0 : \int_0^t \mathbf{1}(X(s) \in A)\ ds > 0\Big\}$$

Much of the analysis here is based on the semigroups $\{\widehat{P}_F^t\}$ considered in Section 2.2 in the special case where $F = \infty \mathbf{1}_{A^c}$ for some $A \in \mathcal{B}$. When F takes this form we denote the semigroup by $\{\widehat{P}_A^t\}$, which can be equivalently expressed,

$$P_A^t g\,(x) = \mathsf{E}_x[g(\mathbb{X}(t))\mathbf{1}(\varrho_{A^c} \geq t)], \qquad g \in L_\infty,\ x \in \mathbb{X},\ t \in \mathbb{T}.$$

Let $\mathcal{C}$ denote the collection of all connected, open subsets of $\mathbb{X}$. In addition to assumption (3) the following assumptions will be imposed throughout:

For each $A \in \mathcal{C}$ the semigroup $\{P_A^t\}$ is ψ_A-irreducible, where ψ_A is Lebesgue measure restricted to A, and every compact subset of A is a small set for $\{P_A^t\}$. (7)

3.1 Exit rates

Our goal in this section is to quantify the rate at which the process moves between elements of $\mathcal{C}$. The motivation for the consideration of transition *rates*, rather than moments, is to set the stage for Markov chain approximations. **Exit rates and metastable sets** are defined as follows

1 The *exit rate* of $A \in \mathcal{C}$ is defined as $\Gamma(A) := -\log(\mathsf{pfe}(A))$, where $\mathsf{pfe}(A)$ denotes the Perron-Frobenius eigenvalue for the semigroup $\{P_A^t : t \in \mathbb{T}\}$ as defined in (6).

2 A set $\mathsf{M} \in \mathcal{C}$ is called *metastable* with exit rate $\Gamma(\mathsf{M})$ if $\Gamma(A) > \Gamma(\mathsf{M})$ for all $A \subset \mathsf{M}$, $A \neq \mathsf{M}$, $A \in \mathcal{C}$.

3 The *V-exit rate* of $A \in \mathcal{B}$ is given by $\Gamma_V(A) := -\log(\mathrm{sr}_V(A))$, where $\mathrm{sr}_V(A)$ denotes the V-spectral radius of $\{P_A^t\}$.

4 For $\mathsf{M} \in \mathcal{B}$ we say that M is *V-metastable* if $\Gamma_V(\mathsf{M}) < \infty$, and $\Gamma_V(A) > \Gamma_V(\mathsf{M})$ for any $A \subset \mathsf{M}$ satisfying $A \in \mathcal{B}$ and $\psi(\mathsf{M}/A) > 0$.

Certain metastable sets are closely related to eigenfunctions:

Theorem 3 *Suppose that $\boldsymbol{X}$ is a diffusion satisfying (3) and (7), and that $\mathsf{M} \in \mathcal{C}$ is both metastable and V-metastable, with common exit rate $\Gamma(\mathsf{M}) = \Gamma_V(\mathsf{M}) < \infty$. Then there exists $h\colon \mathsf{M} \to (0,\infty)$ satisfying the eigenvalue equation*

$$\mathcal{A}h = -\Gamma(\mathsf{M})h \qquad \textit{on } \mathsf{M}. \tag{8}$$

Here, for a set $A \in \mathcal{C}$ and functions $f, g\colon A \to \mathbb{R}$ we write '$g = \mathcal{A}f$ *on* A' if $\{M_f(t \wedge \varrho_{A^c}) : t \in \mathbb{T}\}$ is a local martingale (see (5)).

3.2 The twisted process

Inspired by Theorem 3, this section investigates the consequences of the following **eigenfunction equation**:

For some $\Gamma_0 < \infty$ and some set $\mathsf{M} \in \mathcal{C}$ there exists a function h, positive on M and C^2 in a neighborhood of $\bar{\mathsf{M}}$ such that $\mathcal{D}h\,(x) = -\Gamma_0 h(x)$, for $x \in \mathsf{M}$. (9)

Under (9), for any $x \in \mathbb{X}$ the stochastic process $m_h(t) := h(x)^{-1}h(X(t))e^{\Gamma_0 t}$, $t \in \mathbb{T}$, is a positive martingale up to the stopping time $T_\bullet := \varrho_{\mathsf{M}^c}$. Hence it may serve in a change of measure in the construction of the **twisted process** $\check{\mathbb{X}}$ with state space $\check{\mathbb{X}} := \mathsf{M}$ whose semigroup is defined for any $g \in L_\infty(\mathsf{M})$, and any $x \in \mathsf{M}$ via,:

$$\check{\mathsf{E}}_x[g(\check{X}(t))] := \mathsf{E}_x[m_h(t)g(X(t))\mathbf{1}(T_\bullet > t)]\,. \tag{10}$$

The twisted process is a diffusion. The associated 'twisted generator' is given in Proposition 4 below.

Proposition 4 *Suppose that (3), (7), and (9) hold. Then, The expectation operator $\check{\mathsf{E}}$ defines a diffusion on M, up to the exit time $T_\bullet$. The differential generator is given by,*

$$\check{\mathcal{D}} = I_{h^{-1}}\mathcal{D}I_h + \Gamma_0 I \;=\; \mathcal{D} + \langle \Sigma(\nabla H), \nabla \rangle, \tag{11}$$

where I_h is the multiplication operator: $I_h g = h \cdot g$, and $H(x) = \log(h(x))$ for $x \in \mathsf{M}$. If $\boldsymbol{X}$ is governed by a Smoluchowski equation on $\mathbb{X}$ with potential

U, and if $\Sigma = \sigma^2 I$, then $\mathcal{D}$ is the differential generator for a Smoluchowski equation with potential $U_+ = U - \sigma^2 H$.

The following result characterizes metastability in terms of geometric ergodicity of the twisted process.

Theorem 5 *Assume that (3), (7), and (9) hold. Suppose moreover that the escape-time for the twisted process is infinite a.s., and that the twisted process is $\check{V}$-uniformly ergodic for some $\check{V}\colon \mathsf{M} \to [1,\infty)$, with $h^{-1} \in L_\infty^{\check{V}}$. Then, the set M is both metastable and V_0-metastable, with common exit rate $\Gamma(\mathsf{M}) = \Gamma_{V_0}(\mathsf{M}) = \Gamma_0$ given in (9), and with $V_0 = \check{V}h$.*

We now provide more readily verifiable sufficient conditions under which the conclusions of Theorem 5 will hold.

Theorem 6 *Assume that (3), (7), and (9) hold, and that (V4) is also satisfied for a continuous function $V\colon \mathbb{X} \to [1,\infty)$. Suppose moreover that the Lyapunov function V and the eigenfunction h satisfy the following additional conditions:*

(a) The constant Γ_0 in (9) satisfies $0 < \Gamma_0 < \bar{\Gamma}$.

(b) $h(x) > 0$ for all $x \in \mathsf{M}$, and $h(x) = 0$ for $x \in \partial\mathsf{M} := \bar{\mathsf{M}} \setminus \mathsf{M}$.

(c) $(\nabla h(x))^{\intercal}\Sigma(x)(\nabla h(x)) > 0$ for all $x \in \partial\mathsf{M}$.

(d) $K_n := \{x \in \mathbb{X} : V(x) \le nh(x)\}$ is a compact subset of $\mathbb{X}$ for all $n \ge 1$.

Then,

(i) The escape-time from M for the twisted process is infinite a.s. for $\check{\mathbb{X}}(0) = x \in \mathsf{M}$;

(ii) The twisted process is $\check{V}_1$-uniformly ergodic with $\check{V}_1 = V/h$.

(iii) The set M is both metastable and V-metastable, with exit rate $\Gamma(\mathsf{M}) = \Gamma_V(\mathsf{M}) = \Gamma_0$, where Γ_0 is given in (9).

The following data illustrates the meaning of Theorem 6 for the Smoluchowski equation $\mathrm{d}X = -\nabla U(X)\,\mathrm{d}t + \sigma\,\mathrm{d}W$ in dimension $d = 1$ with the perturbed three well potential U of Figure 1 and $\sigma = 1$. The eigenproblem of the generator has been solved numerically by means of an appropriate finite element solver. The resulting eigenvalues are:

Λ_1	Λ_2	Λ_3	Λ_4	Λ_5	...
0.000	−0.041	−0.069	−0.906	−1.419	...

Thus, the eigenfunctions associated with the low eigenvalues Λ_2 and Λ_3 should allow a decomposition of state space into metastable sets with significantly small exit rates. This eigenfunctions are shown in Figure 2. We observe that the decompositions based on the zeros of the eigenfunctions h_2 or h_3, respectively, are related to the decomposition into the three dominant wells of the potential; deviations of the zeros from the location of the saddle points with maximal barriers are in detail discussed in [7].

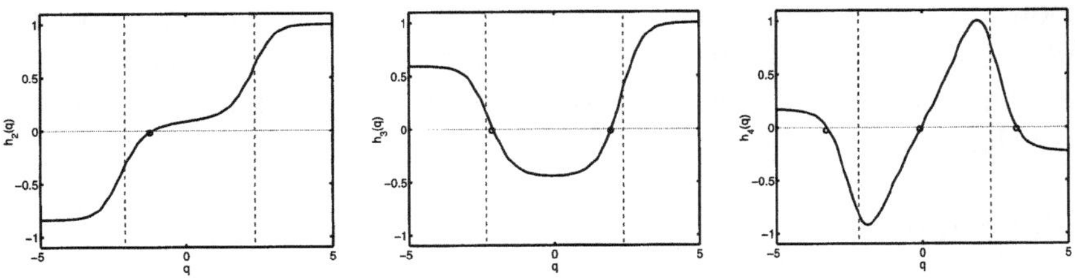

Figure 2. Second to forth eigenfunction $h_2, \ldots, h_4$ of the generator for the Smoluchowski equation with the perturbed three–well potential and $\sigma = 1$. The first eigenfunction h_1 is constant. The dashed lines indicate the main wells of the potential as in Figure 1. Note that the eigenfunctions are remarkable insensitive to large variations of the potential function.

3.3 Consequences for exit times

We show here that $\check{V}$-uniform ergodicity of the twisted process implies that the exit time $T_\bullet := \varrho_{\mathsf{M}^c}$ from a metastable set M is approximately exponentially distributed, with rate $\Gamma(\mathsf{M})$, provided there is a sufficient spectral gap. For the random variable $T_\bullet$ we define the conditional distribution function, and the conditional moment generating function for the residual life at time T by

$$\begin{aligned} F_x(s,T) &= \mathsf{P}_x[(T_\bullet - T) \geq s \mid T_\bullet \geq T], \qquad s \geq 0,\ T \geq 0; \\ M_x(\beta,T) &= \mathsf{E}_x[\exp(\beta(T_\bullet - T)) \mid T_\bullet \geq T], \quad \beta \leq \Gamma,\ T \geq 0. \end{aligned} \tag{12}$$

These quantities are independent of T only for exponential random variables. Theorem 7 states that the rate of decay of the exit time distribution is basically independent of the starting point and of T.

Theorem 7 *Suppose that the conditions of Theorem 6 hold, so that the set* M *is* V*-metastable with exit-rate* $\Gamma > 0$*. Then there exists* $\delta_0 > 0$ *such that for all* $s, T > 0$ *and all* $\beta < \Gamma$*,*

$$\begin{aligned} F_x(s,T) &= e^{-\Gamma s}\left[\frac{1+O\left(e^{-\delta_0 s}V(x)h^{-1}(x)\right)}{1+O\left(e^{-(T+s)\delta_0}V(x)h^{-1}(x)\right)}\right] \\ M_x(\beta,T) &= \tfrac{\Gamma}{\Gamma-\beta} + O(e^{-\delta_0 T}V(x)h^{-1}(x)) \end{aligned}$$

Under the conditions supposed here, an application of Theorem 6 implies that the twisted process is $\check{V}$-uniformly ergodic for some $\check{V}$ satisfying $h^{-1} \in L_\infty^{\check{V}}$. It follows that, for some $\delta_0 > 0$,

$$\check{P}^s h^{-1}(x) = \check{\pi}(h^{-1}) + O(e^{-\delta_0 s} V(x) h^{-1}(x)) \quad s \geq 0,\ x \in \mathbb{X}. \tag{13}$$

Exactly this δ_0 is the constant that enters in Theorem 7. The potential for the twisted process(es) based on the eigenfunction h_3 of our illustrative example is shown in Figure 1. In each of the three parts the twisted potential does not exhibit any significant barriers. Thus, the diffusion is rapidly mixing in all three components and therefore the associated δ_0 is large compared to the low eigenvalues of the full spectrum.

References

[1] S. Balaji and S. P. Meyn, "Multiplicative ergodicity and large deviations for an irreducible Markov chain," *Stochastic Process. Appl.*, vol. 90, no. 1, pp. 123–144, 2000.

[2] A. Bovier, M. Eckhoff, V. Gayrard, and M. Klein, *Metastability in reversible diffusion processes I. Sharp asymptotics for capacities and exit times*, Preprint. Weierstrass-Institut für Angewandte Analysis und Stochastik, Berlin, 2000.

[3] E. G. Coffman, Jr. and V. Misra, "Network resilience: Exploring cascading failures," in *Proceedings of the Fortieth Annual Allerton Conference on Communication, Control, and Computing*, October 2-4, 2002, New York, NY, Plenum Press, 2002.

[4] G. E. Crooks and D. Chandler, "Efficient transition path sampling for nonequilibirum stochastic dynamics," *Phys. Rev. E*, vol. 64, pp. 26–109, 2001.

[5] M. I. Freidlin, "Quasi–deterministic approximation, metastability and stochastic resonance," *Physica D*, vol. 137, pp. 333–352, 2000.

[6] P. W. Glynn and H. Thorisson, "Two-sided taboo limits for Markov processes and associated perfect simulation," *Stochastic Process. Appl.*, vol. 91, no. 1, pp. 1–20, 2001.

[7] W. Huisinga, S. Meyn, and C. Schütte, "Phase transitions and metastability in Markovian and molecular systems," *Ann. Appl. Probab.*, To appear. 2003. Available via http://www.math.fu-berlin.de/~biocomp.

[8] I. Kontoyiannis and S. P. Meyn, "Spectral theory and limit theorems for geometrically ergodic Markov processes," submitted for publication, 2001. Also presented at the 2001 INFORMS Applied Probability Conference, NY, July, 2001.

[9] I. Kontoyiannis and S. P. Meyn, "Spectral theory and limit theorems for geometrically ergodic Markov processes, Part II: Empirical measures & unbounded functionals," submitted for publication, 2001.

[10] S. P. Meyn and R. L. Tweedie, *Markov Chains and Stochastic Stability*, Springer-Verlag, London, 1993.

[11] S. P. Meyn and R. L. Tweedie, "Stability of Markovian processes II: Continuous time processes and sampled chains," *Adv. Appl. Probab.*, vol. 25, pp. 487–517, 1993.

[12] E. Nummelin, *General Irreducible Markov Chains and Nonnegative Operators*. Cambridge University Press, Cambridge, 1984.

[13] L. Rey-Bellet and L. E. Thomas, "Asymptotic behavior of thermal nonequilibrium steady states for a driven chain of anharmonic oscillators," *Comm. Math. Phys.*, vol. 215, no. 1, pp. 1–24, 2000.

[14] C. Schütte, A. Fischer, W. Huisinga, and P. Deuflhard. "A direct approach to conformational dynamics based on hybrid Monte Carlo," *J. Comput. Phys., Special Issue on Computational Biophysics*, vol. 151, pp. 146–168, 1999.

[15] C. Schütte, W. Huisinga, and P. Deuflhard, "Transfer operator approach to conformational dynamics in biomolecular systems," in B. Fiedler, editor, *Ergodic Theory, Analysis, and Efficient Simulation of Dynamical Systems*. Springer, 2001.

[16] E. Seneta and D. Vere-Jones, "On quasi-stationary distributions in discrete-time Markov chains with a denumerable infinity of states," *J. Appl. Probability*, vol. 3, pp. 403–434, 1966.

[17] R. L. Tweedie, "Quasi-stationary distributions for Markov chains on a general state space," *J. Appl. Probability*, vol. 11, pp. 726–741, 1974.

[18] E. Weinan, W. Ren, and E. Vanden Eijnden, "Optimal paths for spatially extended metastable systems driven by noise," Preprint. Available via http://math.nyu.edu/~eve2, 2002.

[19] E. Weinan, W. Ren, and E. Vanden Eijnden, "Probing multi-scale energy landscapes using the string method," Preprint. Available via http://math.nyu.edu/~eve2, 2002.

STOCHASTIC AVERAGING NEAR HOMOCLINIC ORBITS VIA SINGULAR PERTURBATIONS

Richard B. Sowers
Department of Mathematics
University of Illinois at Urbana–Champaign
Urbana, IL 61801
r-sowers@math.uiuc.edu
http://www.math.uiuc.edu/~r-sowers

Abstract We outline a singular-perturbations approach to the graph-valued stochastic averaging results of Freidlin-Wentzell and Freidlin-Weber. We specifically consider the Freidlin-Weber problem (a Newtonian particle in a double-well potential). To show the Freidlin-Weber convergence result, we develop a perturbed test function via a boundary-layer PDE near the homoclinic orbit. Solvability of this PDE is equivalent to the glueing conditions of Freidlin-Wentzell. Details of our calculations will appear elsewhere.

Keywords: Markov processes, stochastic averaging, stratified space

1. Introduction

One of the principal raisons d'être of "applied mathematics" is in *model reduction*; i.e., the development of rigorous methods to replace, often in some limiting regime, a complicated system by a simpler, or lower-dimensional one. We study here a model-reduction problem in *Markov processes*, namely a problem in *stochastic averaging*.

The underpinning of classical averaging is a separation of time scales; there is a coordinate which varies slowly (often thought of as the energy) and a coordinate which quickly varies (typically thought of as an angle). As the ratio of the fast to slow speed increases, it becomes possible to in a sense fix the slowly varying coordinate and carry out a long-term average in the quickly varying component, in the process identifying effective coefficients for the dynamics of the slow motion. A simple example is a Hamiltonian system with small perturbations. The slowly varying coordinate is the value of the Hamiltonian and the quickly varying coordinate is the position (or angle) in the appro-

N. Sri Namachchivaya and Y.K. Lin (eds.),
IUTAM Symposium on Nonlinear Stochastic Dynamics, 83-94.

priate level set of the Hamiltonian. One can seek a reduced model for the slow variable by first rescaling time so that the variation of the slow variable is of order one. As the fast motion becomes faster, the behavior of the slow variable can (often) be described via a closed set of equations (which do not involve the fast variable). Gikhman [9] and Stratonovich [21] developed some of the first averaging results. Completely rigorous arguments involving diffusion processes were given by Khasminskii [11, 12] and several others [2, 3, 17, 18].

The nature and complexity of the reduced or averaged process depends on the complexity of the Hamiltonian H; roughly, the reduced process takes values in the space of orbits of the fast motion. When H is fairly simple, viz., something like a paraboloid with a single isolated elliptic critical point, the reduced process is simply a Markov process on a line—a classical result [12]. This machinery has only recently been extended to handle more complicated Hamiltonians; e.g., Hamiltonians with a finite collection of distinct minima and saddle points; in this case the reduced Markov process takes its values in a *graph* which encodes the topology of the level sets of H [6, 7, 8]; see also the work of Neishtadt [16] and Wolansky [23]. Each leg of the graph corresponds to a region in which the the dynamics are diffeomorphic to a fast deterministic rotation around a cylinder and a slow diffusion and drift in the axial direction of the cylinder. At the end of the legs, however, *glueing conditions* need to be imposed. Heuristically, the glueing conditions generate a family of independent tosses of a coin. When the limiting process hits the vertex of the graph, a new leg is chosen via the toss of the coin. The next excursion out of the vertex occurs on this chosen leg. When the process returns to the vertex, the process repeats with a new coin. See [5] for a more rigorous discussion of this. See also [20] for another example of a process defined on a *stratified space*, which is more general than a graph.

Our goal here is to motivate the glueing conditions of Freidlin, Wentzell, and Weber, via a singular perturbations argument. The basic idea stems from a coordinate change introduced by Khasminskii in [8]. We shall not give the entirety of the proof; see [19] for the details. The original work of Freidlin and Wentzell [8] relied upon a specific type of noise—divergence-free noise, which allowed the invariant measure of the original process to be explicitly identified (Lebesgue measure). Roughly, this allowed one to develop an analysis near homoclinic orbits by comparing probabilities of events where the trajectory starts at different points near homoclinic orbits to integration of such probabilities where initial distribution is distributed according to the known invariant measure. More general noise was considered by Freidlin and Weber in [6]. There, fairly complicated hypoelliptic estimates were used to effect a reduction to the calculations for divergence-free noise. Our arguments, however, are entirely local (near the homoclinic orbits) and nowhere require the knowledge of the globally-defined invariant measure. Our main insight is to identify the proper boundary coordinates (these are motivated by [10]). We then pose a problem involving a coupled collection of partial differential equations. It turns out that the solvability of these coupled PDE's exactly corresponds to the glueing conditions. These coordinates have appeared in related problems (see [20]), and we hope that our efforts here will allow insight into other related problems of interest to the engineering and mathematics communities.

2. Problem Formulation

Fix a double-well potential function U on $\mathbb{R}$. We assume that U is nondegenerate; i.e., U'' is nonzero at the points where $U' = 0$ and that $\lim_{|x|\nearrow\infty} U(x) = \infty$. We also assume that the local maximum of U occurs at 0, and that $U(0) = 0$. Define the energy function

$$H(x,p) \stackrel{\text{def}}{=} \frac{p^2}{2} + U(x)$$

for all $(x,p) \in \mathbb{R}^2$. Consider the stochastic differential equation

$$\begin{aligned} dx_t^\varepsilon &= p^\varepsilon dt \\ dp_t^\varepsilon &= -U'(x_t^\varepsilon)dt + \varepsilon^2\beta(x_t^\varepsilon, p_t^\varepsilon)dt + \varepsilon dW_t \end{aligned}$$

where W is a standard Brownian motion, and where β represents dissipation. Also fix an $(x_\circ, p_\circ) \in \mathbb{R}^2$ as our initial conditions; i.e., $(x_0^\varepsilon, p_0^\varepsilon) = (x_\circ, p_\circ)$. Our aim is to understand the fluctuations of the energy $H(x_t^\varepsilon, p_t^\varepsilon)$. This is the problem studied by Freidlin and Weber in [6] and [7].

A proper understanding of $H(x_t^\varepsilon, p_t^\varepsilon)$ requires looking on the proper time scale. Define $Z_t^\varepsilon \stackrel{\text{def}}{=} (x_{t/\varepsilon^2}^\varepsilon, p_{t/\varepsilon^2}^\varepsilon)$; then in law

$$dZ_t^\varepsilon = \frac{1}{\varepsilon^2}\bar{\nabla} H(Z_t^\varepsilon)dt + b(Z_t^\varepsilon)dt + \sigma(Z_t^\varepsilon)dW_t,$$

where $\bar{\nabla} H(x,p) \stackrel{\text{def}}{=} (\frac{\partial H}{\partial p}, -\frac{\partial H}{\partial x})(x,p)$, $B(x,p) \stackrel{\text{def}}{=} (0, \beta(x,p))$, and $\sigma(x,p) \stackrel{\text{def}}{=} (0,1)$. The initial condition is that $Z_0^\varepsilon = (x_\circ, p_\circ)$. Define the second order operator $\mathscr{L}$, symbol $\langle\cdot,\cdot\rangle_{\mathscr{L}}$, and pseudometric $\|\cdot\|_{\mathscr{L}}$ (on $T^*\mathbb{R}^2$) by

$$\begin{aligned} (\mathscr{L}f)(x,p) &\stackrel{\text{def}}{=} \tfrac{1}{2}\tfrac{\partial^2 f}{\partial p^2}(x,p) + \beta(x,p)\tfrac{\partial f}{\partial p}(x,p) \\ \langle df, dg\rangle_{\mathscr{L}}(x,p) &\stackrel{\text{def}}{=} \tfrac{\partial f}{\partial p}(x,p)\tfrac{\partial g}{\partial p}(x,p) \qquad\qquad (x,p) \in \mathbb{R}^2 \\ \|df\|_{\mathscr{L}}^2(x,p) &\stackrel{\text{def}}{=} \left(\tfrac{\partial f}{\partial p}(x,p)\right)^2 \end{aligned}$$

for all f and g in $C^2(\mathbb{R}^2)$. Then the generator of Z^ε is

$$\mathscr{L}^\varepsilon \stackrel{\text{def}}{=} \frac{1}{\varepsilon^2}\bar{\nabla} H + \mathscr{L}$$

and

$$H(Z_t^\varepsilon) = H(Z_0^\varepsilon) + \int_0^t (\mathscr{L}H)(Z_s^\varepsilon)ds + M_t^\varepsilon$$

where M is a martingale with respect to the filtration of W and with quadratic variation

$$\langle M \rangle_t = \int_0^t \|dH\|^2_{\mathscr{L}}(Z_s^\varepsilon)ds$$

for all $t > 0$.

The dominant part of the dynamics of Z^ε corresponds to the flow

$$\begin{aligned} \dot{\mathfrak{z}}_t(x) &= \bar{\nabla} H(\mathfrak{z}_t(x)) \\ \mathfrak{z}_0(x) &= x. \end{aligned} \qquad t \in \mathbb{R},\ x \in \mathbb{R}^2 \tag{1}$$

The requirements on U imply that there is a homoclinic orbit at $H^{-1}(0)$ which divides the phase space of $\mathfrak{z}$ (viz. $\mathbb{R}^2$) into three regions

$$\begin{aligned} \mathbf{S}_O &\overset{\text{def}}{=} \{(x,p) \in \mathbb{R}^2 : H(x,p) > 0\} \\ \mathbf{S}_L &\overset{\text{def}}{=} \{(x,p) \in \mathbb{R}_- \times \mathbb{R} : H(x,p) < 0\} \\ \mathbf{S}_R &\overset{\text{def}}{=} \{(x,p) \in \mathbb{R}_+ \times \mathbb{R} : H(x,p) < 0\}. \end{aligned}$$

Define for convenience the index set

$$\Lambda \overset{\text{def}}{=} \{O, L, R\}.$$

Inside each of the $\mathbf{S}_\ell$'s, the orbits of $\mathfrak{z}$ are periodic. Thus as long as Z^ε stays inside any given $\mathbf{S}_\ell$, we should be able to asymptotically find closed dynamics for $H(Z^\varepsilon)$. We should (cf. the remarks in [8]) then be able to, in a sense, extend these averaging results to all of $\mathbb{R}^2$ if, in addition to keeping track of the slow variable $H(Z^\varepsilon)$, we also keep track of which of the $\mathbf{S}_\ell$'s Z^ε is in. Define $\mathcal{R}_O \overset{\text{def}}{=} \mathbb{R}_+$ and $\mathcal{R}_L = \mathcal{R}_R = \mathbb{R}_-$; i.e., the space for the limiting process should roughly be the disjoint union

$$E \overset{\text{def}}{=} \cup_{\ell \in \Lambda} (\mathcal{R}_\ell \times \{\ell\})$$

We say roughly since in fact, we want to identify all three points in $\{0\} \times \Lambda$ (they all correspond to the homoclinic orbit $H^{-1}(0)$). We will let $\mathfrak{M} \overset{\text{def}}{=} E/\sim$, where $\sim$ is the equivalence relation such that $(h, \ell) \in E$ is equivalent only to itself if $h \neq 0$, and such that $(0, O) \sim (0, L) \sim (0, R)$; we let $\star \overset{\text{def}}{=} \{0\} \times \Lambda$, the "vertex", be this equivalence class. We equip $\mathfrak{M}$ with the standard quotient topology (see [1, p. 60]). Topologically, $\mathfrak{M}$ is equivalent to a "wye"-shaped subset of $\mathbb{R}^2$; see Figure 1. We define the map $\mathscr{Y} : \mathbb{R}^2 \to \mathfrak{M}$ by setting

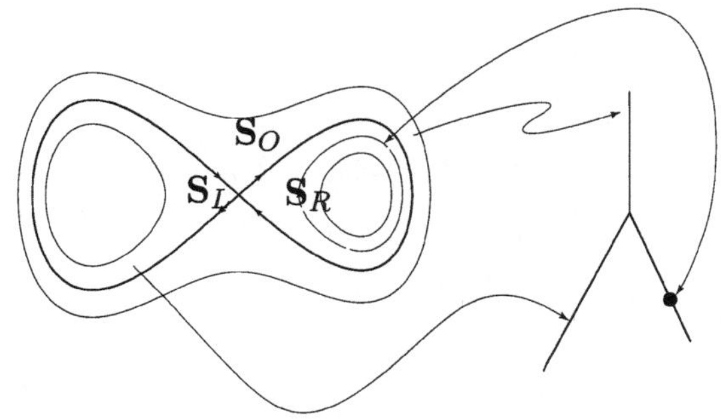

Figure 1. The "Y" graph.

$$\mathscr{Y}(z) \stackrel{\text{def}}{=} (H(z), \ell)$$

if $z \in \mathbf{S}_\ell$, and we define $\mathscr{Y}(z) \stackrel{\text{def}}{=} \star$ if $z \in H^{-1}(0)$. The main result of Freidlin and Wentzell [8] was to show that $\mathscr{Y}(Z^\varepsilon)$ converges in law to an explicitly-defined Markov process on $\mathfrak{M}$ (see also [22]; a diffusion process on such a space is sometimes called a *spider*). Since the limiting process is not on a Euclidean space, we need a formalism more general than that of stochastic differential equations, namely the theory of generators and domains. Define the function spaces

$$\begin{aligned} C(\mathfrak{M}) &\stackrel{\text{def}}{=} \{f \in C(E) : f(0,O) = f(0,L) = f(0,R)\} \\ C^2(\mathfrak{M} \setminus \star) &\stackrel{\text{def}}{=} \{f \in C(\mathfrak{M}) : f(\cdot,\ell) \in C^2(\mathcal{R}_\ell^\circ) \text{ for all } \ell \in \Lambda\}\,; \end{aligned}$$

then $C(\mathfrak{M})$ is (isomorphic to) the collection of continuous functions on $\mathfrak{M}$ (where $\mathfrak{M}$ is endowed with its appropriate topology) and $C^2(\mathfrak{M})$ is (isomorphic to) the collection of continuous functions on $\mathfrak{M}$ which are twice differentiable on the open manifolds corresponding to the legs of the "wye". Note that while $\mathfrak{M}$ is a topological space, it does *not* have a differentiable structure at the vertex $\star$. Define the coefficients

$$
\begin{aligned}
m_2(h,\ell) &\overset{\text{def}}{=} \lim_{T\nearrow\infty}\frac{1}{T}\int_0^T \|dH\|^2_{\mathscr{L}}(\mathfrak{z}_s(x))ds \\
&= \frac{\int_{x'\in\mathbb{R}^2:\mathscr{Y}(x')=(h,\ell)} \frac{\|dH\|^2_{\mathscr{L}}(x')}{\|\bar{\nabla} H(x')\|}\mathscr{H}^1(dx')}{\int_{x'\in\mathbb{R}^2:\mathscr{Y}(x')=(h,\ell)} \|\bar{\nabla} H(x')\|^{-1}\mathscr{H}^1(dx')} \\
m_1(h,\ell) &\overset{\text{def}}{=} \lim_{T\nearrow\infty}\frac{1}{T}\int_0^T (\mathscr{L}H)(\mathfrak{z}_s(x))ds \\
&= \frac{\int_{x'\in\mathbb{R}^2:\mathscr{Y}(x')=(h,\ell)} \frac{(\mathscr{L}H)(x')}{\|\bar{\nabla} H(x')\|}\mathscr{H}^1(dx')}{\int_{x'\in\mathbb{R}^2:\mathscr{Y}(x')=(h,\ell)} \|\bar{\nabla} H(x')\|^{-1}\mathscr{H}^1(dx')}
\end{aligned}
$$

for all $(h,\ell)\in E\setminus\star$, where x is any element of $\mathbb{R}^2\setminus H^{-1}(0)$ such that $\mathscr{Y}(x)=(h,\ell)$ and where $\mathscr{H}^1$ is 1-dimensional Hausdorff measure.

If $f\in C^2(\mathfrak{M})$, we define

$$\mathscr{L}^\dagger f(h,\ell) \overset{\text{def}}{=} \frac{1}{2}m_2(h,\ell)\frac{\partial^2 f}{\partial h^2}(h,\ell) + m_1(h,\ell)\frac{\partial f}{\partial h}(h,\ell)$$

for all $(h,\ell)\in E\setminus\star$. It is fairly easy to show that if $\sup_{(h,\ell)\in E\setminus\star}|\mathscr{L}^\dagger f(h,\ell)|$ is finite, then $\frac{\partial f}{\partial h}(0,\ell)$ exists (for each $\ell\in\Lambda$). We then define the glueing coefficients

$$\mathrm{G}\ell \overset{\text{def}}{=} \int_{x'\in\partial\mathbf{S}_\ell} \|\bar{\nabla} H(x')\|^{-1}\mathscr{H}^1(dx')$$

for all $\ell\in\Lambda$ and we define the generator

$$
\begin{aligned}
\mathscr{L}^\dagger \overset{\text{def}}{=} \Big\{(f,g)\in C(\mathfrak{M})\times C(\mathfrak{M}) : f\in C^2(\mathfrak{M}\setminus\star),\ (\mathscr{L}^\dagger f)=g \text{ on } E\setminus\star \\
\text{and } \mathrm{G}O\tfrac{\partial f}{\partial h}(0,O) = \mathrm{G}L\tfrac{\partial f}{\partial h}(0,L) + \mathrm{G}R\tfrac{\partial f}{\partial h}(0,R)\Big\}.
\end{aligned}
$$

To state the main result of Freidlin and Wentzell devoid of technicalities, let's first restrict the region of interest. Let $\hbar>0$ be small enough that

$$\mathbf{S}_\hbar \overset{\text{def}}{=} \{(x,p)\in\mathbb{R}^2 : |H(x,p)|\le\hbar\}$$

has no critical points of H other than the origin. Let Y be the Markov process which starts at $\mathscr{Y}(x_\circ,p_\circ)$ and has generator $\mathscr{L}^\dagger$ up to the time it leaves $\mathscr{Y}(\mathbf{S}_\hbar)$. In other words, let Y be a stochastic process on some probability triple such that for each $(f,g)\in\mathscr{L}^\dagger$ such that $g\equiv 0$ on $\mathscr{Y}(\mathbb{R}^2\setminus\mathbf{S}_\hbar)$ (this restriction corresponds to killing the process when it leaves $\mathscr{Y}(\mathbf{S}_\hbar)$), there is a martingale M (with respect to the filtration generated by Y) such that

$$f(Y_t) = f(\mathscr{Y}(x_\circ,p_\circ)) + \int_{s=0}^t g(Y_s)ds + M_t$$

for all $t \geq 0$. Define the stopping time

$$\tau^\varepsilon \stackrel{\text{def}}{=} \inf\{t \geq 0 : Z_t^\varepsilon \notin \mathbf{S}_{\hbar}\}$$

Theorem 1 *The processes* $\{\mathscr{Y}(Z_{t\wedge\tau^\varepsilon}^\varepsilon);\, t \geq 0\}$ *converge (as* $\varepsilon \searrow 0$*) in law to* Y.

The glueing conditions roughly mean that when the process hits $\star$, it flips a coin to decide where to make the next excursion. With probability $\mathrm{G}_\ell / \sum_{\ell' \in \Lambda} \mathrm{G}_{\ell'}$, this excursion will be made into leg ℓ. Since transitions from one leg to another correspond to transitions across $H^{-1}(0)$, it is of course natural that the glueing coefficients are given by some statistic of the original generator at the $\partial\mathbf{S}_\ell$'s.

The compactness of the laws of $\mathscr{Y}(Z^\varepsilon)$'s and the uniqueness of the law defined by the generator $\mathscr{L}^\dagger$ are fairly standard. We can carefully use stochastic averaging to locally show that within each leg, $H(Z^\varepsilon)$ evolves according to the generator $\mathscr{L}$. We also need to show that Z^ε does not "stick" at the homoclinic orbit. This can be done too.

Glueing conditions

The main interest, of course, is to understand the glueing conditions.

There are several perspectives. Let's look first at the answer. The glueing coefficients are essentially the numerators of the averaged diffusion coefficient m_2. This is not too surprising. The *diffusion* is the dominant reason why the process moves between the $\mathbf{S}_\ell$'s; a Brownian motion crosses zero many more times than a smooth curve. This brings up a technical issue which gets us started on our analysis. A core of the domain of $\mathscr{L}^\varepsilon$, the generator of Z^ε, is $C_0^2(\mathbb{R}^2)$. In other words, we initially understand the law of Z^ε by looking at test functions which are twice-differentiable near $H^{-1}(0)$. In the limit, however, the dynamics of Y are singular due to the structure of $\mathfrak{M}$. In some way, we need to connect the smoothness requirements of the prelimit process to the singular nature of the limiting process. Another framework in which to study this is continuity conditions and conservation of flux. At each point x of $H^{-1}(0)$, the density of Z^ε (if a density exists) must be continuous and the flux going across $H^{-1}(0)$ at x must sum to zero. The continuity requirement means that if we approach x from within $H^{-1}(0)$, the density has the same limit as if we approach x from outside $H^{-1}(0)$. We need to compress these continuity and conservation laws from all of $H^{-1}(0)$ into a corresponding requirement at $\star$.

Let's focus on how to collapse the generator of Z^ε to $\mathscr{L}^\dagger$. Fix an $(f, g) \in \mathscr{L}^\dagger$. We can then define $\hat{f} : \mathbb{R}^2 \to \mathbb{R}$ by setting

$$\hat{f}(x) \stackrel{\text{def}}{=} \sum_{\ell \in \Lambda} f(H(x), \ell)\chi_{\mathbf{S}_\ell}(x) + f(\star)\chi_{\{0\}}(H(x))$$

We first note that $\hat{f}$ is continuous. However, unless all of the $\frac{\partial f}{\partial h}(0, \ell)$'s are equal (this is clearly allowed since $\mathsf{G}_O = \mathsf{G}_L + \mathsf{G}_R$, but such functions do not form a large enough set to ensure uniqueness of the limiting process), $\hat{f}$ will *not* be differentiable at $H^{-1}(0)$, and thus it will *not* be in the domain of $\mathscr{L}^\varepsilon$. Thus we need to *correct* $\hat{f}$ near $H^{-1}(0)$. This suggests a link with singular perturbations, or more accurately, the notion of *perturbed test functions* (see [4], [13], [14], and [15]). Our main claim is the following.

Proposition 2 *Since the glueing conditions hold, there is a sequence* $\{\Phi^\varepsilon : \varepsilon \in (0,1)\}$ *of bounded functions on* $\mathbb{R}^2$ *such that*

i) $\hat{f} + \Phi^\varepsilon \in C^1(\mathbb{R}^2)$

ii) Φ^ε *and its derivatives tend to zero exponentially as* $|H(x,p)/\varepsilon| \nearrow \infty$.

iii) $\mathscr{L}^\varepsilon \Phi^\varepsilon$ *is* $o(1)$ *as* $\varepsilon \searrow 0$.

iv) Φ^ε *is* $o(1)$ *as* $\varepsilon \searrow 0$.

Thus, the Φ^ε's are approximately harmonic functions (for $\mathscr{L}^\varepsilon$) which smooth $\hat{f}$. It is natural to seek harmonic functions since such harmonic functions naturally describe exit probabilities.

In general, an inner expansion is an approximate solution of a given PDE in a boundary layer near the edge of a boundary. In canonical problems, the role of the inner expansion is to describe the transition between some specified boundary data and the outer expansion (which usually fails to satisfy the specified boundary conditions). In our case, the boundary in question is $H^{-1}(0)$, and the boundary data is implicit; it is given by the "hidden" constraint that $\hat{f} + \Phi^\varepsilon$ is required to be C^1 at $H^{-1}(0)$. This means that at $H^{-1}(0)$, we want the values of $\bar{\Phi}^\varepsilon$ and their appropriate transversal derivatives (i.e., transversal to $H^{-1}(0)$) to match. We also note that effectively we have *three* boundary layers, one in each $\mathbf{S}_\ell$, near $\partial \mathbf{S}_\ell$.

Often the approximate solution of a PDE for an inner expansion is given by using an expanded coordinate in the direction *transversal* to the boundary, and a coordinate in the direction *along* the boundary.

Since $H^{-1}(0)$ is naturally defined as a level set of H, the coordinate in the direction transversal to $H^{-1}(0)$ will naturally be H. Consider for a moment a boundary layer of order ε^α for some $\alpha > 0$. If we apply $\mathscr{L}$ to a to a nonlinear function of H/ε^α, then we get a term of $\varepsilon^{-2\alpha}$ (since $\mathscr{L}$ is a second-order operator). This is due to the diffusion in the direction transversal to $H^{-1}(0)$. We want this to be of the same order as the fast drift, which is of order ε^{-2}. Thus we should take $\alpha = 1$.

Let's next discuss the boundary coordinate. If Θ is a twice-differentiable real-valued function on some open subset $\mathscr{O}$ of $\mathbb{R}^2$ and if we set $\psi^\varepsilon(x) \stackrel{\text{def}}{=}$

$\psi\left(\Theta(x), \frac{H(x)}{\varepsilon}\right)$ for $x \in \mathscr{O}$ where ψ is twice-differentiable, then

$$\begin{aligned}(\mathscr{L}^\varepsilon \psi^\varepsilon)(x) &= \frac{1}{\varepsilon^2}\left\{\frac{\partial \psi}{\partial \theta}(\bar{\nabla} H, \nabla \Theta)(x) + \frac{1}{2}\frac{\partial^2 \psi}{\partial^2 h}\|dH\|^2_{\mathscr{L}}(x)\right\} \\ &\quad + \frac{1}{\varepsilon}\left\{\frac{\partial \psi}{\partial h}(\mathscr{L}H)(x) + \frac{\partial^2 \psi}{\partial \theta \partial h}\langle dH, d\Theta\rangle_{\mathscr{L}}(x)\right\} \\ &\quad + \left\{\frac{\partial \psi}{\partial \theta}(\mathscr{L}\Theta)(x) + \frac{1}{2}\frac{\partial^2 \psi}{\partial \theta^2}\|d\Theta\|^2_{\mathscr{L}}(x)\right\} \end{aligned} \quad (2)$$

for all $x \in \mathscr{O}$, where the various derivatives of ψ are all evaluated at $\left(\Theta(x), \frac{H(x)}{\varepsilon}\right)$. The *Khasminskii* coordinates, which we will denote by Θ_K, equate the coefficients of the two dominant terms; i.e., they solve the transport equation

$$(\bar{\nabla} H, \nabla \Theta_K) = \|dH\|^2_{\mathscr{L}} \quad (3)$$

on some appropriate subset of $\mathbf{S}$.

Remark 3 *Heuristically, the Khasminskii coordinates normalize the diffusion across $H^{-1}(0)$ (this transversal diffusion is responsible for causing the graph-valued limit to choose one leg over the other when it starts at the vertex). Roughly, equal increments of Khasminskii coordinates correspond to equal amounts of transversal diffusion across $H^{-1}(0)$. The glueing conditions are directly related to the relative lengths of the $\partial \mathbf{S}_\ell$'s in the Khasminskii coordinates.*

We are thus led to the coordinate transformation $x \mapsto (\Theta_K(x), H(x))$. One needs to select initial conditions for the PDE (3); these initial conditions will in general be specified along some manifold which is transveral to the level sets of H. In particular, one needs to specify initial data in each of the $\mathbf{S}_\ell$'s. Note that the range of Θ_K in each $\mathbf{S}_\ell$ near $\partial \mathbf{S}_\ell$ is an interval of width G_ℓ; see Figure 2. To make things convenient, we can impose initial data on (3) so that Θ_K starts at 0 just below $(0,0)$, and then increases to G_L as we travel around the left loop (in the direction of $\mathfrak{z}$) of $H^{-1}(0)$ and comes back to $(0,0)$, and then arrange things so that Θ_K further increases around the right loop of $H^{-1}(0)$ (again travelling in the direction of $\mathfrak{z}$) to finally reach $\mathsf{G}_L + \mathsf{G}_R = \mathsf{G}_O$ as we come back to $(0,0)$. Define

$$\mathsf{L}_O = \mathsf{L}_L \stackrel{\text{def}}{=} 0, \quad \mathsf{R}_L = \mathsf{L}_R = \mathsf{G}_L, \quad \text{and} \quad \mathsf{R}_R = \mathsf{R}_O = \mathsf{G}_O$$

and then define

$$\mathsf{I}_\ell \stackrel{\text{def}}{=} (\mathsf{L}_\ell, \mathsf{R}_\ell)$$

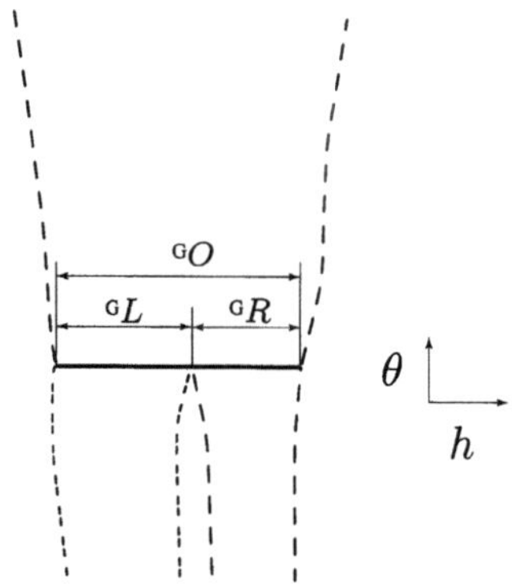

Figure 2. Khasminskii Coordinates (this picture was suggested to the author by N. Sri Namachchivaya, based on some work with Peter Imkeller.)

for all $\ell \in \Lambda$.

The following result, which is the heart of Proposition 2, now has some significance.

Proposition 4 *There is a triplet $(\Psi_O^K, \Psi_L^K, \Psi_R^K)$ of functions such that t he following hold. For each $\ell \in \Lambda$, Ψ_ℓ^K is in $C^\infty(\overline{I_\ell} \times \mathcal{R}_\ell)$ and*

$$\frac{\partial \Psi_\ell^K}{\partial \theta}(\theta, h) + \frac{1}{2}\frac{\partial^2 \Psi_\ell^K}{\partial^2 h}(\theta, h) = 0 \tag{4}$$

for all $(\theta, h) \in I_\ell \times \mathcal{R}_\ell^\circ$ (this is a PDE). Secondly, for every multi-index α such that $|\alpha| \geq 1$, there is a constant $K > 0$ such that

$$\left| D^\alpha \Psi_\ell^K(\theta, h) \right| \leq K e^{-|h|K} \tag{5}$$

for all $\theta \in I_\ell$ and $h \in \mathcal{R}_\ell$ such that $|h| \geq 1$ (these are growth conditions). Thirdly, we have that

$$\Psi_\ell^K({}_{L}\ell, h) = \Psi_\ell^K({}_{R}\ell, h) \tag{6}$$

for all $h \in \mathcal{R}_\ell$ (this is a periodicity condition). Fourthly and finally, we have that

$$\begin{aligned} \lim_{h \searrow 0} \Psi_O^K(\cdot, h) &= \lim_{h \nearrow 0} \Psi_L^K(\cdot, h) \qquad \text{on } I_L \\ \lim_{h \searrow 0} \Psi_O^K(\cdot, h) &= \lim_{h \nearrow 0} \Psi_R^K(\cdot, h) \qquad \text{on } I_R \end{aligned} \tag{7}$$

and that

$$\begin{aligned} \lim_{h\searrow 0} \frac{\partial \Psi_O^K}{\partial h}(\cdot, h) &= \lim_{h\nearrow 0} \frac{\partial \Psi_L^K}{\partial h}(\cdot, h) \qquad \text{on } I_L \\ \lim_{h\searrow 0} \frac{\partial \Psi_O^K}{\partial h}(\cdot, h) &= \lim_{h\nearrow 0} \frac{\partial \Psi_R^K}{\partial h}(\cdot, h) \qquad \text{on } I_R \end{aligned} \tag{8}$$

(*these are matching conditions at the boundaries*).

See [19] for a proof; the sense of convergence in (7) and (8) is also specified there. We then set

$$\Phi^\varepsilon(x) \stackrel{\text{def}}{=} \varepsilon \Psi_\ell^K \left(\Theta_K(x), \frac{H(x)}{\varepsilon} \right) \tag{9}$$

if $x \in \mathbf{S}_\ell$. The PDE condition of (4) should imply that Φ^ε is approximately harmonic. The decay conditions of (5) mean that Φ^ε is a true boundary term. The periodicity conditions of (6) imply that if we approximate the range of Θ_K in $\mathbf{S}_\ell$ (and not just at $\partial\mathbf{S}_\ell$) by an interval of width $\mathfrak{G}_\ell$, then Φ^ε should be almost smooth (in particular where we have specified the initial data for Θ_K). The matching conditions of (7) and (8) are exactly what we need to make $\hat{f} + \Phi^\varepsilon$ continuous and differentiable at $H^{-1}(0)$.

We should note, however, that there are some wrinkles. In general, one has small discontinuities when the orbits of $\mathfrak{z}$ come back to the manifold along which initial data for Θ_K was specified (since the width of the range along the orbits of $\mathfrak{z}$ are exactly the $\mathfrak{G}_\ell$'s only on the homoclinic orbit $H^{-1}(0)$). Secondly, one must consider *singularities* at the origin. There is absolutely no assurance that the terminal data of the backward heat equations is such that (9) is smooth near the origin. These singularities need to be carefully treated.

References

[1] W. M. Boothby, *An Introduction to Differentiable Manifolds and Riemannian Geometry*. Academic Press Inc., Orlando, FL, second edition, 1986.

[2] A. N. Borodin, "A limit theorem for the solutions of differential equations with a random right-hand side," *Teor. Verojatnost. i Primenen.*, vol. 22, no. 3, pp. 498–512, 1977.

[3] A. N. Borodin and M. I. Freidlin, "Fast oscillating random perturbations of dynamical systems with conservation laws," *Ann. Inst. H. Poincaré Probab. Statist.*, vol. 31, no. 3, pp. 485–525, 1995.

[4] L. C. Evans, "The perturbed test function method for viscosity solutions of nonlinear PDE," *Proc. Roy. Soc. Edinburgh Sect. A*, vol. 111, no. (3-4), pp. 359–375, 1989.

[5] S. N. Evans and R. B. Sowers, "Pinching and twisting Markov processes," *Annals of Probability*, to appear.

[6] M. Freidlin and M. Weber, "Random perturbations of nonlinear oscillators," *Ann. Probab.*, vol. 26, no. 3, pp. 925–967, 1998.

[7] M. Freidlin and M. Weber, "A remark on random perturbations of the nonlinear pendulum," *Ann. Appl. Probab.*, vol. 9, no. 3, pp. 611–628, 1999.

[8] M. I. Freidlin and A. D. Wentzell, "Random perturbations of Hamiltonian systems," *Mem. Amer. Math. Soc.*, 109(523):viii+82, 1994.

[9] I. I. Gihman, "Concerning a theorem of N. N. Bogolyubov," *Ukrain. Mat. Ž.*, vol. 4, pp. 215–219, 1952.

[10] R. Z. Khasminskii, "Diffusion processes with a small parameter," *Izv. Akad. Nauk SSSR Ser. Mat.*, vol. 27, pp. 1281–1300, 1963.

[11] R. Z. Khasminskii, "A limit theorem for solutions of differential equations with a random right hand part," *Teor. Verojatnost. i Primenen*, vol. 11, pp. 444–462, 1966.

[12] R. Z. Khasminskii, "Stochastic processes defined by differential equations with a small parameter," *Teor. Verojatnost. i Primenen*, vol. 11, pp. 240–259, 1966.

[13] T. G. Kurtz, "A limit theorem for perturbed operator semigroups with applications to random evolutions," *J. Functional Analysis*, vol. 12, pp. 55–67, 1973.

[14] T. G. Kurtz, "An abstract averaging theorem," *J. Functional Analysis*, vol. 23, no. 2, pp. 135–144, 1976.

[15] H. J. Kushner. *Approximation and Weak Convergence Methods for Random Processes with Applications to Stochastic System Theory*, MIT Press, Cambridge, MA, 1984.

[16] A. I. Neĭshtadt, "Probability phenomena due to separatrix crossing," *Chaos*, vol. 1, no. 1, pp. 42–48, 1991.

[17] G. C. Papanicolaou and W. Kohler, "Asymptotic theory of mixing stochastic ordinary differential equations," *Comm. Pure Appl. Math.*, vol. 27, pp. 641–668, 1974.

[18] G. C. Papanicolaou and W. Kohler, "Asymptotic analysis of deterministic and stochastic equations with rapidly varying components," *Comm. Math. Phys.*, vol. 45, no. 3, pp. 217–232, 1975.

[19] R. B. Sowers, "Stochastic averaging near a homoclinic orbit with multiplicative noise," submitted.

[20] R. B. Sowers, "Stochastic averaging with a flattened Hamiltonian: A Markov process on a stratified space (a whiskered sphere)," *Trans. Amer. Math. Soc.*, vol. 354, no. 3, pp. 853–900, (electronic), 2002.

[21] R. L. Stratonovich, *Topics in the Theory of Random Noise, Vol. I: General Theory of Random Processes. Nonlinear Transformations of Signals and Noise.* Gordon and Breach Science Publishers, New York, 1963. Revised English edition. Translated from the Russian by Richard A. Silverman.

[22] J. Walsh, "A diffusion with discontinuous local time," in *Temps Locaux, Astérisque*, vol. 52–53. Société Mathématique de France, Paris, 1978.

[23] G. Wolansky, "Limit theorem for a dynamical system in the presence of resonances and homoclinic orbits," *J. Differential Equations*, vol. 83, no. 2, pp. 300–335, 1990.

II

PART 2: STABILITY AND BIFURCATIONS

The primary concern in the analysis of *nonlinear* dynamical systems, is the determination and prediction of steady-states or stationary motions (e.g. invariant measures of the local random dynamical systems), and their corresponding stability, which is the focus of the *second* part of the Proceedings. The second *Lead Lecture* of the Symposium given by S. T. Ariaratnam of the University of Waterloo, Canada is the lead paper of this part. The rest of the papers presented during the Symposium, on the general topic of *stability and bifurcations of stochastic systems* are compiled alphabetically, based on the last name of the first author.

In the study of stability of solutions of random dynamical systems, both almost-sure stability and p^{th} moment stability have been widely used. Stability in the almost-sure sense is determined by the sign of the maximal Lyapunov exponent, analogous to the real part of the eigenvalue of the corresponding deterministic system. From an applications standpoint, one may not be satisfied with just almost-sure stability, for a sample process, though stable, may exceed some threshold values or may possess a slow rate of decay. The moment Lyapunov exponent, which describes the exponential growth rate of the pth moment of the solution, gives a more practical measure of the stability. The following four papers deal with this aspect for linear parametrically excited systems.

S. T. Ariaratnam presented results on the almost-sure asymptotic stability of a pair of coupled linear oscillators whose frequencies are equal or nearly equal and parametrically excited by a wide-band stochastic process. Both of the cases when the frequency matrix is of diagonal form and of Jordan-canonical form were considered. N. Sri Namachchivaya presented an approximation for the moment Lyapunov exponent of a similar problem of coupled oscillators in resonance. Volker Wihstutz presented results on stabilization of coupled oscillators by mean zero noise. This work extended the previous results on stabilization by white noise for systems with trace negative linear operators. Wei-Chau Xie presented results on the almost-sure asymptotic stability of a two-dimensional system under combined harmonic and real noise excitations, where the real noise excitation was modeled as an Ornstein-Uhlenbeck process.

In the investigation of stability of nonlinear stochastic systems the challenge has been to evaluate explicitly the top Lyapunov exponents of non-trivial stationary measures. Peter H. Baxendale obtained (asymptotically) the almost sure Lyapunov exponent of the non-trivial stationary measure of the Duffing-van der Pol oscillator. The stochastic averaging techniques were used to estimate the value of the almost sure Lyapunov exponent in the presence of small noise and small viscous damping.

The next three papers provided a phenomenological approach to study the stochastic bifurcation behavior; namely, the qualitative changes in parameterized families of random dynamical systems, e.g. those generated by a family of stochastic differential equations. Adi Bulsara studied the noisy FitzHugh-Nagumo model in the presence of an external sinusoidal driving force, revealing the existence of a stochastic bifurcation. Huw Davies considered a ship model with coupled pitch and roll modes. He presented results of a saturation phenomenon in the pitch mode, with a corresponding large amplitude increase in the roll mode, as the wave excitation amplitude was slowly increased. Rachel Kuske developed a multi-scale analysis for systems particularly sensitive to noise near a critical point, such as a Hopf bifurcation, marking a transition to oscillatory behavior. Both papers of Davies and Kuske provided the insights into the use of multiple-scales in the bifurcation analysis of noisy systems. William Langford presented a deterministic bifurcation result, useful for the investigation of "near-Hamiltonian" or weakly dissipative systems.

STOCHASTIC STABILITY OF COUPLED OSCILLATORS IN INTERNAL RESONANCE

S. T. Ariaratnam and N. M. Abdelrahman
University of Waterloo
Waterloo, Ontario N2L 3G1
CANADA
NMABDELR@engmail.uwaterloo.ca

Abstract The almost-sure asymptotic stability of a pair of coupled linear oscillators whose frequencies are equal or nearly equal and parametrically excited by a wide-band stochastic process is investigated. Using the method of stochastic averaging and a Fourier expansion technique for solving the Fokker-Planck equation associated with the response angles, the maximal Lyapunov exponent is evaluated. The sign of the Lyapunov exponent determines the stability or instability of the oscillators. The cases when the frequency matrix is of diagonal form and of Jordan-canonical form are considered.

1. Introduction

The dynamical behaviour of many linear systems subjected to parametric excitation is described by equations of motion of the form

$$\ddot{\mathbf{q}} + D\dot{\mathbf{q}} + \Omega^2\mathbf{q} + \xi(t)K\mathbf{q} = \mathbf{0} \tag{1}$$

where $\mathbf{q}$ is the vector of modal coordinates, D is the damping matrix, K is the load coupling matrix, Ω is the frequency matrix and $\xi(t)$ represents the parametric excitation. When $\xi(t)$ is a wide-band process of small intensity and the frequency matrix Ω is diagonal with distinct frequencies, the stochastic stability of such a system has been investigated, [1] and [2]. This paper is concerned with the exceptional case of a two degree of freedom system when two frequencies are either equal or nearly equal. In this case, two forms of Ω have to be considered, namely when Ω is diagonal as in conservative systems when the stiffness matrix is derivable from a potential function, and Ω is of Jordan-canonical form as in non-conservative systems.

N. Sri Namachchivaya and Y.K. Lin (eds.),
IUTAM Symposium on Nonlinear Stochastic Dynamics, 97-110.

When the excitation and the damping terms are of sufficiently small intensity, the method of stochastic averaging is employed to obtain expressions for the maximal Lyapunov exponent for each case. The almost-sure stability of the trivial solution is then determined from the sign of the largest Lyapunov exponent and numerical results are presented.

2. Conservative Systems

A conservative system subjected to small intensity excitation and damping forces is described by the equations

$$\ddot{q}_i + 2\varepsilon\beta_i\dot{q}_i + {\omega_i}^2 q_i + \varepsilon^{1/2}\omega_i\xi(t)\sum_{j=1}^{2} k_{ij}q_j = 0, \quad i = 1,2 \tag{2}$$

where β_i and k_{ij}, $i,j = 1,2$ are constants and $0 < \varepsilon \ll 1$. The natural frequencies are such that $\omega_{1,2} = \omega_0(1 \mp \delta_0)$, with $\delta_0 = O(\varepsilon)$, i. e. $\omega_1 \simeq \omega_2$. Transforming to polar coordinates via the transformation

$$q_i = a_i \cos\Theta_i\ , \ \dot{q}_i = -a_i\omega_0 \sin\Theta_i\ , \ \Theta_i = \omega_0 t + \theta_i\ , \ i = 1,2 \tag{3}$$

leads to the following set of first-order equations correct to order $O(\epsilon)$:

$$\begin{aligned}
\dot{a}_i(t) &= -2\varepsilon\beta_i a_i \sin^2\Theta_i + \varepsilon^{1/2}\xi(t)\sum_{j=1}^{2} k_{ij}a_j\cos\Theta_j\sin\Theta_i \\
&\quad + (-1)^i\delta_0\omega_0 a_i \sin 2\Theta_i \\
\dot{\theta}_i(t) &= -\varepsilon\beta_i \sin 2\Theta_i + \varepsilon^{1/2}\xi(t)\sum_{j=1}^{2} k_{ij}\frac{a_j}{a_i}\cos\Theta_j\cos\Theta_i \\
&\quad + 2(-1)^i\delta_0\omega_0\cos^2\Theta_i
\end{aligned} \tag{4}$$

If $\xi(t)$ is taken to be a sufficiently wide-band stochastic process that may be approximated by a white-noise process of spectral density S_0, the above equations may be replaced by the so-called averaged equations following the method of [6] and [4]. As $\varepsilon \to 0$, the solution of the averaged equations converges in the weak sense and up to first order in ϵ to a diffusive Markov process whose govering Itô equations are of the form:

$$da_i = \varepsilon m_{a_i} dt + \varepsilon^{1/2}\sum_{j=1}^{4} \sigma_{a_{ij}} dW_j$$

$$d\theta_i = \varepsilon m_{\theta_i} dt + \varepsilon^{1/2}\sum_{j=1}^{4} \sigma_{\theta_{ij}} dW_j, \ i = 1,2 \tag{5}$$

where W_j are mutually independent unit Wiener processes. The drift coefficients m_{a_i} and m_{θ_i} and the diffusion coefficients $\sigma_{a_{ij}}$, $\sigma_{\theta_{ij}}$ are obtained as:

$$
\begin{aligned}
m_{a_i} &= \frac{1}{16}a_i\left(-16\beta_i + 3k_{ii}{}^2 S_0\right) + \frac{1}{16}\frac{a_j{}^2}{a_i}k_{ij}{}^2 S_0\left(2+\cos 2(\theta_i-\theta_j)\right) \\
&\quad + \frac{1}{8}a_j(k_{ii}-k_{jj})k_{ij}S_0\cos(\theta_i-\theta_j) \\
m_{\theta_i} &= -\frac{1}{8}\frac{a_j}{a_i}(k_{ii}-k_{jj})k_{ij}S_0\sin(\theta_i-\theta_j) \\
&\quad - \frac{1}{8}\frac{a_j{}^2}{a_i^2}k_{ij}{}^2 S_0\sin 2(\theta_i-\theta_j) + (-1)^i(\delta_0/\varepsilon)\omega_0 \quad i\neq j; i,j=1,2
\end{aligned}
$$

$$
\begin{aligned}
[\sigma\sigma^T]_{a_i a_i} &= \frac{1}{8}a_i{}^2 k_{ii}{}^2 S_0 + \frac{1}{8}a_j^2 k_{ij}{}^2 S_0\left(2-\cos 2(\theta_i-\theta_j)\right) \\
[\sigma\sigma^T]_{a_i a_j} &= \frac{1}{8}a_i a_j k_{ij}k_{ji}S_0\cos 2(\theta_i-\theta_j) \quad i\neq j \\
[\sigma\sigma^T]_{a_i\theta_i} &= \frac{1}{4}a_j k_{ii}k_{ij}S_0\sin(\theta_i-\theta_j) + \frac{1}{8}\frac{a_j^2}{a_i}k_{ij}^2 S_0\sin 2(\theta_i-\theta_j) \\
[\sigma\sigma^T]_{a_i\theta_j} &= \frac{1}{4}a_j k_{jj}k_{ij}S_0\sin(\theta_i-\theta_j) + \frac{1}{8}a_i k_{ij}k_{ji}S_0\sin 2(\theta_i-\theta_j) \quad i\neq j \\
[\sigma\sigma^T]_{\theta_i\theta_i} &= \frac{3}{8}k_{ii}{}^2 S_0 + \frac{1}{2}\frac{a_j}{a_i}k_{ii}k_{ij}S_0\cos(\theta_i-\theta_j) \\
&\quad + \frac{1}{8}\frac{a_j^2}{a_i^2}k_{ij}{}^2 S_0\left(2+\cos 2(\theta_i-\theta_j)\right) \\
[\sigma\sigma^T]_{\theta_i\theta_j} &= \frac{1}{4}\left(\frac{a_i}{a_j}k_{ii}k_{ji} + \frac{a_j}{a_i}k_{ij}k_{jj}\right)S_0\cos(\theta_i-\theta_j) + \frac{1}{4}k_{ii}k_{jj}S_0 \\
&\quad + \frac{1}{8}k_{ij}k_{ji}S_0\left(2+\cos 2(\theta_i-\theta_j)\right) \quad i\neq j; i,j=1,2
\end{aligned} \tag{6}
$$

Without loss of generality it is always possible, for a conservative system, to choose a suitable coordinate scaling such that $k_{12} = k_{21}$ and $k_{12}k_{21}$ is invariant under the scaling. A further transformation defined by

$$
\begin{gathered}
\rho = \ln\left(a_1{}^2 + a_2{}^2\right)^{1/2}, \quad \phi = \tan^{-1}(a_2/a_1), \quad \theta = \theta_1 - \theta_2, \\
0 \le \phi \le \pi/2, \quad |\theta| \le \pi,
\end{gathered} \tag{7}
$$

leads to the following Itô equations:

$$d\rho = \varepsilon Q_1(\phi,\theta)dt + \varepsilon^{1/2}\sum_{j=1}^{4}\left(\sigma_{a_{1j}}\frac{\partial\rho}{\partial a_1} + \sigma_{a_{2j}}\frac{\partial\rho}{\partial a_2}\right)dW_j(t)$$

$$d\phi = \varepsilon Q_2(\phi,\theta)dt + \varepsilon^{1/2}\sum_{j=1}^{4}\left(\sigma_{a_{1j}}\frac{\partial\phi}{\partial a_1} + \sigma_{a_{2j}}\frac{\partial\phi}{\partial a_2}\right)dW_j(t)$$

$$d\theta = \varepsilon Q_3(\phi,\theta)dt + \varepsilon^{1/2}\sum_{j=1}^{4}\left(\sigma_{\theta_{1j}} - \sigma_{\theta_{2j}}\right)dW_j(t) \tag{8}$$

where the drift coefficients Q_i are given by

$$\begin{aligned}
&Q_1(\theta,\phi) = \lambda_1\cos^2\phi + \lambda_2\sin^2\phi + \frac{1}{4}k_{12}^2 S_0 \\
&+\frac{1}{32}\left(k_{11}^2 + k_{22}^2 - 4k_{12}^2\right)S_0\sin^2 2\phi \\
&Q_2(\theta,\phi) = -\frac{1}{2}(\lambda_1-\lambda_2)\sin 2\phi + \frac{1}{8}k_{12}^2 S_0\,(2+\cos 2\theta)\cot 2\phi \\
&+\frac{1}{8}(k_{22}-k_{11})k_{12}S_0\cos\theta \\
&+\frac{1}{64}\left(k_{11}^2 + k_{22}^2 - 4k_{12}^2\right)S_0\sin 4\phi \\
&Q_3(\theta,\phi) = -\frac{1}{8}k_{12}^2 S_0\left(\tan^2\phi + \cot^2\phi\right)\sin 2\theta - 2(\delta_0/\varepsilon)\omega_0 \\
&-\frac{1}{4}k_{12}(k_{22}-k_{11})S_0\cot 2\phi\sin\theta
\end{aligned} \tag{9}$$

where $\lambda_i = -\beta_i + \frac{1}{8}k_{ii}^2 S_0$. The angle processes are seen to be independent of the ρ process. Their stationary probability density $p(\theta,\phi)$ is governed by the Fokker-Planck equation

$$\begin{aligned}
\frac{\partial}{\partial\phi}(Q_2 p) + \frac{\partial}{\partial\theta}(Q_3 p) - \frac{1}{2}\frac{\partial^2}{\partial\phi^2}(\Sigma_{22}p) - \frac{1}{2}\frac{\partial^2}{\partial\theta^2}(\Sigma_{33}p)& \\
-\frac{1}{2}\frac{\partial^2}{\partial\phi\partial\theta}(\Sigma_{23}p) = 0&
\end{aligned} \tag{10}$$

where the coefficient Σ_{22}, Σ_{33} and Σ_{23} are found to be

$$\begin{aligned}
\Sigma_{22}(\theta,\phi) &= \frac{1}{32}\left(k_{11}^2 + k_{22}^2 - 4k_{12}^2\right)S_0\sin^2 2\phi + \frac{1}{8}k_{12}^2 S_0\,(2-\cos 2\theta) \\
\Sigma_{33}(\theta,\phi) &= \frac{1}{2}k_{12}^2 S_0\cot^2 2\phi(2+\cos 2\theta) + k_{12}(k_{22}-k_{11})S_0\cot 2\phi\cos\theta \\
&\quad + \frac{1}{8}(k_{11}^2 + k_{22}^2)S_0 + \frac{1}{4}(k_{22}-k_{11})^2 S_0
\end{aligned}$$

$$\Sigma_{23} = \frac{1}{4}(k_{22} - k_{11})k_{12}S_0 \sin\theta + \frac{1}{4}k_{12}^2 S_0 \sin 2\theta \cot 2\phi \tag{11}$$

It is clear from the expressions for the coefficients of the Fokker-Planck equation (10), that ϕ and θ are respectively π and 2π periodic. Hence the solution $p(\theta,\phi)$ of (10) must satisfy the following periodicity and normalizing condition:

$$p(\theta + 2\pi, \phi) = p(\theta, \phi + \pi) = p(\theta, \phi), \quad \int_{-\pi}^{\pi}\int_{0}^{\pi/2} p(\theta,\phi)d\phi d\theta = 1 \tag{12}$$

Because of the presence of the terms $\tan\phi$ and $\cot\phi$, the coefficient functions Q_2, Q_3, Σ_{33} and Σ_{23} are singular at the points $\phi = 0, \pi/2$. To remove this natural singularity the following transformation can be employed:

$$p(\theta,\phi) = \hat{p}(\theta,\phi)\sin^2 2\phi, \quad |\theta| \le \pi, 0 \le \phi \le \pi/2 \tag{13}$$

By employing the transformation, the modified Fokker-Planck equation is given as:

$$\begin{aligned} \frac{\partial}{\partial\phi}\left(\hat{Q}_2\hat{p}\right) + \frac{\partial}{\partial\theta}\left(\hat{Q}_3\hat{p}\right) - \frac{1}{2}\frac{\partial^2}{\partial\phi^2}\left(\hat{\Sigma}_{22}\hat{p}\right) - \frac{1}{2}\frac{\partial^2}{\partial\theta^2}\left(\hat{\Sigma}_{33}\hat{p}\right) \\ - \frac{1}{2}\frac{\partial^2}{\partial\phi\partial\theta}\left(\hat{\Sigma}_{23}\hat{p}\right) = 0 \end{aligned} \tag{14}$$

where the new coefficients are given by

$$\hat{Q}_i = Q_i \sin^2 2\phi, \quad \hat{\Sigma}_{ij} = \Sigma_{ij}\sin^2 2\phi \quad i,j = 2,3 \tag{15}$$

The Fokker-Planck equation after removing the singularity is a regular equation and of bounded and periodic coefficients.

2.1 Double Fourier Expansion

The solution $\hat{p}(\theta,\phi)$ can be sought in the form of a double Fourier expansion as that employed by [3]:

$$\hat{p}(\theta,\phi) = \sum_{m=-\infty}^{\infty}\sum_{l=-\infty}^{\infty}(a_{m,l} + ib_{m,l})\, e^{i(m\theta + 2l\phi)}, \quad i = \sqrt{-1} \tag{16}$$

To obtain real values for the probability density, the following constraints must be applied:

$$a_{m,l} = a_{-m,-l}, \quad b_{m,l} = -b_{-m,-l}, \quad m,l = 0, \pm1, \pm2, \pm3, ... \tag{17}$$

By substituting the Fourier expansion of the probability density $\hat{p}(\theta,\phi)$ into (14) and comparing the coefficients of the same harmonic functions, two recursive relations in terms of the Fourier coefficients can be obtained. If the Fourier expansion converges to the exact density $\hat{p}(\theta, \phi)$ in the mean square sense, the coefficients a_{ml} and $b_{m,l}$ tends to zero for infinitely increasing indices m and l. Therefore, the Fourier expansion can be truncated at the finite indices $m = \pm M$ and $l = \pm L$. By employing the normalizing condition, the following relations can be calculated:

$$\frac{1}{2}\pi^2(a_{0,0} - a_{0,2}) + 8\pi \sum_{\substack{odd \\ l>0}}^{L} \frac{1}{l(l^2-4)} b_{0,l} = 1, \quad b_{0,0} = 0 \tag{18}$$

A system of $2(2M + 1)(2L + 1)$ linear homogeneous equations in terms of the coefficients $a_{m,l}$ and $b_{m,l}$ for $m, l = 0, \pm1, \pm2, \pm3, \ldots$ can be generated by utilizing the two recursive relations. By imposing the relations obtained from the normalizing condition together with those for obtaining real values of the probability density as constraints, unique values for the coefficients $a_{m,l}$ and $b_{m,l}$ $m, l = 0, \pm1, \pm2, \pm3, \ldots$ can be obtained.

2.2 Calculation of Lyapunov Exponent

The largest Lyapunov exponent is defined by

$$\lambda = \lim_{t\to\infty} \frac{1}{t} \ln\left({a_1}^2 + {a_2}^2\right)^{1/2} \tag{19}$$

By substituting for $\rho = \ln\left({a_1}^2 + {a_2}^2\right)^{1/2}$ into (19), the largest Lyapunov exponent can be obtained, following [5], by the following expression:

$$\begin{aligned} \lambda \;=\; & \lim_{t\to\infty} \frac{1}{t} \left\{ \rho(0) + \int_0^t Q_1 dt_1 \right. \\ & \left. + \int_0^t \sum_{j=1}^{4} \left(\sigma_{a_{1j}} \frac{\partial \rho}{\partial a_1} + \sigma_{a_{2j}} \frac{\partial \rho}{\partial a_2} \right) dB_j(t_1) \right\} \end{aligned} \tag{20}$$

The first and third terms in the above equation vanish as $t \to \infty$, since $\rho(0)$ is finite and $B(t)$ grows as $\sqrt{t \ln(\ln t)}$, which is slower than t. The wide-band process $\xi(t)$ is assumed to be stationary and ergodic and therefore the long time average is equal to the ensemble average $w.p.1$. The largest Lyapunov exponent is then given by

$$\lambda = \int_0^{\pi/2} \int_0^{2\pi} Q_1(\theta, \phi) p(\theta, \phi) d\theta d\phi \tag{21}$$

which leads to:

$$
\begin{aligned}
\varepsilon^{-1}\lambda = \sum_{\substack{odd\\ l>0}}^{L} \Bigg[& 2\pi\left(2(\lambda_1+\lambda_2)+k_{12}^2 S_0\right)\frac{1}{l(l^2-4)} \\
& + \frac{\pi}{64}(k_{11}^2+k_{22}^2-4k_{12}^2)S_0\left(-\frac{3}{l}+\frac{4l}{l^2-4}-\frac{l}{l^2-16}\right)\Bigg] b_{0,l} \\
& + 4\pi(\lambda_1-\lambda_2)\sum_{\substack{even\\ l>0}}^{L}\frac{l}{(l^2-1)(l^2-9)}b_{0,l}+\frac{\pi^2}{8}(\lambda_1-\lambda_2)(a_{0,1}-a_{0,3}) \\
& + \frac{\pi^2}{8}(2(\lambda_1+\lambda_2)+k_{12}^2 S_0)(a_{0,0}-a_{0,2}) \\
& + \frac{\pi^2}{256}(k_{11}^2+k_{22}^2-4k_{12}^2)S_0(3a_{0,0}-4a_{0,2}+a_{0,4})
\end{aligned} \tag{22}
$$

It was noticed that the coefficients $a_{m,l}$ become smaller and smaller for large indices and only those of indices $-8 \leq l \leq 8$ and $-2 \leq m \leq 2$ have significant values. It was also found that the coefficients $b_{m,l}$ become smaller and smaller for large indices and only those of indices $-4 \leq l, m \leq 4$ have significant values. By considering various parameter values and using Fourier expansion, a typical plot of the probability density $p(\theta, \phi)$ is shown in Figure (1). Also typical plots of the largest Lyapunov exponent are shown in Figures (2,3).

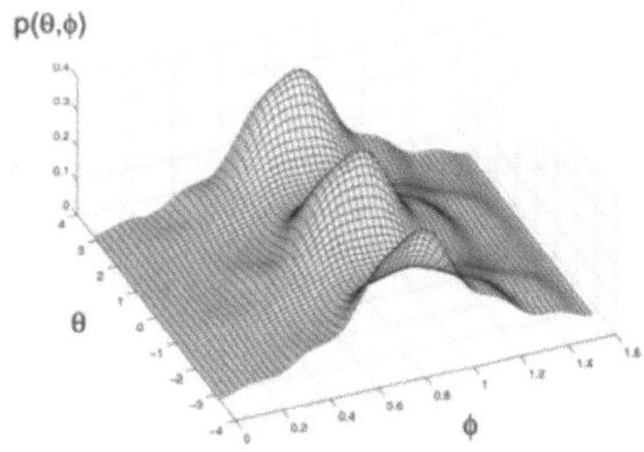

Figure 1. Probability density for a conservative system using Fourier expansion, $\beta_1 = 1.5$, $\beta_2 = 1.2$, $S_0 = 8$, $k_{11} = k_{22} = 0$, $k_{12} = 1.0$, $\omega_0 = 1.0$, $\delta_0/\varepsilon = 0.1$.

It can be inferred from the figures of the largest Lyapunov exponent that the damping terms have a stabilizing effect whereas the excitation spectral density S_0 has a destabilizing effect.

A reasonable analytical approximation of the ϕ marginal of the probability density $p(\theta, \phi)$ can be sought by considering θ to be independent of the fluctuating angle ϕ. As a tribute to the fact that Q_1 is a smooth function

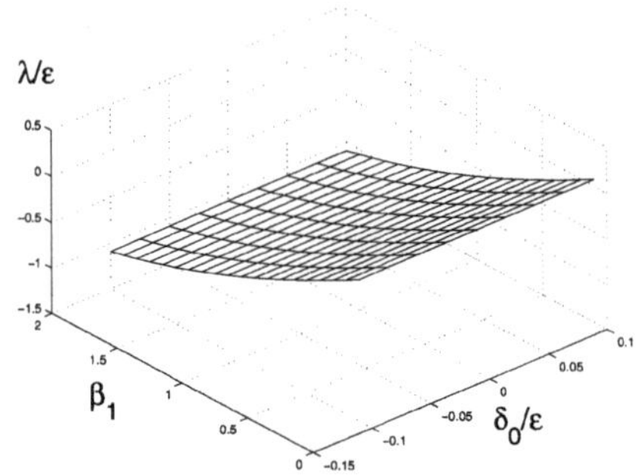

Figure 2. Lyapunov exponent for a conservative system using Fourier expansion, $\beta_2 = 1.5$, $S_0 = 4$, $k_{11} = k_{22} = 0$, $k_{12} = 1.0$, $\omega_0 = 1.0$.

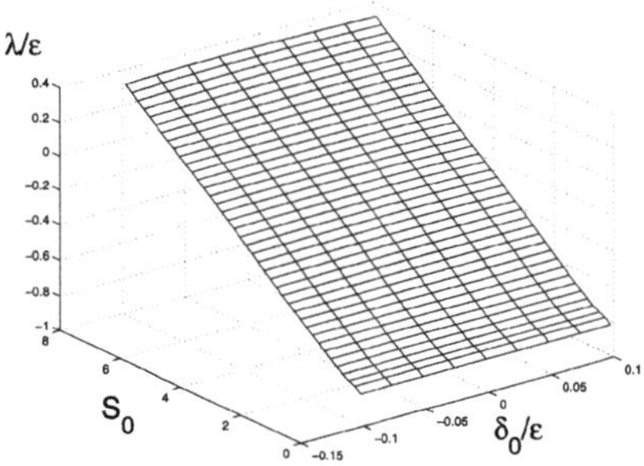

Figure 3. Lyapunov exponent for a conservative system using Fourier expansion, $\beta_1 = 1.0$, $\beta_2 = 0.8$, $k_{11} = k_{22} = 0$, $k_{12} = 1.0$, $\omega_0 = 1.0$.

of ϕ only, it can be verified that a good agreement occurs between the values of the largest Lyapunov exponent obtained by using the Fourier expansion technique and that obtained by using a reasonable analytical approximation.

3. Non-conservative Systems

Consider now a non-conservative two degree of freedom system whose natural frequencies are equal. The motion under parametric excitation is governed by the following non-dimensional linear stochastic differential equations:

$$\ddot{q_1} + 2\beta_1\dot{q_1} + {\omega_0}^2 q_1 + \varepsilon\omega_0\xi(t)\sum_{j=1}^{2} k_{1j}q_j = 0$$

$$\ddot{q_2} + 2\beta_2\dot{q_2} + {\omega_0}^2 q_2 + q_1 + \varepsilon\omega_0\xi(t)\sum_{j=1}^{2} k_{2j}q_j = 0 \qquad (23)$$

where q_i are the non-dimensional generalized coordinates, β_i the non-dimensional viscous damping coefficients, ω_0 the natural frequency, k_{ij} the load coefficients, $0 < \varepsilon \ll 1$. In the absence of damping, β_i, and the parametric excitation, $\xi(t)$, the system of (23) is unstable or said to be in flutter because q_2 contains resonant terms. Letting $q_1 = u_1$, $q_2 = \varepsilon^{-\alpha}u_2$, $\alpha > 0$, (23) are transformed to

$$\ddot{u_1} + 2\beta_1\dot{u_1} + {\omega_0}^2 u_1 + \varepsilon\omega_0\xi(t)[k_{11}u_1 + \varepsilon^{-\alpha}k_{12}u_2] = 0$$
$$\ddot{u_2} + 2\beta_2\dot{u_2} + {\omega_0}^2 u_2 + \varepsilon^{\alpha}u_1 + \varepsilon\omega_0\xi(t)[\varepsilon^{\alpha}k_{21}u_1 + k_{22}u_2] = 0 \qquad (24)$$

In order to include the effect of the coupling terms k_{12}, k_{21} and of u_1 in the second of (24), for a wide-band excitation one must take $\alpha = 2/3$. Introducing the transformation

$$u_i = a_i\cos\Theta_i,\quad \dot{u_i} = -a_i\omega_0\sin\Theta_i\quad \Theta_i = \omega_0 t + \theta_i,\quad i = 1,2 \qquad (25)$$

and applying the method of variation of parameters (24) transforms, to terms of first order in $\epsilon^{2/3}$, to the following:

$$\dot{a_1}(t) = -2\varepsilon^{2/3}{\beta_1}^* a_1\sin^2\Theta_1 + \varepsilon^{1/3}\xi(t)k_{12}a_2\cos\Theta_2\sin\Theta_1$$
$$\dot{\theta}_1(t) = -\varepsilon^{2/3}{\beta_1}^*\sin 2\Theta_1 + \frac{\varepsilon^{1/3}}{a_1}\xi(t)k_{12}a_2\cos\Theta_2\cos\Theta_1$$
$$\dot{a_2}(t) = -2\varepsilon^{2/3}{\beta_2}^* a_2\sin^2\Theta_2 + \varepsilon^{2/3}a_1\cos\Theta_1\sin\Theta_2$$
$$\dot{\theta}_2(t) = -\varepsilon^{2/3}{\beta_2}^*\sin 2\Theta_2 + \frac{\varepsilon^{2/3}a_1}{a_2}\cos\Theta_1\cos\Theta_2 \qquad (26)$$

where $\beta_i = \varepsilon^{2/3}\beta_i{}^*$, $i = 1,2$. Again when $\xi(t)$ is approximated by a white noise process with spectral density S_0, the method of stochastic averaging applied to these equations results in the following Itô equations:

$$da_i = \varepsilon^{2/3}m_{a_i}dt + \varepsilon^{1/3}\sum_{j=1}^{4}\sigma_{a_{ij}}dW_j$$

$$d\theta_i = \varepsilon^{2/3}m_{\theta_i}dt + \varepsilon^{1/3}\sum_{j=1}^{4}\sigma_{\theta_{ij}}dW_j,\quad i = 1,2 \qquad (27)$$

where W_j are mutually independent unit Wiener processes. The drift coefficients m_{a_i}, m_{θ_i} and the diffusion coefficients $\sigma_{a_{ij}}$, $\sigma_{\theta_{ij}}$ are obtained by using the averaging procedure of [6].

$$
\begin{aligned}
m_{a_1} &= -\beta_1{}^* a_1 + \frac{k_{12}{}^2}{16}\frac{a_2^2}{a_1} S_0 \left(2 + \cos 2(\theta_1 - \theta_2)\right), \\
m_{\theta_1} &= -\frac{k_{12}{}^2}{8}\frac{a_2^2}{a_1^2} S_0 \sin 2(\theta_1 - \theta_2) \\
m_{a_2} &= -\beta_2{}^* a_2 - \frac{1}{2\omega_0} a_1 \sin(\theta_1 - \theta_2), \; m_{\theta_2} = \frac{1}{2\omega_0}\frac{a_1}{a_2}\cos(\theta_1 - \theta_2) \\
[\sigma\sigma^T]_{a_1 a_1} &= \frac{k_{12}{}^2}{8} a_2^2 S_0 \left(2 - \cos 2(\theta_1 - \theta_2)\right), \\
[\sigma\sigma^T]_{a_1 \theta_1} &= \frac{k_{12}{}^2}{8}\frac{a_2^2}{a_1} S_0 \sin 2(\theta_1 - \theta_2) \\
[\sigma\sigma^T]_{\theta_1 \theta_1} &= \frac{k_{12}{}^2}{8}\frac{a_2^2}{a_1^2} S_0 \left(2 + \cos 2(\theta_1 - \theta_2)\right)
\end{aligned}
$$

$$
\begin{aligned}
&[\sigma\sigma^T]_{a_1 a_2} = [\sigma\sigma^T]_{a_2 a_2} = [\sigma\sigma^T]_{a_1 \theta_2} = [\sigma\sigma^T]_{a_2 \theta_1} = 0 \\
&[\sigma\sigma^T]_{a_2 \theta_2} = [\sigma\sigma^T]_{\theta_1 \theta_2} = [\sigma\sigma^T]_{\theta_2 \theta_2} = 0
\end{aligned} \tag{28}
$$

By introducing the transformation of (7) and making use of Itô's differential rule, the following Itô equations for the function ρ and the angle processes ϕ and θ can be obtained:

$$
\begin{aligned}
d\rho &= \varepsilon^{2/3} Q_1(\phi, \theta) dt + \varepsilon^{1/3} \sum_{j=1}^{4} \left(\sigma_{a_{1j}} \frac{\partial \rho}{\partial a_1} + \sigma_{a_{2j}} \frac{\partial \rho}{\partial a_2}\right) dW_j(t) \\
d\phi &= \varepsilon^{2/3} Q_2(\phi, \theta) dt + \varepsilon^{1/3} \sum_{j=1}^{4} \left(\sigma_{a_{1j}} \frac{\partial \phi}{\partial a_1} + \sigma_{a_{2j}} \frac{\partial \phi}{\partial a_2}\right) dW_j(t) \\
d\theta &= \varepsilon^{2/3} Q_3(\phi, \theta) dt + \varepsilon^{1/3} \sum_{j=1}^{4} \left(\sigma_{\theta_{1j}} - \sigma_{\theta_{2j}}\right) dW_j(t)
\end{aligned} \tag{29}
$$

where the drift coefficients Q_i are given by:

$$
\begin{aligned}
Q_1(\theta,\phi) &= \frac{k_{12}^2}{32} S_0 \cos 2\theta \sin^2 2\phi - \frac{1}{4\omega_0} \sin\theta \sin 2\phi \\
&+ \frac{k_{12}^2}{16} S_0 (1-\cos 2\phi)^2 - \frac{1}{2}({\beta_1}^* + {\beta_2}^*) - \frac{1}{2}({\beta_1}^* - {\beta_2}^*) \cos 2\phi \\
Q_2(\theta,\phi) &= \frac{k_{12}^2}{8} S_0 \sin^2\phi (\sin 2\phi - \tan\phi) \\
&- \frac{k_{12}^2}{16} S_0 \cos 2\theta \sin^2\phi (\sin 2\phi + \tan\phi) - \frac{1}{2\omega_0} \sin\theta \cos^2\phi \\
&+ \frac{1}{2}({\beta_1}^* - {\beta_2}^*) \sin 2\phi \\
Q_3(\theta,\phi) &= -\frac{k_{12}^2}{8\omega_0^2} S_0 \sin 2\theta \tan^2\phi - \frac{1}{2} \cos\theta \cot\phi
\end{aligned} \tag{30}
$$

The Fokker-Planck equation for the stationary probability density $p(\theta,\phi)$ associated with the angle processes θ and ϕ is similar to (10) with

$$
\begin{aligned}
\Sigma_{22}(\theta,\phi) &= \frac{k_{12}^2}{16} S_0 (1+2\sin^2\theta)(1-\cos 2\phi)^2 \\
\Sigma_{33}(\theta,\phi) &= \frac{k_{12}^2}{8} S_0 (1+2\cos^2\theta)\tan^2\phi \\
\Sigma_{23}(\theta,\phi) &= -\frac{k_{12}^2}{8} S_0 \sin 2\theta \sin^2\phi \tan\phi
\end{aligned} \tag{31}
$$

The probability density must satisfy the same periodicity and normalizing conditions of (12). Again the singularity in the coefficients Q_2, Q_3, Σ_{23} and Σ_{33} are removed by taking $p(\theta,\phi) = \hat{p}(\theta,\phi)\sin^2 2\phi$ so that $\hat{p}(\theta,\phi)$ satisfies an equation similar (14). As before the solution for $p(\theta,\phi)$ is sought in the form of a truncated Fourier series with the coefficents satisfying the constraint conditions of (17) and (18). The solution procedure is similar to that of the previous case where now the largest Lyapunov exponent is evaluated by using (21) with Q_1 being given by (30)

$$
\begin{aligned}
\varepsilon^{-2/3}\lambda = &\tfrac{k_{12}^2}{128} S_0 \pi^2 [5a_{0,0} - 4(a_{0,1} + a_{0,2} - a_{0,3}) - a_{0,4}] \\
&+ \tfrac{1}{4}({\beta_1}^* + {\beta_2}^*)\pi^2 (a_{0,2} - a_{0,0}) + \tfrac{1}{8}({\beta_1}^* - {\beta_2}^*)\pi^2 (a_{0,3} - a_{0,1}) \\
&+ \tfrac{k_{12}^2}{512} S_0 \pi^2 [6a_{2,0} - 4(a_{2,2} + a_{2,-2}) + a_{2,4} + a_{2,-4}]
\end{aligned}
$$

$$
\begin{aligned}
&-\tfrac{1}{32\omega_0}\pi^2[3(a_{1,-1}-a_{1,1})-a_{1,-3}+a_{1,3}] \\
&+k_{12}^2 S_0\pi \sum_{\substack{odd\\ l>0}}^{L} \tfrac{l^2-10}{l(l^2-4)(l^2-16)} b_{0,l} - k_{12}^2 S_0 \pi \sum_{\substack{even\\ l>0}}^{L} \tfrac{l}{(l^2-1)(l^2-9)} b_{0,l} \\
&-4\pi(\beta_1{}^* + \beta_2{}^*) \sum_{\substack{odd\\ l>0}}^{L} \tfrac{1}{l(l^2-4)} b_{0,l} - 4\pi(\beta_1{}^* - \beta_2{}^*) \sum_{\substack{even\\ l>0}}^{L} \tfrac{l}{(l^2-1)(l^2-9)} b_{0,l} \\
&-\tfrac{3\pi k_{12}^2}{2} S_0 \sum_{\substack{odd\\ l>0}}^{L} \tfrac{1}{l(l^2-4)(l^2-16)} (b_{-2,l}+b_{2,l}) \\
&-\tfrac{3\pi}{\omega_0} \sum_{\substack{even\\ l>0}}^{L} \tfrac{1}{(l^2-1)(l^2-9)} (b_{-1,l}-b_{1,l}) \qquad (32)
\end{aligned}
$$

It was observed that the coefficients $a_{m,l}$ and $b_{m,l}$ become smaller and smaller for large indices and only those of indices $-8 \leq l \leq 8$ and $-4 \leq m \leq 4$ have significant values. Again typical plot of the probability density, $p(\theta, \phi)$, is shown in Fig.(4) and typical plots of the largest Lyapunov exponent, λ, are

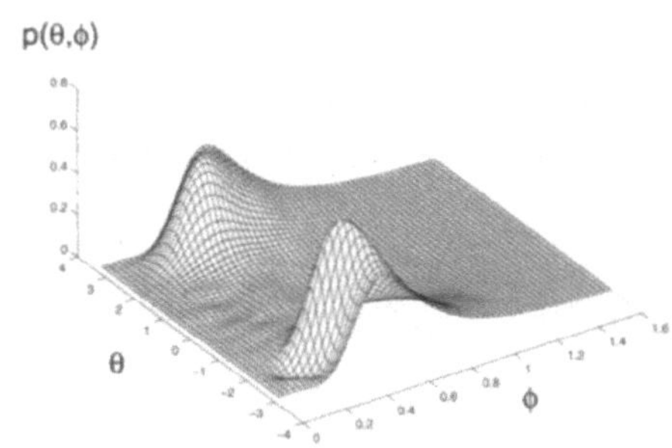

Figure 4. Probability density for a non-conservative system using Fourier expansion with $\beta_1{}^* = 1.5$, $\beta_2{}^* = 1.2$, $S_0 = 8$, $k_{12} = 2.0$, $\omega_0 = 1.0$.

shown in Figs(5,6). It can be inferred from the corresponding figures of the largest Lyapunov exponent that the damping terms, $\beta_1{}^*$ and $\beta_2{}^*$ have a stabilizing effect whereas the spectral density S_0 has a destabilizing effect.

4. Conclusions

The stochastic stability of two degree of freedom systems whose natural frequencies are commensurate and parametrically excited by a white noise process was investigated. Two cases were studied, namely, conservative systems having nearly equal frequencies and non-conservative systems whose frequencies are equal. The stability of these systems was examined by evaluating the largest Lyapunov exponent. The method of stochastic averaging together with a Fourier expansion technique for solving the associated Fokker-Planck equation governing the stationary probability density of the angle vari-

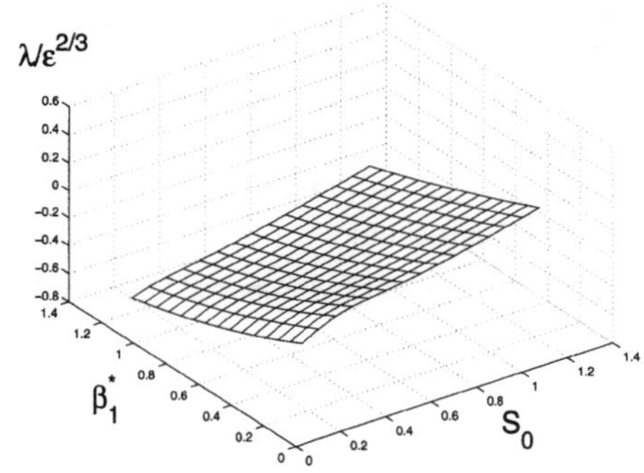

Figure 5. Lyapunov exponent for a non-conservative system using Fourier expansion, $\beta_2{}^* = 0.7$, $k_{12} = 5.0$, $\omega_0 = 1.0$.

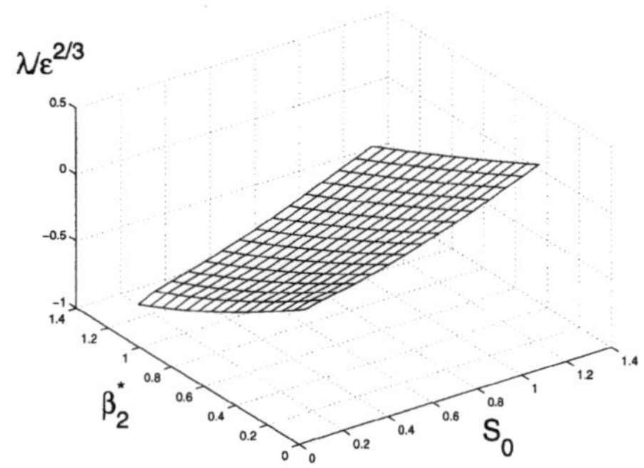

Figure 6. Lyapunov exponent for a non-conservative system using Fourier expansion, $\beta_1{}^* = 1.0$, $k_{12} = 6.0$, $\omega_0 = 1.0$.

ables was employed to evaluate the largest Lyapunov exponent. Numerical results are presented to show the effect of the damping terms and the excitation spectral density on values of the largest Lyapunov exponent.

References

[1] S. T. Ariaratnam, D. S. F. Tam, and W.-C. Xie, "Lyapunov exponents and stochastic stability of coupled linear systems," *Probabilistic Eng. Mech.*, vol. 6, no. 2, pp. 51–56, 1990.

[2] S. T. Ariaratnam and W.-C. Xie, Wie-Chau (1992) "Lyapunov exponents and stochastic stability of coupled linear systems under real noise excitation," *Journal of Applied Mechanics*, vol. 59, pp. 664–673, 1992.

[3] R. Griesbaum and W. Wedig, "Generalized Fourier analysis for stochastic stability problems," *Proceedings of the IUTAM Symposium on Nonlinearity and Stochastic Structural Dynamics*, ed. S. Narayanan, R. N. Iyengar, Kluwer Academic Publishers, pp. 77–94, 1999.

[4] R. Z. Khasminskii, "A limit theorem for the solutions of differential equations with random right-hand sides ," *Theory of Probability and its Application*, vol. 11, no. 3, pp. 390–406, 1966. (English Translation).

[5] R. Z. Khasminskii, "Necessary and sufficient conditions for the asymptotic stability of linear stochastic systems," *Theory of Probability and its Application*, vol. 12, no. 1, pp.144–147, 1967. (English Translation).

[6] R. L. Stratonovich, *Topics in the Theory of Random Noise*, vol. 1, Gordon and Breach, New York, 1963.

DYNAMICS OF GLOBALLY COUPLED NOISY EXCITABLE ELEMENTS: THE FITZHUGH-NAGUMO CASE

J. A. Acebrón
Department of Physics
University of California
San Diego, La Jolla, CA 92093, USA

A. R. Bulsara
SPAWAR Systems Center Code D363
49590 Lassing Road, RM A341
San Diego, CA 92152-6147

W.-J. Rappel
Department of Physics
University of California
San Diego, La Jolla, CA 92093, USA
acebron@physics.ucsd.edu

Abstract We study the noisy FitzHugh-Nagumo model in the presence of an external sinusoidal driving force. We derive a Fokker-Planck equation for both the single element and for the globally coupled system. We introduce an efficient way to numerically solve this Fokker-Planck equation and show that the external driving force leads to a classical resonance when its frequency matches the underlying systems frequency. This resonance is also investigated analytically by exploiting the different timescales in the problem. Agreement between the analytical results and numerical results is excellent and reveals the existence of a stochastic bifurcation.

Keywords: noise, FitzHugh-Nagumo, excitable media, Fokker-Planck

N. Sri Namachchivaya and Y.K. Lin (eds.),
IUTAM Symposium on Nonlinear Stochastic Dynamics, 111-124.

1. Introduction

The FitzHugh-Nagumo model (FHN) is a simplified version of the celebrated Hodgkin-Huxley model [4], which describes the firing mechanism in a excitable nerve cell. In the FHN, the dynamics of the nerve cell has been reduced to two variables: a fast, activation, variable and a slow, recovery, variable [5]. Due to its relative simplicity, the FHN and its extensions has been studied extensively. Both single neurons and populations of diffusively coupled neurons have been investigated. In addition, the periodically driven FHN, where either the slow or the fast equation contains a time-periodic driving term, has received considerable attention [8, 9, 7, 10, 6]. Moreover, as the FHN displays a rich phase diagram that includes excitable, oscillatory and bistable regimes, it has become a "workhorse" in the field of pattern formation (see e.g., [14, 3]).

In this paper, we investigate the FHN in the presence of noise and a probe signal. In addition to studying the single element, we examine the effect of coupling the FHN elements in a global fashion. Particular attention is paid to the classical resonance effect that can arise when a system with an underlying frequency is driven by a probe signal [9, 1]

2. Model Equations

Let us start with the most general form of the FHN model:

$$\begin{aligned} \frac{dx}{dt} &= Ax^3 + Bx^2 + Cx + Hy + I + \xi, \\ \frac{dy}{dt} &= Ex + Fy + G, \end{aligned} \tag{1}$$

where ξ's are Gaussian white noises, with $<\xi(t)>= 0$, $<\xi(t)\xi(t')>= 2D\delta(t-t')$ and where A through G are parameters that determine the dynamics of the system. To make the treatment in this paper as general as possible all relevant expressions will be derived using the above set of equations. However, when presenting the results of numerical calculations, we have chosen to limit ourselves here to the investigation of the FHN in its more conventional notation:

$$\begin{aligned} \alpha\frac{dx}{dt} &= x(x-a)(1-x) - y + \xi, \\ \frac{dy}{dt} &= x - py - b. \end{aligned} \tag{2}$$

where α is typically taken to be small.

In order to study the properties of the stochastic differential equation in (1), we start by deriving the Fokker-Planck equation (FPE) for the density probability, which is given by:

$$\frac{\partial \rho}{\partial t} = D\frac{\partial^2 \rho}{\partial x^2} - \frac{\partial}{\partial x}\left[(A\,x^3 + B\,x^2 + C\,x + H\,y + I)\rho\right] \\ -\frac{\partial}{\partial y}\left[(E\,x + F\,y + G)\rho\right], \qquad (3)$$

which has to be accompanied by initial and boundary data (decay to zero as $x \to \pm\infty, y \to \pm\infty$, with sufficiently high rate), and the normalization condition

$$\int_{-\infty}^{+\infty}\int_{-\infty}^{+\infty} dx\,dy\,\rho(x,y,t) = 1. \qquad (4)$$

In the following, we are interested in finding solutions of the FPE for large time. This search is greatly facilitated by the fact that the FPE has a unique stationary solution. This can be seen by noting that there exists a Lyapunov function (see [11] and references therein). It then follows that such a stationary solution is unique and globally stable.

For parameters values for which analytical progress is difficult to achieve one has to resort to numerics. Direct simulation of the Langevin equations (1), as has been commonly done in the FHN repertoire, can be computationally intensive. For reasonably accurate results one typically has to average over many realizations. This is particularly the case for systems close to a bifurcation point where one has to distinguish between different stable solutions and for systems where the noise is large. Numerical solutions of the FPE, on the other hand, can be obtained much faster. Rather then using a finite difference scheme we have used an efficient spectral method for which we expand the density probability ρ using a basis of Hermite polynomials

$$\rho(x,y,t) = \sum_{n=0}^{\infty} r_n^m(t) H_n(x)\, H_m(y) e^{-x^2} e^{-y^2}. \qquad (5)$$

Note that this expansion satisfies the boundary conditions, and the normalization condition with $r_0^0 = 1/\pi$.

Let us insert (5) into the FPE (3). We then obtain the following hierarchy of coupled ordinary differential equations for $r_n^m(t)$.

$$\dot{r}_n^m = (\frac{3}{2}A\,n^2 + C\,n + F\,m)r_n^m + \left[B(n-\frac{1}{2}) + I\right] r_{n-1}^m \\ + \left[D + \frac{3}{4}A(n-1) + \frac{1}{2}C\right] r_{n-2}^m + \frac{B}{4}r_{n-3}^m + \frac{A}{8}r_{n-4}^m \\ +B\,n(m+1)\,r_{n+1}^m + A\,n(n+1)(n+2)r_{n+2}^m \\ +G\,r_n^{m-1} + \frac{F}{2}r_n^{m-2} + \frac{1}{2}(H+E)r_{n-1}^{m+1} + E(n+1)r_{n+1}^{m-1} \qquad (6) \\ n = 0, 1\ldots,\infty, m = 0, 1, \ldots, \infty,$$

where $< x >$, and $< y >$ are given by

$$< x >= \int_{-\infty}^{+\infty} \int_{-\infty}^{+\infty} dx\, dy\, x\, \rho(x, y, t) = \pi r_1^0, \tag{7}$$

$$< y >= \int_{-\infty}^{+\infty} \int_{-\infty}^{+\infty} dx\, dy\, y\, \rho(x, y, t) = \pi r_0^1. \tag{8}$$

The numerical method consists of truncating the infinite hierarchy of first-order, coupled nonlinear differential equations, for a reasonable number of modes $n = 0, \ldots, N$, and $m = 0, \ldots, M$, setting $r_{N+1}^{M+1} = 0$. We have compared the numerical solution obtained via the Fokker-Planck approach, to the solution of the Langevin equations, obtained by averaging over a large number of realizations. In Figure 1, we plot the first moment $< x >$ as a function of time, obtained numerically by solving the Langevin equations (1) and by solving the FPE using the above-described spectral method. The spectral method (with $N = M = 7$ moments) is seen to provide excellent agreement with the more conventional and time-consuming technique based on numerically integrating the coupled stochastic differential equations (1).

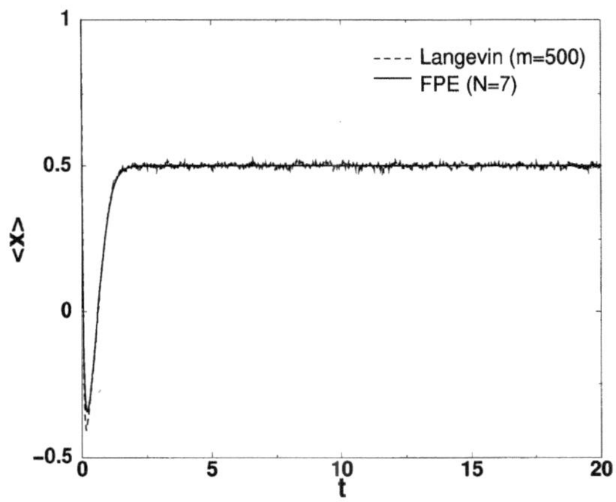

Figure 1. Comparison between the numerical solution of the Langevin equations (2) (averaged over $m = 500$ realizations) and the solution of the Fokker-Planck equation by the spectral method with $N = M = 7$ coefficients. Parameters are $D = 0.01, b = 0.5, p = 1, \alpha = 0.05$.

Following this preamble, we start with an extension of the model equations (1) to describe a system of globally linearly coupled FitzHugh-Nagumo elements. Global coupling is also most amenable (of all the possible coupling schemes) to theoretical treatment. We will couple the elements in the following global fashion:

$$\frac{d x_i}{dt} = A x_i^3 + B x_i^2 + C x_i + H y_i + I + \frac{K}{N} \sum_{j=1}^{N} (x_j - x_i) + \xi_i, \tag{9}$$

$$\frac{d y_i}{dt} = E x_i + F y_i + G, \quad i = 1, \cdots, N. \tag{10}$$

With this type of coupling, the FPE for the perfectly synchronized system is identical to the FPE of a single element.

We are interested in the analytical investigation of the Langevin dynamics above, for the case of very large N. A neat picture of such a case can be given by the limiting-model obtained when $N \to \infty$ (thermodynamic limit). In this limit, it is well known [2, 12] that models with mean-field coupling are described by an evolution equation for the one-particle probability density. This can be seen by noting that the hierarchy of equations for all the multiparticle probability densities can be closed by assuming molecular chaos. In such a way, the one-system probability density $\rho(x, y, t)$ is asymptotically in the limit, $N \to \infty$, the solution of the following nonlinear Fokker-Planck equation:

$$\frac{\partial \rho}{\partial t} = D \frac{\partial^2 \rho}{\partial x^2} - \frac{\partial}{\partial x} [(A x^3 + B x^2 + C x + H y + K(\bar{x} - x) + I)\rho] - \frac{\partial}{\partial y} [(E x + F y + G)\rho], \tag{11}$$

where

$$\bar{x} = \int_{-\infty}^{+\infty} \int_{-\infty}^{+\infty} dx\, dy\, x\, \rho(x, y, t). \tag{12}$$

The hierarchy (6) now becomes:

$$\begin{aligned} \dot{r}_n^m = {} & (\frac{3}{2} A n^2 + C n + F m - K n) r_n^m + \left[B(n - \frac{1}{2}) + I + \pi K r_1^0 \right] r_{n-1}^m \\ & + \left[D + \frac{3}{4} A(n-1) + \frac{1}{2} C - \frac{K}{2} \right] r_{n-2}^m + \frac{B}{4} r_{n-3}^m + \frac{A}{8} r_{n-4}^m \\ & + B n(m+1) r_{n+1}^m + A n(n+1)(n+2) r_{n+2}^m + G r_n^{m-1} \\ & + \frac{F}{2} r_n^{m-2} + \frac{1}{2}(H + E) r_{n-1}^{m+1} + E(n+1) r_{n+1}^{m-1} \\ & n = 0, \ldots, \infty, m = 0, \ldots, \infty, \end{aligned} \tag{13}$$

Note that now the hierarchy consists of a system of coupled first-order nonlinear differential equations.

In Figure 2, we have compared the numerical solution obtained via the non-

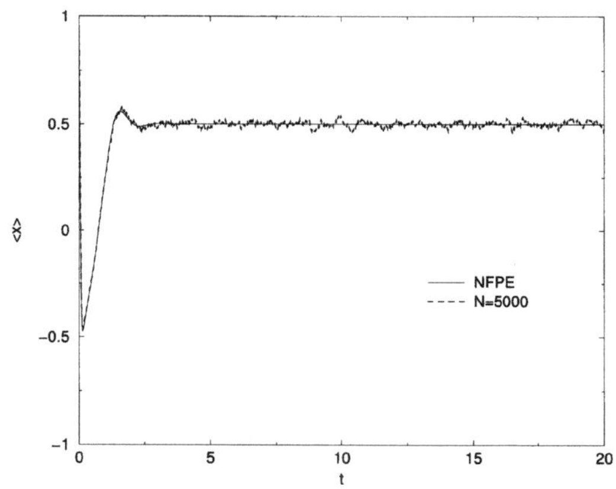

Figure 2. Comparison between solution obtained by means of FPE, and direct numerical simulation of the Langevin equations for $N = 5000$ oscillators. Parameters are as in Figure 1, and $K = 1$.

linear Fokker-Planck approach to the solution of the Langevin equations for a large number of FHN oscillators ($N = 5000$). The solution of the FPE, corresponding to $N \to \infty$, provides excellent agreement with the finite N case and shows that $N = 5000$ is already close to infinity for all practical purposes.

3. Bifurcation Analysis

To investigate the bifurcations in the noisy case it is worthwhile to determine the underlying frequency of the system. One way of determining this frequency is to compute $< x >$ from the Langevin equations and evaluate its time dependence. Unfortunately, this is computationally very costly. On the other hand, $< x >$ calculated from the FPE, which offers a computationally superior way to characterize the system, does not display a time-dependent behavior for a single FHN oscillator. Therefore, an alternative way of finding the frequency needs to be employed. Fortunately, as we will see below, including an external time-sinusoidal "probe" signal leads to a classical resonance which can be used to determine the underlying frequency [9]. We will consider an external signal that has a time-sinusoidal component $G = G_0 + q\sin(\omega_p t)$ in (1) (or, equivalently, $b = b_0 + q\sin(\omega_p t)$ in (2)).

To illustrate the effect of the probe signal, we first performed Langevin simulations and calculated $< y >$. In Figure 3 we have plotted the power spec-

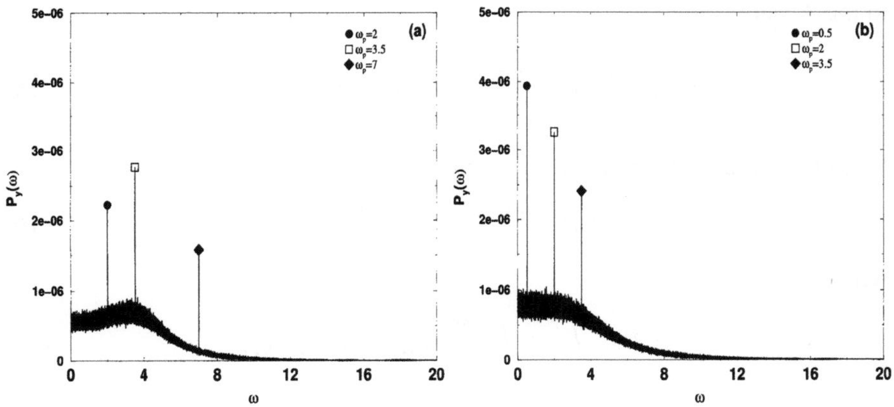

Figure 3. Power spectrum of the variable y for three different values of the probe signal, and noise strength; (a) $D = 0.003$, and (b) $D = 0.005$. Simulations done by using the Langevin equations. Other parameters are $b_0 = 0.5, p = 1, q = 0.01$, and $\alpha = 0.05$.

trum of this quantity, for three different probe signals; two of them with frequencies ω_p that differs significantly from and one that is very close to the underlying frequency. The power spectrum was obtained by averaging 100 time-series of 2^{23} timesteps each. The figure illustrates clearly that for a probe signal frequency that matches the broad peak (corresponding to the, in general, non-sinusoidal running oscillations) in the power spectrum of the unprobed system, the signal is amplified. Thus, adding a probe signal gives us a tool to investigate the dynamics of the noisy system. On the other hand, it is worthwhile to exploit such a result to study the bifurcations in our noisy system. In Figure 3a, and b, we show the results for two different values of the noise strength. For small noise strength (Figure 3a), the peak in the power spectrum reaches a maximum for non-zero values of the probing frequency while for larger noise strengths (Figure 3b) this peak is reached for $\omega_p = 0$. Thus, there is a qualitative change in the power spectrum and in the dynamics of the system, which can be interpreted as the signature of a bifurcation.

Analytical progress by means of the Fokker-Planck equation can be made if we consider a small amplitude signal, $q = \varepsilon Q$, where $\varepsilon \ll 1$. The resulting FPE can then be analyzed via the method of multiple scales. Thus, (3) contains terms with two different time scales. It is then to be expected that an appropriate asymptotic method will be able to capture the long-time behavior of ρ. This may be achieved by introducing fast and slow timescales as follows:

$$\tau = \frac{t}{\varepsilon}, t = t. \tag{14}$$

We look for a distribution function satisfying the boundary condition according to the Ansatz:

$$\rho(x,y,t;\varepsilon) = \sum_{n=0}^{2} \rho^{(n)}(x,y,t,\tau)\varepsilon^n + O(\varepsilon^3) \tag{15}$$

From (15), the average of x is given by

$$< x > = < x >^{(0)} + \varepsilon < x >^{(1)} + O(\varepsilon^2), \tag{16}$$

where

$$< x >^{(j)} = \int_{-\infty}^{+\infty} \int_{-\infty}^{+\infty} dx\, dy\, x\, \rho^{(j)}(x,y,t). \tag{17}$$

The average of y is given by similar equations. Inserting (15) into (3), we obtain the following hierarchy of equations for $\rho^{(j)}$:

$$\frac{\partial \rho^{(0)}}{\partial \tau} = 0, \tag{18}$$

$$\begin{aligned} \frac{\partial \rho^{(1)}}{\partial \tau} &= D\frac{\partial^2 \rho^{(0)}}{\partial x^2} - \frac{\partial}{\partial x}\left[(A\,x^3 + B\,x^2 + C\,x + H\,y + I)\rho^{(0)}\right] \\ &- \frac{\partial}{\partial y}\left[(E\,x + F\,y + G_0)\rho^{(0)}\right] - \frac{\partial \rho^{(0)}}{\partial t}, \end{aligned} \tag{19}$$

$$\begin{aligned} \frac{\partial \rho^{(2)}}{\partial \tau} &= D\frac{\partial^2 \rho^{(1)}}{\partial x^2} - \frac{\partial}{\partial x}\left[(A\,x^3 + B\,x^2 + C\,x + H\,y + I)\rho^{(1)}\right] \\ &- \frac{\partial}{\partial y}\left[(E\,x + F\,y + G_0)\rho^{(1)}\right] - \frac{\partial \rho^{(1)}}{\partial t} \\ &- Q\sin(\omega_p t)\frac{\partial \rho^{(0)}}{\partial y}, \end{aligned} \tag{20}$$

where the normalization conditions

$$\int_{-\infty}^{+\infty} \int_{-\infty}^{+\infty} \rho^{(n)}(x,y,t)\, dx\, dy = \delta_{0n} \tag{21}$$

follows from (4). Equation (18) implies that $\rho^{(0)}$ is independent of τ. Then, the terms in the right side of (19) which do not have τ-dependent coefficients give rise to secular terms (unbounded on the τ-time scale). The condition that no secular terms should appear is

$$\begin{aligned} D\frac{\partial^2 \rho^{(0)}}{\partial x^2} &- \frac{\partial}{\partial x}\left[(A\,x^3 + B\,x^2 + C\,x + H\,y + I)\rho^{(0)}\right] \\ &- \frac{\partial}{\partial y}\left[(E\,x + F\,y + G_0)\rho^{(0)}\right] - \frac{\partial \rho^{(0)}}{\partial t} = 0. \end{aligned} \tag{22}$$

This equation should be solved for $\rho^{(0)}$ together with the normalization condition and initial condition data. Note that this problem is equivalent to solving the FPE (3) without the probe signal as the effects of the probe signal appear first when calculating the first-order correction, $\rho^{(1)}$.

To calculate these first-order corrections, we again impose the condition that no secular terms appear and that the right-hand side of (20) vanishes. The resulting equation is:

$$D\frac{\partial^2 \rho^{(1)}}{\partial x^2} - \frac{\partial}{\partial x}\left[(A\,x^3 + B\,x^2 + C\,x + H\,y + I)\rho^{(1)}\right] \\ -\frac{\partial}{\partial y}\left[(E\,x + F\,y + G_0)\rho^{(1)}\right] \\ -\frac{\partial \rho^{(1)}}{\partial t} - Q\sin(\omega_p t)\frac{\partial \rho^{(0)}}{\partial y} = 0 \tag{23}$$

The analysis of the equation above can be readily accomplished in Fourier space. Fourier transforming (23), we obtain

$$i\omega\hat{\rho}^{(1)} = D\frac{\partial^2 \hat{\rho}^{(1)}}{\partial x^2} - \frac{\partial}{\partial x} \\ \left[(A\,x^3 + B\,x^2 + C\,x + H\,y + I)\hat{\rho}^{(1)}\right] \\ -\frac{\partial}{\partial y}\left[(E\,x + F\,y + G_0)\hat{\rho}^{(1)}\right] \\ -i\frac{Q}{2}\frac{\partial}{\partial y}\left[\hat{\rho}^{(0)}(\omega + \omega_p) - \hat{\rho}^{(0)}(\omega - \omega_p)\right], \tag{24}$$

where

$$\hat{\rho}^{(j)}(x, y, \omega) = \int_{-\infty}^{\infty} dt\, e^{-i\omega t}\rho^{(j)}(x, y, t), \tag{25}$$

$$< \hat{x} >^{(j)} = \int_{-\infty}^{+\infty}\int_{-\infty}^{+\infty} dx\, dy\, x\, \hat{\rho}^{(j)}(x, y, t), \tag{26}$$

$$j = 0, 1. \tag{27}$$

The equation (24) should be solved for $\hat{\rho}^{(1)}$ together with $\int_{-\infty}^{+\infty}\int_{-\infty}^{+\infty} dx\, dy\, \hat{\rho}^{(1)} = 0$. Since $\rho^{(0)}$ evolves to a stationary solution for long-time (i.e., $\hat{\rho}^{(0)} = \delta(\omega)f(\delta_1, \delta_2)$), we find that $\hat{\rho}^{(1)} = 0$ is the only solution of (24), unless $\omega = \pm\omega_p$. Then, (24),(27) imply that

$$\hat{\rho}^{(1)} = \eta^{+}(x, y)\delta(\omega - \omega_p) + \eta^{-}(x, y)\delta(\omega + \omega_p). \tag{28}$$

Inserting (28) in (24), we obtain two uncoupled equations for η^+, and η^-. These can be solved, by expanding $\eta^\pm$ in Hermite polynomials,

$$\eta^\pm(x,y) = \sum_{n=0}^{\infty}\sum_{m=0}^{\infty}(T^\pm)_n^m H_n(x)\,H_m(y)e^{-x^2}e^{-y^2}, \tag{29}$$

and solving the corresponding nonlinear systems of equations for the coefficients $(T^\pm)_n^m$. Once we obtain $(T^\pm)_n^m$, we can calculate $<\hat{x}>^{(1)}$ from (27). Notice that $\hat{\rho}(+\omega_p) = \hat{\rho}^*(-\omega_p)$, by taking the complex conjugate in (24), and (27). Then it follows from (28), and (29) that $(T^+)_n^m = \left((T^-)_{-n}^{-m}\right)^*$. Therefore we conclude that $<\hat{x}>^{(1)}(-\omega_p) = \left(<\hat{x}>^{(1)}\right)^*(+\omega_p)$, and the inverse Fourier transform yields

$$\begin{aligned}<x>^{(1)}(t) = {} & 2\,\mathrm{Re}\left(<\hat{x}>^{(1)}(\omega_p)\right)\cos(\omega_p t)\\ & -2\,\mathrm{Im}\left(<\hat{x}>^{(1)}(\omega_p)\right)\sin(\omega_p t).\end{aligned} \tag{30}$$

Knowing $<x>^{(1)}(t)$, its amplitude can be readily computed, and the result is

$$A_{<x>} = 2\sqrt{<\hat{x}>^{(1)}(<\hat{x}>^{(1)})^*} + O(\varepsilon^2). \tag{31}$$

In Figure 4, we plotted the numerical solution for two different values of the

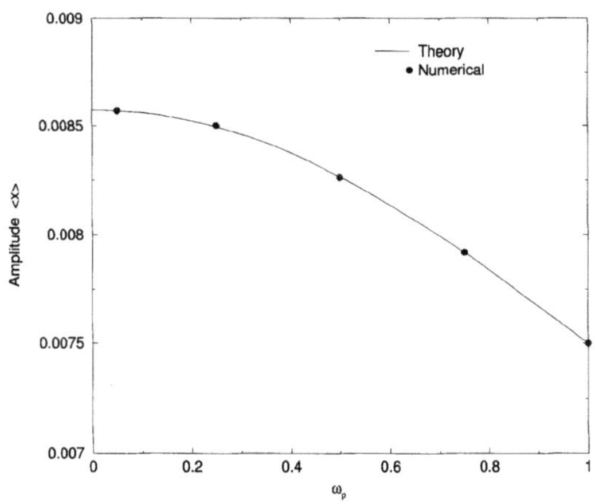

Figure 4. Comparison between the theoretical results and the numerical simulations, marked by symbols. Parameters are $D = 0.05$, $b_0 = 0.3, p = 1, q = 0.01$, and $\alpha = 0.05$.

coupling and the theoretical approximation (31), showing a remarkable agreement with the theoretical results corresponding to the first-order expansion. It should be noticed, however, that the amplitude of the probe signal considered here is small, $q = 0.01$. For increasing strength of the amplitude, higher

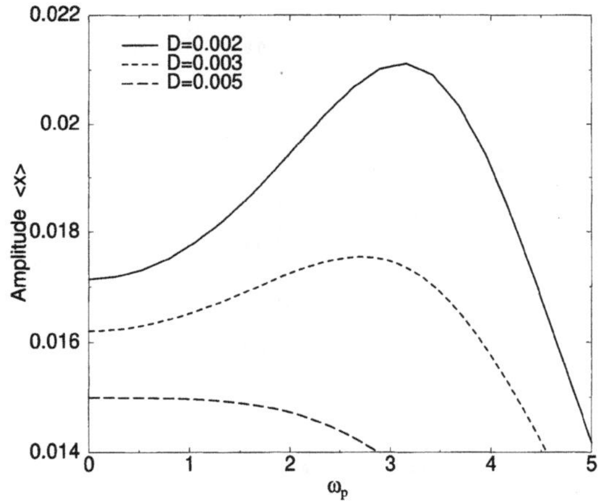

Figure 5. $< x >$ versus the frequency of the probe signal for different values of the noise strength. Results obtained by means of the theory for a single FHN. Parameters are $b_0 = 0.5, p = 1, q = 0.01$, and $\alpha = 0.05$.

orders in the expansion may be required. Once $\rho^{(1)}$ is known, it is also straightforward to find the successive terms in the expansion. Without entering into a detailed study, some general features can easily be drawn from the hierarchy of equations for $\rho^{(j)}$. Similarly to the analysis for $\hat{\rho}^{(1)}$, and by taking into account that $\hat{\rho}^{(1)}$ is a function exclusively of $\omega \pm \omega_p$, it is straightforward to prove that $\hat{\rho}^{(2)} = 0$ is the only solution, unless $\omega = 0, \pm 2\omega_p$. In general, successive terms will depend on higher harmonics of the main frequency ω_p.

By applying the theory above, we were able to obtain the amplitude of $< x >$ as function of the frequency of the probe signal, shown in Figure 5. It should be noticed that for increasing noise levels, the peak moves to smaller values of ω and then disappears. This was already observed in Figure 3 and can be interpreted as a sign of a stochastic bifurcation.

We now investigate the case of coupled FHN elements for which the dynamical response exhibits bifurcations, even in the absence of a probe signal. The bifurcation, of the Hopf-type, is shown in Figure 6 where we have plotted the amplitude of $< x >$ vs. coupling for a fixed level of noise. Below some critical coupling strength, the system is not synchronized and the solution of the FPE is stationary. On the other hand, above this critical coupling strength, the system synchronizes and exhibits a time-dependent behavior.

The inclusion of a probe signal will elicit a time-dependent solution of the FPE, even when the system *without* the probe signal has a stationary solution. The amplitude of the response ($< x >$) depends critically on the frequency

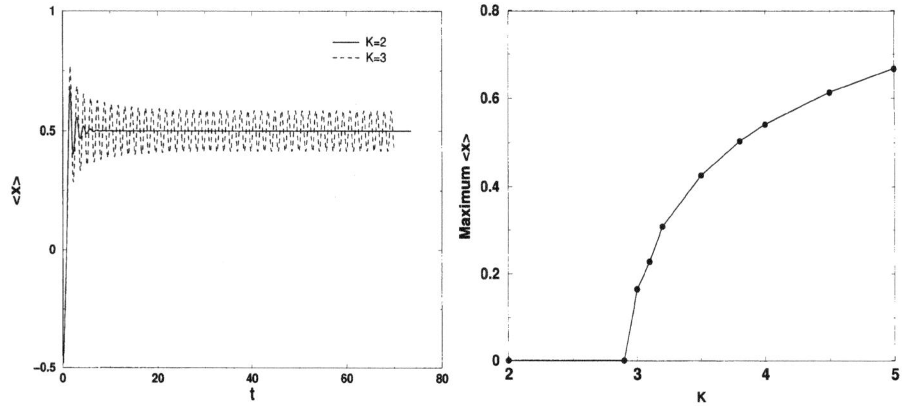

Figure 6. (a) $< x >$ versus time for two different values of the coupling showing a clear bifurcation. (b) Amplitude of $< x >$ vs. coupling for a fixed level of noise. Parameters are as in Figure 5

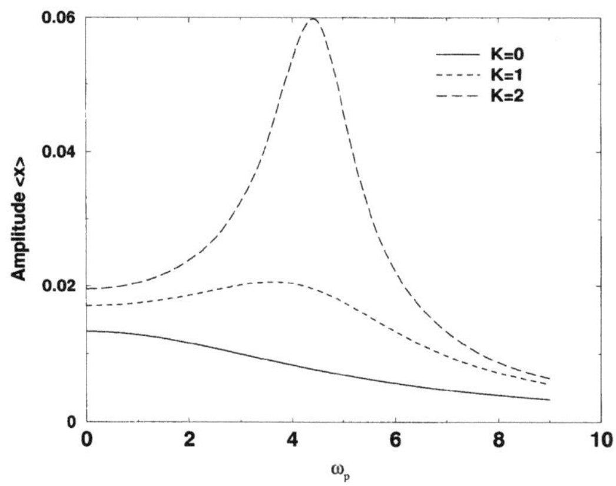

Figure 7. Amplitude of $< x >$ versus the frequency of the probe signal for three different values of the coupling strength. Parameters are as in Figure 5

of the probe signal as is shown in Figure 7. In contrast to similar coupled systems (see e.g., [1]), increasing the coupling does not lead to the "death" of the oscillatory region and the optimal frequency actually increases as the coupling is increased. The position and amplitude of the peak in Figure 7 depends on the coupling strength. For $K = 0$ the response curve does not exhibit a peak showing that there is no underlying frequency in the problem. Increasing K produces an underlying frequency which appears as a peak in the curve. Notice that for $K > 2.9$ the system will synchronize in the absence of a probe signal. This, then, leads to a response that has two principal frequencies: the frequency arising from the Hopf bifurcation and the probe frequency.

4. Conclusion

In this paper, we have investigated the single and globally coupled FHN model in the presence of noise and an injection signal. We have derived a FPE for the system and have shown that we can solve this FPE efficiently by using a suitably chosen expansion. We find that there is a classical resonance effect when the frequency of the probe signal approaches the one of the underlying system. We also characterize this resonance by separating the fast and slow time scales in the problem and find that, for small driving amplitudes, the agreement between numerical and analytical results is excellent. Finally, we reveal the existence of a stochastic bifurcation (see Figure 3 and Figure 5), manifested by the qualitative change of the peak location in the curves of $A_{<x>}$ vs. ω_p.

Future work will include the further characterization of the bifurcation we observed. Note that the bifurcation we found is different from the one found in earlier work [13]. We plan to address these differences in a future publication. In addition, we plan to investigate further the response of the globally coupled system to a probe signal. Particular attention will be paid to the possibility that, upon inclusion of an input signal, a population can become synchronized and can produce a large output signal. By varying the intrinsic parameters, including the coupling constant, the response can thus be "tuned" at different frequencies. Whether or not real neurons make use of this mechanism remains to be seen.

Acknowledgments

This work has been supported by the Office of Naval Research (Code 331). We also thank the National Partnership for Advanced Computational Infrastructure at the San Diego Supercomputer Center for computing resources.

References

[1] J. A. Acebrón, A. R. Bulsara, and W.-J. Rappel, Fluctuation and noise letters, to appear.

[2] R. C. Desai and R. Zwanzing, *J. Stat. Phys*, vol. 19, p. 1, 1978.

[3] A. Hagberg and E. Meron, *Nonlinearity*, vol. 7, p. 805, 1994.

[4] A. L. Hodgkin and A. F. Huxley, *J. Physiol.* (London), vol. 117, p. 500, 1952.

[5] J. Keener and J. Sneyd, *Mathematical Physiology*, Springer-Verlag, Berlin, 1998.

[6] C. Kurrer and K. Schulten, *Phys. Rev. E*, vol. 51, p. 6213, 1995.

[7] B. Lindner and L. Schimansky-Geier, *Phys. Rev. E*, vol. 60, p. 7270, 1999.

[8] A. Longtin, *Chaos Soliton Fractal*, vol. 11, p. 1835, 2000.

[9] S. R. Massanés and C. J. Pérez Vicente, *Int. J. Bif. Chaos*, vol. 9, p. 2295, 1999; *Phys. Rev. E*, vol. 59, p. 4490, 1999.

[10] A. S. Pikovsky and J. urths, *Phys. Rev. Lett.*, vol. 78, p. 775, 1997.

[11] H. Risken, *The Fokker-Planck equation: Methods of Solution and Applications*, (Springer Verlag, Berlin, 1996).

[12] M. Shiino, *Phys. Lett.*, vol. 112A, p. 302, 1985; *Phys. Rev.*, vol. A36, p. 2393, 1987.

[13] S. Tanabe and K. Pakdaman, *Phys. Rev. E*, vol. 63, 031911, 2001.

[14] A. T. Winfree, *Chaos*, vol. 1, p. 303, 1991.

LYAPUNOV EXPONENTS AND STABILITY FOR THE STOCHASTIC DUFFING-VAN DER POL OSCILLATOR

Peter H. Baxendale

Department of Mathematics

University of Southern California

Los Angeles, CA 90089-1113

USA

baxendal@math.usc.edu

Abstract Let λ denote the almost sure Lyapunov exponent obtained by linearizing the stochastic Duffing-van der Pol oscillator

$$\ddot{x} = -\omega^2 x + \beta\dot{x} - Ax^3 - Bx^2\dot{x} + \sigma x \dot{W}_t$$

at the origin $x = \dot{x} = 0$ in phase space. If $\lambda > 0$ then the process $\{(x_t, \dot{x}_t) : t \geq 0\}$ is positive recurrent on $\mathbf{R}^2 \setminus \{(0,0)\}$ with stationary probability measure μ, say. For $\lambda > 0$ let $\widetilde{\lambda}$ denote the almost sure Lyapunov exponent obtained by linearizing the same equation along a typical stationary trajectory in $\mathbf{R}^2 \setminus \{(0,0)\}$. The sign of $\widetilde{\lambda}$ is important for stability properties of the two point motion associated with the original equation. We use stochastic averaging techniques to estimate the value of $\widetilde{\lambda}$ in the presence of small noise and small viscous damping.

Keywords: stochastic Duffing-van der Pol oscillator, stability, Lyapunov exponent, stochastic averaging

1. Introduction

The stochastic Duffing-van der Pol oscillator is currently the object of much study in the theory of stability and bifurcations for stochastic dynamical systems. See for example Arnold [2], Arnold, Sri Namachchivaya and Schenk-Hoppé [4], Keller and Ochs [13], and Schenk-Hoppé [18]. In this paper we study the system with multiplicative white noise. To be specific we consider the stochastic Duffing-van der Pol oscillator

$$\ddot{x} = -\omega^2 x + \beta\dot{x} - Ax^3 - Bx^2\dot{x} + \sigma x \dot{W}_t$$

N. Sri Namachchivaya and Y.K. Lin (eds.),
IUTAM Symposium on Nonlinear Stochastic Dynamics, 125-135.

with $\omega > 0$ and $A \geq 0$ and $B \geq 0$. Here $-\beta$ is the coefficient of viscous damping and σ is the intensity of the white noise forcing. To ensure that the system is truly nonlinear we assume that at least one of A and B is strictly positive; if $B = 0$ we assume also that $\beta < 0$ so as to obtain some dissipation in the system.

Putting $y = -\dot{x}/\omega$ we get the 2-dimensional stochastic differential equation

$$\left\{\begin{array}{rcl} dx_t &=& -\omega y_t\, dt, \\ dy_t &=& \left(\omega x_t + \beta y_t + (A/\omega)x_t^3 - Bx_t^2 y_t\right) dt - (\sigma/\omega)x_t\, dW_t. \end{array}\right. \tag{1}$$

Notice that the system has multiplicative noise and hence that $(0,0)$ is a fixed point for (1). Linearizing at $(0,0)$ gives

$$du_t = \begin{bmatrix} 0 & -\omega \\ \omega & \beta \end{bmatrix} u_t dt - \frac{\sigma}{\omega}\begin{bmatrix} 0 & 0 \\ 1 & 0 \end{bmatrix} u_t\, dW_t. \tag{2}$$

The (top) almost sure Lyapunov exponent $\lambda = \lambda(\beta, \sigma, \omega)$ for the system (1) linearized at $(0,0)$ is defined as the almost sure limit

$$\lambda(\beta, \sigma, \omega) = \lim_{t\to\infty} \frac{1}{t} \log \|u_t\| \tag{3}$$

where $\{u_t : t \geq 0\}$ is a solution of (2). It is easy to show that the limit exists almost surely and does not depend on u_0 so long as $u_0 \neq 0$.

Clearly the sign of λ determines the almost sure stability or instability of the linearized process $\{u_t : t \geq 0\}$. It also determines the behavior near $(0,0)$ of the nonlinear process (x_t, y_t). This in turn controls the behavior of the process (x_t, y_t) considered on the state space $\mathbf{R}^2 \setminus \{(0,0)\}$. See Section 2 for details.

If $\lambda > 0$ then the solution $\{(x_t, y_t) : t \geq 0\}$ of (1) is positive recurrent on $\mathbf{R}^2\setminus\{(0,0)\}$ with a unique invariant probability measure $\mu = \mu(\beta, \sigma, \omega, A, B)$, say. In this case we are interested in the question of stability along trajectories. Let (x_0, y_0) and $(\tilde{x}_0, \tilde{y}_0)$ be distinct initial points in $\mathbf{R}^2 \setminus \{(0,0)\}$, and let $\{(x_t, y_t) : t \geq 0\}$ and $\{(\tilde{x}_t, \tilde{y}_t) : t \geq 0\}$ be the corresponding solutions of (1) generated by the same noise $\{W_t : t \geq 0\}$. This produces a coupling of the trajectories $\{(x_t, y_t) : t \geq 0\}$ and $\{(\tilde{x}_t, \tilde{y}_t) : t \geq 0\}$. We wish to consider the behavior of $\|(\tilde{x}_t, \tilde{y}_t) - (x_t, y_t)\|$ as $t \to \infty$. In this paper we shall consider the distinct but closely related question of linearized stability along trajectories.

Linearizing the system along a trajectory $\{(x_t, y_t) : t \geq 0\}$ in $\mathbf{R}^2$ gives

$$dv_t = \begin{bmatrix} 0 & -\omega \\ \omega + \frac{3A}{\omega}x_t^2 - 2Bx_t y_t & \beta - Bx_t^2 \end{bmatrix} v_t dt - \begin{bmatrix} 0 & 0 \\ \frac{\sigma}{\omega} & 0 \end{bmatrix} v_t\, dW_t. \tag{4}$$

In the same way that the process u_t gives information about the behavior of (x_t, y_t) near $(0,0)$, we expect that the process v_t will give information about the behavior of $(\tilde{x}_t, \tilde{y}_t) - (x_t, y_t)$ when $\|(\tilde{x}_t, \tilde{y}_t) - (x_t, y_t)\|$ is small. Thus we consider the stability of the solution v_t of (4) when $(x_0, y_0) \neq (0, 0)$. In particular we consider the (top) almost sure Lyapunov exponent $\widetilde{\lambda} = \widetilde{\lambda}(\beta, \sigma, \omega, A, B)$, say, for the system (1) linearized along a stationary trajectory $\{(x_t, y_t) : t \geq 0\}$ in $\mathbf{R}^2 \setminus \{0\}$, defined as the almost sure limit

$$\widetilde{\lambda}(\beta, \sigma, \omega, A, B) = \lim_{t \to \infty} \frac{1}{t} \log \|v_t\|. \tag{5}$$

Notice that the almost sure limit is with respect to the product measure $\mathbf{P} \times \mu$ where $\mathbf{P}$ represents the probability measure for $\{W_t : t \geq 0\}$ and μ is the distribution of $(x_0, y_0) \in \mathbf{R}^2 \setminus \{(0, 0)\}$. Recall that μ and thus $\widetilde{\lambda}$ exist only when $\lambda > 0$.

An exact formula for $\lambda(\beta, \sigma, \omega)$ is given by Imkeller and Lederer [12]; earlier numerical calculations by Kozin and Prodromou [15] gave criteria in terms of β, σ and ω for $\lambda(\beta, \sigma, \omega)$ to be positive or negative. However the evaluation of $\widetilde{\lambda}(\beta, \sigma, \omega, A, B)$ is much more complicated, see Section 3, and to date there is no closed form formula.

In this paper we use stochastic averaging techniques to obtain estimates for $\widetilde{\lambda}$ in the case of small noise and small viscous damping. We replace σ by $\varepsilon\sigma$ and β by $\varepsilon^2\beta$. A result of Auslender and Milstein [5] gives

$$\lambda(\varepsilon^2\beta, \varepsilon\sigma, \omega) \sim \varepsilon^2 \left(\frac{\beta}{2} + \frac{\sigma^2}{8\omega^2} \right)$$

as $\varepsilon \to 0$, so we look for estimates on $\widetilde{\lambda}(\varepsilon^2\beta, \varepsilon\sigma, \omega, A, B)$ when $\beta > -\sigma^2/4\omega^2$.

In Section 4 we consider the case when $B = 0$. In this case the system can be regarded as a small perturbation of a Hamiltonian system, and we have

$$\widetilde{\lambda}(\varepsilon^2\beta, \varepsilon\sigma, \omega, A, 0) \sim \varepsilon^{2/3}\widetilde{\lambda}_0$$

where $\widetilde{\lambda}_0 > 0$. This implies (linearized) instability along trajectories. In Section 5 we consider the case when $A > 0$ and $B > 0$. Now we have

$$\widetilde{\lambda}(\varepsilon^2\beta, \varepsilon\sigma, \omega, A, 0) \sim \varepsilon^2\widetilde{\lambda}_{\text{ave}}$$

where $\widetilde{\lambda}_1$ is the top Lyapunov exponent along trajectories for a stochastically averaged version of the original system. The value of $\widetilde{\lambda}_{\text{ave}}$ is given by a two dimensional integral formula. The sign of $\widetilde{\lambda}_{\text{ave}}$ depends only on the dimensionless quantities $\beta\omega^2/\sigma^2 > -1/4$ and $\omega B/A$. Numerical simulations presented in Section 5.1 show that for $-\sigma^2/4\omega^2 < \beta < 0$ the Lyapunov exponent $\widetilde{\lambda}_{\text{ave}}$ is positive if $\omega B/A \leq 1$ and is negative if $\omega B/A \geq 1.5$.

2. Behavior Near 0

The following result makes precise the idea that the behavior of the linearization u_t at $(0,0)$ determines the behavior of (x_t, y_t) near $(0,0)$. It is essentially an application of Theorems 2.12 and 2.13 of Baxendale [7], see also Theorem 2.8 of Baxendale [8]. The verification of condition (2.3) for [7] (or condition $H1(z)$ of [8]) is carried out in [8] for the case $B = 0$, $\beta < 0$, and in [9] for the case $B > 0$.

Theorem 1 *For the system (1)*

(i) *If $\lambda < 0$ then*

$$\limsup_{t\to\infty} \frac{1}{t} \log \|(x_t, y_t)\| \leq \lambda$$

almost surely, for all $(x_0, y_0) \neq 0$.

(ii) *If $\lambda > 0$ there exists a unique probability measure μ on $\mathbf{R}^2 \setminus \{(0,0)\}$ such that*

$$\lim_{t\to\infty} \frac{1}{t} \int_0^t \phi(x_s, y_s)\, ds = \int \phi\, d\mu$$

almost surely, for all bounded measurable $\phi : \mathbf{R}^2 \setminus \{(0,0)\} \to \mathbf{R}$ and all $(x_0, y_0) \neq (0,0)$. In particular μ is the unique invariant measure for $\{(x_t, y_t) : t \geq 0\}$ on $\mathbf{R}^2 \setminus \{(0,0)\}$.

Notice that if $\lambda > 0$ then the linearized process $u_t \to \infty$ almost surely, whereas the original process (x_t, y_t) is affected by the cubic nonlinearity and does not tend to infinity.

3. Khasminskii-Carverhill Formula

In this section we assume that $\lambda > 0$, so that the invariant probability measure μ on $\mathbf{R}^2 \setminus \{(0,0)\}$ exists. We apply the method developed by Khasminskii [14] for linear stochastic differential equations and extended to the nonlinear setting by Carverhill [11]. For v_t given by (4) write $v_t = \|v_t\| \begin{bmatrix} \cos\theta_t \\ \sin\theta_t \end{bmatrix}$. Then

$$\begin{aligned} d\theta_t &= \Big(\omega + (\frac{3A}{\omega}x_t^2 - 2Bx_t y_t)\cos^2\theta_t + (\beta - Bx_t^2)\sin\theta_t\cos\theta_t \\ &\quad -\frac{\sigma^2}{\omega^2}\sin\theta_t\cos^3\theta_t\Big)\, dt - \frac{\sigma}{\omega}\cos^2\theta_t dW_t. \end{aligned} \tag{6}$$

Moreover, Itô's formula for $\log\|v_t\|$ and the ergodic theorem give

$$\widetilde{\lambda} = \lim_{t\to\infty} \frac{1}{t}\log\|v_t\| = \int_{(\mathbf{R}^2\setminus\{0\})\times\mathbf{S}^1} Q(x,y,\theta)\, d\nu(x,y,\theta) \tag{7}$$

where

$$Q(x,y,\theta) = \left(\frac{3A}{\omega}x^2 - 2Bxy\right)\sin\theta\cos\theta + (\beta - Bx^2)\sin^2\theta + \frac{\sigma^2}{2\omega^2}\cos 2\theta\cos^2\theta$$

and ν is the invariant probability measure for the diffusion process $\{(x_t, y_t, \theta_t) : t \geq 0\}$ on $(\mathbf{R}^2 \setminus \{(0,0)\}) \times \mathbf{S}^1$ given by (1) and (6). The existence of ν is implied by the existence of its marginal μ, which in turn is guaranteed by $\lambda > 0$. The uniqueness of ν can be verified using the methods of Arnold and San Martin [3].

The formula for λ is obtained similarly, using just the original setting of Khasminskii. Putting $x = y = 0$ reduces the three dimensional problem in (7) to the one dimensional problem handled by [12] and [15].

4. The Duffing Case $B = 0$, $\beta < 0$

In this section we consider the system

$$\begin{cases} dx_t &= -\omega y_t\,dt \\ dy_t &= \left(\omega x_t + \varepsilon^2\beta y_t + \frac{A}{\omega}x_t^3\right)dt - \frac{\varepsilon\sigma}{\omega}x_t\,dW_t \end{cases}$$

which may be regarded as a small perturbation of the Hamiltonian system with

$$H(x,y) = \frac{\omega}{2}\left(x^2 + y^2\right) + \frac{A}{4\omega}x^4.$$

Theorem 2 (Baxendale and Goukasian [10]) *If* $-\sigma^2/4\omega^2 < \beta < 0$ *then*

$$\widetilde{\lambda}(\varepsilon^2\beta, \varepsilon\sigma, \omega, A, 0) = \varepsilon^{2/3}\widetilde{\lambda}_0 + O(\varepsilon^{4/3}) \quad \textit{as } \varepsilon \to 0$$

where $\widetilde{\lambda}_0 = \widetilde{\lambda}_0(\beta, \sigma, \omega) > 0$.

There are three main ideas in the proof. The first one is to write the equation for v_t in the moving frame given by the vectors $\nabla^{\perp}H(x,y)$ and $\nabla H(x,y)/\|\nabla H(x,y)\|^2$ (where $\nabla^{\perp}H(x,y)$ denotes the symplectic gradient of H). The resulting equation has the form

$$dv_t = \begin{bmatrix} 0 & J(x_t,y_t) \\ 0 & 0 \end{bmatrix} v_t\,dt + \varepsilon^2 M_0(x_t,y_t)v_t\,dt + \varepsilon M_1(x_t,y_t)v_t\,dW_t \quad (8)$$

where $J(x,y)$ is a scalar function and $M_0(x,y)$ and $M_1(x,y)$ are matrix-valued functions.

The second main idea is to use a technique due to Pinsky and Wihstutz [17] for small perturbations of nilpotent linear systems. Their method shows that the constant coefficient linear stochastic differential equation

$$dv_t = \begin{bmatrix} 0 & a \\ 0 & 0 \end{bmatrix} v_t dt + \varepsilon^2 \begin{bmatrix} a_{11} & a_{12} \\ a_{21} & a_{22} \end{bmatrix} v_t dt + \varepsilon \begin{bmatrix} b_{11} & b_{12} \\ b_{21} & b_{22} \end{bmatrix} v_t \, dW_t \tag{9}$$

has Lyapunov exponent

$$\lambda(\varepsilon) = \varepsilon^{2/3} \gamma_0 |a|^{2/3} |b_{21}|^{2/3} + O(\varepsilon^{4/3}) \quad \text{as } \varepsilon \to 0$$

so long as $a \neq 0$ and $b_{21} \neq 0$, where

$$\gamma_0 = \frac{\pi}{2^{1/3} 3^{1/6} [\Gamma(\frac{1}{3})]^2} \approx 0.29.$$

The exact value of γ_0 is due to Ariaratnam and Xie [1].

The third main idea is to use stochastic averaging techniques to pass from the variable coefficient equation (8) to the constant coefficient equation (9). The stochastic averaging involves three different time scales, so that the original three dimensional integral formula for $\widetilde{\lambda}$ reduces to a one dimensional integral formula for $\widetilde{\lambda}_0$ which involves a strictly positive integrand.

5. The Case $A > 0$, $B > 0$

Following Arnold, N. Sri Namachchivaya & Schenk-Hoppé [4] we rescale spatially $x \to x/\varepsilon$, $y \to y/\varepsilon$ to obtain

$$\begin{cases} dx_t &= -\omega y_t \, dt \\ dy_t &= \omega x_t \, dt + \varepsilon^2 \left(\beta y_t + \frac{A}{\omega} x_t^3 - B x_t^2 y_t \right) dt - \frac{\varepsilon \sigma}{\omega} x_t \, dW_t. \end{cases}$$

This system is a small perturbation of the rigid rotation with constant angular velocity ω. Write $z_t = \begin{bmatrix} \cos \omega t & \sin \omega t \\ -\sin \omega t & \cos \omega t \end{bmatrix} \begin{bmatrix} x_t \\ y_t \end{bmatrix}$, then

$$dz_t = \varepsilon^2 F(z,t) dt + \varepsilon G(z,t) dW_t \tag{10}$$

where the vector fields $F(z, \cdot)$ and $G(z, \cdot)$ have period $2\pi/\omega$. Applying the method of stochastic averaging to the k-point motion of this SDE (rescale time by factor $1/\varepsilon^2$ and take weak limit), we get

$$\begin{aligned} dz_t &= \left(\frac{\beta}{2} - \frac{B \|z_t\|^2}{8} \right) z_t dt - \frac{3A}{8\omega} \|z_t\|^2 J z_t dt \\ &\quad + \frac{\sigma}{2\sqrt{2}\omega} \left(K_1 z_t \, dW_t^1 + K_2 z_t \, dW_t^2 + \sqrt{2} J z_t \, dW_t^3 \right) \end{aligned} \tag{11}$$

where

$$J = \begin{bmatrix} 0 & 1 \\ -1 & 0 \end{bmatrix}, \quad K_1 = \begin{bmatrix} 1 & 0 \\ 0 & -1 \end{bmatrix}, \quad K_2 = \begin{bmatrix} 0 & 1 \\ 1 & 0 \end{bmatrix}.$$

It is important to observe that the stochastic averaging done in the passage from (10) to (11) is carried out at the level of the k-point motions, for all $k \geq 1$. Related results are contained in Section 5.6 of Kunita [16]. If the stochastic averaging is done only for the one point motion of (10) the resulting information is only the generator for the one point motion of (11). There are many ways of choosing a stochastic differential equation (SDE) with a given generator for the one point motion, and these different choices of SDE can have many different stability behaviors, and many different values for the Lyapunov exponent along trajectories, see Baxendale [6]. However by doing stochastic averaging for the k-point motions, we gain information about the local characteristics of the stochastic flow associated with (11). In particular the law of the stochastic flow associated with (11) is uniquely determined.

In this setting Arnold, N. Sri Namachchivaya and Schenk-Hoppé [4] choose the simpler SDE

$$\begin{aligned} dz_t &= \left(\frac{\beta}{2} - \frac{B\|z_t\|^2}{8}\right) z_t dt - \frac{3A}{8\omega}\|z_t\|^2 J z_t dt \\ &\quad + \frac{\sigma}{2\sqrt{2}\omega}\left(z_t\, dW_t^1 + \sqrt{3} J z_t\, dW_t^2\right) \end{aligned} \tag{12}$$

which generates the same one-point motion as (11). This SDE has $\lambda = \beta/2 + \sigma^2/8\omega^2$ and $\widetilde{\lambda} = 0$ (when $\lambda > 0$). Moreover the (one-point) amplitude process $r_t = \|z_t\|$ given by (11) or (12) satisfies an SDE of the form

$$dr_t = \left(\frac{\beta r_t}{2} - \frac{B r_t^3}{8} + \frac{3\sigma^2 r_t}{16\omega^2}\right) dt + \frac{\sigma}{2\sqrt{2}\omega} r_t\, dW_t \tag{13}$$

which has $\lambda = \beta/2 + \sigma^2/8\omega^2$ and $\widetilde{\lambda} = -2\lambda$ (when $\lambda > 0$). Thus equations (11), (12), and (13) give three different values for the Lyapunov exponent along trajectories. The following theorem tells us that (11) is the correct SDE to study.

Theorem 3 (Baxendale [9]) *Assume $A > 0$, $B > 0$. Let λ_{ave} and $\widetilde{\lambda}_{\text{ave}}$ be the Lyapunov exponents for the averaged SDE (11). Then as $\varepsilon \to 0$,*

$$\lambda(\varepsilon^2\beta, \varepsilon\sigma, \omega) \sim \varepsilon^2 \lambda_{\text{ave}} = \varepsilon^2\left(\frac{\beta}{2} + \frac{\sigma^2}{8\omega^2}\right) \tag{14}$$

and if $\beta > -4\sigma^2/\omega^2$ then

$$\widetilde{\lambda}(\varepsilon^2\beta, \varepsilon\sigma, \omega, A, B) \sim \varepsilon^2 \widetilde{\lambda}_{\text{ave}}. \tag{15}$$

In fact the first result (14) is due to Auslender and Milstein [5]; the explicit formula for λ_{ave} comes from the rotational symmetry of (11). The result (15) is plausible because the Lyapunov exponent is not affected by spatial rescaling and rigid rotations. However since stochastic averaging guarantees only convergence on finite intervals of time there is work to be done. The proof of (15) uses the Khasminskii-Carverhill formula (7) and the weak convergence of the invariant measures ν on $(\mathbf{R}^2 \setminus \{0\}) \times \mathbf{S}^1$ as $\varepsilon \to 0$. It is here that the condition $B > 0$ is used. If $B = 0$ then the sequence of invariant measures is not tight and the estimate (15) fails, see Section 4. The condition $A > 0$ ensures that the limit measure is unique. It will also ensure that the measure $\widetilde{\nu}$ in (16) is unique.

It remains to calculate the Lyapunov exponent $\widetilde{\lambda}_{\text{ave}}$. The rotational symmetry in (11) reduces the calculation in the Khasminskii-Carverhill formula to a two-dimensional problem. We have

$$\widetilde{\lambda}_{\text{ave}} = -\frac{\beta}{2} - \frac{\sigma^2}{8\omega^2} + \int \left(\frac{3Ar^2}{8\omega} \sin 2\psi - \frac{Br^2}{8} \cos 2\psi \right) d\widetilde{\nu}(r,\psi) \tag{16}$$

where $\widetilde{\nu}$ is the invariant probability on $(0,\infty) \times \mathbf{S}^1$ for

$$dr_t = \left(\frac{\beta r_t}{2} - \frac{Br_t^3}{8} + \frac{3\sigma^2 r_t}{16\omega^2} \right) dt + \frac{\sigma}{2\sqrt{2\omega}} r_t \, dB_t^1$$

and

$$\begin{aligned} d\psi_t &= \left(\frac{B}{8} r_t^2 \sin 2\psi_t + \frac{3Ar_t^2}{4\omega} \cos^2 \psi_t \right) dt \\ &\quad - \frac{\sigma}{2\sqrt{2\omega}} \sin 2\psi_t \, dB_t^1 + \frac{\sigma}{2\sqrt{2\omega}} (-1 + \cos 2\psi_t) \, dB_t^2. \end{aligned}$$

The processes r_t and ψ_t are related to the processes (x_t, y_t) and θ_t used in the derivation of the formula (7) by the formulas $\begin{bmatrix} x_t \\ y_t \end{bmatrix} = r_t \begin{bmatrix} \cos\phi_t \\ \sin\phi_t \end{bmatrix}$ and $\psi_t = \theta_t - \phi_t$. Moreover, the dependence of $\widetilde{\lambda}_{\text{ave}}$ on the parameters β, σ, ω, A and B satisfies

$$\widetilde{\lambda}_{\text{ave}}(\beta,\sigma,\omega,A,B) = \frac{\sigma^2}{\omega^2} \widetilde{\lambda}_{\text{ave}}\left(\frac{\omega^2\beta}{\sigma^2}, 1, 1, 1, \frac{\omega B}{A}\right).$$

We see that the sign of $\widetilde{\lambda}_{\text{ave}}$ depends only on the dimensionless quantities $\omega^2\beta/\sigma^2$ and $\omega B/A$. The first quantity measures the strength of the viscous damping relative to the noise intensity. The second quantity measures the relative sizes of the cubic dissipation $Bx^2\dot{x}$ and cubic restoring force Ax^3.

5.1 Numerical calculations for $\widetilde{\lambda}_{\text{ave}}$

Write $s_t = r_t^{-2}$, then s_t satisfies the affine equation

$$ds_t = \left(-\beta s_t + \frac{B}{4}\right) dt - \frac{\sigma}{\sqrt{2}\omega} s_t \, dB_t^1$$

The following numerical results used a first order Euler scheme to simulate the process (s_t, ψ_t) on some time interval $[0, T]$, and this was used to approximate the integral in (16) by the corresponding discrete time average. See Talay [19] for a theoretical discussion of this method. The simulations all had $\omega = \sigma = 1$ and $T \approx 7.8 \times 10^5$. The value of β was increased from -0.245 to 0.005 in steps of 0.005. In Figures 1 and 2 the dotted line shows the exact value

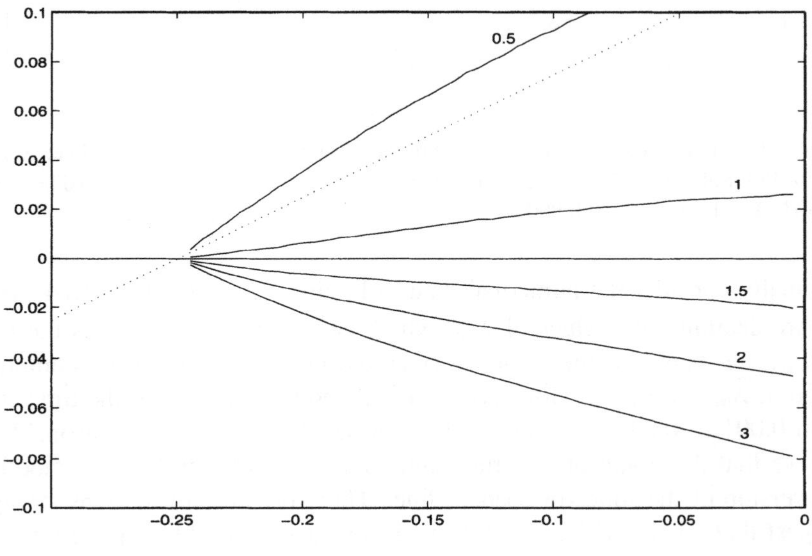

Figure 1. Lyapunov exponent as a function of β when $\sigma^2/\omega^2 = 1$. Dotted line: $\lambda = \beta/2 + 1/8$. Solid lines: $\widetilde{\lambda}$ for $\omega B/A = 0.5$, 1, 1.5, 2 and 3.

$\lambda_{\text{ave}} = \beta/2 + 1/8$. The solid lines show the results of simulation of $\widetilde{\lambda}_{\text{ave}}$ with time step $h = 0.001$. The dashed line shows the results of simulation of $\widetilde{\lambda}_{\text{ave}}$ with time step $h = 0.0005$. Figure 1 shows $\widetilde{\lambda}_{\text{ave}}$ for the values $\omega B/A = 0.5$, 1, 1.5, 2 and 3. Figure 2 shows $\widetilde{\lambda}_{\text{ave}}$ for the values $\omega B/A = 1$, 1.24 and 1.5.

From the numerical data in Figure 1 we see in the range $-\sigma^2/4\omega^2 < \beta < 0$ we have

$$\left.\begin{array}{l}\widetilde{\lambda}_{\text{ave}} > 0 \\ \widetilde{\lambda}_{\text{ave}} < 0\end{array}\right\} \text{ if } \left\{\begin{array}{l}\omega B/A \leq 1 \\ \omega B/A \geq 1.5\end{array}\right.$$

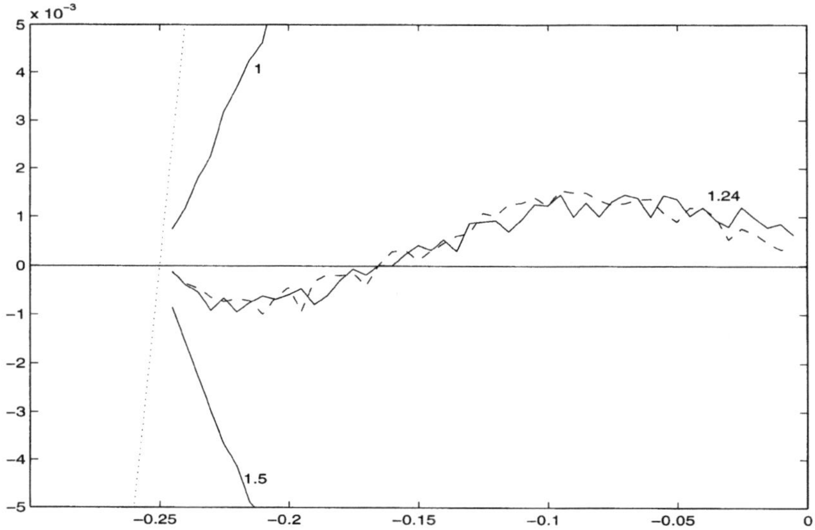

Figure 2. Lyapunov exponent as a function of β when $\sigma^2/\omega^2 = 1$. Dotted line: $\lambda_{\text{ave}} = \beta/2 + 1/8$. Solid lines: $\widetilde{\lambda}_{\text{ave}}$ for $\omega B/A = 1$, 1.24 and 1.5 with $h = 0.001$. Dashed line: $\widetilde{\lambda}_{\text{ave}}$ for $\omega B/A = 1.24$ with $h = 0.0005$.

Finally consider the numerical data in Figure 2 for $\omega B/A = 1.24$. Here we are dealing with values of $\widetilde{\lambda}_{\text{ave}}$ which are of the same order as the time step h, and hence of the same order as the possible error in the computed value of $\widetilde{\lambda}_{\text{ave}}$, see [19]. However, Figure 2 shows that halving the time step, from 0.001 to 0.0005, has little effect on the data. Thus it is reasonable to believe that the graph of the true value of $\widetilde{\lambda}_{\text{ave}}$ will be close to a smoothed out version of the solid (or dashed) line. Therefore our calculations strongly suggest that when $\omega B/A = 1.24$ the averaged system (11) experiences two dynamic bifurcations as β is increased from $-\infty$ to 0. One occurs when $\beta = -\sigma^2/4\omega^2$ as the fixed point (0,0) loses stability, but the system still has stability along trajectories. The other occurs at approximately $\beta = -0.17\sigma^2/\omega^2$, as the property of stability along trajectories is lost.

Acknowledgments

I would like to thank Levon Goukasian (University of Southern California) for his assistance with the numerical computations in Section 5.1.

References

[1] S. T. Ariaratnam and W. C. Xie, "Lyapunov exponent and rotation number of a two-dimensional nilpotent stochastic system," *Dynam. Stability Systems*, vol. 5, pp. 1–9, 1990.

[2] L. Arnold, *Random dynamical systems.* Springer, Berlin Heidelberg New York, 1998.

[3] L. Arnold and L. San Martin, "A control problem related to the Lyapunov spectrum of stochastic flows," *Matemática Aplicada e Computacional*, vol. 5, pp. 31–64, 1986.

[4] L. Arnold, N. Sri Namachchivaya, and K. Schenk-Hoppé, "Toward an understanding of the stochastic Hopf bifurcation: a case study," *Internat. J. Bifur. Chaos*, vol. 6, pp. 1947–1975, 1996.

[5] E. Auslender and G. Mil'shtein, "Asymptotic expansions of the Liapunov index for linear stochastic systems with small noise," *J. Appl. Math. Mech.*, vol. 46, pp. 277–283, 1982.

[6] P. Baxendale, "Asymptotic behaviour of stochastic flows of diffeomorphisms," in *Stochastic processes and their applications. Proc. Nagoya 1985.* (K. Itô and T. Hida, eds) Lect. Notes Math., vol. 1203, pp. 1–19. Springer, Berlin Heidelberg New York, 1986.

[7] P. Baxendale, "Invariant measures for nonlinear stochastic differential equations," in: *Lyapunov Exponents. Proc. Oberwolfach 1990.* (L. Arnold, H. Crauel and J.-P. Eckmann, eds) Lect. Notes Math., vol. 1486, pp. 123–140. Springer, Berlin Heidelberg New York, 1991.

[8] P. Baxendale, "A stochastic Hopf bifurcation," *Probab. Th. Rel. Fields*, vol. 99, pp. 581–616, 1994.

[9] P. Baxendale, "Stochastic averaging and asymptotic behavior of the stochastic Duffing - van der Pol equation," Preprint. 2002.

[10] P. Baxendale and L. Goukasian, "Lyapunov exponents for small random perturbations of Hamiltonian systems," *Ann. Probab.*, vol. 30, pp. 101–134, 2002.

[11] A. Carverhill, "A formula for the Lyapunov numbers of a stochastic flow. Application to a perturbation theorem," *Stochastics*, vol. 14, pp. 209–226, 1985.

[12] P. Imkeller and C. Lederer, "An explicit description of the Lyapunov exponents of the noisy damped harmonic oscillator," *Dynam. Stability Systems*, vol. 14, pp. 385–405, 1999.

[13] H. Keller and G. Ochs, "Numerical approximation of random attractors," in *Stochastic dynamics* (H. Crauel, M. Gundlach, eds) pp. 93–115. Springer, Berlin Heidelberg New York, 1999.

[14] R. Khasminskii, "Necessary and sufficient conditions for the asymptotic stability of linear stochastic systems," *Theory Probab. Appl.*, vol. 12, pp. 144–147, 1967.

[15] F. Kozin and S. Prodromou, "Necessary and sufficient conditions for almost sure sample stability of linear Itô equations," *SIAM J. Appl. Math.*, vol. 21, pp. 413–424., 1971.

[16] H. Kunita, *Stochastic flows and stochastic differential equations*, Cambridge University Press, 1990.

[17] M. Pinsky and V. Wihstutz, "Lyapunov exponents of nilpotent Itô systems," *Stochastics*, vol. 25, pp. 43–57., 1988.

[18] K. Schenk-Hoppé, "Bifurcation scenarios of the noisy Duffing-van der Pol oscillator," *Nonlinear dynamics*, vol. 11, pp. 255–274, 1996.

[19] D. Talay, "The Lyapunov exponent for the Euler scheme for stochastic differential equations," in: *Stochastic dynamics* (H. Crauel, M. Gundlach, eds.) pp. 241–258, Springer, Berlin Heidelberg New York, 1999.

SLOW SWEEP THROUGH A PERIOD-DOUBLING CASCADE: AN EXAMPLE OF A NOISY PARAMETRIC BIFURCATION

Huw G. Davies and Philotas Kyriakidis

University of New Brunswick

Fredericton, N.B., Canada, E3B 5A3

davies@unb.ca

Abstract The response of a nonlinear system when a parameter is slowly varied through a bifurcation point may be influenced strongly by very-low-level noise. Noise, which is inherent in any system, both in experiments and in numerical simulations, is actually necessary in some cases to trigger the change in response from the initial form to the bifurcated form. This is the case for example when the initial form of the solution continues to exist even though in an unstable state after the bifurcation point. We analyse two such systems involving two different types of bifurcations, one involving a continuous system and the other a discrete map, and show the critical effect of noise in each case.

(i) A ship model with coupled pitch and roll modes exhibits a saturation phenomenon of the pitch mode, with a corresponding large amplitude increase of the roll mode as the wave excitation amplitude is slowly increased. This is a form of transcritical bifurcation. During a slow sinusoidal variation of the excitation amplitude, we show that noise determines whether and when the possibly catastrophic increase in roll amplitude occurs.

(ii) Quadratic maps exhibit cascades of period-doubling bifurcations. We use a renormalisation scheme to describe high-period orbits. Low-period orbits can be stabilised by slow sinusoidal sweep of a control parameter. Increasing the rate of sweep has a stabilising effect, but increasing the noise level destabilises the orbits by triggering period-doubling.

We analyse all these systems using matched asymptotic expansions in terms of the small rate of variation of the parameter. A nested set of three expansions is needed to describe the jump phenomenon from one type of behaviour to another, that is, from the now unstable original form to the stable bifurcated form. The innermost expansion in each case describes how noise is necessary to trigger exponential growth for example of period-2 response away from the locally unstable period-1 response; the mean-square period-2 response is calculated. This expansion matches to an inner expansion that describes the rapid change from one form of response behaviour to the other, and this in turn matches to the outer ex-

N. Sri Namachchivaya and Y.K. Lin (eds.),
IUTAM Symposium on Nonlinear Stochastic Dynamics, 137-146.

pansion that describes the bifurcated response. Comparisons of the analytic estimates with numerical simulations of the describing equations are excellent in all cases.

Keywords: nonstationary bifurcations, parametric variation, noise

1. Introduction

Nonlinear systems may exhibit a variety of bifurcations. We examine the behaviour of several systems as a control parameter is varied slowly through a bifurcation. We consider those cases where after the bifurcation the original form of the response continues to exist, although now in an unstable state, simultaneously with the new stable bifurcated form. Typically, the emergence of the bifurcated form of the response appears to be delayed beyond the bifurcation point found in the autonomous or stationary state. The original form of the response persists into the parameter region where it is locally unstable; either noise or an applied perturbation is necessary to trigger the bifurcated form.

A number of such systems have been discussed in the literature; they are referred to as having noise-controlled dynamics. Lythe and Proctor [5, 6] have studied a number of such problems including intermittent destabilization caused by shear of a fluid in cellular convective motion. Celet et al.[1] used experiments with a laser to show that slow passage through a resonance with bistable response curves is strongly dependent on noise. Pisarchik did numerical simulations of the Henon map and of the laser rate equations and showed that period doubling bifurcations could be destroyed by sufficiently rapid periodic variation. This was confirmed by [3] who used matched asymptotic expansions to quantify the effect for a general unimodal map. These examples cover a wide variety of bifurcations of nonlinear systems described by partial and ordinary differential equations and discrete maps.

In what follows we first describe two examples of noise-controlled systems, showing the competing effects of rate of parameter change through the bifurcation and noise. With linear parametric variation, the appearance of the bifurcating form is more delayed as the rate of variation increases. With sinusoidal variation through the bifurcation the bifurcated form may not appear at all if the variation is rapid; sinusoidal variation can stabilize the initial form of the response. In both cases, the addition of noise is destabilizing; an increase in noise level tends to generate the bifurcated form by accelerating the repulsion away from the locally unstable initial form. The noise levels that do this may be extremely small. We then analyse these systems and show how matched asymptotic expansions can be used to describe the response and to evaluate and quantify these competing effects.

2. A Coupled-modes Model

Nayfeh [7] described how his earlier model of coupled pitch and roll modes of ship motion could explain some types of capsize. A simplified form of Nayfeh's model that describes the main features of the motion is given by the coupled equations:

$$y_1'' + \mu_1 y_1' + \omega_1^2 y_1 = y_1 y_1 \tag{1}$$
$$y_2'' + \mu_2 y_2' + \omega_2^2 y_2 = y_1^2 + F\cos(\Omega t) \tag{2}$$

where y_1 and y_2 are the amplitudes of the roll and pitch modes, respectively. The quadratic coupling terms describe fluid inertia effects. Here the direct excitation is only of the pitch mode. Nayfeh showed that as the excitation amplitude F increases beyond a certain level, the pitch mode saturates, and the nonlinear coupling causes parametric excitation of the roll mode.

Davies and Pan [2] used a simplified form of Nayfeh's model to discuss nonstationary motion. After a multiple time scale analysis and a change of variables, a restricted set of motions is described by the equations:

$$a' = -a - b + F \tag{3}$$
$$b' = ab + \sigma f \tag{4}$$

where now the excitation amplitude $F = B_0 + B_1\cos(\epsilon t)$, where ϵ is a small frequency of variation. This model describes excitation at basic frequency Ω but with slow amplitude modulation at frequency ϵ. The variable f represents a Gaussian noise with zero mean and unit variance, while σ is the typically very small noise amplitude. It is convenient to continue to call the variables a and b the pitch and roll modes, respectively, even though there has been a change of variables from the original equations of Nayfeh. With no modulation ($B_1 = 0$) and no noise, this model describes a bifurcation at the point $B_0 = 0$; the solution $(a, b) = (B_0, 0)$ is stable for $B_0 < 0$, while $(a, b) = (0, B_0)$ is stable for $B_0 > 0$.

Figures 1 and 2 show the case of $\epsilon = 0.05$, and very low noise. In Figure 1 the amplitudes of the modal variables a and b are plotted against time. To show the sensitivity of this model to perturbations or noise, there is an added small perturbation of the roll mode b of magnitude 0.02 at $t = 200$. As can be seen the effect of the perturbation initially seems to decay, but the perturbation actually generates a large roll amplitude b, which indeed then decays during the next cycle. A similar perturbation applied to the pitch mode, amplitude a, has no such effect.

In Figure 2 the amplitudes are plotted against the excitation amplitude F to show the form of the bifurcation at $F = 0$; no perturbation has been added in Figure 2. The linear response of a for $F > 0$ is unstable. But the sinusoidal variation is sufficiently rapid that it stabilizes the pitch mode; no roll mode is

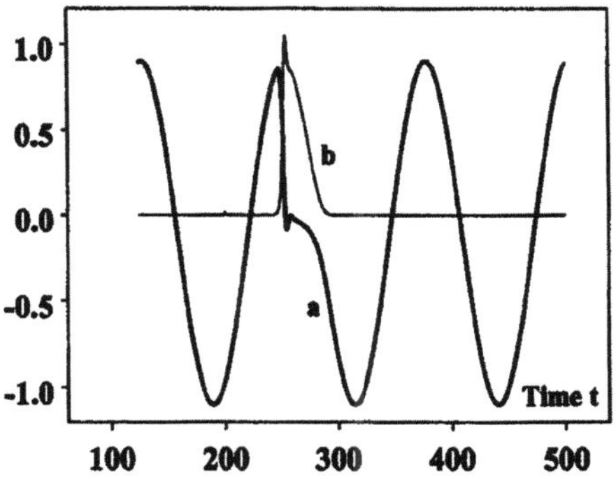

Figure 1.

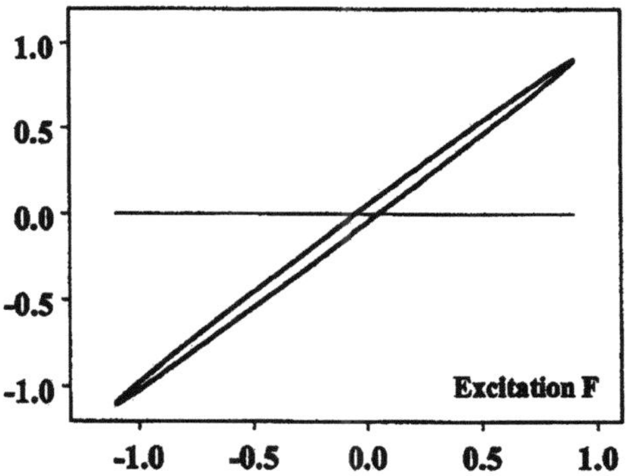

Figure 2.

generated; the amplitude of b is negligibly small for all time. The response curve for a is not a straight line because of the lag caused by the variation.

Figure 3 has the same modulation frequency of $\epsilon = 0.05$ as in Figure 2, but with a noise amplitude of $\sigma = \exp(-15)$, still a very small value. Now a large roll mode is excited; and the randomness of the jump in amplitude for each cycle is clearly seen.

Figure 4 has the same noise level as in Figure 3, but the frequency has been increased to $\epsilon = 0.07$. There is only a low amplitude roll mode response. A further increase to $\epsilon = 0.08$ effectively eliminates the roll mode completely.

The competing effects of stabilization by increasing the frequency of variation and destabilization by increasing the noise level can be seen clearly. It is also of interest to note that if the noise term is applied instead to (1) rather than (2), even noise levels of amplitude $\sigma = 1$ generate of course a noisy amplitude of pitch mode a, but no roll amplitude b. The noise-model itself is also of importance.

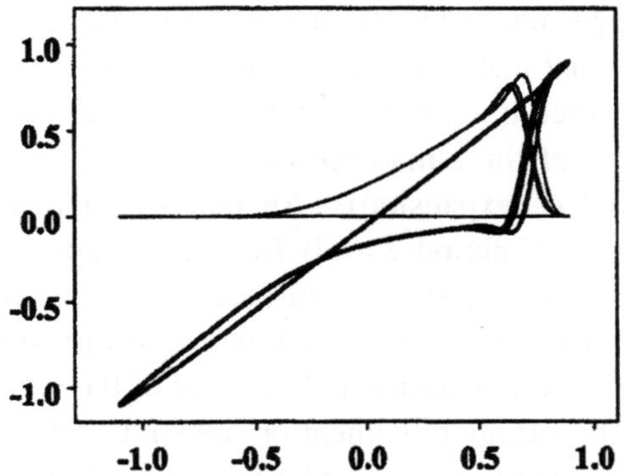

Figure 3.

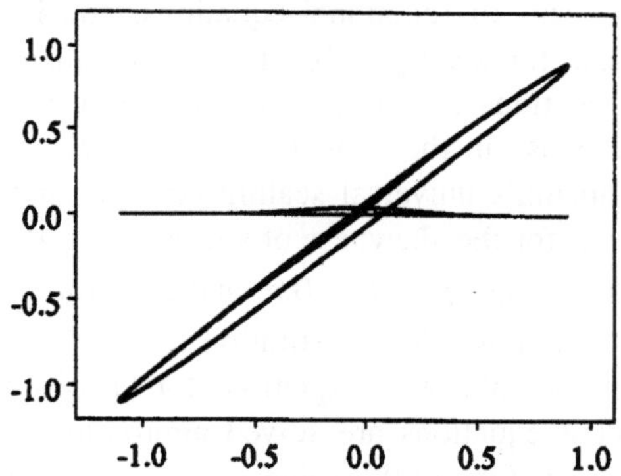

Figure 4.

3. Period-doubling of a Map

Davies and Rangavajhula [3] studied a unimodal map with slow parametric variation and additive noise. They again showed the competing effects of rate of variation and noise level. They used the logistic map as an example to compare the analysis with numerical simulation. Here we use instead the related quadratic map:

$$y_{n+1} \quad = \quad y_n^2 - y_n - x_n + \sigma f_n \tag{5}$$

with either slow linear or sinusoidal variation (variation of $O(\epsilon)$ where $\epsilon \lll 1$) of the parameter x. The stationary map with no noise has a bifurcation from period-1 to period-2 at the point $(y, x) = (0, 0)$. The stationary period-2 response has the simple form $y = \pm x^{\frac{1}{2}}$, stable in the region $0 < x < 0.5$. Again a noise term has been added; f_n is a Gaussian random variable with unit variance and zero mean, and the amplitude of the noise $\sigma \ll 1$.

Davies and Rangavajhula [3] discussed the period-1 to period-2 bifurcation. They showed that matched asymptotic expansions in terms of the small parameter ϵ can be used to describe the nonstationary transition from period-1 to period-2. A set of three matched expansions is needed to describe this transition (a triple deck expansion). An inner-inner expansion shows how noise causes repulsion of period-2 away from the locally unstable period-1 response in the region of x above the bifurcation. This matches in an overlap region to an inner expansion, which again in turn matches in the standard [8] manner to the outer solution. The order of the various expansions depends on whether the variation is linear or sinusoidal. We now show how this analysis can be extended to higher order period-doublings using a renormalization scheme to suitably scale the estimates.

Feigenbaum [4] described a universal renormalization for unimodal maps. His analysis is in terms of functional equations, and is not easy to apply. We look here instead for a simple algebraic renormalization based on comparing the exact equations for the period-4 response with a scaled version of the period-2 response in the form $(y + \beta)/\alpha$, where in the limit, $\alpha \approx -2.5029$ is Feigenbaum's universal scaling constant and β is a very small correction to account for the skewness of successive period-doubling curves. Write $z = y - (0.5)^{\frac{1}{2}}$ and $\eta = x - 0.5$, and look for (z, η) values that describe the period-4 window. The substitution $z = (x^{\frac{1}{2}} + \beta)/\alpha$ into the iterate of (5) yields two simultaneous equations for (x/α^2) and (β/α) in terms of η. These implicit equations are solved approximately by series expansions up to third order. Finally the value of α is obtained by forcing the solution to give the known values of the first few bifurcation points. The result is an explicit solution in the form:

$$\begin{aligned} \alpha &= C_0 + C_{11}x + C_{12}x^2 + C_{13}x^3 + \cdots \\ \beta &= C_{21}x + C_{22}x^2 + C_{23}x^3 + \cdots \\ \eta &= C_{31}x + C_{32}x^2 + C_{33}x^3 + \cdots \end{aligned} \tag{6}$$

The value $z = (x^{\frac{1}{2}} + \beta)/\alpha$ plotted against $(\eta + 0.5)$ for $0 < x < 0.5$ gives the period-4 window with range $0.5 < x < 0.61\cdots$. This solution can then be used iteratively for a succession of period doublings. The result is shown in Figure 5; these of course are the values for the stationary case.

Note that the amplitudes are plotted against $-\ln(x \inf - x)$, where $x \inf \approx 0.651156$ is the accumulation point, the value of x where chaotic response begins. On this scale, each period-doubling window has the same 'length'. Note also that only the principal periodic responses that scale successively as $1/\alpha$ are plotted; these are the regions closest to the amplitude $y = 0.5$, the extremum of the stationary version of the map (5).

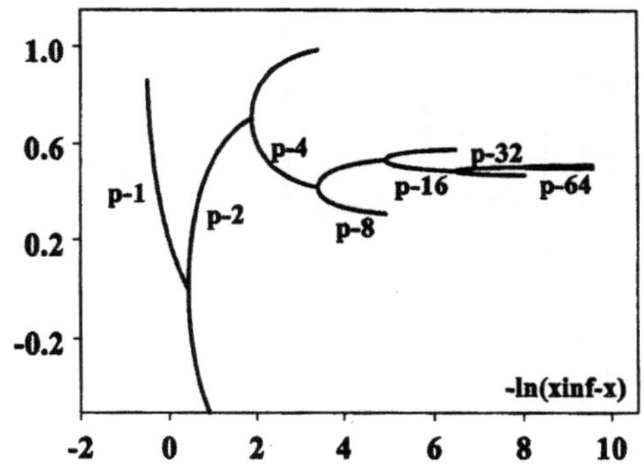

Figure 5.

Next consider the nonstationary response with added noise. Davies and Rangavajhula [3] showed how matched asymptotic expansions can be used to describe the response. The inner-inner expansion, the innermost of a nested set of three expansions, describes the influence of noise on the repulsion of a period-2 response from the locally unstable period-1 response. For linear parameter variation it is given by

$$y^{ii} \quad = \quad y^{p-1} + y_{rms} \tag{7}$$

where

$$y^2_{rms} \quad = \quad (\sigma^2/4)(\pi/4\epsilon)^{\frac{1}{2}}(\exp(x^2/\epsilon)\text{erfc}((x^2/\epsilon)^{\frac{1}{2}}) \tag{8}$$

where erfc represents the complementary error function and y^{p-1} is the period-1 solution. The response generated by the noise has Gaussian distribution with variance y_{rms}. The estimate (8) can be used to determine where the period-doubling becomes evident, that is where the canard occurs. The noise term y_{rms} is negligibly small until the canard; then y_{rms} becomes $\approx O(1)$ when $x = (2\epsilon|ln\sigma|)^{\frac{1}{2}}$, a scaling that has been reported for many similar problems.

The nonstationary response for higher period-doublings can be obtained with suitable scaling, namely in the form: amplitude as $1/\alpha$, x-scaling as $1/\delta$, where $\delta \approx 4.66$ is the Feigenbaum universal constant associated with the values of x at which period-doubling occurs (values of which can also be found from the cubic solution, above), linear rate ϵ as (2δ), and noise as k, where $k \approx 7$ is a value obtained during the comparison of the exact period-4 and scaled period-2 that was described earlier. Feigenbaum's universal constants are the limiting values at the accumulation point; the analysis just described yields slightly modified values for low periodic orbits.

In Figure 6, both data from a numerical simulation of a linear variation with $\epsilon = 0.0001$, $\sigma = \exp(-15)$ and appropriate values of y^{ii} from (8) are

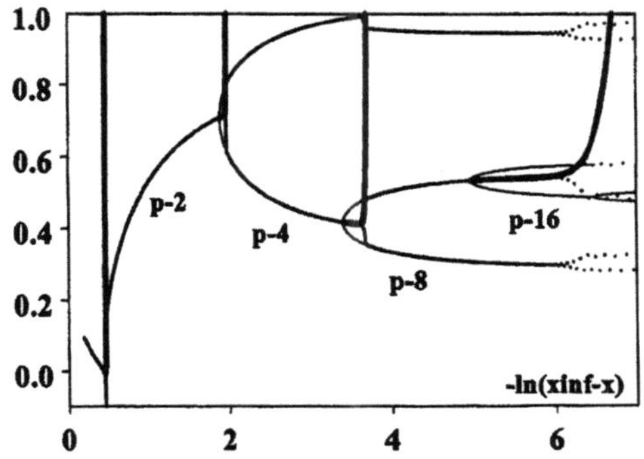

Figure 6.

superimposed on the renormalization data of Figure 5. The fine lines are the stationary renormalization results; the dots are the nonstationary simulation, and the heavy lines are values of y^{ii}. y^{ii} shows where the jump occurs within each period-doubling window. Note that the simple renormalization scheme is very accurate. The nonstationary simulation is $O(\epsilon)$ away from the stationary values except near bifurcation points. Note also that y^{ii} indeed gives a good estimate of where the jump occurs. The value of ϵ is kept constant throughout the sweep; the effective ϵ in the calculation of y^{ii} for each period must be scaled by 2δ. The noise (with κ) and amplitude (with α) scale as before. The jump for higher periods is calculated using the scaled values of ϵ and σ in the y^{ii} expression for period-1 to period-2. The relative position of the jump to period-doubling is seen to occur later in each successive window, until the effective ϵ is too high for the subsequent period-doubling to show up. The scaling indicates that at the nth period-doubling the jump occurs at the effective value (as it would be calculated within the period-2 window) given by $x = (2\epsilon(2\delta)^n|ln\kappa^n\sigma|)^{\frac{1}{2}}$; the nth period-doubling window only appears during the nonstationary sweep if this quantity is less than 0.5.

Figure 7 shows five separate sinusoidal variations. The ranges of variation are scaled using δ so that the bifurcation is at the same relative position within the window for each successive period-doubling. Now the frequency ϵ must be scaled just by the factor 2, with amplitude and noise scaling as before. In this case the response curves at each successive period-doubling are similar, with successive scaling just of the amplitude by the factor α. An appropriate change is made at each stage to the y^{p-1} term in (7) using the cubic (6).

4. Conclusion

Two examples have been given showing the critical effect of very low level noise in triggering the onset of a bifurcated form of response. The discus-

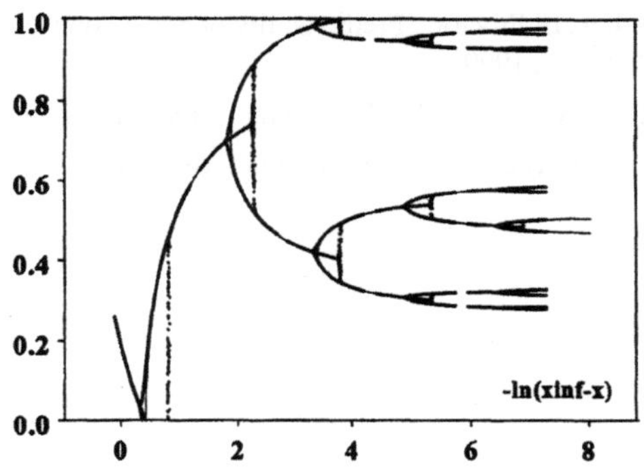

Figure 7.

sion applies to bifurcations where the initial form of the response persists beyond the bifurcation into a parameter region where it is locally unstable. Examples of transcritical and period-doubling bifurcations have been discussed. The analysis of a period-doubling bifurcation of a discrete map has been extended to higher order doublings using a simple renormalization scheme. The values and the appropriate scaling obtained from the scheme compare very well with numerical simulations. The scaled estimate of the inner-inner expansion (y^{ii}) obtained from previous work gives a good estimate of the canard from one period to another.

The important general conclusion to be noted is that even what might be thought of as negligibly low noise levels in a system may in fact be critical in determining how the system responds during slow passage through bifurcations.

Acknowledgment

This work was supported by the Natural Sciences and Engineering Research Council of Canada.

References

[1] J. C. Celet, *et al.* "Slowly passing through resonance strongly depends on noise," *Phys. Rev. Lett.*, vol. 81, pp. 975–978, 1999.

[2] H. G. Davies and R. Pan, "Nonstationary bifurcations about a simple bifurcation," *J. Sound Vibn.*, vol. 172, pp. 155–170, 1994.

[3] H. G. Davies and K. Rangavajhula, "A period-doubling bifurcation with slow parametric variation and additive noise," *Proc. Roy. Soc. Lond. A*, vol. 457, pp. 2965–2982, 2001.

[4] M. J. Feigenbaum, "Quantitative universality for a class of nonlinear transformations," *J. Stat. Phys.*, vol. 19, pp. 25–52, 1978.

[5] G. D. Lythe and M. R. E. Proctor, "Noise and slow-fast dynamics in a three-wave resonance problem," *Phys. Rev. E*, vol. 47, pp. 3122–3127, 1993.

[6] G. D. Lythe and M. R. E. Proctor, "Predictability of noise-controlled dynamics," *Physica D*, vol. 133, pp 362–370, 1999.

[7] A. H. Nayfeh, "Plenary Lecture. ASME Biennial Vibration Conference. Pittsburg, 2001.

[8] M. van Dyke, M. "Perturbation Methods in Fluid Mechanics," Academic Press, 1964.

MULTI-SCALE ANALYSIS OF NOISE-SENSITIVITY NEAR A BIFURCATION

R. Kuske
University of British Columbia
#121-1984 Mathematics Road
Vancouver, BC V6T 1Z2
rachel@math.ubc.ca

Abstract We develop a multi-scale analysis for stochastic differential equations. Such models are particularly sensitive to noise when the system is near a critical point, such as a Hopf bifurcation, which marks a transition to oscillatory behavior. In particular, we are interested in the case when the combined effects of the noise and the bifurcation amplify oscillations which would decay in the deterministic system. The derivation of reduced equations for the envelope of the oscillations provides an efficient analysis of the dynamics by separating the influence of the noise from the intrinsic oscillations over long time scales.

Keywords: Stochastic, bifurcation, Duffing-van der Pol, multiple scales, amplitude equations

1. Introduction

In this paper we give an asymptotic analysis of the effect of noise near a Hopf bifurcation for the stochastic Duffing-van der Pol equation. In particular, we are interested in the case where the sensitivity to the noise is demonstrated through a resonance. In the presence of noise, oscillations are sustained in the subcritical region, where oscillations die out in the absence of noise, as shown in Figure 1. In the absence of external periodic forcing this phenomenon has been called autonomous stochastic resonance [6], where the noise excites the oscillations intrinsic to the deterministic dynamics. It has been observed in many systems with delays, including models of neurons, lasers, and a variety of oscillators with delayed feedback (see [7]-[4], and references therein). It has also been observed in systems without delays, where it has been studied in the context of stochastic bifurcations [15].

N. Sri Namachchivaya and Y.K. Lin (eds.),
IUTAM Symposium on Nonlinear Stochastic Dynamics, 147-156.

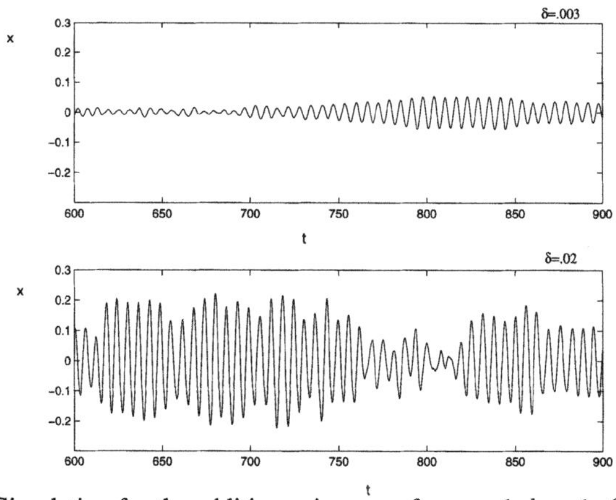

Figure 1. Simulation for the additive noise case, for $a = 1$, $b = 1$, $\beta = -\epsilon^2$, $\epsilon = 0.07$. In the top figure $\delta = 0.003$, and the bottom figure $\delta = 0.02$. Even though ϵ is the same for both, the bottom figure has larger amplitude oscillations.

In this paper we consider a canonical example of a system, the Duffing-van der Pol equation, which has been studied both in the context of deterministic and stochastic cases. We will consider both additive and multiplicative noise cases. Using multi-scale analysis, we derive envelope equations for the stochastic amplitude of the oscillations. We see that this approach leads to results similar to those in [2]- [1]. The multiscale approach provides a new viewpoint of the dynamics in which the oscillations have the deterministic frequency associated with the Hopf bifurcation and are modulated by a slowly varying stochastic amplitude. It also has an attractive physical interpretation which is directly related to resonance with Fourier-type components.

2. Multi-scale analysis

We begin with the Duffing-van der Pol equation with additive noise, which we write as a system of stochastic differential equations.

$$dx = y\,dt \tag{1}$$

$$dy = \left[-\omega^2 x + \beta y - a x^3 - b x^2 y\right] dt + \delta dw. \tag{2}$$

For comparison, we also consider multiplicative noise, replacing (2) with

$$dy = \left[-\omega^2 x + \beta y - a x^3 - b x^2 y\right] dt + \delta x dw. \tag{3}$$

We are interested in the case where β is near zero, since $\beta = 0$ corresponds to a Hopf bifurcation in the absence of noise. That is, for $\beta < 0$ oscillations decay over time and for $\beta > 0$ there is a stable oscillatory behavior. We will explore the noise-sensitivity of the system for small values of β, restricting our analysis to $\delta \ll 1$ in order to understand where small noise can play a significant role in the dynamics. We will obtain relevant scalings related to this noise-sensitivity. In order to focus on these parameters, we set all other parameters to unity, $\omega = a = b = 1$.

It is well known that in the absence of noise ($\delta = 0$), one can give an asymptotic approximation to the solution when the parameters correspond to close proximity to the Hopf bifurcation. The method of multiple scales is useful for giving a description in which the natural mode of oscillation associated with the bifurcation appears explicitly in the approximation [8]. The form is

$$x \sim A(T)\cos\omega t + B(T)\sin\omega t, \quad T = \epsilon^2 t, \tag{4}$$

where ϵ^2 is the parameter measuring the proximity to the bifurcation. Here $A(T)$ and $B(T)$ are functions of a slow time T, which are treated as constants with respect to the fast oscillations with frequency ω on the t time scale. By deriving equations on the slow T scale for the amplitude, or envelope, described by $A(T)$ and $B(T)$, one gets an asymptotic approximation for the process near the Hopf bifurcation, that is, for $\epsilon^2 \ll 1$. In the remainder of this paper, we derive such envelope or amplitude equations in the stochastic case $\delta \neq 0$.

2.1 Additive noise: $\beta < 0$

We begin by considering the subcritical case with additive noise (2). In the absence of noise, oscillations decay exponentially in time. In the case of additive noise, the noise causes a resonance-type effect in exciting the natural modes of the system, even when the noise is small (See Figure 1). This is particularly evident when $|\beta| \ll 1$, that is, β takes values near the Hopf bifurcation point of the deterministic system.

We borrow techniques from the derivation of envelope equations via multiscale analysis, used in studying the bifurcations of patterns near critical [8]-[12]. The analysis starts with the identification of a slow time scale $T = \epsilon^2 t$, which is related to the proximity of β to zero, $\beta = \epsilon^2\beta_2$, for $\beta_2 = O(1)$ with respect to ϵ. Note that $\beta_2 < 0$ in the subcritical case. Then we look for an asymptotic approximation to the solution in the form of a slowly varying amplitude, or envelope, and for the critical mode corresponding to the Hopf bifurcation. A standard linear analysis of the deterministic system shows that this critical mode has frequency ω. Therefore, we expect approximations of the form

$$\hat{x}(t) = \epsilon A(T)\cos\omega t + \epsilon B(T)\sin\omega t. \tag{5}$$

Here we have assumed that $A(T)$ and $B(T)$ are slowly varying functions of time, which is appropriate when $\epsilon \ll 1$ and $\delta \ll 1$. Below we also determine the relationship between δ and ϵ for the validity of the asymptotic approximation.

We derive equations for $A(T)$ and $B(T)$ of the form

$$dA = \psi_A dT + \sigma_A d\xi_1(T), \quad dB = \psi_B dT + sigma_B d\xi_2(T), \tag{6}$$

with $\xi_1(T)$ and $\xi_2(T)$ as independent standard Brownian motions.

In order to get the drift coefficients ψ_A, ψ_B and the diffusion coefficients σ_A, σ_B, we use two identities expressing $d\hat{x}$ and $d\hat{y}$ in terms of A and B. In the Appendix we use multi-scale analysis to obtain the leading order approximation for the coefficients,

$$\psi_A = \frac{1}{2}\beta_2 A + \frac{3}{8}aB(A^2 + B^2) - \frac{1}{8}bA(A^2 + B^2), \tag{7}$$

$$\psi_B = \frac{1}{2}\beta_2 B - \frac{3}{8}aA(A^2 + B^2) - \frac{1}{8}bB(A^2 + B^2),$$

$$\sigma_A = -\frac{\delta}{2\epsilon^2}, \quad \sigma_B = \frac{\delta}{2\epsilon^2}. \tag{8}$$

Note that ψ_A and ψ_B are identical to the result of the multiscale analysis in the deterministic case.

From (8) we can conclude that the multiscale analysis is valid when the magnitude of the noise is $\delta = O(\epsilon^2)$ or smaller. For $\delta \gg \epsilon^2$, we see that the noise dominates the dynamics of the envelope. Then it is inappropriate to use the multi-scale approximation in which the envelope is a slowly varying stochastic process. This scaling relationship $\delta \sim \epsilon^2$ is not unexpected, since it appears in other analyses of the non-trivial behavior of this system near the Hopf bifurcation [2]- [1].

Using (6)-(8) we numerically simulate the solution for $A(T)$ and $B(T)$ for $\delta = O(\epsilon^2)$, and thus obtain the approximation $\hat{x}$ for $x(t)$. Since this approximation holds in an averaged or weak sense, we compare the results by comparing simulations which approximation the probability density, as shown in Figure 2. Note that the simulation of the multi-scale approximation is approximately $O(1/\epsilon^2)$ faster than the simulation of the original model (2), since the approximation (5) requires the simulation of (6) on the T time scale rather than the t time scale.

2.2 Multiplicative noise: $\beta < 0$

The procedure for approximating x in (3) for this case is similar to that described in the previous section. We again use the form (5) for $\hat{x}$.

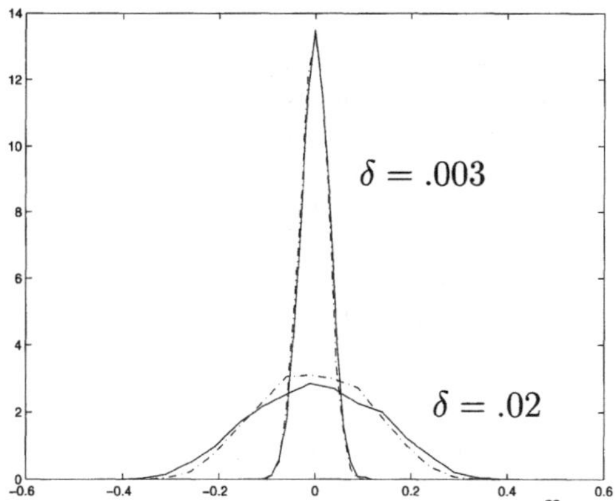

Figure 2. The probability density function $p(x)$ and $p(\hat{x})$, obtained from 5000 numerical simulations of the full system (2) (dash-dotted lines) and the multi-scale approximation (solid line) for $a = 1$, $b = 1$, $\beta = -\epsilon^2$, $\epsilon = 0.07$ for large t. Here t is chosen sufficiently large in order to approximate the invariant density for additive noise. The two sets of curves correspond to two different noise levels, $\delta = 0.003$ and $\delta = 0.02$, giving more and less concentrated densities, respectively.

As in Baxendale [2]- [3], we assume the form of equations for A and B

$$\begin{pmatrix} dA \\ dB \end{pmatrix} = \begin{pmatrix} \psi_A \\ \psi_B \end{pmatrix} dT + \sum_{j=1}^{3} \Sigma_j \begin{pmatrix} A \\ B \end{pmatrix} d\mathcal{W}_j(T). \tag{9}$$

Here the drift coefficients ψ_A and ψ_B are the same as in the additive noise case, but the noise is in terms of three independent Brownian motions $\mathcal{W}_j$. In [3] it is shown that the averaged equations for the stochastic van der Pol-Duffing equation have this form, in particular for describing the two-point motion and the associated Lyapunov exponent. The constant matrices Σ_j follow from a consistency comparison of the infinitesimal generators for the original and averaged processes.

$$\Sigma_0 = \frac{c_0}{2}\begin{bmatrix} 0 & 1 \\ -1 & \end{bmatrix} \quad \Sigma_1 = \frac{c_1}{2\sqrt{2}}\begin{bmatrix} 1 & 0 \\ 0 & -1 \end{bmatrix} \quad \Sigma_2 = \frac{c_2}{2\sqrt{2}}\begin{bmatrix} 0 & 1 \\ 1 & 0 \end{bmatrix}. \tag{10}$$

Here c_j depends on the small parameters δ and ϵ.

Our multi-scale analysis gives the same form for the noise terms in the stochastic amplitude equations for the delay equation with multiplicative noise. This can be seen by writing the noise as a type of Fourier series representation,

$$\delta S x(t) dw = \delta \mathcal{M}_0 \overline{x} dw_0 + \delta \mathcal{M}_1 \overline{x} \cos 2\omega t dw_1 + \delta \mathcal{M}_2 \overline{x} \sin 2\omega t dw_2. \tag{11}$$

where

$$S = \begin{bmatrix} 0 & 0 \\ 1 & 0 \end{bmatrix}, \quad \overline{x} = (x\ y)^T, \quad \mathcal{M}_j = \begin{bmatrix} m_{j1} & m_{j2} \\ m_{j3} & m_{j4} \end{bmatrix}, \tag{12}$$

w_j are independent Brownian motions, and $\mathcal{M}_j$ are constant matrices that must be determined. Note that we could include other modes in (11), such as $\mathcal{M}\overline{x}\sin k\omega t$ for $k > 2$. However, under the multi-scale projection onto $\cos\omega t$ and $\sin\omega t$ as shown in the Appendix (10), these terms do not contribute to the stochastic amplitude equations. In addition, there are certain restrictions on the coefficients m_{jk}, as discussed below. This forces us to express the term $Sx(t)dw$ in more than one term of such a Fourier representation. Here we retain all terms which would possibly contribute to the leading order behavior of the envelope, due to a resonance with the dominant mode $\cos\omega t, \ \sin\omega t$.

As in the case of additive noise, we write $w_j(t) = \mathcal{W}_j(T)/\epsilon$ and use the multi-scale projection (10) onto $\cos\omega t$ and $\sin\omega t$, treating $\mathcal{W}_j(T)$ as independent of t. This gives m_{jk} in terms of the entries for Σ_j.

$$m_{01} + m_{04} = 0, \quad m_{12} - m_{13} = c_0\frac{\epsilon}{\delta} \tag{13}$$

$$m_{j1} - m_{j4} = \frac{c_j\epsilon\sqrt{2}}{\delta}, \quad m_{j2} + m_{j3} = 0, \quad j = 1, 2 \tag{14}$$

Note that there is not a unique solution to this system, so we have some freedom in choosing the matrices $\mathcal{M}_j$. For example, we can take $m_{03} = -1$, $m_{14} = m_{24} = -\sqrt{2}$, and all other $m_{jk} = 0$, so that $c_0 = c_1 = c_2 = \delta/\epsilon$ agrees with Σ_j [3].

For the subcritical case with multiplicative noise, the deterministic decay plays a role in both the deterministic and stochastic aspects of the dynamics, and thus the noise does not dominate the behavior as in the case of additive noise. In Figure 3 we compare the time dependent

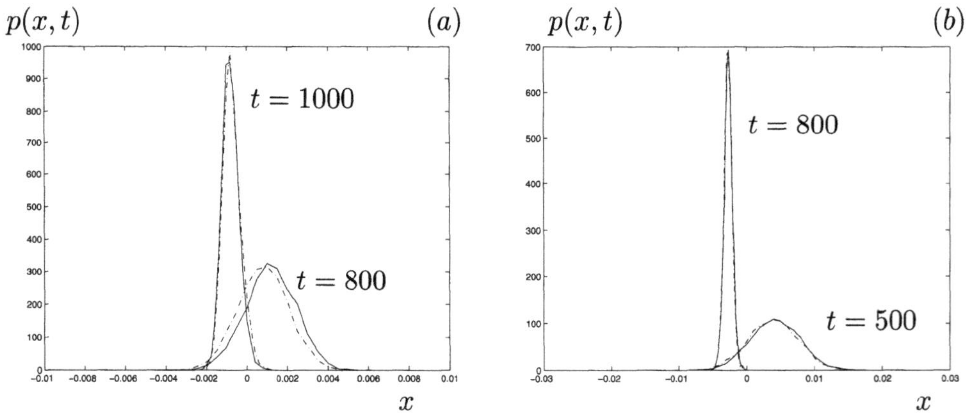

Figure 3. The probability density function $p(x,t)$ and $p(\hat{x},t)$, obtained from the simulation of the full system (3) (dash-dotted lines) and the multi-scale approximation $\hat{x}$ (solid line) for $a = 1$, $b = 1$, $\beta = -\epsilon^2$, $\epsilon = .1$, and $\delta = .02$. In (a) the results are graphed for two different times, $t = 800$ and $t = 1000$ and initial condition, $x(0) = .5, y(0) = 0$.

probability density function for the original and multi-scale systems, observing how the shape of the density changes over time. In [2]- [1] the scaling $\delta = \epsilon$ is also used to obtain averaged equations of the same form as the envelope equations for A and B, and they show that an explicit expression for the probability density can be obtained for the averaged system. In [1], only two terms are used to represent the noise in the averaged system.

3. Supercritical Case: $\beta > 0$

For the supercritical case, we take $\beta = \epsilon^2\beta_2$ with $\beta_2 > 0$. In the absence of noise, oscillations are stable, which are approximated by

$$x \sim \epsilon\rho\cos((1+\epsilon^2\omega_2)t), \quad \rho^2 = 4\frac{\beta_2}{b}, \quad \omega_2 = \frac{3}{8}a\rho^2. \tag{15}$$

Then the multiscale approximation has the form

$$\hat{x} = \epsilon(\rho + A(T))\cos((1+\epsilon^2\omega_2)t) + \epsilon B(T)\sin((1+\epsilon^2\omega_2)t). \tag{16}$$

We restrict our attention to the case of additive noise, noting that the multiplicative noise can be treated in a similar way.

Using the same multi-scale method to obtain the multi-scale approximation as in Section 2.1, we find that $A(T)$ and $B(T)$ in (16) follow the stochastic system of the form (6). In this case, the coefficients are

$$\begin{aligned}
\psi_A &= -\beta_2 A - \omega_2 B + \frac{3}{8}a\rho^2 B + \frac{3}{4}a\rho AB - \frac{1}{8}b\rho\left(\frac{3}{4}A^2 + B^2\right) + \\
&\quad \frac{3}{8}aB(A^2+B^2) - \frac{1}{8}bA(A^2+B^2), \\
\psi_B &= \omega_2 A - \frac{9}{8}a\rho^2 A - \frac{1}{4}b\rho AB - \frac{3}{8}a\rho\left(\frac{3}{4}A^2 + B^2\right) - \\
&\quad \frac{3}{8}aA(A^2+B^2) - \frac{1}{8}bB(A^2+B^2).
\end{aligned} \tag{17}$$

and σ_A and σ_B are given by (8). Once again ψ_A and ψ_B are identical to the result of the multiscale analysis in the deterministic case. In Figure 4 we compare the probability densities

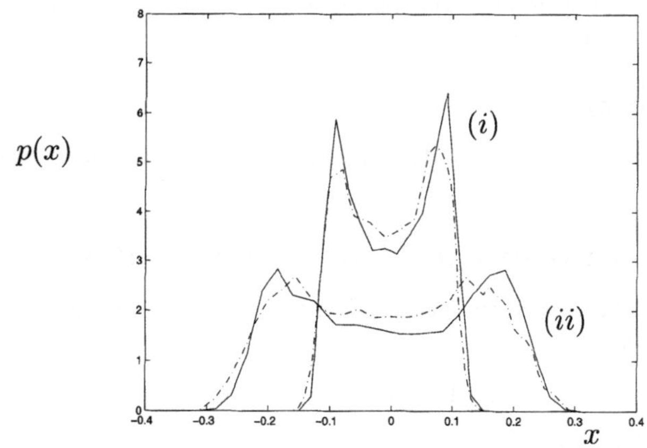

Figure 4. The probability density function $p(x)$ and $p(\hat{x})$, obtained from the numerical simulations (24,000 realizations) of the full system (dash-dotted lines) and the multi-scale approximation (16) (solid line) for $a = 1$, $b = 1$, $\beta = \epsilon^2$, for two cases: (i) $\epsilon = .05$ and $\delta = .002$ and (ii) $\epsilon = .1$ and $\delta = .01$. Here t is chosen sufficiently large so that it is anticipated that $p(x)$ is the invariant density for the case when the noise is additive. Note that a very large number of simulations is needed to completely resolve the density.

obtained from simulation of the original variable x, and the multi-scale approximation using (16) for $\hat{x}$ with (6), (8), and (17).

4. Conclusion

We apply a multiscale analysis to analyze the effect of noise on the canonical model of the stochastic Duffing-van der Pol equation. This model is particularly sensitive to noise when the delay is near a critical value, corresponding to a Hopf bifurcation. In the subcritical case, oscillations which would decay in the absence of noise are amplified by the interaction of the noise and the bifurcation. We derive stochastic equations for the slow evolution of the envelope of these oscillations at the Hopf frequency. The analysis leads to a description which separates the deterministic and stochastic elements of the dynamics. Advantages of this approximation include efficiency of simulations over large times and relatively simple dynamics described by the envelope equations. The analysis also leads to a natural scaling between the magnitude of the noise δ and the proximity to the bifurcation ϵ^2. Note it is not necessary to assume this scaling relationship at the beginning of the analysis; it appears naturally as a result of the analysis, from which it is clear that the analysis is valid for $\delta = O(\epsilon^2)$ for additive noise, and $\delta = O(\epsilon)$ for multiplicative noise.

The method has also been applied to stochastic systems with delay: a linear system and the logistic model, both with delay and additive small noise [11], and a canonical model with delayed negative feedback [10]. For the linear system the approximation is compared with a result obtained from the Fourier analysis. For the logistic model both the sub- and super-critical cases are considered; that is, the analysis is applied for delays above and below the critical delay. The method is particularly useful for systems with delays since it avoids complications due to the memory in the system, which can make other analyses impractical.

The phenomenon demonstrated here occurs in many applications with similar features, including neuronal models, optics, and information systems [7]- [4]. The analysis exploits the interaction of the noise with the Hopf bifurcation, features which are common to these other models. Thus the analysis can be applied to a wide variety of stochastic models, in order to understand the effects of noise near a bifurcation.

Acknowledgments

The author wishes to thank Prof. Salah Mohammed for encouragements and useful discussions for this work.

Appendix: Details of the Multi-scale Analysis

We start with (2) and the ansatz for the multi-scale approximation (5)-(6). We also need the expression for $\hat{y}(t)$ which is

$$\hat{y}(t) = -\epsilon\omega\left[A(T)\sin\omega t - B(T)\cos\omega t\right]. \tag{1}$$

First, from Itô's formula, we have

$$\begin{aligned} d\hat{x} &= \frac{\partial \hat{x}}{\partial t}dt + \frac{d\hat{x}}{dA}dA + \frac{d\hat{x}}{dB}dB && (2)\\ &= \hat{y}dt + \epsilon\cos\omega t(\psi_A dT + \sigma_A d\xi_1) + \epsilon\sin\omega t(\psi_B dT + \sigma_B d\xi_2) \\ d\hat{y} &= \frac{\partial \hat{y}}{\partial t}dt + \frac{d\hat{y}}{dA}dA + \frac{d\hat{y}}{dB}dB && (3)\\ &= \left[-\omega^2\hat{x}\right]dt - \epsilon\omega\sin\omega t(\psi_A dT + \sigma_A d\xi_1) + \epsilon\omega\cos\omega t(\psi_B dT + \sigma_B d\xi_2). \end{aligned}$$

Here we have left out terms such as $\frac{d^2\hat{x}}{dA^2}$ and $\frac{d^2\hat{y}}{dB^2}$, since these quantities vanish, according to (5) and (1). The second set of expressions for $d\hat{x}$, $d\hat{y}$ comes from substituting (5) and (1) into (2),

$$\begin{aligned} d\hat{x} &= [-\epsilon(A(T)\sin\omega t + B(T)\cos\omega t)]dt \quad (4) \\ d\hat{y} &= \epsilon\left[-\omega^2(A(T)\cos\omega t + B(T)\sin\omega t) - \epsilon^2\omega\beta_2(A(T)\sin\omega t - B(T)\cos\omega t)\right. \\ &\quad -a\epsilon^2(A(T)\cos\omega t + B(T)\sin\omega t)^3 \quad (5) \\ &\quad \left.-b\omega\epsilon^2(A(T)\cos\omega t + B(T)\sin\omega t)^2(-A(T)\sin\omega t + B(T)\cos\omega t)\right] dt + \delta dw. \end{aligned}$$

Now we set (4) and (5) equal to (2) and (3), and collect coefficients of like powers of ϵ, noting the $dT = \epsilon^2 dt$. We find that the $O(\epsilon)$ terms cancel. Then the next terms are $O(\epsilon^3)$, and for convenience we write them in terms of the slow time T.

$$(\cos\omega t\,\psi_A + \sin\omega t\,\psi_B)dT + \cos\omega t\,\sigma_A\,d\xi_1 + \sin\omega t\,\sigma_B\,d\xi_2 = 0 \quad (6)$$

$$\begin{aligned} \epsilon(-\sin\omega t\,\psi_A + \cos\omega t\,\psi_B)dT + \epsilon(-\sin\omega t\,\sigma_A\,d\xi_1 + \cos\omega t\,\sigma_B\,d\xi_2) &= \\ \epsilon[-\beta_2(A(T)\sin\omega t - B(T)\cos\omega t) - a(A(T)\cos\omega t + B(T)\sin\omega t)^3 & \\ -b(A(T)\cos\omega t + B(T)\sin\omega t)^2(-A(T)\sin\omega t + B(T)\cos\omega t)]dT + \delta dw. & \quad (7) \end{aligned}$$

Here we have used $\omega = 1$ in the coefficients. Then we equate the drift and diffusion terms. For the diffusion terms we have

$$\begin{aligned} \cos\omega t\,\sigma_A\,d\xi_1 + \sin\omega t\,\sigma_B\,d\xi_2 &= 0 \quad (8) \\ \epsilon(-\sin\omega t\,\sigma_A\,d\xi_1 + \cos\omega t\,\sigma_B\,d\xi_2) &= \delta dw \quad (9) \\ &= \frac{\delta}{\epsilon}\left[\cos\omega t\,dw_1(T) + \sin\omega t\,dw_2(T)\right] \end{aligned}$$

Here we have used well-known identities, first to express dw in terms of $\sin\omega t$, $\cos\omega t$ and two independent Brownian motions w_2 and w_1, and second to write $w_2(t)$ and $w_1(t)$ as functions of the slow time T.

At this stage we employ the multi-scale assumption, that is, that the fast time scale t and the slow time scale T are independent. This is a useful approximation when $\epsilon \ll 1$. In order to obtain σ_A and σ_B, we eliminate the fast oscillations in t, integrating over one period $2\pi/\omega$, using the orthogonality of $\sin\omega t$ and $\cos\omega t$, and treating t and T as independent. For example,

$$\int_0^{2\pi/\omega} [\epsilon\cos\omega t\cdot(8) - \epsilon\sin\omega t\cdot(9)]\,dt \quad \Rightarrow \sigma_A = -\frac{\delta}{2\epsilon^2}. \quad (10)$$

Similarly, $\sigma_B = \frac{\delta}{2\epsilon^2}$.

We use the same method to solve for the drift coefficients, substituting the drift terms from (6) and (7) for (8) and (9), respectively, into (10). Again treating functions of T as independent from t yields (7).

References

[1] L. Arnold, N. Sri Namachchivaya, and K. Schenk-Hoppe, "Toward an understanding of the stochastic Hopf bifurcation: a case study," *Int. J. Bif. Chaos*, vol. 6, pp. 1947–1975, 1996.

[2] P. Baxendale,"Lyapunov exponents and stability for the stochastic Duffing-van der Pol oscillator," this proceedings.

[3] P. Baxendale,"Stochastic averaging and asymptotic behavior of the stochastic Duffing-van der Pol equation, in review.

[4] A. Beuter, J. Belair, and C. Labrie, "Feedback and delays in neurological diseases: a modeling study using dynamical systems," *Bull. Math. Bio.*, vol. 55, pp. 525–541, 1993.

[5] J. L. Cabrera, J. Gorronogoitia, and F. J. de la Rubia, "Noise-correlation-time-mediated localization in random nonlinear dynamical systems," *Phys. Rev. Lett.*, vol. 82, pp. 2816–2819, 1999.

[6] H. Gang, T. Ditzinger, C. Z. Ning, and H. Haken, *Phys. Rev. Lett.*, vol. 71, p. 807, 1993.

[7] J. Garcia-Ojalvo and R. Roy, "Noise amplification in a stochastic Ikeda model," *Phys. Lett. A*, vol. 224, pp. 51–56, 1996.

[8] J. Kevorkian and J. D. Cole, *Multiple Scale and Singular Perturbation Methods*, Springer-Verlag, New York, 1996.

[9] S. Kim, S. H. Park, and H.-B. Pyo, "Stochastic resonance in coupled oscillator systems with time delay," *Phys. Rev. Lett.*, vol. 82, pp. 1620–1623, 1999.

[10] R. Kuske, "Asymptotic analysis of noise-amplified oscillations for subcritical delays," *Differential Equations and Dynamical Systems*, to appear.

[11] R. Kuske and M. Malgorzata, "Multi-scale analysis of stochastic delay differential equations," preprint.

[12] P. Manneville, *Dissipative Structures and Weak Turbulence*, 1990, Academic Press, San Diego, 1990.

[13] N. Sri Namachchivaya, R. Sowers, and L. Vedula, "Nonstandard Reduction of Noisy Duffing-van der Pol Equation," *J. Dyn. Sys.*, vol. 16, pp. 223–245, 2001.

[14] T. Ohira and Y. Sato, "Resonance with Noise and Delay," *Phys. Rev. Lett.*, vol. 82, pp. 2811–2815, 1999.

[15] *Stochastic Dynamics*, H. Crauel and M. Gundlach, eds. Springer-Verlag, New York, 1999.

HOPF MEETS HAMILTON UNDER WHITNEY'S UMBRELLA

William F. Langford
Department of Mathematics and Statistics
University of Guelph
Guelph ON Canada N1G 2W1 *
wlangfor@uoguelph.ca

Abstract In Hamiltonian mechanics, the classical Hopf bifurcation theorem is not directly applicable. Instead, there is an analogous "Hamiltonian-Hopf bifurcation theorem" in which *two pairs* of complex conjugate eigenvalues approach the imaginary axis symmetrically from the left and right, then merge in double purely imaginary eigenvalues and separate along the imaginary axis (or the reverse). This phenomenon has codimension one within the class of Hamiltonian systems. In the general case of non-Hamiltonian vector fields, the occurrence of double imaginary eigenvalues has codimension three. This paper presents a first investigation of the interface between these two cases. They meet in an interesting topological singularity known as Whitney's umbrella. We show that the Hamiltonian case lies on the "handle" of Whitney's umbrella. This allows us to investigate near-Hamiltonian or weakly dissipative systems that lie in a tubular neighborhood of the handle of Whitney's umbrella.

Keywords: Hopf bifurcation, Hamiltonian mechanics, Whitney umbrella, versal deformation, weak dissipation.

1. Introduction

In [8], W. R. Hamilton introduced the basic ideas of what we now call *Hamiltonian mechanics*. He postulated that in mechanics the action is stationary under small variations in the motion. Expressing a mechanical system in terms of coordinates q and momenta p, and the total energy as a function $H(q,p)$, the evolution of the system in time t satisfies

$$\frac{dH(q(t),p(t))}{dt} = 0 \quad \forall\, t, \tag{1}$$

*Research supported in part by the Natural Sciences and Engineering Research Council of Canada.

N. Sri Namachchivaya and Y.K. Lin (eds.),
IUTAM Symposium on Nonlinear Stochastic Dynamics, 157-165.

that is, $H(q,p)$ is stationary along a trajectory. Any trajectory $(q(t),p(t))$ is a solution of the system of differential equations

$$\frac{dq}{dt} = \frac{\partial H}{\partial p}(q,p), \quad \frac{dp}{dt} = -\frac{\partial H}{\partial q}(q,p), \tag{2}$$

which implies (1). To this day, Hamiltonian mechanics remains fundamentally important in theoretical and applied mechanics. However, Hamilton's assumptions are an idealization, often not completely satisfied by "real world" mechanical systems. Hamiltonian systems are both *conservative* and *deterministic*, while most real systems involve both *dissipation* of energy and *stochastic* effects. If the dissipation and stochasticity are very small, then the Hamiltonian model may give a sufficiently accurate description of the system. However, often this is not the case. Several papers in this volume address the issue of stochastic perturbations of Hamiltonian systems. This paper is concerned with the effects of weak dissipation on a particular class of Hamiltonian systems, those undergoing the *Hamiltonian–Hopf bifurcation.*

2. Classical Hopf Bifurcation Theorem

This section presents a brief summary of the classical Hopf bifurcation theorem, see [9] or [6]. The Hopf bifurcation theorem describes the generic mechanism by which a periodic orbit is created or disappears, in a small neighborhood of an equilibrium point. Let us consider a parametrized family of differential equations

$$\dot{y} = f(y,\mu), \quad f(0,\mu) = 0, \tag{3}$$

where $f : \mathbb{R}^n \times \mathbb{R} \to \mathbb{R}^n$ is a sufficiently smooth ($f \in C^k, k \geq 3$) function of the state variables y and a real parameter μ. We assume the existence of an equilibrium point, translated to the origin for convenience; and at this equilibrium assume that the derivative with respect to y

$$A = D_y f|_{(0,0)} \tag{4}$$

has a simple pair of imaginary eigenvalues $\pm i\omega_0$, and no others with zero real part. Then A is nonsingular, and the implicit function theorem guarantees that the equilibrium point of (3) has a unique smooth extension $y(\mu)$ for μ near 0. Without loss of generality, we assume that this branch of solutions has been translated to $y \equiv 0$. Now the derivative A has a smooth extension

$$A(\mu) = D_y f|_{(0,\mu)}, \tag{5}$$

and the simplicity of the imaginary eigenvalues implies that they too have unique smooth extensions along the branch, for μ near 0,

$$\sigma(\mu) \pm i\omega(\mu), \quad \sigma(0) = 0, \quad \omega(0) = \omega_0. \tag{6}$$

Next we assume Hopf's transversality or crossing condition, with respect to the parameter μ

$$\frac{d\sigma}{d\mu}(0) \neq 0. \tag{7}$$

Under these assumptions, the main conclusion of Hopf's theorem is that there is a unique (up to phase shift) smooth branch of periodic solutions, in any small neighborhood of $(y, \mu) = (0, 0)$. This branch may be parametrized by the amplitude of the periodic solution, here denoted r, and furthermore there is a smooth relationship between the parameters μ and r of the form

$$r[c_0\mu + c_2 r^2 + c_4 r^4 + \cdots] = 0, \tag{8}$$

where $\cdots$ denotes higher order terms in r. The coefficients c_0, c_2, etc. are smooth functions of μ. Hopf [9] showed that (7) is equivalent to $c_0(0) \neq 0$, so that, by the Implicit Function Theorem, (8) can be solved uniquely for μ as a function of $r^2 > 0$ near the origin. He also presented a formula for the calculation of $c_2(0)$ in terms of Taylor coefficients of f. In the case $c_2(0) \neq 0$, the branch of periodic solutions exists (locally) as a curve on one side only of $\mu = 0$, and the direction (supercritical or subcritical) and stability of the bifurcating branch is determined by the sign of $c_2(0)/c_0(0)$. This curve defines the classical *Hopf bifurcation diagram.*

3. Hamiltonian-Hopf Bifurcation

First consider the behavior of the eigenvalues of the linearization, at an equilibrium point, of Hamiltonian system (2). Any linear Hamiltonian system has the form

$$\frac{dy}{dt} = Ay, \quad A \equiv JS, \quad J = \begin{pmatrix} 0 & I \\ -I & 0 \end{pmatrix}, \tag{9}$$

where I is the $n \times n$ indentity matrix and S is a symmetric $2n \times 2n$ matrix. It follows that the characteristic polynomial $p(\lambda)$ of A is an even real function, $p(-\lambda) = p(\lambda)$, and thus for a Hamiltonian system the spectrum in the complex plane of the linearization at a fixed point is symmetric with respect to both the real and imaginary axes. Simple eigenvalues that occur on the imaginary axis are "trapped" there, and cannot move off to either side as a parameter varies. Therefore the classical Hopf bifurcation, in which a complex conjugate pair of *simple* eigenvalues crosses the imaginary axis transversely, *cannot* occur in Hamiltonian systems. The nearest analogue to classical Hopf bifurcation, for linear Hamiltonian systems, involves four eigenvalues as follows. Assume the Hamiltonian $H(q, p, \mu)$ depends smoothly on a real parameter μ. Consider two simple imaginary eigenvalues of $A(\mu)$ with $\mathrm{Im}(\lambda) > 0$, together with their two complex conjugates, as shown on the left in Figure 1. These are restricted to moving along the imaginary axis as the param-

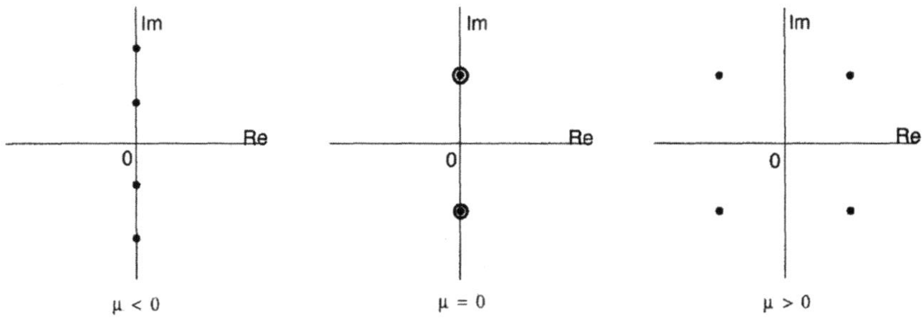

Figure 1. Eigenvalues of a 4×4 Hamiltonian matrix $A(\mu)$: splitting of double purely imaginary eigenvalues at $\mu = 0$.

eter varies, until they meet and coalesce in double purely imaginary eigenvalues (simultaneously with their conjugates) as in the middle diagram of Figure 1. These double eigenvalues may now split off of the imaginary axis, forming a symmetric quartet of simple complex eigenvalues $\{\lambda, \bar{\lambda}, -\lambda, -\bar{\lambda}\}$, as on the right in Figure 1. Of course, this scenario also may play in reverse as μ varies. The merging of eigenvalues on the imaginary axis (called 1:1 resonance) has codimension one in the class of Hamiltonian vector fields. This behaviour occurs in many physical applications.

Since the classical Hopf bifurcation theorem does not apply here, it is natural to ask whether there is a different but analogous theorem giving rise to periodic solutions for the Hamiltonian context in Figure 1. This question was answered first by [13]; see also [12]. Their theorem gives sufficient conditions under which either zero or two families of periodic solutions exist in a neighborhood of an equilibrium point, for which the linearization has eigenvalues that behave as in Figure 1. Because their theorem describes the generic analog of the Hopf bifurcation theorem for Hamiltonian systems, it has come to be called the Hamiltonian-Hopf bifurcation. Subsequently, this work has been generalized and extended, for example by van der Meer (1985). In addition to the periodic orbits one can find quasiperiodic orbits, invariant tori, stable and unstable manifolds, homoclinic orbits and chaos. In the present paper we consider only periodic orbits. The Hamiltonian-Hopf bifurcation theorem for periodic solutions, first presented by [13], is summarized here as follows.

Proposition. **([13])** Consider a parametrized family of Hamiltonian systems in $\mathbb{R}^4$,

$$\frac{dq}{dt} = \frac{\partial H}{\partial p}(q, p, \mu), \quad \frac{dp}{dt} = -\frac{\partial H}{\partial q}(q, p, \mu), \tag{10}$$

where $q, p \in \mathbb{R}^2$, $\mu \in \mathbb{R}$, and H is smooth in all its variables. Assume that the linearization $A(\mu)$ as in (9) has *double* nonsemisimple imaginary eigenvalues $\pm i\omega_0$ at $\mu = 0$, that split as in Figure 1 as μ varies through zero. Assume further a nondegeneracy condition on the nonlinear terms in H, that takes the form of the nonvanishing of a certain cubic coefficient

$$C_3 \neq 0. \tag{11}$$

Then, in a neighborhood of the equilibrium point $(y, \mu) = (0, 0)$, there exist two families of periodic solutions (centers). Depending on the sign of C_3, the two familes are either bounded and connected or unbounded and not connected.

4. Whitney's Umbrella

An observation of ([1], page 221) fits the current situation very well:

> "...the natural object of study is not the individual object (say, a vector field with a complicated singular point), but a family so large that the singularity of the type under consideration does not disappear under a small deformation of the family. This simple argument of Poincaré shows the futility of such a large number of studies in the theory of differential equations and in other areas of analysis that it is always somewhat dangerous to mention it."

The Hamiltonian-Hopf bifurcation described above represents a highly degenerate case when viewed in the setting of general *dissipative* dynamical systems. In general, double imaginary eigenvalues are not restricted as in Figure 1, but may split and leave the imaginary axis in any two independent directions. One is led naturally to ask whether it is possible to characterize, for example as a parametrized family, *all possible* dissipative systems that are close to the Hamiltonian-Hopf bifurcation, perhaps after a change of coordinates. For the linear case, an affirmative answer to this question is given in a classic paper of [1]. Arnold called such a parametrized family a *versal deformation*. The minimum number of parameters in a versal deformation is the *codimension* of the singularity. Although there may be additional physical parameters in a model, those other parameters can not lead to any new local behaviour which is not already included in the versal deformation. For the case at hand, the codimension is found to be 3.

The linear Hamiltonian system (9) with $n = 2$ and double imaginary eigenvalues as above may be rewritten as a two dimensional *complex* system in Jordan normal form

$$\frac{d}{dt}\begin{bmatrix} z_1 \\ z_2 \end{bmatrix} = \begin{bmatrix} i & 1 \\ 0 & i \end{bmatrix}\begin{bmatrix} z_1 \\ z_2 \end{bmatrix}, \tag{12}$$

after rescaling t so that $\omega_0 = 1$. Applying Arnold's versal deformation to the matrix in (12) gives the 3-parameter family of linear (non-Hamiltonian, dissipative) systems with coefficient matrix in "Arnold-Jordan normal form"

$$\frac{d}{dt}\begin{bmatrix} z_1 \\ z_2 \end{bmatrix} = \begin{bmatrix} i+\epsilon & 1 \\ \mu+i\nu & i+\epsilon \end{bmatrix}\begin{bmatrix} z_1 \\ z_2 \end{bmatrix}. \tag{13}$$

The eigenvalues of (13) are the roots of the characteristic polynomial

$$(\lambda - i - \epsilon)^2 - (\mu + i\nu) = 0, \tag{14}$$

thus are given by

$$\lambda_\pm = (i+\epsilon) \pm \sqrt{\mu + i\nu}. \tag{15}$$

To these must be added their two complex conjugates for the original four dimensional system.

Now consider the classical Hopf bifurcation of a periodic solution from the trivial equilibrium. One condition, from Section 2, is that the linearized system (13) has a purely imaginary pair of *simple* eigenvalues. Note that the two eigenvalues in (15) may cross the imaginary axis *independently*. In terms of the parameters (ν, μ, ϵ), the necessary condition for Hopf bifurcation, $\mathrm{Re}(\lambda_\pm) = 0$, yields

$$\mathcal{H}: \quad \nu^2 + 4\epsilon^2(\mu - \epsilon^2) = 0. \tag{16}$$

As remarked by several authors, these equations define topologically a *Whitney umbrella* in the (ν, μ, ϵ) parameter space, see [15] and Figure 2 and similar figures in [14], [4], and [7]. The standard form of the Whitney umbrella, as given by [3] or [5], is

$$y^2 = zx^2, \tag{17}$$

which follows from (16) with the transformation $y = \nu$, $x = 2\epsilon$, $z = \epsilon^2 - \mu$. (This transformation to standard form turns the umbrella shown in Figure 2 upside down.) For future reference, we will call the singular variety in (16) and Figure 2 the *umbrella* of primary Hopf bifurcations, denoted by the symbol $\mathcal{H}$. We call the entire μ-axis the *handle* of the umbrella, as indicated by a vertical line in Figure 2. The umbrella self-intersects transversely along the handle for $\mu < 0$. The crossing angle at this intersection goes to zero, as $\mu \to 0^-$.

At every point on the umbrella but not on the μ-axis or handle, the matrix $A(\mu, \nu, \epsilon)$ of (13) has exactly one purely imaginary eigenvalue and one with nonzero real part (to which must be added their two complex conjugates). Along the negative μ-axis, there are two distinct purely imaginary eigenvalue pairs, as on the left side of Figure 1. They coalesce as double purely imaginary eigenvalues for parameters (μ, ν, ϵ) at the origin of Figure 2. Along the positive μ-axis (on the handle above the umbrella) the four eigenvalues are symmetrically arranged, one in each quadrant, as on the right side of Figure 1.

The umbrella separates $\mathbb{R}^3$ into three disjoint open regions in Figure 2, characterized by the signs of the nonzero real parts of the complex eigenvalues $\mathrm{Re}(\lambda_\pm)$ and their congugates, as follows.

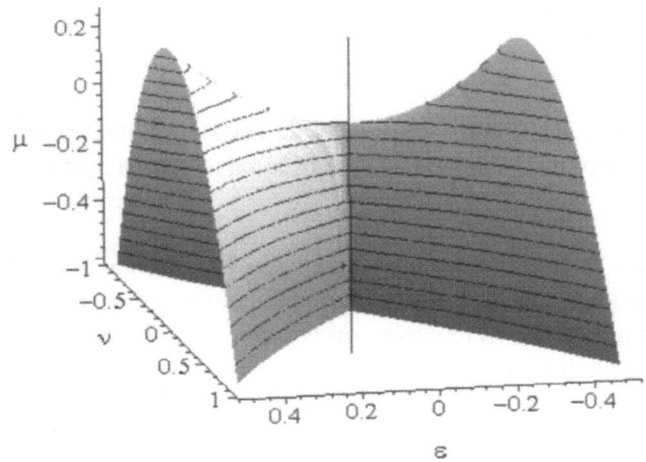

Figure 2. The Whitney umbrella of Hopf bifurcations.

- Under the umbrella, in the region with $\epsilon < 0$, both real parts are negative.
- Under the umbrella, in the region with $\epsilon > 0$, both real parts are positive.
- Everywhere above the umbrella the two real parts have opposite signs.

After comparison with the Hamiltonian case, we obtain the following new result.

Proposition. (Linearization) In the codimension-3 unfolding of the 1:1 resonant Hopf bifurcation with double imaginary eigenvalues $\pm i\omega_0$ as shown in Figure 2, the Hamiltonian-Hopf bifurcation corresponds to the special case, restricted to the *handle* of Whitney's umbrella (represented as a vertical line in Figure 2). Along this handle in Figure 2, the parameter μ is the same as the parameter μ in the Hamiltonian case shown in Figure 1.

5. Near-Hamiltonian Hopf Bifurcation

In studying nonlinear systems near a singular equilibrium, it is often advantageous to bring them to Poincaré-Birkhoff normal form. This normal form is not unique, but may be applied in the same way to both of the general nonlinear Hamiltonian and non-Hamiltonian systems, of which (9) and (13) are the linear parts, respectively, to permit direct comparison. The general nonlinear non-Hamiltonian normal form is given, as in [14] or [10], by

$$\begin{bmatrix} \dot{z}_1 \\ \dot{z}_2 \end{bmatrix} = \begin{bmatrix} i+\epsilon & 1 \\ \mu + i\nu & i+\epsilon \end{bmatrix} \begin{bmatrix} z_1 \\ z_2 \end{bmatrix} + \begin{bmatrix} \phi_1 & 0 \\ \phi_2 & \phi_1 \end{bmatrix} \begin{bmatrix} z_1 \\ z_2 \end{bmatrix}. \tag{18}$$

Here ϕ_1, ϕ_2 are complex-valued functions of the invariant polynomials $z_1\bar{z}_1$ and $\mathrm{Im}[\bar{z}_1 z_2]$; this means that they have the form

$$\phi_j = \phi_j(|z_1|^2, \mathrm{Im}[\bar{z}_1 z_2], \alpha), \quad j = 1, 2, \tag{19}$$

and furthermore, $\phi_j(0,0,0) = 0$, $j = 1, 2$. The Hamiltonian normal form is a restricted case of this, obtained by setting equal to zero all of the following:

$$\mathrm{Re}[\phi_1] = 0, \quad \mathrm{Im}[\phi_2] = 0, \quad \epsilon = 0, \quad \nu = 0. \tag{20}$$

It was shown by [14] that for the general non-Hamiltonian normal form (18), subject to a nondegeneracy condition on the nonlinear terms, a classical Hopf bifurcation of a limit cycle occurs on crossing the Whitney umbrella at any point not on the handle of the umbrella. The direction of bifurcation and stability of the cycle were determined. This implies that in a tubular neighborhood of the handle for $\mu < 0$, where the two "sheets" of the umbrella cross transversely, two limit cycles bifurcate from the equilibrium point in close proximity. This paper announces the following result. The mathematical proof will be published elsewhere.

Proposition. (Weak Dissipation Limit) In the non-Hamiltonian normal form (18) in a tubular neighborhood of the handle of the Whitney umbrella, as the "dissipation" terms listed in (20) tend to 0, the periodic solutions that persist correspond directly to those found by [13] in the Hamiltonian case. There are two such families of periodic solutions, and they are either bounded or unbounded depending on the sign of a computable coefficient, provided that it is nonzero as in the theorem of Meyer and Schmidt.

6. Conclusion

This analysis shows that the codimension-1 Hamiltonian-Hopf bifurcation of [13] can be viewed as a singular limit of the codimension-3 dissipative 1:1 resonant normal form of [14]. The essential singularity in which these two cases meet is topologically equivalent to Whitney's umbrella. For applications, this work is a step towards providing a justification for the use of Hamiltonian equations to describe real-world systems that are in fact only near-Hamiltonian.

The new results presented here raise more questions than they answer. Little is known about systems at the interface between Hamiltonian and dissipative. What is the fate of the normally hyperbolic invariant tori found in [14], as the system approaches the Hamiltonian limit? Do they give rise to KAM tori and chaos as found in conservative systems? What is the behaviour of the hyperbolic stable and unstable manifolds, above the umbrella near the handle? Do heteroclinic orbits persist? Is there a possibility of pinched tori and nontrivial monodromy? Finally, of particular interest for this IUTAM Symposium, what are the effects of *stochastic* perturbations, perhaps together with dissipation, on the Hamiltonian-Hopf bifurcation? Work in progress is beginning to shed light on these and other questions, and will be published elsewhere.

Acknowledgments

The author wishes to thank G. Iooss, K. Meyer, and R. Cushman for their suggestions and encouragement of this work.

References

[1] V. I. Arnold, "On matrices depending on parameters," *Russian Math. Surveys*. vol. 26, pp. 29–43, 1971.

[2] V. I. Arnold, *Geometrical Methods in the Theory of Ordinary Differential Equations*, New York: Springer-Verlag, 1983.

[3] V. I. Arnold, S. M. Gussein-Zade, and A.N. Varchenko, *Singularities of Differential Maps, Vol. 1*. Monographs in Mathematics, vol. 82. Boston: Birkhäuser, 1985.

[4] J. E. Furter, "Hopf bifurcation at non-semisimple eigenvalues: a singularity theory approach," in *Bifurcation and Symmetry*, Editors E. Allgower, K. Böhmer, and M. Golubitsky. ISNM vol. 104, pp. 135–145. Basel: Birkhäuser Verlag, 1992.

[5] M. Golubitsky and V. Guillemin, *Stable Mappings and their Singularities*, Graduate Texts in Mathematics, vol. 70, New York: Springer-Verlag, 1973.

[6] M. Golubitsky and W. F. Langford, "Classification and unfoldings of degenerate Hopf bifurcations," *J. Differential Equations*, vol. 41, pp. 375–415, 1981.

[7] W. Govaerts, J. Guckenheimer, and A. Khibnik, "Defining functions for multiple Hopf bifurcations," *SIAM J. Numer. Anal.*, vol. 34, pp. 1269–1288, 1997.

[8] W. R. Hamilton, "On a General Method of Expressing the Paths of Light, and of Planets, by the Coefficients of a Characteristic Function," Reprinted in The Mathematical Papers of Sir William Rowan Hamilton, Vol. 1. [1931]. Cambridge: Cambridge University Press, 1833.

[9] E. Hopf, "Abzweigung einer periodischen Lösung von einer stationären Losüng eines Differentialsystems," *Ber. Math.-Phys. Kl. Sächs Akad. Wiss. Leipzig*, vol. 94, pp. 3–22, 1942.

[10] G. Iooss and M. Adelmeyer, *Topics in Bifurcation Theory and Applications*, Advanced Series in Nonlinear Dynamics, vol. 3, Singapore: World Scientific, 1992.

[11] J. C. van der Meer, *The Hamiltonian Hopf Bifurcation*, Lecture Notes in Mathematics, vol. 1160, New York: Springer-Verlag, 1985.

[12] K. R. Meyer and G. R. Hall, *Introduction to Hamiltonian Dynamical Systems and the N-Body Problem*, Applied Mathematical Sciences, vol. 90, New York: Springer-Verlag, 1990.

[13] K. R. Meyer and D. S. Schmidt, "Periodic orbits near L_4 for mass ratios near the critical mass ratio of Routh." *Celest. Mech.*, vol. 4, pp. 99–109, 1971.

[14] S. A. van Gils, M. Krupa, and W. F. Langford, "Hopf bifurcation with non-semisimple 1:1 resonance," *Nonlinearity*, vol. 3, pp. 825–850, 1990.

[15] H. Whitney, "The general type of singularity of a set of $2n-1$ smooth functions of n variables," *Duke Math. J.*, vol. 10, pp. 161–172, 1943.

STOCHASTIC STABILITY OF TWO COUPLED OSCILLATORS IN RESONANCE

N. Sri Namachchivaya
Department of Aeronautical and Astronautical Engineering
University of Illinois at Urbana–Champaign
Urbana, IL 61801, USA
sri@nsgsun.aae.uiuc.edu

N. Ramakrishnan
Department of Aeronautical and Astronautical Engineering
University of Illinois at Urbana–Champaign
Urbana, IL 61801, USA

H. J. Van Roessel
Department of Mathematical and Statistical Sciences
University of Alberta,
Edmonton, Alberta T6G-2G1, Canada
henry.vanroessel@ualberta.ca

L. Vedula
Department of Aeronautical and Astronautical Engineering
University of Illinois at Urbana–Champaign
Urbana, IL 61801, USA

Abstract We obtain an approximation for the moment Lyapunov exponent of two coupled oscillators with commensurable frequencies driven by a small intensity real noise. A stochastic averaging scheme along with a Girsanov transformation and the Feynman-Kac formula are used to explicitly obtain the eigenvalue problem for the moment Lyapunov exponent. An orthogonal expansion for the eigenvalue problem based on Galerkin method is used to derive the stability results.

N. Sri Namachchivaya and Y.K. Lin (eds.),
IUTAM Symposium on Nonlinear Stochastic Dynamics, 167-177.

1. Introduction

An important measure of the stability of random dynamical systems is provided by the moment Lyapunov exponent which is defined as

$$g(p; x_0) = \lim_{t\to\infty} \frac{1}{t} \ln \mathbb{E}\|x(t; x_0)\|^p,$$

where $x(t; x_0)$ is the solution of a linear stochastic system. If $g(p; x_0) < 0$, then, $\mathbb{E}\|x(t; x_0)\| \to 0$ as $t \to \infty$, and the linear stochastic system is said to be stable in the pth moment. The connection between the moment Lyapunov exponent and the associated large-deviation problem for linear systems perturbed by white noise was first established in [9]. An alternative approach to the study of moment Lyapunov exponents and large deviations was developed by [3] in which estimates were obtained for the probability with which an almost-sure stable system exceeds a threshold. This probability is controlled by the *stability index* which is the unique nonzero solution of the equation $g(p) = 0$.

An asymptotic approximation for the moment Lyapunov exponent was obtained using a perturbative approach by [6] and [2]. These results are, however, limited to the case when the moment p is small. An asymptotic expansion for the moment Lyapunov exponent and stability index valid for all p was obtained for a two-dimensional linear stochastic system by [5]. This was extended to the case of two coupled oscillators by [7] where the moment Lyapunov exponent was obtained separately using both a perturbative approach as well as using the stochastic averaging. The two methods yielded the same generator $L(p)$ from which the moment Lyapunov exponent is obtained as the principal eigenvalue. A key assumption made in [7] is that the frequencies of the two oscillators are non-commensurable. The first extension of these results to coupled oscillators with commensurable frequencies was carried out by [8] using a perturbation method.

In this work, we shall extend the procedure used in [7] to obtain the moment Lyapunov exponent using *stochastic averaging* for the case of two coupled oscillators in *resonance*. By resonance, we mean that a relationship of the form $m_1\omega_1(\eta) = m_2\omega_2(\eta)$ exists for some integer values of m_1 and m_2 and system parameters η. The resonance effects would in this case, persist in all regions of the phase-space, and are due to a particular choice of the system parameters η. In the presence of resonance, energy given to one of the modes of vibration is continuously exchanged among all the modes involved in that resonance and strongly influences the overall response of the system.

2. Problem Formulation

Consider a two-degrees-of-freedom system of coupled harmonic oscillators parametrically excited by noise with small viscous damping

$$\ddot{q}_i + \omega_i^2 q_i + 2\varepsilon^2 \zeta_i \omega_i \dot{q}_i + \varepsilon k_{ij} q_j f(\xi(t)) \;=\; 0, \quad i = 1,2, \tag{1}$$

where the repeated subscripts implies summation and $\xi(t)$ is a stationary noise process with strong mixing properties and f is smooth. Our interest is in the case when the two frequencies are commensurable, i.e., there exists a relation of the form $m_1\omega_1 = m_2\omega_2$ where, m_1 and m_2 are integers. We transform (1) to amplitude-phase form using the change of variables,

$$q_i = a_i \cos\phi_i, \qquad \dot{q}_i = -a_i\omega_i \sin\phi_i, \qquad i = 1,2, \tag{2}$$

where, $\phi_i = \psi_i + \omega_i t$. Using (2) in (1), we obtain the following set of equations in amplitudes, a_i and phase, ϕ_i

$$\dot{a}_i = \varepsilon^2 b_{a_i} + \varepsilon F_{a_i}, \qquad \dot{\psi}_i = \varepsilon^2 b_{\psi_i} + \varepsilon F_{\psi_i}, \qquad i = 1,2, \tag{3}$$

where $b_{a_i} = -\zeta_i\omega_i a_i(1-\cos 2\phi_i)$, $b_{\psi_i} = -\zeta_i\omega_i \sin 2\phi_i$ and the functions F_{a_i}, F_{ϕ_i} can be obtained through some simple calculations. Equation (3) is in the so-called *standard form* for averaging. Our goal is to study (3) and show that as ε tends to zero, the dynamics of $x_t^\varepsilon = (a_t^\varepsilon, \psi_t^\varepsilon)$ tends to a lower-dimensional Markov process and to identify its generator. The mathematical problem is to rigorously characterize the limiting dynamics of the state as the fluctuations of the noise become more and more rapid. To carry out this program, certain strong mixing assumptions are enforced upon the noise process. Asymptotically, the state process becomes diffusive and Markovian; the noise becomes extraneous, effecting a dimensional reduction of the original system to the dimension of the state. Using the results of [4] on stochastic averaging, (3) can be reduced to the following set of Itô stochastic differential equations

$$\begin{aligned} da_i &= \eta_{a_i}(a,\phi)\,dt + \sigma_{a_{ij}}(a,\phi)\,dW_j(t), \\ d\psi_i &= \nu_{\psi_i}(a,\phi)\,dt + \sigma_{\psi_{ij}}(a,\phi)\,dW_j(t), \end{aligned} \tag{4}$$

where $i,j = 1,2$ and $\phi = \psi_1 - \psi_2$. Equivalently, (3) reduces to a diffusive Markov process with infinitesimal generator

$$\begin{aligned} \widetilde{L} = {} & \eta_{a_i}(a,\phi)\frac{\partial}{\partial a_i} + \nu_{\psi_i}(a,\phi)\frac{\partial}{\partial \psi_i} + \frac{1}{2}\Sigma_{a_i a_j}(a,\phi)\frac{\partial^2}{\partial a_i \partial a_j} \\ & + \frac{1}{2}\Sigma_{\psi_i\psi_j}(a,\phi)\frac{\partial^2}{\partial \psi_i \partial \psi_j} + \frac{1}{2}\Sigma_{a_i\psi_j}(a,\phi)\frac{\partial^2}{\partial a_i \partial \psi_j} \end{aligned} \tag{5}$$

where $\Sigma = \sigma\sigma^T$. The drift and diffusion terms in the averaged equations (4) are given by

$$\eta_i(a,\phi) = \mathbb{M}\{\int_{-\infty}^{0} \mathbb{E}[\frac{\partial F_{a_i}}{\partial a_j}(a,\psi,t,\xi(t))F_{a_j}(a,\psi,t+\tau,\xi(t+\tau))]d\tau$$
$$+ \int_{-\infty}^{0} \mathbb{E}[\frac{\partial F_{a_i}}{\partial \psi_j}(a,\psi,t,\xi(t))F_{\psi_j}(a,\psi,t+\tau,\xi(t+\tau))]d\tau + b_{a_i}(a,\psi,t)\},$$

$$\nu_i(a,\phi) = \mathbb{M}\{\int_{-\infty}^{0} \mathbb{E}[\frac{\partial F_{\psi_i}}{\partial a_j}(a,\psi,t,\xi(t))F_{a_j}(a,\psi,t+\tau,\xi(t+\tau))]d\tau$$
$$+ \int_{-\infty}^{0} \mathbb{E}[\frac{\partial F_{\psi_i}}{\partial a_j}(a,\psi,t,\xi(t))F_{\psi_j}(a,\psi,t+\tau,\xi(t+\tau))]d\tau + b_{\psi_i}(a,\phi,t)\},$$
$$\Gamma_{ij}(a,\phi) = \mathbb{M}\{\int_{-\infty}^{\infty} \mathbb{E}\left[F_i(x,t,\xi(t))F_j(x,t,\xi(t+\tau))\right]d\tau\},$$

where $\mathbb{M}(\cdot) := \lim_{T\to\infty} \frac{1}{T}\int_0^T(\cdot)dt$.

Stochastic averaging is at the heart of the problem of weak convergence of stochastic processes, or more exactly, weak convergence of the laws of Markov processes. Hence, it is not clear, what can be said about the sample stability of trajectories of (3) if the Lyapunov exponent of (4) is negative. However, it has been shown by [7] that the asymptotic expansion of both moment and maximal Lyapunov exponent are the same whether we use a perturbative approach or an averaging approach to study the stability of (1).

Equation (4) is the *starting point* for the rest of our analysis. It is worth noting that in the non-resonant case (see [7]) one of the advantages of the above approximate Itô equations (4) is that the amplitudes a_i are decoupled from the phases ϕ_i and they form a diffusive Markov process. However, in the resonant case the amplitude equations are coupled to the phase equations through the phase difference $\phi = \psi_1 - \psi_2$.

3. Moment Lyapunov Exponents

We transform the Itô equations for the amplitudes in terms of the norm, $r = \|a\|$ and an angle θ using the following change of variables

$$r = \sqrt{a_1^2 + a_2^2}, \qquad \theta = \tan^{-1}\left(\frac{a_2}{a_1}\right), \tag{6}$$

which results in

$$g(p;x_0) = g(p;r_0) = \lim_{t\to\infty} \frac{1}{t}\log\mathbb{E}[r^p].$$

The transformation is singular at $\theta = 0, \pi/2$. Applying Itô's rule and using (4), the Itô equations for r and the angle θ yields

$$\begin{aligned} dr &= \eta(\theta,\phi)\, r\, dt + \sigma_j(\theta,\phi)\, r\, dW_j(t), \\ d\theta &= \mu(\theta,\phi)\, dt + \Psi_j(\theta,\phi)\, dW_j(t), \\ d\phi &= \nu(\theta,\phi)\, dt + \chi_j(\theta,\phi)\, dW_j(t). \end{aligned} \tag{7}$$

The coefficients $\eta(\theta,\phi)$, $\sigma_j(\theta,\phi)$, $\mu(\theta,\phi)$, $\Psi_j(\theta,\phi)$, $\nu(\theta,\phi)$ and $\chi_j(\theta,\phi)$ in (7) are related to the coefficients $\eta_{a_i}(a,\phi)$, $\Sigma_{a_i a_j}(a,\phi)$, $\nu_{\psi_1}(a,\phi)$, $\Sigma_{\psi_i\psi_j}(a,\phi)$, $\sigma_{a_i a_j}(a,\phi)$ and are easily obtained using Itô's rule. The solution $r(t;r_0)$ can be explicitly written as

$$r(t) = r_0\; e^{\int_0^t Q_1(\theta_s,\phi_s)ds + \int_0^t \sigma_j(\theta_s,\phi_s)dW_j(s)}, \tag{8}$$

where

$$Q_1(\theta,\phi) = \eta(\theta,\phi) - \frac{1}{2}|\sigma|^2(\theta,\phi).$$

The pth moment of the normed response is then given by

$$\begin{aligned} \mathbb{E}\left[r^p(t)\right] &= r_0^p\; \mathbb{E}[e^{p\int_0^t Q_1(\theta_s,\phi_s)ds + p\int_0^t \sigma_j(\theta_s,\phi_s)dW_j(s)}] \\ &= r_0^p\; \mathbb{E}[e^{p\int_0^t \sigma_j(\theta_s,\phi_s)dW_j(s) - \frac{p^2}{2}\int_0^t |\sigma|^2(\theta_s,\phi_s)ds} \\ &\qquad \cdot\; e^{p\int_0^t Q_1(\theta_s,\phi_s)ds + \frac{p^2}{2}\int_0^t |\sigma|^2(\theta_s,\phi_s)ds}]. \end{aligned} \tag{9}$$

Now applying the Girsanov theorem, and making use of stochastic differential equations in the phases, θ and ϕ, we calculate the pth moment of the normed response. Hence, we obtain

$$\mathbb{E}[r^p(t)] = r_0^p\; \mathbb{E}[e^{p\int_0^t Q_1(\theta_s,\phi_s)ds + \frac{p^2}{2}\int_0^t |\sigma|^2(\theta_s,\phi_s)ds}]. \tag{10}$$

The expectation on the right hand side is with respect to the new process generated by

$$\begin{aligned} L(p) =& \Big(p\,\sigma(\theta,\phi)\Psi(\theta,\phi) + \mu(\theta,\phi)\Big)\frac{\partial}{\partial\theta} + \Big(p\,\sigma(\theta,\phi)\chi(\theta,\phi) + \nu(\theta,\phi)\Big)\frac{\partial}{\partial\phi} \\ &+ \frac{1}{2}\Psi^2(\theta,\phi)\frac{\partial^2}{\partial\theta^2} + \frac{1}{2}\chi^2(\theta,\phi)\frac{\partial^2}{\partial\phi^2} + \Big(\Psi(\theta,\phi)\chi(\theta,\phi)\Big)\frac{\partial^2}{\partial\theta\partial\phi} \end{aligned} \tag{11}$$

Now, to compute $\mathbb{E}[r^p(t)]$ from 10 we make use of the Feynman-Kac formula, according to which the expectation $\mathbb{E}[r^p(t)] = \psi(\theta,\phi,p,t)$ is the solution of

$$\left(-\frac{\partial}{\partial t} + L + pQ_1 + \frac{p^2}{2}\sigma^2\right)\psi(\theta,\phi,p,t) = 0, \qquad \psi(\theta,\phi,p,0) = 1, \tag{12}$$

where L is the generator of the θ and ϕ-process given by (11). Finally, using Theorem 5.2.1 from [1], the moment Lyapunov exponent is obtained as the principal eigenvalue of the operator

$$\widetilde{L}(p) = L(p) + p\,Q_1 + \frac{p^2}{2}\sigma^2 = L(p) + p\eta + \frac{p(p-1)}{2}|\sigma|^2$$

with non-negative eigenfunction $\psi_0(p)$, normalized so that $\|\psi_0(p)\| = 1$, along with appropriate boundary conditions, i.e.,

$$\widetilde{L}(p)\psi_0 = g_2(p)\psi_0. \tag{13}$$

After a lengthy calculation, we get

$$\begin{aligned}\widetilde{L}(p)\psi_0 = {} & \frac{1}{2}\sigma^2_{\theta\theta}(\theta,\phi)\frac{\partial^2\psi_0}{\partial\theta^2}(\theta,\phi) + \frac{1}{2}\sigma^2_{\phi\phi}(\theta,\phi)\frac{\partial^2\psi_0}{\partial\phi^2}(\theta,\phi) \\ & + \sigma^2_{\phi\theta}(\theta,\phi)\frac{\partial^2\psi_0}{\partial\theta\partial\phi}(\theta,\phi) + [\mu(\theta,\phi) + p\,\widehat{\mu}(\theta,\phi)]\frac{\partial\psi_0}{\partial\theta}(\theta,\phi) \\ & + [\nu(\theta,\phi) + p\,\widehat{\nu}(\theta,\phi)]\frac{\partial\psi_0}{\partial\phi}(\theta,\phi) + [p\,Q(\theta,\phi) + \frac{p^2}{2}\widehat{Q}(\theta,\phi)]\psi_0(\theta,\phi),\end{aligned}$$

where the drift and diffusion coefficients are given in the Appendix. The boundary conditions need to be determined by considering the diffusion process with generator $L(p)$. For physical systems with symmetry, one-to-one resonance is the rule rather than the exception. Hence, while averaging these coefficients in (3), we have made use of this specific resonance condition called *one-to-one* semi-simple resonance, where we define

$$\omega = \omega_1 = \omega_2.$$

The results for the other resonance cases can also be easily obtained. In addition, we have made use of the cosine and sine spectrum given, respectively, by

$$S_\xi(\omega) = 2\int_0^\infty \mathcal{R}_\xi(\tau)\,\cos(\omega\tau)\,d\tau \quad \text{and} \quad T_\xi(\omega) = 2\int_0^\infty \mathcal{R}_\xi(\tau)\,\sin(\omega\tau)\,d\tau,$$

where $\mathcal{R}_\xi$ is the correlation function of f. Equation (13) was derived without any restriction on the size of p. This equation along with appropriate boundary conditions forms the eigenvalue problem for which $g_2(p)$ is the principal eigenvalue. In order to reduce the number of cases to be evaluated, we can simplify the coefficients $p_{ij} := k_{ij}/\omega_i$ by a suitable scaling of the state variables x. It can be shown as in [7] that it is always possible to take $p_{12} = \pm p_{21} = \kappa$. Since the semi-simple linear form ($\omega_1 = \omega_2 = \omega$) naturally occurs in conservative systems, we shall only consider $p_{12} = p_{21} = \kappa$, in which case the coefficients are given in the Appendix. Except for some special cases, the general solution of (13) cannot be obtained explicitly for $g_2(p)$.

The domain for the diffusion process in θ and ϕ is

$$\mathcal{D} = \{(\theta, \phi) : \ \theta \in [0, \pi/2], \ \phi \in [0, 2\pi]\}.$$

Since all the coefficients in (13) are periodic in ϕ, ψ_0 is $2\pi-$ periodic in ϕ. In general it is possible to have singularities in θ but *only the nonsingular cases are considered here.* In order to obtain the boundary conditions at the boundaries $\theta = 0$ and $\frac{\pi}{2}$, we investigate the asymptotic behavior of the diffusion process near these boundaries. Simple calculations show that $\sigma^2_{\theta\theta}$ is finite at $\theta = 0, \pi/2$, whereas

$$\mu(\theta, \phi) \sim \frac{\kappa^2}{16\theta} \{ S_\xi(2\,\omega) + S_\xi(0)\,(1 + \cos(2\phi))\}, \qquad (\theta \to 0^+)$$

$$\mu(\theta, \phi) \sim \frac{\kappa^2}{16(\theta - \pi/2)} \{S_\xi(2\,\omega) + S_\xi(0)(1 + \cos(2\phi))\}. \qquad (\theta \to \frac{\pi}{2}^-)$$

Therefore, the diffusion process near the boundary is characterized by the drift term. It follows that, for $\kappa \neq 0$, the trajectories close to the boundaries $\theta = 0, \frac{\pi}{2}$ are repelled. Thus for the case when both the degrees of freedom are coupled, *i.e,* $p_{12} \neq 0$ and $p_{21} \neq 0$, the boundaries at $\theta = 0$ and $\theta = \frac{\pi}{2}$ are not attainable states. Also, trajectories starting at either of these boundaries are pushed into the interior of the domain $\mathcal{D}$. For the remaining part of this study, we shall make the assumption that $\kappa \neq 0$. This assumption is valid and appropriate for most of the practical situations to which these results can be applied. The results deduced above imply that the probability flux in a direction normal to the boundaries at $\theta = 0$ and $\theta = \frac{\pi}{2}$, is identically equal to zero and so we *assume a Neumann boundary condition* for ψ_0.

4. The Eigenvalue Problem

As in [7, 10], the solution of (13) can be calculated from an orthogonal expansion. The nature of the coefficients of the equation suggests that a double Fourier series is appropriate. Since ψ_0 is 2π-periodic in ϕ and because of the assumed Neumann boundary conditions

$$\frac{\partial \psi_0}{\partial \theta}(0, \phi) = \frac{\partial \psi_0}{\partial \theta}(\frac{\pi}{2}, \phi) = 0 \tag{14}$$

we may express ψ_0 as

$$\psi_0(\phi, \theta) = \sum_{n=0}^{\infty} \sum_{m=0}^{\infty} (a_{mn} \cos m\phi + b_{mn} \sin m\phi) \cos(2n\theta). \tag{15}$$

Insertion of (15) into (13) leads to an equation of the form:

$$\sum_{r=0}^{\infty} \sum_{s=0}^{\infty} \left(\begin{bmatrix} \widehat{C}_{rsmn} & \overline{C}_{rsmn} \\ \widehat{D}_{rsmn} & \overline{D}_{rsmn} \end{bmatrix} - g(p) \begin{bmatrix} \widehat{E}_{rsmn} & \overline{E}_{rsmn} \\ \widehat{F}_{rsmn} & \overline{F}_{rsmn} \end{bmatrix} \right) \begin{pmatrix} a_{rs} \\ b_{rs} \end{pmatrix} = 0, \tag{16}$$

where the coefficients $\widehat{C}_{rsmn}$ etc. are easily obtained expressions involving a_{mn} and b_{mn}. The existence of a nontrivial solution of (16) requires that the determinant of the coefficient matrix be zero. Thus to evaluate $g(p)$, the leading eigenvalue of the coefficient matrix, we construct a sequence of approximations by truncating the sums. An approximate solution to (16) is obtained by numerically solving the truncated equation

$$\sum_{r=0}^{N}\sum_{s=0}^{N}\left(\begin{bmatrix}\widehat{C}_{rsmn} & \overline{C}_{rsmn}\\ \widehat{D}_{rsmn} & \overline{D}_{rsmn}\end{bmatrix} - g(p)\begin{bmatrix}\widehat{E}_{rsmn} & \overline{E}_{rsmn}\\ \widehat{F}_{rsmn} & \overline{F}_{rsmn}\end{bmatrix}\right)\begin{pmatrix}a_{rs}\\ b_{rs}\end{pmatrix} = 0.$$

The set of approximate eigenvalues obtained by this procedure converges to the corresponding true eigenvalues as $N \to \infty$. However, the amount of calculation increases drastically with an increase in the number of terms considered.

5. Numerical Results

We now present some numerical results to illustrate the convergence of various orders of approximations. For this purpose, we consider the numerical values $p_{11} = 1$, $p_{22} = 2$, $\kappa = 1$, $\delta_1 = 1$, $\delta_2 = 2$, $S_\xi(2\omega) = S_\xi(0) = 1$, $T_\xi(2\omega) = 0$. We obtain several sets of results corresponding to $N = 0, 1, 2, 3$ giving rise respectively to 1, 6, 15 and 28 terms in the Fourier expansion. From the re-

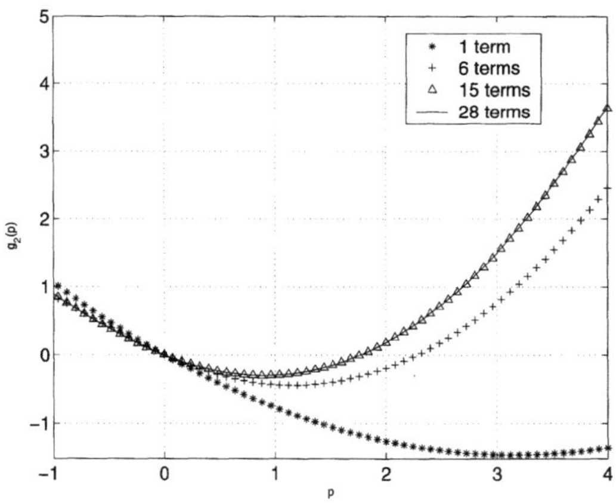

Figure 1. Variation of Moment Lyapunov Exponent, $g_2(p)$, with p

sults shown in Figure 1, it is clear that the results for $N = 2$ (15 terms) and $N = 3$ (28 terms) are essentially the same. Thus, it suffices to use a 15 term expansion in order to obtain a sufficiently accurate expression for $g_2(p)$. For the parameter values given above, the system is almost-surely stable but moment unstable beyond $p \approx 1.77$.

In conclusion, a method to compute the moment Lyapunov exponent of a two degree-of-freedom coupled linear system, with commensurable eigenvalues, under random parametric excitation was developed. We derive the generator $L(p)$ for finite p, whose principal eigenvalue is the moment Lyapunov exponent, based on stochastic averaging. Except for some special cases the general solution of (13) cannot be obtained explicitly for $g_2(p)$. In this paper we have obtained approximate solutions based on Fourier analysis and it is shown that a fifteen term expansion is sufficiently accurate.

Acknowledgments

The authors would like to acknowledge the support of the Office of Naval Research under grant number N001-15-005, the National Science Foundation under grant number CMS 00-84944 and NSERC of Canada.

Appendix: Drift and Diffusion Coefficients

The drift and diffusion coefficients of $\widetilde{L}(p)$ given by (13) are evaluated below.

$$\begin{aligned}
\mu(\theta,\phi) =& [\sigma^2_{\theta\theta}(\theta,\phi) + \frac{1}{4}\kappa^2 S_\xi(0)\cos(2\,\phi)]\cot 2\theta - \frac{1}{2}\Lambda\sin 2\theta \\
&+ \frac{1}{8}\,S_\xi(2\,\omega)(p_{11}+p_{22})\kappa\,\cos(\phi)\cos 2\,\theta \\
&+ \frac{1}{8}\,T_\xi(2\,\omega)(p_{11}+p_{22})\kappa\,\sin(\phi) - \frac{1}{8}\,S_\xi(0)(p_{11}-p_{22})\kappa\,\cos(\phi)
\end{aligned}$$

$$\begin{aligned}
\widehat{\mu}(\theta,\phi) =& \{[A_0 + \frac{1}{16}p_{11}p_{22}S_\xi(2\omega)\cos(2,\phi)]\cos 2\theta - \frac{1}{2}E_0\}\sin 2\theta \\
&+ \frac{1}{16}\,\kappa\,[2\,(p_{11}-p_{22})\cos^2 2\theta \\
&\qquad + (p_{11}+p_{22})\cos 2\theta - (p_{11}-p_{22})]\,S_\xi(2\,\omega)\,\cos(\phi)
\end{aligned}$$

$$\begin{aligned}
\nu(\theta,\phi) =& -\frac{1}{8}\,(p_{11}-p_{22})(p_{11}+p_{22})T_\xi(2\,\omega) - \frac{1}{2}\,S_\xi(0)\sin(2\,\phi)\kappa^2\csc^2 2\,\theta \\
&+ \frac{1}{4}\,\kappa\,\{[(p_{11}+p_{22})T_\xi(2\,\omega)\cos(\phi) + (p_{11}-p_{22})\,S_\xi(0)\sin(\phi)]\cos 2\,\theta \\
&- (p_{11}+p_{22})\,S_\xi(2\,\omega)\,\sin(\phi)\}\csc 2\,\theta + \frac{1}{4}\,S_\xi(0)\sin(2\,\phi)\kappa^2
\end{aligned}$$

$$\begin{aligned}
\widehat{\nu}(\theta,\phi) =& -\frac{1}{8}\,\kappa\,[\{(p_{11}-p_{22}) + (p_{11}+p_{22})\}\,S_\xi(2\,\omega)\,\sin(\phi)]\csc 2\,\theta \\
&- \frac{1}{8}p_{11}\,p_{22}\,S_\xi(0)\sin(2\,\phi)
\end{aligned}$$

$$\begin{aligned}
\sigma^2_{\theta\theta}(\theta,\phi) =& [A_0 + \frac{1}{16}\,p_{11}p_{22}\,S_\xi(2\,\omega)\cos(2\,\phi)]\cos^2 2\theta \\
&+ C_0 - \frac{1}{16}\,\{\,p_{11}p_{22}S_\xi(2\,\omega) + 2\,\kappa^2\,S_\xi(0)\}\cos(2\,\phi) \\
&- \frac{1}{8}\,\kappa\,(p_{11}-p_{22})\,S_\xi(2\,\omega)\,\cos(\phi)\cos 2\,\theta\,\sin 2\,\theta
\end{aligned}$$

$$\sigma^2_{\theta\phi}(\theta,\phi) = \frac{1}{4}\,\kappa^2\,S_\xi(0)\sin(2\,\phi)\cot 2\,\theta - \frac{1}{4}\,\kappa\,(p_{11}-p_{22})\,S_\xi(0)\sin(\phi)$$

$$\begin{aligned}
\sigma^2_{\phi\phi}(\theta,\phi) =& \frac{1}{2}\omega^2\kappa^2\{S_\xi(0) + S_\xi(2\,\omega) + S_\xi(0)\cos(2\,\phi)\}\cot^2(2\,\theta) \\
&- \frac{1}{2}\omega^2\,\kappa(p_{11}-p_{22})\,\{2\,S_\xi(0) +\,S_\xi(2\,\omega)\}\cos(\phi)\cot(2\,\theta) \\
&+ \frac{1}{8}\omega^2[({p_{11}}^2 + {p_{22}}^2)S_\xi(2\,\omega) + 2(p_{11}-p_{22})^2 S_\xi(0) \\
&- 2p_{11}p_{22}S_\xi(2\,\omega)\cos(2\,\phi)]
\end{aligned}$$

$$\begin{aligned}
Q(\theta,\phi) =& \sigma^2(\theta,\phi) + \frac{1}{2}\Lambda\cos 2\theta + D + \frac{1}{8}\,S_\xi(0)\kappa^2\cos(2\,\phi) \\
&+ \frac{1}{8}\,S_\xi(2\,\omega)(p_{11}+p_{22})\kappa\,\cos(\phi)\sin 2\,\theta
\end{aligned}$$

$$\begin{aligned}
\widehat{Q}(\theta,\phi) =& -[A_0 + \frac{1}{16}\,p_{11}p_{22}\,S_\xi(2\,\omega)\cos(2\,\phi)]\cos^2 2\theta E_0\cos 2\theta + F_0 \\
&+ \frac{1}{16}\,S_\xi(2\,\omega)p_{11}p_{22}\cos(2\,\phi) + \frac{1}{8}\,\kappa(p_{11}+p_{22})\,S_\xi(2\,\omega)\,\cos(\phi)\sin 2\,\theta \\
&+ \frac{1}{8}\,\kappa\,(p_{11}-p_{22})\,S_\xi(2\,\omega)\,\cos(\phi)\,\cos 2\,\theta\sin 2\,\theta
\end{aligned}$$

with the constants given by

$$
\begin{aligned}
A_0 &= \frac{\kappa^2}{8} S_\xi(\omega) - \frac{p_{11}^2 + p_{22}^2}{32} S_\xi(2\omega), \quad C_0 = \frac{\kappa^2}{8} S_\xi(0) + \frac{p_{11}^2 + p_{22}^2}{32} S_\xi(2\omega), \\
D_0 &= \frac{1}{2}(\Lambda_1 + \Lambda_2) + \frac{\kappa^2}{8}[S_\xi(\omega) - S_\xi(0)], \\
E_0 &= \frac{1}{16}(p_{11}^2 - p_{22}^2) S_\xi(2\omega_2), \quad F_0 = \frac{\kappa^2}{8} S_\xi(\omega) + \frac{p_{11}^2 + p_{22}^2}{32} S_\xi(2\omega) \\
\Lambda &= \Lambda_1 - \Lambda_2, \quad \Lambda_i = -\delta_i + \frac{1}{8} p_{ii}^2 S_\xi(2\omega), \quad i = 1, 2.
\end{aligned}
\tag{1}
$$

References

[1] L. Arnold, "A formula connecting sample and moment stability of linear stochastic systems," *SIAM Journal of Applied Mathematics*, vol. 44, no. 4, pp. 793–802, 1984.

[2] L. Arnold, M. M. Doyle, and N. Sri Namachchivaya, "Small noise expansion of moment Lyapunov exponents for general two dimensional systems," *Dynamics and Stability of Systems*, vol. 12, no. 3, pp. 187–211, 1997.

[3] P. H. Baxendale, *Moment stability and large deviations for linear stochastic differential equations*, in *Taniguchi Symposium PMPP*, 1987.

[4] R. Z. Khasminskii, "A limit theorem for solutions of differential equations with random right-hand-side," *Theory of Probability and Applications*, vol. 11, no. 3, pp. 390–406, 1966.

[5] R. Z. Khasminskii and N. Moshchuk, "Moment Lyapunov exponent and stability index for linear conservative system with small random perturbation," *SIAM Journal of Applied Mathematics*, vol. 58, no. 1, pp. 245–256, 1998.

[6] N. Sri Namachchivaya, H. J. Van Roessel, and M. M. Doyle, "Moment Lyapunov exponent for two coupled oscillators driven by real noise," *SIAM Journal of Applied Mathematics*, vol. 56, pp. 1400–1423, 1996.

[7] N. Sri Namachchivaya and H. J. Van Roessel, "Moment Lyapunov exponent and stochastic stability of two coupled oscillators driven by real noise," *ASME Journal of Applied Mechanics*, vol. 68, pp. 1400–1412, 2001.

[8] N. Sri Namachchivaya and H. J. Van Roessel, "Stochastic Stability of Two Coupled Oscillators in Resonance," (in preparation).

[9] D. W. Stroock, "On the rate at which a homogeneous diffusion approaches a limit, an application of the large deviation theory of certain stochastic integrals," *Annals of Probability*, vol. 14, pp. 840–859, 1986.

[10] W. V. Wedig, "Lyapunov exponents of stochastic systems and related bifurcation problems," in *Stochastic Structural Dynamics: Progress in Theory and Applications*, S. T. Ariaratnam, G. I. Schuëller, and I. Elishakoff, Eds., Elsevier Applied Science, London & New York, 1988.

ON STABILIZING THE DOUBLE OSCILLATOR BY MEAN ZERO NOISE

Volker Wihstutz
Department of Mathematics
University of North Carolina at Charlotte
University Blvd
Charlotte, NC 28223, USA
wihstutz@uncc.edu

Abstract By similar arguments as for the *single* inverted pendulum it is shown that stabilization by random vibration is also possible for a family of linear double oscillators which includes the inverted *double* pendulum with a certain configuration for the parameters of damping, restoring and coupling.

Keywords: Stabilization by noise, random vibration, inverted double pendulum, stabilization of the linear double oscillator, double oscillator

1. Introduction

Research on parametrically excited dynamical systems like the harmonic oscillator or coupled oscillators has involved many engineers, physicists and mathematicians in the last century. In the case of random excitement (parametric noise, multiplicative noise), since the fundamental work of H. Furstenberg [10], Khasminskii [11], and Oseledec [17], Lyapunov exponents have become the main tools of investigation, especially for problems of stability.

One of these problems is the question whether or not parametric noise can stabilize an otherwise unstable equilibrium. In the case of the inverted pendulum answers are known by experiment (Bogdanoff [8], Popp [18]), as well as, by theory (see e.g., Kapitsa [14], Meerkov [16], Bellman, Bentsman and Meerkov [7], Kao and Wihstutz [12], and Wihstutz [22]). Lyapunov exponents associated with the double oscillator have been studied among others by Ariaratnam, Tam and Xie [1], Ariatnam and Xie [2], and Wedig [21], and more recently by the group around N. Sri Namachchivaya that approaches the problem mostly by means of moment Lyapunov exponents. (See, for example, N. Sri Namachchivaya, Van Roessel and Doyle [19], Doyle, N. Sri Na-

N. Sri Namachchivaya and Y.K. Lin (eds.),
IUTAM Symposium on Nonlinear Stochastic Dynamics, 179-190.

machchivaya and van Roessel [9].) Especially N. Sri Namachchivaya and Vedula in [20] address the problem of stabilization by noise.

In this paper, almost sure or pathwise Lyapunov exponents are considered, that is the exponential growth rate $\lambda(x, \omega) = \lim_{t\to\infty} \frac{1}{t} \log |x_t(\omega)|$ of each trajectory $t \to x_t(\omega)$ of the stochastic process x_t. By naturally generalizing the method developed for the single oscillator to higher dimensions it is shown as a first result that under certain constraints on the parameters the double oscillator can be stabilized by the same class of noise that stabilizes a single oscillator. This result covers the special case of stabilizing an inverted double pendulum.

In Section 2 we recall the method for stabilizing a single inverted pendulum. We describe the admissible noise in Section 3. The stabilizability result is presented in Section 4 and proved in Section 5.

2. Stabilizing the Inverted Pendulum

We recall that in order for a noise $\dot{F}_{t/\epsilon}$ to stabilize the single inverted pendulum

$$\ddot{y} + (\omega d)\dot{y} + (-\omega^2 + \dot{F}_{t/\epsilon} b)y = 0 \tag{1}$$

or with $x := [y, \dot{y}]'$,

$$dx = Axdt + Bx \circ dF_{t/\epsilon}, \quad A = \begin{bmatrix} 0 & 1 \\ -q & -e \end{bmatrix}, B = \begin{bmatrix} 0 & 0 \\ -b & 0 \end{bmatrix} \tag{2}$$

(where $q = -\omega^2$ and $e = \omega d > 0$), two basic properties are required for the stochastic process F_t (see [12]).

First, $F_{t/\epsilon}$ must give rise to a time-dependent and random linear transformation $T(F_{t/\epsilon})$ of $\mathbf{R}^2$ which preserves the Lyapunov exponent of (2) and transforms the stochastic differential equation (2) — after reduction to trace zero — to a random family of ODEs,

$$d\tilde{z} = [\tilde{A} + \tilde{B}(F_{t/\epsilon})]\tilde{z}dt, \tag{3}$$

where $\tilde{B}(F_{t/\epsilon})$ is a fast oscillating matrix with mean zero and $\tilde{A}$ is a deterministic skew-symmetric matrix (for which the dot product $s \cdot \tilde{A}s = 0$, $s = \tilde{z}/|\tilde{z}|$).

Second, $F_{t/\epsilon}$ is required to be such that a certain averaging principle (over an infinite time horizon) can be applied to the system (3) yielding

$$\lambda^{\tilde{z}}(\epsilon) \to s \cdot \tilde{A}s = 0(\text{as } \epsilon \to 0). \tag{4}$$

In order to obtain the Lyapunov exponent of (2) one has to undo the reduction to zero, that is add $\frac{1}{2}\text{trace}A = -\frac{1}{2}\omega d$. The result is

$$\lambda(\epsilon) = \lambda^{x}(\epsilon) = \lambda^{\tilde{z}}(\epsilon) - \frac{1}{2}\omega d \to -\frac{1}{2}\omega d < 0 \tag{5}$$

(as $\epsilon \to 0$).

Turning our attention to the *pair of coupled oscillators*

$$\ddot{y}_1 + \omega_1 d_{11}\dot{y}_1 + \omega_1 d_{12}\dot{y}_2 + q_1 y_1 + \dot{F}_{t/\epsilon}(b_{11}y_1 + b_{12}y_2) \quad (6)$$

$$\ddot{y}_2 + \omega_2 d_{21}\dot{y}_1 + \omega_2 d_{22}\dot{y}_2 + q_2 y_2 + \dot{F}_{t/\epsilon}(b_{21}y_1 + b_{22}y_2), \quad (7)$$

where $q_i = \pm\omega_i^2$, $\omega_i > 0$ $(i = 1, 2)$, we represent the double oscillator in analogy to (2) by putting

$$\mathbf{x} = \begin{bmatrix} x_1 \\ x_2 \end{bmatrix}, x_1 = [y_1, y_2]', x_2 = [\dot{y}_1, \dot{y}_2]' \in \mathbf{R}^2$$

(rather than keeping together the position and velocity of each of the single oscillators as is often done: $x_i = [y_i, \dot{y}_i]', i = 1, 2$). Then (6) and (7) reads

$$d\mathbf{x} = \mathbf{A}\,\mathbf{x}\,dt + \mathbf{B}\mathbf{x} \circ dF_{t/\epsilon}, \quad (8)$$

with

$$\mathbf{A} = \begin{bmatrix} 0 & I \\ -Q & -E \end{bmatrix}, \mathbf{B} = \begin{bmatrix} 0 & 0 \\ -B & 0 \end{bmatrix}, \quad (9)$$

where $Q = \text{diag}(q_1, q_2)$, $E = ((e_{ij})) = \Omega D$ and $\Omega = \text{diag}(\omega_1, \omega_2)$. Note that the class of systems modelled by the equations (6) and (7), or equivalently (8), contains the system of two pendula, one hinging at the other, referred to as *double pendulum.* Here y_i denotes the angle that measures the (not too large) deviation of the ith pendulum from the vertical, and $q_i = \pm\omega_i^2$ is determined by gravity.

If both parameters q_i are negative, we have the *inverted double pendulum.* If $q_1 q_2 > 0$, then either the first pendulum is inverted ($q_1 = -\omega_1^2$), while the second pendulum is suspended at the first one ($q_2 = \omega_2^2$), or the first pendulum is normally suspended ($q_1 = \omega_1^2$) and supports the inverted second one ($q_2 = -\omega_2^2$). Of course the model includes also the stable system of the two pendula in normal suspension.

In the case of the double pendulum with at least one pendulum inverted we are interested in stabilizing the system with only one noise source: that is by randomly vibrating the support (or suspension point) of only the first pendulum - a situation which is modelled with help of vanishing $detB$.

3. Admissible Noise

Concerning the noise, we consider the same two families of stochastic processes F_t whose derivatives dF_t can stabilize the single (inverted) pendulum. That is

(I) F_t is stationary and ergodic with mean zero and finite variance and P-almost surely differentiable with respect to t,

$$dF_t = f_0(t)dt; \quad (10)$$

or

(II) F_t is defined with help of a background noise as a generalization of the Ornstein-Uhlenbeck process (see example (ii) below). Let ξ_t be a non-degenerate stationary and ergodic diffusion process satisfying the Stratonovich stochastic differential equation

$$d\xi_t = X_0(\xi_t)dt + \sum_{k=1}^{r} X_k(\xi_t) \circ dW_t^k, \xi_0 \in M, \tag{11}$$

where $X_0, X_1, \ldots, X_r$ are smooth vectorfields on a smooth manifold M and $W_t^1, \ldots, W_t^r$ are independent standard Wiener processes. Let $\mathcal{G}$ denote the infinitesimal generator of the Markov process ξ_t. To define F_t, let φ_0 be a smooth function on M with $\mathbf{E}\varphi_0(\xi) = 0$ and $\mathbf{E}\varphi_0^2 < \infty$ and $\psi = \mathcal{G}^{-1}(\varphi_0)$; then

$$F_t = \psi(\xi_t). \tag{12}$$

There is an equivalent description of the second family:

(II') Given the background noise ξ_t on the manifold M and a smooth function φ_0 with $\mathbf{E}\varphi_0(\xi) = 0$ and $\mathbf{E}\varphi_0^2 < \infty$ as above, we choose further smooth functions $f_1, \ldots, f_r$ on M and consider

$$F_t = F_0 + \int_0^t \varphi_0(\xi_s)ds + \sum_{k=1}^{r} \int_0^t f_k(\xi_s)dW_s^k \qquad \text{(Itô)}, \tag{13}$$

or in Stratonovich form

$$F_t = F_0 + \int_0^t f_0(\xi_s)ds + \sum_{k=1}^{r} \int_0^t f_k(\xi_s) \circ dW_s^k \qquad \text{(Stratonovich)}, \tag{14}$$

where $f_0 = \varphi_0 + \frac{1}{2}\sum_{k=1}^{r}[X_k(\xi) \cdot \nabla f_k(\xi)]$. We now *require*

$$lim_{t\to\infty} \frac{\mathbf{E}F_t^2}{t} = 0. \tag{15}$$

Examples and counter examples.

(i) $F_t = \sigma cos(\alpha t + \theta)$ $\sigma, \alpha > 0$, and θ uniformly distributed over $[0, 2\pi]$; $dF_t = -\sigma\alpha\sin(\alpha t + \theta)dt$.

(ii) $F_t = \xi_t$ = Ornstein-Uhlenbeck process = background noise: $d\xi_t = -\gamma_0\xi_t + \gamma_1 dW_t$, ξ_0 normally distributed with mean 0 and variance $\gamma_1^2/(2\gamma_0)$, $\varphi_0(\xi) = -\gamma_0\xi$, $F_t = \psi(\xi_t) = \xi_t$ (or $f_0(\xi) = \varphi_0(\xi) = -\gamma_0\xi, r = 1, f_1(\xi) = const = \gamma_1$, $dF_t = d\xi_t$).

(iii) Note that white noise, $dF_t = dW_t$, does not belong to the families described above, since $\frac{\mathbf{E}F_t^2}{t} = \frac{\mathbf{E}W_t^2}{t} = \frac{t}{t} = 1$.

Remark.

(i) As to the equivalency of the definitions (II) and (II') of the process F_t above, note that (II') follows from (II) immediately by Itô's formula. Putting $f_k = X_k(\xi) \cdot \nabla_\xi \psi(\xi)$, together with $\mathcal{G}(\psi) = \varphi_0$ we have

$$dF_t = d\psi(\xi_t) = \varphi_0(\xi_t)dt + \sum_{k=1}^{r} f_k(\xi_t)dW_t^k$$

with $\mathbf{E}\varphi_0 = 0$, $\mathbf{E}\varphi_0^2 < \infty$, thus $\mathbf{E}F(\xi_t)^2 < \infty$, and therefore $\frac{\mathbf{E}F(\xi_t)^2}{t} = \frac{const}{t} \to 0$ as $t \to \infty$.

The proof of the implication (II') $\Rightarrow$ (II), that is to say that the functional F_t of $\xi_\cdot$ in (II') is actually a function of ξ_t is more involved (see Kao and Wihstutz [13]).

(ii) Note that in any case F_t is stationary and ergodic with mean zero. Therefore $\mathbf{E}(dF) = 0$ and the random vectorfield $\mathbf{x} \to [\mathbf{A} + \mathbf{B}\dot{F}_t]\mathbf{x}$ or $\mathbf{x} \to [(\mathbf{A} + \varphi_0\mathbf{B})dt + \sum_{k=1}^{r} f_k\mathbf{B}dW_t^k]\mathbf{x}$, respectively, averages out to $\mathbf{x} \to \mathbf{A}\mathbf{x}$ of the unperturbed system. So dF can be regarded as *random vibration*. It is a stochastic analogue of what Meerkov ([15], [16]) and Bellman et. al ([7]) call *vibrational control.*

4. A Stabilization Result

The following stabilization result will be shown true.

Lemma 1 *Given the system (8), (9) that is*

$$d\mathbf{x} = \mathbf{A}\mathbf{x}dt + \mathbf{B}\mathbf{x} \circ dF_{t/\epsilon}, \tag{16}$$

where

$$\mathbf{A} = \begin{bmatrix} 0 & I \\ -Q & -E \end{bmatrix}, \mathbf{B} = \begin{bmatrix} 0 & 0 \\ -B & 0 \end{bmatrix}, \tag{17}$$

and $F_{t/\epsilon}$ is an admissible speeded up stochastic process from (I) or (II) with $\sigma^2 := \mathbf{E}F_t^2$, put

$$G = G(\sigma^2) = Q - \frac{1}{4}E^2 + \sigma^2 B^2. \tag{18}$$

If

$$traceG > 0, detG > 0 \text{ and } (traceG)^2 - 4detG > 0, \tag{19}$$

then the top Lyapunov exponent $\lambda(\epsilon)$ of (16) satisfies

$$\lambda(\epsilon) \to -\frac{1}{2}traceE \text{ as } \epsilon \to 0. \tag{20}$$

Therefore, if $traceE > 0$*, for* $\epsilon > 0$ *sufficiently small, the system (16) is almost surely globally exponentially stable.*

We will show that the assumptions (19) of the lemma are satisfied for σ^2 sufficiently large, if the matrix $B = ((b_{ij}))$ satisfies

$$\det B > 0, \text{trace}B \neq 0 \text{ and } b_{12}b_{21} > 0 \tag{21}$$

or if ($detB = 0$, but for instance)

$$B = b_1 \begin{bmatrix} 1 & \eta \\ \frac{c}{\eta} & c \end{bmatrix}, \tag{22}$$

where $b_1 \neq 0$, $c > 0$, and: $\eta > 0$ sufficiently small and $e_{12}\text{trace}E > 0$ or $\eta > 0$ sufficiently large and $e_{21}\text{trace}E > 0$. So we obtain

Theorem 2 *The double oscillator (16) with* $Q = diag(\pm\omega_1^2, \pm\omega_2^2)$ *that satisfies the conditions (21) or (22) can be stabilized by the mean zero noise* $dF_{t/\epsilon}$ *with* F_t *from family (I) or (II).*

To emphazise that this result includes stabilizability of an inverted double pendulum we state the following corollary.

Corollary 3 *The inverted double pendulum satisfying (22) can be stabilized with help of one noise source that is, more precisely, by randomly vibrating the supporting point with mean zero noise* $dF_{t/\epsilon}$*, where the stochastic process* F_t *is taken from family (I) or family (II).*

5. Proofs

5.1 Proof of the lemma

The proof of Lemma 5.1 follows the principles established for the single oscillator. It has 4 steps: Reduction to trace zero, random transformation, skew-symmetrization and homogeneization (averaging out the fast variables).

Reduction to trace zero. The linear transformation $\mathbf{T}_E = \begin{bmatrix} I & 0 \\ \frac{1}{2}E & I \end{bmatrix}$ (which preserves the Lyapunov exponents) and subsequent addition of the matrix $\mathbf{D}_E = \begin{bmatrix} \frac{1}{2}E & 0 \\ 0 & \frac{1}{2}E \end{bmatrix}$ to the drift matrix at the right hand side (which changes the Lyapunov exponents) yields the trace zero system

$$d\hat{\mathbf{x}} = \hat{\mathbf{A}}\,\hat{\mathbf{x}}\,dt + \hat{\mathbf{B}}\,\hat{\mathbf{x}} \circ dF_{t/\epsilon}, \tag{23}$$

where $\hat{\mathbf{B}} = \mathbf{T}_E \mathbf{B} \mathbf{T}_E^{-1} = \mathbf{B}$ and

$$\hat{\mathbf{A}} = \mathbf{T}_E \mathbf{A} \mathbf{T}_E^{-1} + \mathbf{D}_E = \begin{bmatrix} -\frac{1}{2}E & I \\ -Q + \frac{1}{4}E^2 & -\frac{1}{2}E \end{bmatrix} + \begin{bmatrix} \frac{1}{2}E & 0 \\ 0 & \frac{1}{2}E \end{bmatrix} = \begin{bmatrix} 0 & I \\ -Q + \frac{1}{4}E^2 & 0 \end{bmatrix} \tag{24}$$

and the top Lyapunov exponent

$$\lambda(\epsilon) = \lambda^{\mathbf{x}}(\epsilon) = \lambda^{\hat{\mathbf{x}}}(\epsilon) - \frac{1}{4} traceE. \tag{25}$$

So, if $traceE$ is positive, then for ϵ small $\lambda^{\mathbf{x}}(\epsilon)$ becomes negative, *provided*

$$\lambda^{\hat{\mathbf{x}}}(\epsilon) \to 0 \text{ as } \epsilon \to 0, \tag{26}$$

which will be shown in the remaining steps.

Random transformation. Since the stochastic process F_t from section 3 satisfies $lim_{t\to\infty} \frac{1}{t} F_t = 0$, its pathwise Lyapunov exponent $\lambda(F) = lim_{t\to\infty} \frac{1}{t} \log |F_t|$ vanishes P-almost surely (see Kao and Wihstutz [12]). Therefore the random transformation $\mathbf{z} = \mathbf{T}(t/\epsilon)\hat{\mathbf{x}}$ of $\mathbf{R}^4$ with

$$\mathbf{T}(t/\epsilon) = \begin{bmatrix} I & 0 \\ F_{t/\epsilon} B & I \end{bmatrix} \tag{27}$$

preserves the Lyapunov exponents, that is to say

$$\lambda^{\hat{\mathbf{x}}}(\epsilon) = \lambda^{\mathbf{z}}(\epsilon). \tag{28}$$

By Itô's formula the noisy vector field in (23), $\hat{\mathbf{B}}\hat{\mathbf{x}} \circ dF_{t/\epsilon}$, vanishes so that we obtain the random family of ODEs

$$d\mathbf{z} = \mathbf{T}(t/\epsilon)\hat{\mathbf{A}}\mathbf{T}(t/\epsilon)^{-1}\mathbf{z}dt = \begin{bmatrix} -F_{t/\epsilon}B & I \\ -Q + \frac{1}{4}E^2 - F_{t/\epsilon}^2 B^2 & F_{t/\epsilon}B \end{bmatrix} \mathbf{z}dt, \tag{29}$$

which we rewrite as

$$d\mathbf{z} = \{\mathbf{C}(G) + \mathbf{F}(t/\epsilon, B)\}\mathbf{z}dt, \tag{30}$$

with constant matrix

$$\mathbf{C}(G) = \begin{bmatrix} 0 & I \\ -G & 0 \end{bmatrix}, \tag{31}$$

$G = Q - \frac{1}{4}E^2 + \sigma^2 B^2$, $\sigma = EF_0^2$, and fast oscillating matrix of mean zero

$$\mathbf{F}(t/\epsilon, B) = \begin{bmatrix} -F_{t/\epsilon}B & 0 \\ -F_{t/\epsilon}^2 B^2 + \sigma^2 B^2 & F_{t/\epsilon}B \end{bmatrix}. \tag{32}$$

Skew symmetry. There exists a constant linear transformation $\mathbf{T}_{skew}$ of $\mathbf{R}^4$ which renders $\mathbf{C}(G)$ skew-symmetric (while preserving the properties of $\mathbf{F}(t/\epsilon, B)$), if all eigenvalues α of $\mathbf{C}(G)$ are purely imaginary and distinct. The four roots $\alpha_1, ..., \alpha_4$ of the characteristic equation of $\mathbf{C}(G)$, which is computed as

$$\alpha^4 + (traceG)\alpha^2 + detG = 0,$$

are given by $\rho = \alpha^2$, where

$$\rho_{1,2} = -\frac{traceG}{2} \pm \frac{1}{2}\sqrt{(traceG)^2 - 4detG} \tag{33}$$

and $\alpha_{1,2} = \pm\sqrt{\rho_1}$, $\alpha_{3,4} = \pm\sqrt{\rho_2}$. It follows that the eigenvalues α_i, $(i = 1, 2, 3, 4)$ are purely imaginary and distinct if ρ_1 and ρ_2 are negative and distinct. The latter is the case, if

$$\text{trace}G > 0 \text{ and } 0 < (\text{trace}G)^2 - 4\det G < (traceG)^2 \tag{34}$$

or, equivalently,

$$\text{trace}G > 0, \det G > 0 \text{ and } (\text{trace}G)^2 - 4\det G > 0. \tag{35}$$

Given (35), we consider the transformation $\mathbf{z} \to \tilde{\mathbf{z}} = \mathbf{T}_{skew}\mathbf{z}$,

$$d\tilde{\mathbf{z}} = \{\tilde{\mathbf{C}} + \tilde{\mathbf{F}}(t/\epsilon, B)\}\tilde{\mathbf{z}}dt, \tag{36}$$

with skew-symmetric 4×4 -matrix $\tilde{\mathbf{C}} = \mathbf{T}_{skew}\mathbf{C}(G)\mathbf{T}_{skew}^{-1}$ and the stationary and ergodic fast oscillating mean-zero matrix

$$\tilde{\mathbf{F}}(t/\epsilon, B) = \mathbf{T}_{skew} \begin{bmatrix} -F_{t/\epsilon}B & 0 \\ -F_{t/\epsilon}^2 B^2 + \sigma^2 B^2 & F_{t/\epsilon}B \end{bmatrix} \mathbf{T}_{skew}^{-1}. \tag{37}$$

Remark:. We remark that under our assumptions on the 2×2-matrix G, this matrix has two distinct real positive eigenvalues, is thus diagonalizable, and $\tilde{G} = T_{diag}GT_{diag}^{-1} = diag(\gamma_1, \gamma_2)$, T_{diag} a suitable transformation of $\mathbf{R}^2$, has an invertible root $\sqrt{\tilde{G}}$. Therefore we can choose the transformation

$$\mathbf{T}_{skew} = Diag(I, \sqrt{\tilde{G}}^{-1}) \circ Diag(T_{diag}, T_{diag}),$$

to obtain

$$\tilde{\mathbf{C}} = \begin{bmatrix} 0 & diag(\sqrt{\gamma_1}, \sqrt{\gamma_2}) \\ -diag(\sqrt{\gamma_1}, \sqrt{\gamma_2}) & 0 \end{bmatrix}.$$

Averaging. Applying the averaging principle from [12] with respect to the infinite horizon $[0, \infty)$ yields the desired relation (26),

$$\lambda^{\hat{\mathbf{x}}}(\epsilon) = \lambda^{\mathbf{z}}(\epsilon) = \lambda^{\tilde{\mathbf{z}}}(\epsilon) \to \mathbf{s} \cdot \tilde{\mathbf{C}}(\tilde{G})\mathbf{s} = 0, \text{ as } \epsilon \to 0.$$

The lemma is proved.

5.2 Proof of the lemma

We first note that the entries of $G = Q - \frac{1}{4}E^2 + \sigma^2 B^2 = ((g_{ij}))$ are

$$\begin{aligned} g_{11} &= q_{11} - \frac{1}{4}(e_{11}^2 + e_{12}e_{21}) + \sigma^2(b_{11}^2 + b_{12}b_{21}), \\ g_{12} &= q_{12} - \frac{1}{4}(e_{12}\text{trace}E) + \sigma^2(b_{12}\text{trace}B), \\ g_{21} &= q_{21} - \frac{1}{4}(e_{21}\text{trace}E) + \sigma^2(b_{21}\text{trace}B), \\ g_{22} &= q_{22} - \frac{1}{4}(e_{22}^2 + e_{12}e_{21}) + \sigma^2(b_{22}^2 + b_{12}b_{21}). \end{aligned}$$

So, if $b_{12}b_{21} > 0$, then for σ^2 sufficiently large: $g_{11} > 0$ and $g_{22} > 0$, thus trace$G > 0$. If, in addition, trace$B$ does not vanish, then (again for sufficiently large σ^2) g_{12} and g_{21} have the same sign $\neq 0$, thus $g_{12}g_{21} > 0$. Therefore $(\text{trace}G)^2 - 4detG = (g_{11} - g_{22})^2 + 4g_{12}g_{21} > 0$ (and G has distinct positive eigenvalues $\gamma_1 \neq \gamma_2$ and is thus diagonalizable.) It remains to show that $\det G > 0$.

Now, due to bi-linearity (of det as function of the columns), $\det G$ is of the form

$$\det G = \det G(\sigma^2) = const + \sigma^2 const + \sigma^4 detB^2. \tag{38}$$

So, *in case* $detB \neq 0$, $detG > 0$ follows readily from this representation for σ^2 large enough.

In case $detB = 0$, $B \neq 0$, note that with $b = \begin{bmatrix} b_1 \\ Kb_1 \end{bmatrix} \in \mathbf{R}^2$, without loss of generality $B = [b, \eta b] = b_1 \begin{bmatrix} 1 & \eta \\ K & \eta K \end{bmatrix}, b_1 \neq 0, B^2 = BtraceB, traceB = b_1(1 + \eta K)$.

Let us denote the first and second column of $Q - \frac{1}{4}E^2$ by v and w, respectively. Then due to the bi-linearity of the determinant and since $detB^2 = (traceB)^2 \det[b, \eta b] = 0$,

$$\begin{aligned} detG &= det\{[v, w] + \sigma^2(traceB)[b, \eta b]\} \\ &= det[v + \sigma^2(traceB)b, w + \sigma^2(traceB)\eta b] \\ &= det[v, w] + \sigma^2(traceB)det[\eta v - w, b]. \end{aligned}$$

Since for $Q = diag(q_1, q_2)$, $(q_i = \pm\omega_i^2)$, $v = \begin{bmatrix} q_1 & - & \frac{1}{4}(e_{11}^2 + e_{12}e_{21}) \\ & - & \frac{1}{4}(e_{21}traceE) \end{bmatrix}$

and $w = \begin{bmatrix} & - & \frac{1}{4}(e_{12}traceE) \\ q_2 & - & \frac{1}{4}(e_{22}^2 + e_{12}e_{21}) \end{bmatrix}$, we have

$$\det[\eta v - w, b](traceB)$$
$$= \{\eta[q_1 - \frac{1}{4}(e_{11}^2 + e_{12}e_{21})] + \frac{1}{4}(e_{12}traceE)\}\{Kb_1(traceB)\}$$
$$-\{\eta[-\frac{1}{4}(e_{21}traceE)] - [q_2 - \frac{1}{4}(e_{22}^2 + e_{12}e_{21})]\}\{b_1(traceB)\}$$
$$= \{(\eta K)q_1 + q_2 - \frac{1}{4}[(\eta K)(e_{11}^2 + e_{12}e_{21}) + (e_{22}^2 + e_{12}e_{21})]$$
$$+\frac{1}{4}K(e_{12}traceE) + \frac{1}{4}\eta(e_{21}traceE)\}b_1(traceB).$$

It follows that $det[\eta v - w, b](traceB) > 0$, whence $detG > 0$ for σ^2 sufficiently large, *if* $b_1 \neq 0, c > 0, \eta K = c > 0$ and: $K = c/\eta > 0$ sufficiently large and $e_{12}(traceE) > 0$ or $\eta > 0$ sufficiently large and $e_{21}(traceE) > 0$. This proves theorem 2.

Remark.

(i) On skew-symmetrization. If $\mathbf{C}(G) = \begin{bmatrix} 0 & I \\ -G & 0 \end{bmatrix}$ cannot be linearly transformed into a skew-symmetric matrix, a closer look at the averaging principle in [12] is necessary. However, it is doubtful that stabilization by random vibration is possible in this case, the reason being that the averaging probably is not of such a kind that all directions of the state space are more or less equally often visited, so that the modes are not "uniformly mixed".

(ii) On averaging. The averaging as discussed in this paper is imposed on the system by the noise in a much more indirect way than , for instance, in case of the system $dx = Axdt + \frac{1}{\epsilon}Bx \circ dW_t$, A and B 2×2- matrices with $B = \begin{bmatrix} 0 & 1 \\ -1 & 0 \end{bmatrix}$ skew-symmetric, which has been discussed in an inspiring paper of L. Arnold [3], 1979 (that kicked off quite a few studies on stability of dynamical systems in presence of noise; for higher dimensions see e.g., L. Arnold [4](1990) and L. Arnold, et. al [5](1983), [6](1996)). In L. Arnold [3] the white noise forces the system directly to move through all directions - the more "uniformly," the smaller ϵ. In the situation discussed in this paper, the mixing of the modes is only seen in a randomly moving coordinate system, that is for system (36). Since

F_t is stationary and ergodic, Oceledec's Multiplicative Ergodic Theorem [17] holds for this system. There are p distinct Lyapunov exponents $\lambda_i^{\tilde{\mathbf{z}}}(\epsilon), i = 1, 2, \ldots, p$, $(1 \leq p \leq 4)$ which sum up to $trace[\tilde{\mathbf{C}} + \tilde{\mathbf{F}}] = 0$. This implies that there are both stable and unstable modes. The averaging principle tells us that all Lyapunov exponents go to zero with ϵ tending to zero, not only the top one, meaning that the stable modes become less stable,while the unstable ones become stable, truly an averaging or mixing effect.

References

[1] S. T. Ariaratnam, D. S. F. Tam, and Wei-Xau Xie, "Lyapunov exponents and stochastic stability of coupled linear systems under white noise excitation," *Prob. Engineer. Mechan.*, vol. 6, no. 2, pp. 51–56, 1991.

[2] S. T. Ariaratnam and Wei-Xau Xie, "Lyapunov exponents and stochastic stability of coupled linear systems under real noise excitation," *Transaction of the ASME*, vol. 59, pp. 664–673, 1992.

[3] L. Arnold, "A new example of an unstable system being stabilized by random parameter noise," *Inform. Comm. Math. Chem.*, pp. 133–140, 1979.

[4] L. Arnold, "Stabilization by noise revisited," *Z. Angew. Math. Mech.*, vol. 70, pp. 235–246, 1990.

[5] L. Arnold, H. Crauel, and V. Wihstutz, "Stabilization of linear systems by noise," *SIAM J. Control Optim.*, vol. 21, pp. 451–461, 1983.

[6] L. Arnold, A. Eisenberg, and V. Wihstutz, "Large noise asymptotics of invariant measures, with application to Lyapunov exponents," *Stochastics and Stochastic Reports*, vol. 59, pp. 71–142, 1996.

[7] R. Bellman, J. Bentsman, and S. Meerkov, "Vibrational control of nonlinear systems," *IEEE Trans. Automat. Control*, vol. 31, pp. 710–724, 1986.

[8] J. L. Bogdanoff and S. J. Criton, *Experiments with an inverted pendulum subject to random parametric excitation*, Report Purdue University, Lafayette, 1965.

[9] M. M. Doyle, N. Sri Namachchivaya, and H. J. Van Roessel, "Asymptotic stability of structural systems based on Lyapunov exponents and moment Lyapunov exponents," *Int. J. Non-Linear Mechanics*, vol. 32, no. 4, pp. 681–692, 1997.

[10] H. Furstenberg, "Noncommutating random products," *Trans. Amer. Math. Soc.*, vol. 108, pp. 377–428, 1963.

[11] R. Z. H'asminsky (Khasminskii), *Stochastic Stability of Differential Equations*, Russ.: Nauka, Moscow, 1969, Engl.: Alphen, Sijthoff and Noordhoff, 1980.

[12] J. Kao and V. Wihstutz, "Stabilization of companion form systems by mean zero noise," *Stochastics and Stochastic Reports*, vol. 49, pp. 1–25, 1994.

[13] J. Kao and V. Wihstutz, "Characterization of stochastic processes which stabilize linear companion form systems," *Stoch. Proc. Appl.*, vol. 89, pp. 49–68, 2000.

[14] P. L. Kapitsa, "Dynamic stability of a pendulum with an oscillating support point," *Zh. Eksperim. Teor. Fiz.*, vol. 21, no. 5, 1951.

[15] S. M. Meerkov, "Vibrational control," *Automation and Remote Control*, vol. 34, pp. 201–209, 1973.

[16] S. M. Meerkov, "Principle of vibrational control: Theory and Application," *IEEE Trans. Automat. Control*, vol. 25, pp. 755–762, 1980.

[17] V. I. Oseledec, "A multiplicative ergodic theorem: Lyapunov characteristic numbers for dynamical systems," *Trans. Moscow Math. Soc.*, vol. 19, pp. 197–231, 1968.

[18] K. Popp, Experiments of stabilizing the inverted pendulum, Technical University Hannover, Germany (1995)

[19] N. Sri Namachchivaya, H. J. Van Roessel, and M. M. Doyle, "Moment Lyapunov exponent for two coupled oscillators driven by real noise," *SIAM J. Appl. Math.*, vol. 56, no. 5, pp. 1400–1433, 1996.

[20] N. Sri Namachchivaya and Lalit Vedula, "Real noise stabilization of linear systems: Application to flow induced oscillations," *J. Dynamics and Stabilization of Systems*, vol. 15, no. 2, pp. 185–208, 2000.

[21] W. V. Wedig, "Lyapunov exponents of stochastic systems and related bifurcation problems," in: *Stochastic Structural Dynamics: Progress in theory and Applications*, S. T. Ariaratnam, G. I. Schueller, and I. Elishakoff (eds.), Elsevier Applied Science, London and New York, 1988.

[22] V. Wihstutz, "Stabilization by random vibration," *Design Engineering (DE)*, vol. 84, no. 1, pp. 881–892, 1995.

STABILITY OF A TWO-DIMENSIONAL SYSTEM UNDER COMBINED HARMONIC AND REAL NOISE PARAMETRIC EXCITATIONS

Wei-Chau Xie
Solid Mechanics Division, Faculty of Engineering
University of Waterloo, Waterloo, Ontario, Canada N2L 3G1
xie@uwaterloo.ca

Abstract The dynamic stability of a two-dimensional system under combined excitations of harmonic and real noise, which is modelled as an Ornstein-Uhlenbeck process, is studied through the determination of the moment Lyapunov exponents and the Lyapunov exponents. The influence of the real noise excitation on the primary parametric resonance due to the harmonic excitation is investigated.

Keywords: Dynamic stability, Moment Lyapunov exponents, Lyapunov exponents, combined harmonic and real noise excitations

1. Introduction

The investigation of the dynamic stability of elastic systems frequently leads to the study of the dynamical behaviour of the solutions of a parameterized family of differential equations of the form

$$\dot{\mathbf{x}} = \mathbf{f}(\mathbf{x}, \gamma), \quad \mathbf{x} = (x_1, x_2, \ldots, x_n)^T \in \mathbb{R}^n, \quad \mathbf{f}(\mathbf{0}, \gamma) = \mathbf{0}, \tag{1}$$

where γ is a scalar parameter characterizing the load condition.

In many practical situations, the loading may be subjected to fluctuations of a stochastic nature, such as those arising from earthquakes, wind, and ocean waves. The Ornstein-Uhlenbeck process given by $\mathrm{d}\xi(t) = -\alpha\,\xi(t)\,\mathrm{d}t + \sigma\,\mathrm{d}W(t)$, where $W(t)$ is a standard Wiener process, is a simple, Gaussian, explicitly representable stationary process, and is often used to model a realizable noise process. There are also engineering situations where the loading on the system contains both periodic components and stochastic fluctuations. An example of such a system is the uncoupled flapping motion of rotor blades in forward flight under the effect of atmospheric turbulence.

N. Sri Namachchivaya and Y.K. Lin (eds.),
IUTAM Symposium on Nonlinear Stochastic Dynamics, 191-201.

The sample or almost-sure stability of the trivial solution of system (1) is determined by the Lyapunov exponent of its linearized system, which characterizes the average exponential rates of growth of the solutions of the system for time parameter t large, defined as

$$\lambda = \lim_{t\to\infty} \frac{1}{t} \log \|\mathbf{x}(t)\|, \tag{2}$$

where $\|\cdot\|$ denotes the Euclidean vector norm. The trivial solution of the dynamical system is stable with probability one (w.p.1) if the top Lyapunov exponent is negative, whereas it is unstable w.p.1 if the top Lyapunov exponent is positive.

On the other hand, the stability of the pth moment, $E[\|\mathbf{x}(t)\|^p]$, of the trivial solution of the dynamical system is determined by the moment Lyapunov exponent defined as

$$\Lambda(p) = \lim_{t\to\infty} \frac{1}{t} \log E\left[\|\mathbf{x}(t)\|^p\right], \tag{3}$$

where $E[\cdot]$ denotes expected value. If $\Lambda(p) < 0$, then $E[\|\mathbf{x}(t)\|^p] \to 0$ as $t \to \infty$. The pth moment Lyapunov exponent $\Lambda(p)$ is a convex analytic function in p with $\Lambda(0) = 0$ and the top Lyapunov exponent $\lambda = \Lambda'(0)$.

To have a complete picture of the dynamic stability of a dynamical system, it is important to study both the sample and moment stability and to determine both the top Lyapunov exponent and the pth moment Lyapunov exponent. A systematic presentation of the theory of random dynamical systems and a comprehensive list of references can be found in [1].

Although the moment Lyapunov exponents are important in the study of dynamic stability of stochastic systems, the actual evaluations of the moment Lyapunov exponents are very difficult. Using the analytic property of the moment Lyapunov exponents, [2] obtained weak noise expansions of the moment Lyapunov exponents of a two-dimensional system in terms of εp, where ε is a small parameter, under both white noise and real noise excitations. Khasminskii and Moshchuk [3] obtained an asymptotic expansion of the moment Lyapunov exponent of a two-dimensional system under white noise parametric excitation in terms of the small fluctuation parameter ε. Sri Namachchivaya and Van Roessel [5] studied the moment Lyapunov exponents of two coupled oscillators driven by real noise. Xie [7] obtained weak noise expansions of the moment Lyapunov exponent in terms of the small fluctuation parameter, of a two-dimensional system exhibiting pitch-fork bifurcation under real noise excitation.

The almost sure stability of dynamical systems under combined harmonic and stochastic excitations was first studied by [4] by evaluating the largest Lyapunov exponents.

In this paper, the dynamic stability of a two-dimensional system under combined parametric excitations of harmonic and real noise is studied through the determination of the moment Lyapunov exponents and Lyapunov exponents.

2. Formulation

Consider a two-dimensional system under combined harmonic and real noise excitations as follows

$$q''(\tau) + 2\varepsilon\beta q'(\tau) + \left[\Omega_0^2 + \varepsilon\mu_0 \sin\nu_0\tau + \varepsilon^{1/2}\gamma_0\xi(\tau)\right]q(\tau) = 0, \tag{4}$$

where $\varepsilon > 0$ is a small parameter, and $\xi(\tau)$ is a real noise process in time τ, modelled as an Ornstein-Uhlenbeck process, given by

$$\mathrm{d}\xi(\tau) = -\alpha_0\xi(\tau)\,\mathrm{d}\tau + \sigma_0\,\mathrm{d}W(\tau). \tag{5}$$

Applying the transformation $q(\tau) = x(\tau)\,\mathrm{e}^{-\varepsilon\beta\tau}$ and time scaling $t = \Omega\tau$, (4) and (5) become

$$\ddot{x}(t) + \left[1 + \varepsilon\mu\sin\nu t + \varepsilon^{1/2}\gamma\zeta(t)\right]x(t) = 0, \tag{6}$$

$$\mathrm{d}\zeta(t) = -\alpha\zeta(t)\,\mathrm{d}t + \sigma\,\mathrm{d}W(t), \tag{7}$$

where $\mu = \mu_0/\Omega^2$, $\nu = \nu_0/\Omega$, $\gamma = \gamma_0/\Omega^2$, $\alpha = \alpha_0/\Omega$, $\sigma = \sigma_0/\sqrt{\Omega}$, $\Omega^2 = \Omega_0^2 - \varepsilon^2\beta^2$.

The Lyapunov exponents and the moment Lyapunov exponents of systems (4) and (6) are related as follows:

$$\lambda_{q(\tau)} = -\varepsilon\beta + \Omega\lambda_{x(t)}, \qquad \Lambda_{q(\tau)}(p) = -\varepsilon p\beta + \Omega\Lambda_{x(t)}(p).$$

The eigenvalue problem governing the moment Lyapunov exponent of system (6) can be established using an approach originally employed by [6] for a two-dimensional linear Itô stochastic system.

Letting $x_1 = x$, $x_2 = \dot{x}$, $\theta = \nu t$, applying the transformation $\cos\varphi = x_1/\|\mathbf{x}\|$, $\sin\varphi = x_2/\|\mathbf{x}\|$, $\|\mathbf{x}\| = (x_1^2 + x_2^2)^{1/2}$, and defining the pth norm $P = \|\mathbf{x}\|^p$, the Itô differential equations for P and φ can be obtained using Itô's Lemma:

$$\mathrm{d}\begin{Bmatrix} P \\ \varphi \end{Bmatrix} = \begin{Bmatrix} -pP(\varepsilon\mu\sin\theta + \varepsilon^{1/2}\gamma\zeta)\cos\varphi\sin\varphi \\ -\left[1 + (\varepsilon\mu\sin\theta + \varepsilon^{1/2}\gamma\zeta)\cos^2\varphi\right] \end{Bmatrix}\mathrm{d}t. \tag{8}$$

Applying a linear transformation $S = T(\zeta,\varphi,\theta)P$, $P = T^{-1}(\zeta,\varphi,\theta)S$, $0 \leqslant \varphi < \pi$, $0 \leqslant \theta < 2\pi$, the Itô differential equation for the transformed pth norm process S can also be obtained using Itô's Lemma:

$$\mathrm{d}S = \Big\{\tfrac{1}{2}\sigma^2 T_{\zeta\zeta} - \alpha\zeta T_\zeta + \nu T_\theta - \left[1 + (\varepsilon\mu\sin\theta + \varepsilon^{1/2}\gamma\zeta)\cos^2\varphi\right] T_\varphi - p(\varepsilon\mu\sin\theta + \varepsilon^{1/2}\gamma\zeta)\cos\varphi\sin\varphi\, T\Big\} P\,\mathrm{d}t + \sigma T_\zeta P\,\mathrm{d}W. \quad (9)$$

For bounded and non-singular transformation $T(\zeta,\varphi,\theta)$, both processes P and S are expected to have the same stability behaviour. Therefore, $T(\zeta,\varphi,\theta)$ is chosen so that the drift term of the Itô differential equation (9) is independent of the processes ζ, φ, and θ, i.e.

$$\mathrm{d}S = \Lambda S\,\mathrm{d}t + \sigma T_\zeta T^{-1} S\,\mathrm{d}W. \quad (10)$$

Comparing the drift terms in equations (9) and (10), it is seen that such a transformation $T(\zeta,\varphi,\theta)$ is the eigenfunction of the eigenvalue problem

$$\tfrac{1}{2}\sigma^2 T_{\zeta\zeta} - \alpha\zeta T_\zeta + \nu T_\theta - \left[1 + (\varepsilon\mu\sin\theta + \varepsilon^{1/2}\gamma\zeta)\cos^2\varphi\right] T_\varphi - p(\varepsilon\mu\sin\theta + \varepsilon^{1/2}\gamma\zeta)\cos\varphi\sin\varphi\, T = \Lambda T, \quad (11)$$

with Λ being the eigenvalue. From (10), it is clear that Λ is the Lyapunov exponent of the transformed pth norm process S; hence Λ is the moment Lyapunov exponent of system (6), i.e. $\Lambda = \Lambda_{x(t)}(p)$.

3. Moment Lyapunov Exponents

In this section, a method of regular perturbation is applied to solve (11) to obtain weak noise expansions of $\Lambda_{x(t)}(p)$.

It is well known that, in the absence of the real noise excitation, system (6) is in primary parametric resonance when the harmonic excitation frequency ν is in the vicinity of 2. Hence, let $\nu = 2 + \varepsilon\Delta$, where Δ is the mistune parameter.

Applying the transformation $\theta = \psi - 2\varphi$, the differential operator in (11) becomes $\mathcal{L}(p) = L_0 + \varepsilon^{1/2}L_1 + \varepsilon L_2$, where

$$L_0 = \tfrac{1}{2}\sigma^2\frac{\partial^2}{\partial\zeta^2} - \alpha\zeta\frac{\partial}{\partial\zeta} - \frac{\partial}{\partial\varphi},$$

$$L_1 = -\gamma\zeta\Big[\cos^2\varphi\Big(2\frac{\partial}{\partial\psi} + \frac{\partial}{\partial\varphi}\Big) + p\cos\varphi\sin\varphi\Big],$$

$$L_2 = \Delta\frac{\partial}{\partial\psi} - \mu\sin(\psi - 2\varphi)\Big[\cos^2\varphi\Big(2\frac{\partial}{\partial\psi} + \frac{\partial}{\partial\varphi}\Big) + p\cos\varphi\sin\varphi\Big].$$

Expanding $\Lambda_{x(t)}(p)$ and $T(\zeta,\varphi,\psi)$ as $\Lambda_{x(t)}(p) = \sum_{i=0}^{\infty}\varepsilon^{i/2}\Lambda_i(p)$, $T(\zeta,\varphi,\psi)$ $=\sum_{i=0}^{\infty}\varepsilon^{i/2}T_i(\zeta,\varphi,\psi)$, where $T_i(\zeta,\varphi,\psi)$ are periodic functions in φ of period π and in ψ of period 2π, substituting into (11), and equating terms of equal power of ε results in the perturbation equations,

$$\begin{array}{ll} O(1): & L_0T_0 = \Lambda_0 T_0, \\ O(\varepsilon^{1/2}): & L_0T_1 + L_1T_0 = \Lambda_1 T_0 + \Lambda_0 T_1, \\ O(\varepsilon^1): & L_0T_2 + L_1T_1 + L_2T_0 = \Lambda_2 T_0 + \Lambda_1 T_1 + \Lambda_0 T_2, \\ \vdots & \vdots \end{array} \tag{12}$$

3.1 Zeroth-order perturbation

The zeroth-order perturbation equation is $L_0T_0 = \Lambda_0 T_0$. It is well known that $\Lambda_{x(t)}(0) = 0$, which implies $\Lambda_i(0) = 0$, $i = 0, 1, \ldots$. Since the perturbation equation does not contain the parameter p explicitly, $\Lambda_0(0) = 0$ suggests that $\Lambda_0(p) = 0$ for all values of p.

Seeking a solution in the form $T_0(\zeta,\varphi,\psi) = Z_0(\zeta)\Phi_0(\varphi)\Psi_0(\psi)$ and substituting into the perturbation equation. It is easy to show that $\Phi_0(\varphi) = C_1$, $Z_0(\zeta) = C_2$, and $T_0(\zeta,\varphi,\psi) = \Psi_0(\psi)$, where $\Psi_0(\psi)$ is a periodic function of period 2π.

The adjoint equation of the zeroth-order perturbation equation is

$$\tfrac{1}{2}\sigma^2 \frac{\partial^2 T_0^*}{\partial \zeta^2} + \alpha\zeta \frac{\partial T_0^*}{\partial \zeta} + \frac{\partial T_0^*}{\partial \varphi} + \alpha T_0^* = 0. \tag{13}$$

Similarly, seeking a solution of the form $T_0^*(\zeta,\varphi,\psi) = Z_0^*(\zeta)\Phi_0^*(\varphi)\Psi_0^*(\psi)$ and substituting into (13) leads to

$$\tfrac{1}{2}\sigma^2 \frac{\ddot{Z}_0^*}{Z_0^*} + \alpha\zeta \frac{\dot{Z}_0^*}{Z_0^*} + \alpha = -\frac{\dot{\Phi}_0^*}{\Phi_0^*} = \kappa,$$

and $\Psi_0^*(\psi)$ can be taken as an arbitrary periodic function of period 2π. Since the coefficients of the differential operator $\mathfrak{L}(p)$ are periodic functions in φ of period π, the Φ_0^* equation leads to

$$\Phi_0^*(\varphi) = 1/\pi, \qquad 0 \leqslant \varphi < \pi, \tag{14}$$

which is the probability density function of a random number uniformly distributed in $(0,\pi)$. The Z_0^* equation can be solved to yield

$$Z_0^*(\zeta) = \frac{1}{\sqrt{2\pi}\sigma_\zeta} \exp\Big(-\frac{\zeta^2}{2\sigma_\zeta^2}\Big), \tag{15}$$

where $\sigma_\zeta = \sigma/\sqrt{2\alpha}$. Equation (15) indicates that the real noise process $\zeta(t)$ is a normally distributed random variable with mean value $\mu_\zeta = 0$ and standard deviation σ_ζ.

3.2 First-order perturbation

The first-order perturbation equation becomes $L_0 T_1 = \Lambda_1 T_0 - L_1 T_0$. It is easy to show that $L_1 T_0 = -g_1^{(1)}(\varphi, \psi) \cdot \zeta$, where

$$g_1^{(1)}(\varphi, \psi) = \gamma\big[2\cos^2\varphi\, \dot{\Psi}_0(\psi) + p\cos\varphi \sin\varphi\, \Psi_0(\psi)\big].$$

The solvability condition of the first-order perturbation equation is given by, from Fredholm Alternative, $(\Lambda_1 T_0 - L_1 T_0, T_0^*) = 0$, where (f, g) denotes the inner product of functions $f(\zeta, \varphi, \psi)$ and $g(\zeta, \varphi, \psi)$ defined as

$$(f, g) = \int_{\psi=0}^{2\pi} \int_{\varphi=0}^{\pi} \int_{\zeta=-\infty}^{+\infty} f(\zeta, \varphi, \psi)\, g(\zeta, \varphi, \psi)\, d\zeta\, d\varphi\, d\psi.$$

Hence,

$$\Lambda_1 = \frac{(L_1 T_0, T_0^*)}{(T_0, T_0^*)} = 0.$$

The first-order perturbation equation becomes $L_0 T_1 = g_1^{(1)}(\varphi, \psi) \cdot \zeta$. From [7], a solution of $L_0 T(\zeta, \varphi, \psi) = f(\zeta) g(\varphi, \psi)$ is given by

$$T(\zeta; \hat{\varphi}, s; \psi) = \int_0^s g(\hat{\varphi} - r, \psi)\, E\big[f(z(r))\big]\, dr, \tag{16}$$

where $E\big[f(z(r))\big]$ is the expected value of the random variable $f(z(r))$, in which $z(r)$ is the normally distributed random variable with mean value $\mu_{z(r)} = \zeta e^{-\alpha(r-s)}$ and variance $\sigma^2_{z(r)} = \sigma^2\big[1 - e^{-2\alpha(r-s)}\big]/2\alpha$. The solution $T(\zeta, \varphi, \psi)$ is then obtained by replacing $\varphi = \hat{\varphi} - s$ and passing the limit $s \to -\infty$.

A solution of T_1 is obtained as $T_1(\zeta, \varphi, \psi) = G_1^{(1)}(\varphi, \psi) \cdot \zeta$, where

$$G_1^{(1)}(\varphi, \psi) = -\frac{\gamma}{2\alpha(\alpha^2 + 4)}\big[2(\alpha^2 \cos 2\varphi + 2\alpha \sin 2\varphi + \alpha^2 + 4)\, \dot{\Psi}_0(\psi)$$
$$+ p\alpha(\alpha \sin 2\varphi - 2\cos 2\varphi)\, \Psi_0(\psi)\big].$$

3.3 Second-order perturbation

Since $\Lambda_0(p) = \Lambda_1(p) = 0$, the second-order perturbation equation becomes $L_0 T_2 = \Lambda_2 T_0 - L_1 T_1 - L_2 T_0$. From Fredholm Alternative, the solvability condition is $(\Lambda_2 T_0 - L_1 T_1 - L_2 T_0,\ T_0^*) = 0$, which can be shown to be

$$\int_{\psi=0}^{2\pi} \Big\{ \int_{\varphi=0}^{\pi} \Big[\Lambda_2 \Psi_0(\psi) + \langle \zeta^2 \rangle\, S_2^{(1)}(\varphi, \psi) - \Delta\, \dot{\Psi}_0(\psi) + \mu \sin(\psi - 2\varphi)$$
$$\times \big(2\cos^2\varphi\, \dot{\Psi}_0(\psi) + p\cos\varphi \sin\varphi\, \Psi_0(\psi)\big)\Big]\, d\varphi \Big\}\, \Psi_0^*(\psi)\, d\psi = 0. \tag{17}$$

where

$$S_2^{(1)}(\varphi,\psi)=\gamma\Big[\cos^2\varphi\Big(2\frac{\partial G_1^{(1)}}{\partial\psi}+\frac{\partial G_1^{(1)}}{\partial\varphi}\Big)+p\cos\varphi\sin\varphi\, G_1^{(1)}\Big].$$

Since (17) is valid for an arbitrary function $\Psi_0^*(\psi)$, the solvability condition leads to

$$a\,\ddot{\Psi}_0(\psi)+(b+2q\,\sin\psi)\,\dot{\Psi}_0(\psi)+\big[\Lambda_2+c\,p(p+2)-pq\cos\psi\big]\,\Psi_0(\psi)=0, \quad (18)$$

where $q=\mu/4$, and

$$a=-\frac{\gamma^2\sigma^2(3\alpha^2+8)}{4\alpha^2(\alpha^2+4)},\quad b=-\frac{8\alpha\Delta+2\alpha^3\Delta+\gamma^2\sigma^2}{2\alpha(\alpha^2+4)},\quad c=-\frac{\gamma^2\sigma^2}{16(\alpha^2+4)}.$$

Hence $\Lambda_2(p)$ is the eigenvalue of the eigenvalue problem (18) with a second-order ordinary differential operator and $\Psi_0(\psi)$ is the associated eigenfunction.

3.4 Determination of Λ_2

Since the coefficients of (18) are periodic functions of period 2π, a series expansion of the eigenfunction $\Psi_0(\psi)$ may be taken as

$$\Psi_0(\psi)=C_0+\sum_{k=1}^{N}(C_k\cos k\psi+S_k\sin k\psi), \quad (19)$$

where C_0, C_k, S_k, $k=1,2,\ldots,N$, are constants to be determined. Substituting (19) into (18), multiplying the resulting equation by $\cos m\psi$, $\sin m\psi$, $m=0,1,\ldots,N$, respectively, and integrating with respect to ψ from 0 to 2π results in a set of $2N+1$ homogeneous linear equations for the unknown coefficients C_0, C_k, S_k, $k=1,2,\ldots,N$. In order to have a nontrivial solution, the determinant of the coefficient matrix of these $2N+1$ linear algebraic equations $\Delta^{(N)}$ must be zero. The determinantal equation $\Delta^{(N)}$ leads to a polynomial equation of degree $2N+1$ in Λ_2 of the form

$$\Lambda_2^{2N+1}+d_{2N}\Lambda_2^{2N}+\cdots+d_1\Lambda_2+d_0=0. \quad (20)$$

An approximate result of Λ_2 can be obtained by solving equation (20). Note that the exact result of Λ_2 is obtained when the number of terms $N\to\infty$.

When $N=1$, (20) is a cubic equation and the solution is given by

$$\Lambda_2^{(1)}=\frac{1}{6}(A_2-2d_2)-\frac{2}{3}\frac{3d_1-d_2^2}{A_2}, \quad (21)$$

where

$$A_2=\big(-108\,d_0+36\,d_1d_2-8\,d_2^3+12A_1\big)^{1/3},$$

$$A_1=\big(81\,d_0^2-54\,d_0d_1d_2+12\,d_1^3+12\,d_0d_2^3-3\,d_1^2d_2^2\big)^{1/2}.$$

When $N > 1$, there is no analytical solution available for (20) and numerical approach has to be employed to find an approximate root $\Lambda_2^{(N)}$. In Figure 1, the analytical result $\Lambda_2^{(1)}$ is plotted along with the numerical result $\Lambda_2^{(8)}$. It is

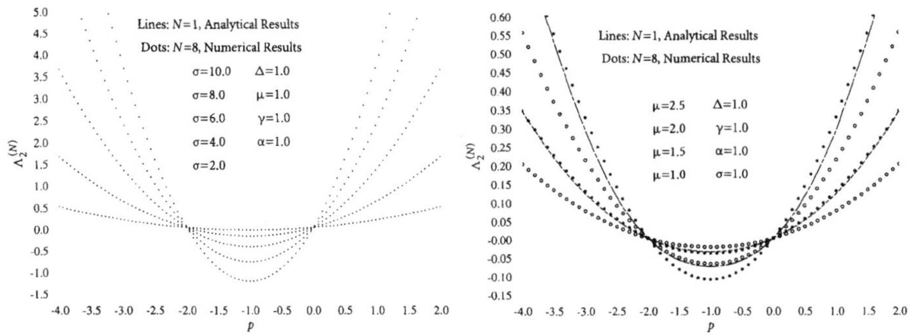

Figure 1. Second-Order Perturbation of Moment Lyapunov Exponent $\Lambda_2^{(N)}$.

seen that, for a given value of $\mu = 1$, the two results agree extremely well for all values of σ. On the other hand, for a given value of $\sigma = 1$, $\Lambda_2^{(1)}$ and $\Lambda_2^{(8)}$ agree very well for smaller values of μ up to 2. When the value of μ increases, discrepancies exist between $\Lambda_2^{(1)}$ and $\Lambda_2^{(8)}$. $3D$-surface plots of $\Lambda_2^{(8)}$ are shown in Figure 2 for various combinations of the parameters μ, γ, α, and σ to study the influence of the real noise excitation and the harmonic excitation on the parametric resonance and the moment Lyapunov exponent.

It is seen that for small values of the noise intensity σ, small values of the amplitude of real noise excitation γ, or large values of the amplitude of the harmonic excitation μ, the influence of the real noise excitation is small and the harmonic excitation is dominant; hence the effect of the primary parametric resonance is significant. On the other hand, when the noise intensity σ is increased, the amplitude of the real noise excitation γ is large, or the amplitude of the harmonic excitation μ is small, the impact of the real noise excitation intensifies and the prominence of the primary parametric resonance is decreased.

3.5 Determination of λ_2

The Lyapunov exponent is given by

$$\lambda_{x(t)} = \lim_{p \to 0} \frac{\Lambda_{x(t)}(p)}{p} \approx \varepsilon \lambda_2^{(N)} + o(\varepsilon), \qquad \lambda_2^{(N)} = \lim_{p \to 0} \frac{\Lambda_2^{(N)}(p)}{p}. \tag{22}$$

Equation (22) implies $\Lambda_2^{(N)} = O(p)$ when $p \to 0$, and hence $[\Lambda_2^{(N)}]^k = o(p)$ for $k > 1$. From (20), the second-order perturbation of the Lyapunov exponent is given by $\lambda_2^{(N)} = \lim_{p \to 0} d_0/d_1$. Since the determination of $\lambda_2^{(N)}$ does not

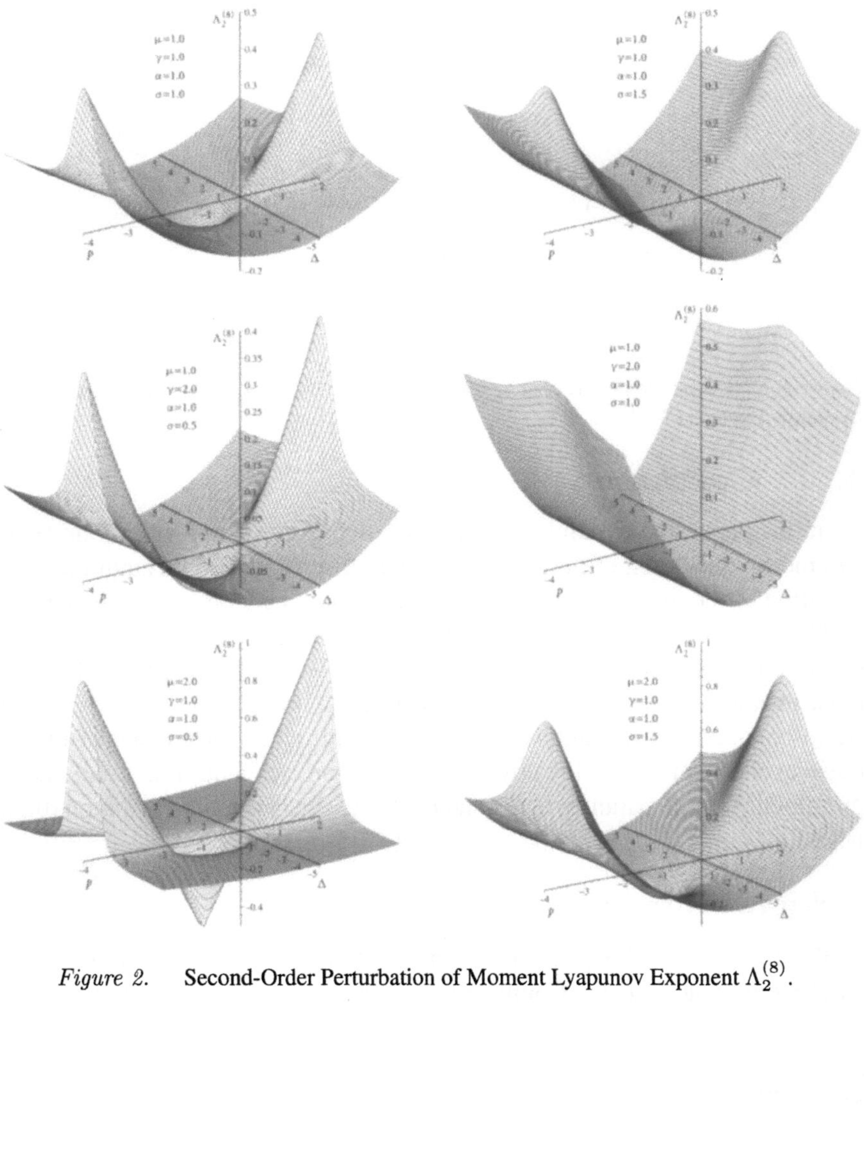

Figure 2. Second-Order Perturbation of Moment Lyapunov Exponent $\Lambda_2^{(8)}$.

require the solution of the polynomial equation (20), a larger value of N can be taken. Typical plots of $\lambda_2^{(N)}$ are shown in Figure 3 for $\sigma = 1.0$.

In order to check the validity of the perturbation results, a digital simulation is performed. Equations (6) and (7) are discretized using the Euler scheme and solved numerically. Numerical results of $\lambda_{x(t)}$ from simulation are plotted in Figure 4 along with the perturbation results $\lambda_{x(t)} \approx \varepsilon\lambda_2^{(14)}$ for various values of μ. It is observed that both results agree very well.

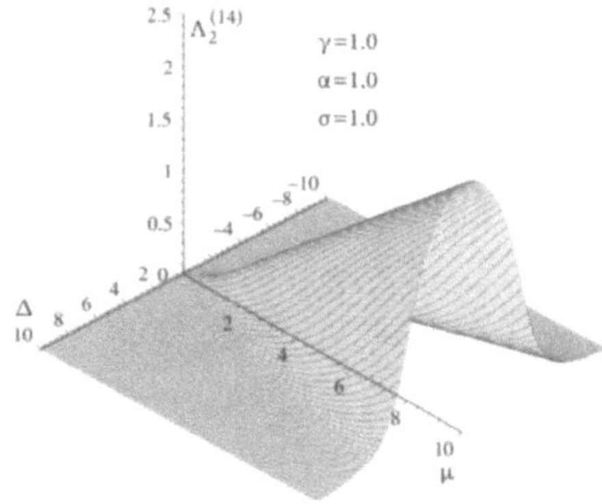

Figure 3. Second-Order Perturbation of Lyapunov Exponent $\lambda_2^{(14)}$.

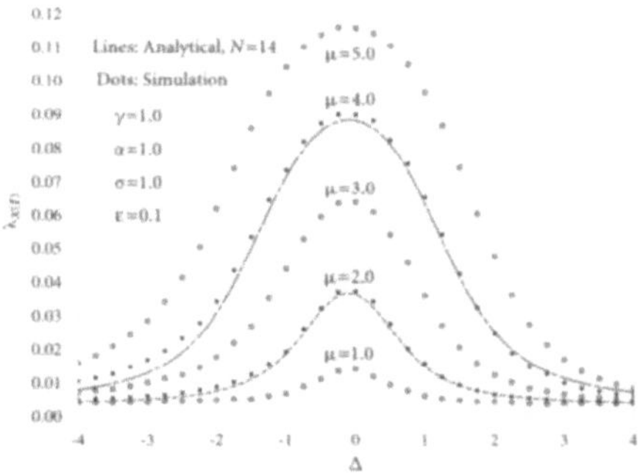

Figure 4. Lyapunov Exponent $\lambda_{x(t)}$.

4. Conclusion

In this paper, the dynamic stability of a two-dimensional system under the parametric excitation of harmonic and real noise is studied through the determination of the moment Lyapunov exponents and the Lyapunov exponents.

The effect of the real noise excitation on the primary parametric resonance due to the harmonic excitation is investigated. When the influence of the real noise excitation is small, the harmonic excitation is dominant and the parametric resonance is significant. On the other hand, when the impact of the real noise excitation is large, the effect of the harmonic excitation is small and the prominence of the parametric resonance due to the harmonic excitation diminishes.

Acknowledgments

The research for this paper was supported, in part, by the Natural Sciences and Engineering Research Council of Canada through Grant No. OGPO131355.

References

[1] L. Arnold, *Random Dynamical Systems*, Berlin: Springer-Verlag, 1998.

[2] L. Arnold, M. M. Doyle, and N. Sri Namachchivaya, "Small Noise Expansion of Moment Lyapunov Exponents for Two-Dimensional Systems," *Dynamics and Stability of Systems*, vol. 12, no. 3, pp. 187–211, 1997.

[3] R. Khasminskii and N. Moshchuk, "Moment Lyapunov Exponent and Stability Index for Linear Conservative System with Small Random Perturbation," *SIAM Journal of Applied Mathematics*, vol. 58, no. 1, pp. 245–256, 1998.

[4] N. Sri Namachchivaya, "Almost Sure Stability of Dynamical Systems under Combined Harmonic and Stochastic Excitations," *Journal of Sound and Vibration*, vol. 151, no. 1, pp. 77–90, 1991.

[5] N. Sri Namachchivaya and H. J. Van Roessel, "Moment Lyapunov Exponent and Stochastic Stability of Two Coupled Oscillators Driven by Real Noise," *Transaction ASME Journal of Applied Mechanics*, vol. 68, no. 6, pp. 903–914, 2001.

[6] W. Wedig, "Lyapunov Exponent of Stochastic Systems and Related Bifurcation Problems," *Stochastic Structural Dynamics–Progress in Theory and Applications*. Editors: S. T. Ariaratnam, G. I. Schuëller, and I. Elishakoff, Elsevier Applied Science, pp. 315–327, 1988.

[7] W.-C. Xie, "Moment Lyapunov Exponents of a Two-Dimensional System under Real Noise Excitation," *Journal of Sound and Vibration*, vol. 239, no. 1, pp. 139–155, 2001.

III

PART 3: CONTROL AND ESTIMATION

Stochastic control, a common concept in many engineering disciplines, is to determine the best strategy for controlling the motion of a dynamical system in the presence of stochastic uncertainties. Modeling of engineering systems includes model identification, and parameter and state estimation. Interplay among stochastic control, random dynamical systems, and design and analysis of large, complex engineering systems is the focus of the *third* part of the Proceedings. The lead paper of this part was authored by P. R. Kumar of the University of Illinois at Urbana-Champaign, USA, the third *Lead Lecturer* of the Symposium. The rest of the papers presented during the Symposium, on the general topic of *control and estimation of nonlinear stochastic systems* are compiled alphabetically, based on the last name of the first author.

Control of system networks, subjected to stochastic uncertainty such as noise and/or delays, is an important area of current interest. P. R. Kumar presented a survey of queuing network models of manufacturing systems, as well as wired and wireless networks used for communication. He presented results on stability, stabilizability, performance analysis, and optimal control of such systems. Agnessa Kovaleva considered a new class of problems associated with control against large deviation in a weakly perturbed system. An averaging procedure was developed for deriving an approximate solution of the risk-sensitive control problem in the small noise limit. Erik Johnson considered semiactive control of structures using "smart" damping devices for vibration mitigation in the development of an efficient control algorithm which accounts for the dissipative characteristics of the device. W. Q. Zhu considered the stability of controlled quasi partially integrable Hamiltonian systems with pure parametric stochastic excitations.

Estimation and identification of uncertain nonlinear systems are of primary concern throughout aerospace, computer, electrical, and mechanical engineering applications. Rafail Khasminskii presented results on parameter estimation in linear and nonlinear systems which are stable in probability. Some simple mechanical examples were presented and discussed. Marcello Vasta presented methods to estimate unknown parameters in non-linear, randomly excited dynamic systems, which have applications in studies of the response of structures to random environmental loads, such as wind and ocean wave forces. Fa Ma re-examined the Bouc-Wen differential model of hysteresis which is routinely used in the characterization of nonlinear damping and in system identification. Patricia Davies considered the effects of noise in nonlinear system identification with applications to structures with viscoelastic elements. It was shown that the bias of estimates changes when different numbers of data sets are used in the parameter estimation.

THE DISTRIBUTED NONLINEAR STOCHASTIC WORLD OF NETWORKS

P. R. Kumar
University of Illinois at Urbana-Champaign
Department of Electrical and Computer Engineering, and
Coordinated Science Laboratory
1308 West Main Street
*Urbana, IL 61801-2307, USA**
prkumar@uiuc.edu

Abstract Currently there is much interest in networks. This includes queueing network models of manufacturing systems, as well as wired and wireless networks used for communication. At their heart, networks too are just dynamical systems, often stochastic, nonlinear, and distributed. Many of the issues of interest for such systems can be said to be what they have always been for dynamical and controlled dynamical systems – stability, stabilizability, performance analysis, and optimal control. We illustrate this with respect to problems of much contemporary attention, by examining the stability problem for re-entrant lines, and the determination of the information theoretic capacity of wireless networks. We illustrate how the modern emphasis in the former area is on computational proofs of stability, and in the latter area is in the establishment of scaling laws.

Keywords: Queueing networks, communication networks, manufacturing systems, wireless networks, stability, stabilizability, performance analysis, optimal control, capacity, information theory.

1. Introduction

Networks are once again in vogue with the current interest in communication networks, both wired and wireless. This is actually the fifth such phase of interest in networks. The first was sparked in the early 1900's by the growth of telephony, and was influenced by the work of Erlang. The second phase of in-

*This material is based upon work partially supported by USARO under Contract Nos. DAAD19-00-1-0466 and DAAD19-01010-465, DARPA under Contract No. N00014-01-1-0576, AFOSR under Contract No. F49620-02-1-0217, DARPA/AFOSR under Contract No. F49620-02-1-0325, and NSF under Contract No. NSF ANI 02-21357. Any opinions, findings, and conclusions or recommendations expressed in this publication are those of the authors and do not necessarily reflect the views of the above agencies.

N. Sri Namachchivaya and Y.K. Lin (eds.),
IUTAM Symposium on Nonlinear Stochastic Dynamics, 205-216.

terest was in the 1950's, sparked by the interest in modeling job shops for manufacturing products, and a key result concerning the steady-state distribution of certain networks of queues was obtained by [6]. The third phase occurred in the 1970's, sparked by the interest in the performance evaluation of computer systems. A key result was the work of Baskett, Chandy, Muntz and Palacios [1] on the steady-state distribution of networks somewhat more general than Jackson's. The fourth phase in the 1990's was again sparked by manufacturing, though this time it was the growth of the electronics industry and the concomitant growth of wafer fabrication plants whose "re-entrant" topology was significantly different to elude exact analysis. An account of this work can be found in [8]. The fifth phase, which is actually proceeding in parallel with the fourth phase, since semiconductor manufacturing is still of great interest, was sparked by the wired internet, and the more recent wireless networks which may be on the cusp of a takeoff.

Networks are, at their core, just dynamical systems. They are stochastic, since they suffer from the vagaries of randomness. They are nonlinear, with interactions often being discrete. Finally, in many cases, they are spatially dispersed and informationally decentralized, with the consequence that actions are taken at different nodes or locations based on local events – that is, they are distributed.

While the technological focus of networks nowadays may appear to distinguish them from other more classical systems, at their heart the basic issues are essentially the same, though the models may be different and challenging. The purpose of this paper is to relate problems studied in modern networking to problems traditionally studied in classical dynamical systems, and to highlight their essential similarities, while at the same time bringing attention to the considerable challenges that lie in analyzing such systems.

2. Some Questions of Perhaps Permanent Interest

For the purpose of seeing how networking issues can be classified, consider first the following sequence of issues of interest for differential equations or controlled differential equations.

Stability

Give a system

$$\frac{dx(t)}{dt} = f(x(t)), \quad f(0) = 0,$$

often the first question of interest is whether the origin is stable in some appropriate sense, e.g., asymptotically stable. This question is often addressed by Lyapunov's second method. One searches for a Lyapunov function $V(x)$ satisfying $\dot{V}(x) \leq 0$ or stronger properties, along with suitable growth conditions on V.

Stabilizability

When one turns to controlled systems, the issue of synthesizing stable control laws arises. Given a controlled differential equation

$$\frac{dx(t)}{dt} = f(x(t), u(t)), \qquad f(0,0) = 0,$$

does there exist a *state feedback control law*

$$u(t) = g(x(t)),$$

such that the origin is an asymptotically stable equilibrium point of

$$\frac{dx(t)}{dt} = f(x(t), g(x(t)))?$$

One should note that inquiring whether a *given* system is stable is a *descriptive question*, but synthesizing a stabilizing control law is a *prescriptive matter.*

Performance

Stability is a binary notion. Either an equilibrium point of a certain system is stable in a certain sense, or it is not. Given that a system is stable, one can then ask finer questions. For example, for a particular control law $u(t) = g(x(t), t)$, what is the value of the *cost*

$$\int_0^T c(x(t), u(t))dt \tag{1}$$

that is incurred? Such questions lie in the realm of *performance analysis* or *evaluation*. Performance analysis is also a descriptive issue.

Optimal Control

Rather than inquiring what is the performance of an already specified control law, one could try to determine an *optimal control law*. That is, one wants to find a control law $u(t) = g(x(t), t)$ which minimizes the cost function (1). The Pontryagin Minimum principle may be employed to determine such an optimal control for deterministic systems, while dynamic programming (see [10]) is the preferred approach when the system is stochastic.

The theory of networks can be perfectly well defined as one which attempts to obtain answers to questions such as these, in the specific context of networks. In the rest of this paper we will develop this theme and illustrate the particular issues that arise, as well as delineating some questions which are at the cutting edge of research in the field.

3. A Single Queue

It is simplest to begin with the case of a single queue, shown in Figure 1. Suppose that fluid arrives to it at rate λ, and the server can pump fluid out at

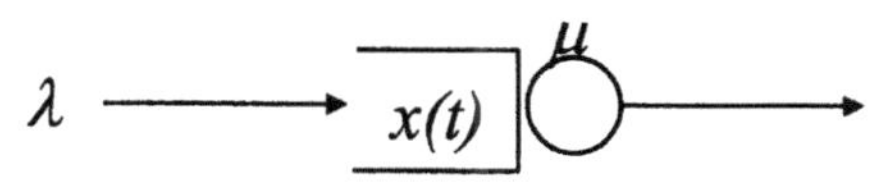

Figure 1. A single queue with arrival rate λ and service rate μ.

maximum rate μ. Let $x(t)$ denote the amount of fluid in the buffer. Clearly, such a queue is stable in the sense that $x(t) \to 0$ irrespective of the initial condition $x(0)$, if and only if $\lambda < \mu$, in which case

$$\begin{aligned} \dot{x}(t) &= \lambda - \mu \text{ if } x(t) > 0, \\ &= 0 \qquad \text{for } x(t) = 0. \end{aligned}$$

A typical trajectory is shown in Figure 2. When $\lambda < \mu$, $x(t)$ hits zero in finite

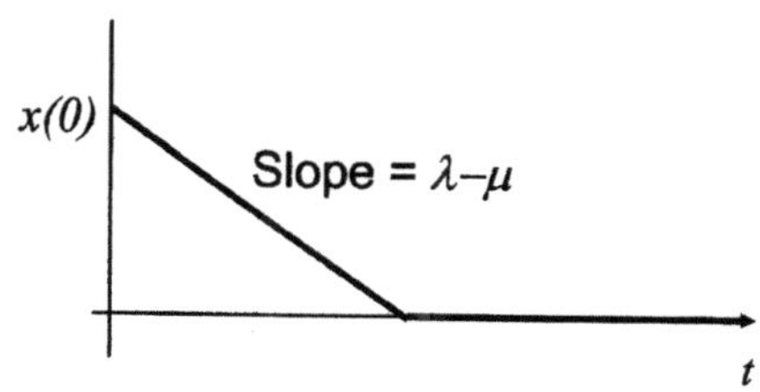

Figure 2. A typical trajectory of the fluid queue when $\lambda < \mu$.

time and thereafter stays at zero. One can say that the *stability region* of the system (in parameter space) is the set of all parameters λ strictly less than μ, i.e., $\lambda < \mu$.

Now let us consider a stochastic version of this queue where "customers" arrive as a Poisson process of rate λ, while service times are i.i.d., and exponentially distributed with mean $\frac{1}{\mu}$ (or rate μ).

Even though it does not make a difference to the overall system's stability, let us suppose the service policy, i.e., the *control policy*, is to serve the customer at the head of the queue. This is the First Come First Serve (FCFS) policy.

Then $x(t)$, the number of customers in the queue, is a Markov chain of the Birth-Death variety. Its infinitesimal generator has transition rates $q_{i,i+1} = \lambda$ for all $i \geq 0$, and $q_{i,i-1} = \mu$ for all $i \geq 1$.

Typically, as one moves to more complex systems, it is difficult to determine the transient behavior of the system, and one focuses solely on the steady-state behavior.

In this system, if $\lambda < \mu$, then the Markov chain is positive recurrent – a form of *stability*. One can also calculate that the invariant probability distribution of x is geometric, i.e., $\pi(x) = \rho^x(1-\rho)$ where $\rho := \frac{\lambda}{\mu}$. The mean number of customers in the system is $L := \frac{\rho}{1-\rho}$.

A well known theorem in queueing theory, called Little's theorem, asserts that under fairly general conditions, L, the mean number of customers in the system, is related to W, the mean waiting time or delay of a customer, by $L = \lambda W$, where λ is the arrival rate of customers.

In this case, we thus see that

$$\text{Mean Delay } W = \frac{\rho}{\lambda(1-\rho)} = \frac{1}{\mu - \lambda}.$$

The determination of the mean delay is typical of what is meant by *performance analysis*.

If our cost criterion is indeed to minimize the mean delay then it does not really matter which customer the server serves, as long as the server is "non-idling," that is, it does not stay idle when there is any waiting customer. It is a simple matter to show that idling only increases the mean delay in this example.

In our simple context, it is a trivial matter that the FCFS policy is *optimal* with respect to the delay cost function.

Thus, already in this simple system, we see the notions of stability, the set of parameters for which certain policies are stable, the notion of performance evaluation, and the notion of an optimal policy.

4. Queueing Networks

We now turn to networks of queues. Such systems are of interest in several applications. One example of much importance in this electronic age is semiconductor manufacturing, see [7]. Figure 3 illustrates the flow of wafers in a wafer fabrication plant, see [12]. One starts with wafers of silicon, and builds several "layers" of chemical patterns on it. Each layer requires several steps. Thus wafers return to the same machines again and again for the individual processing steps of subsequent layers. This gives rise to the "re-entrant" flow line shown in Figure 3.

The first question, see [13], that one can ask of such systems is: Can they indeed be unstable even under some non-idling scheduling policies?

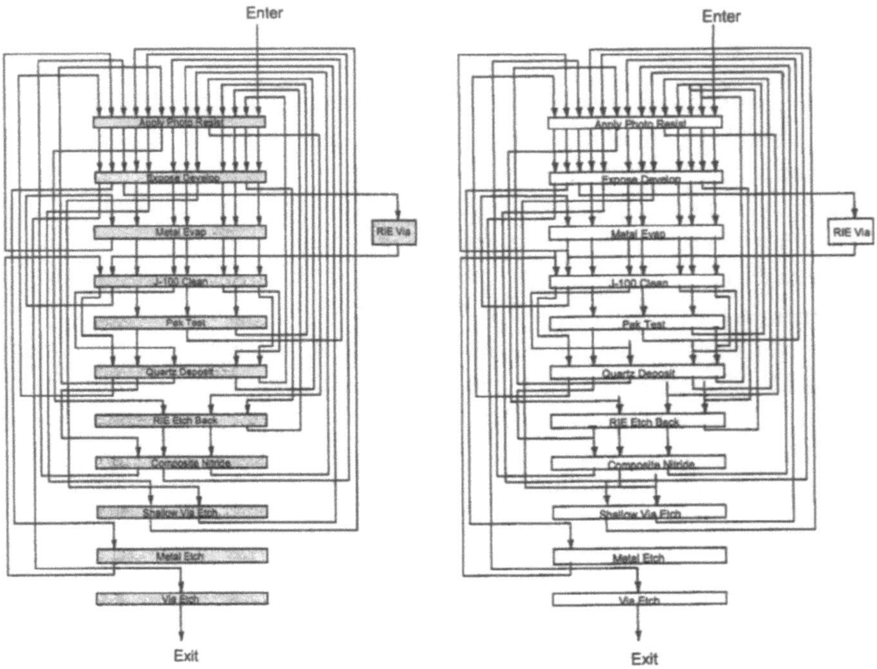

Figure 3. A semiconductor manufacturing plant.

Example: An unstable system

Consider the simple system shown in Figure 4. Customers arrive to the

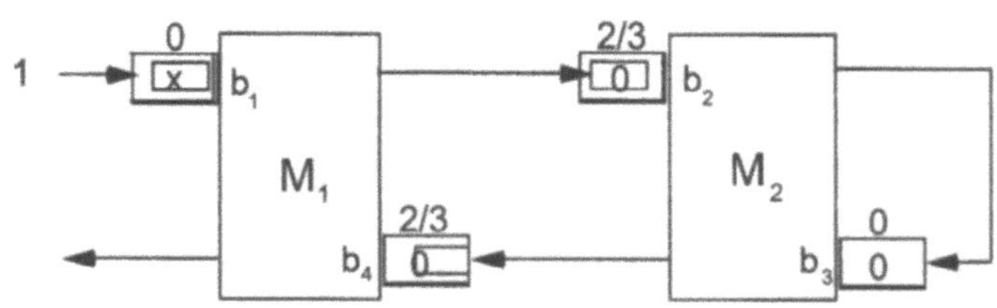

Figure 4. An unstable queueing network.

system periodically, with an interarrival period of exactly one hour. They first visit buffer b_1 at machine 1, where they require ϵ hours of attention, where ϵ is a small number. Then they visit buffer b_2, located at machine 2, where they require $\frac{2}{3}$ hours of service, then on to buffer b_3, also located at machine 2 requiring ϵ hours of service, and finally to buffer b_4 located back at machine 1 (hence a re-entrant line), from which they require $\frac{2}{3}$ hours of service, before exiting the system.

Consider the priority policy at machine 1 which favors customers in buffer b_4 over b_1, and serves b_1 only when b_4 is empty. Similarly at machine 2, consider the policy which favors buffer b_2 over b_3. This is an example of a *distributed* policy; nodes make local decisions based only on locally available information.

Note that each customer requires a total of $\left(\frac{2}{3} + \epsilon\right)$ hours of service at machine 1, while they arrive one hour apart. Hence, one may expect that machine 1 would not be unstable. Similarly for machine 2.

However, it turns out that the interactions between machine 1 and 2 due to the priority policies in place result in

$$\limsup_{t \to \infty} \sum_{i=1}^{4} x_i(t) = +\infty,$$

where $x_i(t)$ is the content of buffer b_i at time.

Thus the system is unstable!

5. Computational Proofs of Stability for Queueing Networks

We have seen that stability is indeed a property that needs to be carefully checked in large systems.

The next issue that arises is how to check for stability in large systems. Lyapunov methods again come to the fore, which in the context of positive recurrence for Markov chains has been studied by [4]. We will investigate though the stronger notion of whether the system has a steady-state distribution with a finite mean.

At this point one difference emerges with respect to the classical study of stability. Since the systems involved can be huge, we wish to automate the search for a Lyapunov function. That is, we would like the computer to generate a Lyapunov function which proves stability. Of course, what we need is really an algorithm to generate a Lyapunov function (though it may not always be successful).

Consider a queueing network where customers arrive to a buffer b_1 as a Poisson process of rate λ, and thereafter proceed to buffers $b_2, b_3, \ldots, b_L$, exiting from b_L. Suppose there are M machines, with every buffer served by one of the machines, and write "$i \in \sigma$" if buffer b_i is served by machine σ. Let the service time required by customers in buffer b_i be exponentially distributed with rate μ_i $\left(\text{i.e., mean } \frac{1}{\mu_i}\right)$.

One can render this a discrete time system by sampling at arrival times as well as at service completion epochs, assuming that all buffers are being worked on, if not on real customers then on virtual customers. Let $x(n)$ be the state of the system at time n, i.e., the vector of the number of customers in the L buffers. Also let e_i be the i-th coordinate vector with a one in the i-th place.

Let us consider a quadratic Lyapunov function of the form $V(x) := x^T Qx$ with $Q = Q^T$. To ensure that V is nonnegative it suffices to ensure that all entries q_{ij} of Q are nonnegative, since the components x are all nonnegative. The evolution of V is simply as follows, with $\mathfrak{F}_n$ denoting the past σ-algebra up to the n-th sample; and $u_i(n)$ the indicator function of whether a real customer from b_i is actually in service:

$$\begin{aligned} E[V(x(n+1)) \mid \mathfrak{F}_n] &= \lambda(x(n)+e_1)^T Q(x(n)+e_1) \\ &\quad + \sum_{i=1}^{L-1} \mu_i u_i(n)(x(n)+e_{i+1}-e_i)^T \\ &\quad Q(x(n)+e_{i+1}-e_i) \\ &\quad + \mu_L u_L(n)(x(n)-e_L)^T Q(x(n)-e_L) \\ &\quad + \left(1-\lambda-\sum_{i=1}^{L} \mu_i u_i(n)\right) x^T(n) Q x(n). \end{aligned}$$

From this it follows that if for some $\gamma > 0$,

$$\lambda q_{1i} + \sum_{\sigma} \sum_{j \in \sigma} \mu_j (q_{j+1,i} - q_{ji}) u_j(n) \;\; \leq \;\; -\gamma \text{ whenever } x_i(n) \geq 1, \quad (2)$$

(defining $q_{L+1,i} = 0$) then

$$Ex^T(n+1)Qx(n+1) \;\; \leq \;\; Ex^T(n)Qx(n) - \gamma E \sum_i x_i(n) + c,$$

where c is a constant. By telescoping the inequality it is an easy matter to establish that $\limsup_{N\to\infty} \frac{1}{N+1} \sum_{n=0}^{N} E \sum_i x_i(n)$ is bounded.

The inequality (2), however, involves a random quantity $u_j(n)$, the control variable. Now suppose that we are considering only non-idling policies under which a server is never idled when there are any customers waiting for it. Then if b_i is served by machine $\bar{\sigma}$, and $x_i(n) \geq 1$, then $\sum_{j\in\bar{\sigma}} u_j(n) = 1$. Then it can be shown (see [9]) that (2) will follow if for every i, and $\bar{\sigma}$ serving i,

$$\lambda q_{1i} + \text{Max}_{j\in\bar{\sigma}} \mu_j (q_{j+1,i} - q_{ji}) + \sum_{\sigma \neq \bar{\sigma}} \sum_{j \in \sigma} \mu_j (q_{j+1,i} - q_{ji})^+ \leq \gamma. \qquad (3)$$

Thus we obtain a test of stability: Find $Q = Q^T = [q_{ij}]$ such that (3) is satisfied for some $\gamma > 0$. Noting that (3) can be scaled, γ can be taken to be 1.

Whether (3) is feasible for some choice of $q_{ij} = q_{ji} \geq 0$ can be determined by linear programming. Thus, we have obtained a linear programming test to establish that the queueing network is stable, in this case to mean that the queue lengths have a finite mean in steady-state, see [9].

In fact, one can also obtain bounds on performance too by studying the evolution of quadratic forms. We refer the reader to [2] and [11], for details of this approach.

The interesting points to note are that we do not seek to obtain closed form expressions for the performance, or to analytically establish that a certain network operating under a certain scheduling policy is stable. Rather we wish to develop algorithms which can provide such proofs of stability or bounds on performance. This is the modern direction in which research is proceeding.

6. Determining the Capacity of a Wireless Network

In the $M/M/1$ queue studied in Section 3, it was shown that for any value of the arrival rate λ strictly less than μ, there exists a policy which stabilizes the system. The supremum of all the values of λ for which the system is stabilizable is thus μ. We say that μ is the *capacity* of the system.

Similarly, in the re-entrant line considered in Section 5, it is fairly easy to see that the capacity of the system is determined by the load imposed on the most bottlenecked server. Note that when a customer visits b_i, it consumes $\frac{1}{\mu_i}$ time units, on average, from the machine serving that buffer. Hence, each customer consumes a total mean time of $\sum_{i\in\sigma}\frac{1}{\mu_i}$ from machine σ. If customers arrive on average every $\frac{1}{\lambda}$ time units, then a necessary condition for stability is

$$\sum_{i\in\sigma}\frac{1}{\mu_i} \quad < \quad \frac{1}{\lambda} \text{ for every } \sigma. \tag{4}$$

In fact, through a "time-sharing" policy, the system can be stabilized whenever (4) is satisfied. Thus

$$\text{Capacity of system} \quad = \quad \frac{1}{\text{Max}_\sigma \sum_{i\in\sigma}\frac{1}{\mu_i}},$$

where the capacity is defined as the supremum of the arrival rates for which the system is stabilizable.

Generally, in queueing networks of the above type, it is easy to deduce what the capacity of the system is.

However, when one turns to wireless networks, the center of much attention these days, even the capacity of the system is unknown! The problem is that wireless networks have no links. Nodes simply radiate energy, see Figure 5.

When several nodes transmit simultaneously, a receiver receives the superposition of all the transmitted signals, attenuated according to their distances with respect to their transmitters. In addition there is ambient noise. Information has to be encoded in signals, and decoded from them. Clearly,

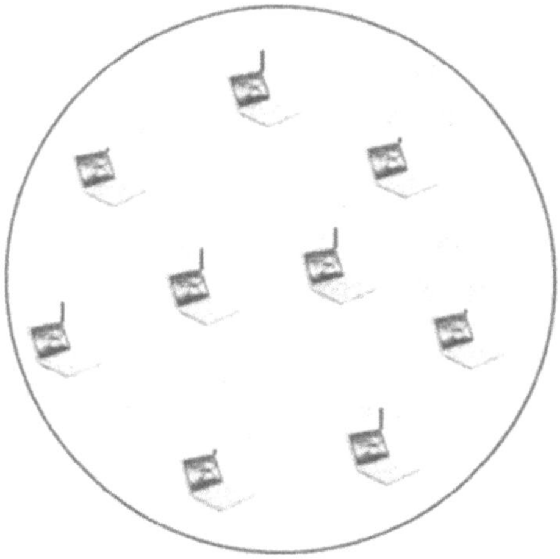

Figure 5. A wireless network.

this is a highly nonlinear (coding certainly is), stochastic, distributed system. From a system theoretic viewpoint, it can be regarded as a team theoretic problem where different agents have different informations. Such problems are among the most intractable problems there are, see [15].

In this particular communication systems context, this is a problem lying in the realm of network information theory, see [3], which seeks to generalize the seminal work of [14].

The problem can be modeled as follows. Consider n nodes located on a two-dimensional plane, with Euclidean distance $\rho_{ij} \geq \rho_{\min} > 0$ between any two nodes i and j. Each node's power for transmission is limited to P. Transmitted signals get attenuated over distance ρ according to $\frac{e^{-\gamma\rho}}{\rho^{\delta}}$, where $\gamma \geq 0$ measures absorption by the medium, and $\delta \geq 0$ is called the path loss exponent. Suppose node i transmits $x_i(t)$ at discrete time t. Node j receives $y_j(t) = \sum_{j \neq i} \frac{e^{-\gamma\rho_{ij}}}{\rho_{ij}^{\delta}} x_i(t) + w_j(t)$, where the noise $w_j(t) \sim N(0, \sigma^2)$ and i.i.d. for all j and t.

From the observations $\{y_j(t)\}_{t=1}^{T}$ that node j receives, and knowledge of its own private information, node j needs to deduce the information sent to it by other nodes. Let $\{R_{ij} : i \neq j\}$ be a *feasible vector of rates* that the network can support. This is defined in the usual Shannon theoretic way, see [3]. Define the *transport capacity* of the network as

$$C_T \quad := \quad \sup_{\text{Feasible}(R_{ij})} \sum_{(i,j)} \rho_{ij} R_{ij}.$$

This is the supremal distance-weighted sum of rates that the network can support. Its unit of measurement is bit-meters/sec, and it is analogous to the metric of man-miles/year that airlines use.

Over the past fifty years, ever since the seminal work of Shannon, there has been much effort in trying to characterize (the closure of) the set of feasible rate vectors. However, even the case of a system with three nodes, where one node is a sender, one is a receiver, and a third node is there to help by serving as a "relay," the so called "relay channel," is unsolved.

However, by incorporating distances explicitly, both in modeling node locations and distances between them, as well as in the metric of transport capacity, one can make progress.

There is one additional shift in emphasis that permits the theory to develop in a satisfactorily general way. We shall not try to determine the exact transport capacity, but will try to determine *scaling laws* for it, as the number of nodes n increases. It is shown in [16] that:

Theorem 1 *The transport capacity can at best grow linearly in the number of nodes:*

$$C_T \leq \frac{c(\gamma, \delta, \rho_{min})P}{\sigma^2} \cdot n.$$

It can also be shown that by operating the network in a "multi-hop" manner, that is, by relaying packets from node to node, one can often attain a transport capacity of this linear order; see [5]. This shows that in such scenarios: (i) the transport capacity indeed scales exactly linearly in n, i.e., $C_T = \Theta(n)$ using Knuth's notation, and (ii) the multi-hop mode of operating wireless networks is order-optimal.

Results such as these form a bridge between the theory of networks, and the world of protocol development for wireless networks – a rather hectic world these days.

7. Conclusion

The theme of this paper is to show that problems which have traditionally drawn the attention of researchers in dynamical systems and their control – stability, stabilizability, performance and optimal control – are also the main issues in the modern world of networks. Due to the complex nature of models of networks, these age old problems continue to challenge us. Even issues such as stability of large networks, and the set of arrival rates at which they can be stabilized – the capacity problem, are at the cutting edge of research in the area.

References

[1] F. Baskett, K. M. Chandy, R. R. Muntz, and F. G. Palacios, "Open, closed and mixed networks of queues with different classes of customers," *J. Assoc. Comput. Mach.*, vol. 22, no. 2, pp., 1975.

[2] D. Bertsimas, I. C. Paschalidis, and J. N. Tsitsiklis, "Optimization of multiclass queueing networks: Polyhedral and nonlinear characterizations of achievable performance," *Annals of Applied Probability*, vol. 4, pp. 43–75, 1994.

[3] T. Cover and J. Thomas, *Elements of Information Theory*. Wiley and Sons, New York, 1991.

[4] F. G. Foster, "On Markoff chains with an enumerable infinity of states," *Proc. Cambridge Phil. Soc.*, vol. 47, pp. 587–591, 1952.

[5] P. Gupta and P. R. Kumar, "The capacity of wireless networks," *IEEE Transactions on Information Theory*, vol. IT-46, no. 2, pp. 388–404, 2000.

[6] J. R. Jackson, "Networks of waiting lines," *Mathematics of Operations Research*, vol. 5, pp. 518–521, 1957.

[7] P. R. Kumar, "Re–entrant lines," *Queueing Systems: Theory and Applications: Special Issue on Queueing Networks*, vol. 13, no. 1–3, pp. 87–110, 1993.

[8] P. R. Kumar, "Scheduling queueing networks: Stability, performance analysis and design," in F. P. Kelly and R. Williams, editors, *IMA Volumes in Mathematics and its Applications*, pp. 21–70. Springer-Verlag, New York, Volume 71, 1995..

[9] P. R. Kumar and S. P. Meyn, "Stability of queueing networks and scheduling policies," *IEEE Transactions on Automatic Control*, vol. 40, no. 2, pp. 251–260, 1995.

[10] P. R. Kumar and P. P. Varaiya, *Stochastic Systems: Estimation, Identification and Adaptive Control*, Prentice-Hall, Englewood Cliffs, NJ, 1986.

[11] S. Kumar and P. R. Kumar, "Performance bounds for queueing networks and scheduling policies," *IEEE Transactions on Automatic Control*, vol. AC-39, pp. 1600–1611, 1994.

[12] S. C. H. Lu, D. Ramaswamy, and P. R. Kumar, "Efficient scheduling policies to reduce mean and variance of cycle–time in semiconductor manufacturing plants," *IEEE Transactions on Semiconductor Manufacturing*, vol. 7, no. 3, pp. 374–385, 1994.

[13] S. H. Lu and P. R. Kumar, "Distributed scheduling based on due dates and buffer priorities," *IEEE Transactions on Automatic Control*, vol. AC-36, no. 12, pp. 1406–1416, 1991.

[14] C. E. Shannon, "A mathematical theory of communication," *Bell Syst. Tech. J.*, vol. 27, pp. 379–423, 1948.

[15] H. S. Witsenhausen, "Separation of estimation and control for discrete time systems," *IEEE Proceedings*, vol. 59, pp. 1557–1566, 1971.

[16] L. Xie and P. R. Kumar, "A network information theory for wireless communication: Scaling laws and optimal operation," Submitted to *IEEE Transactions on Information Theory*, 2002.

NOISE EFFECTS IN NONLINEAR SYSTEM IDENTIFICATION WITH APPLICATIONS TO STRUCTURES AND SYSTEMS WITH VISCOELASTIC ELEMENTS

Patricia Davies, Rong Deng, Timothy Doughty and Anil K. Bajaj
Ray. W. Herrick Laboratories,
School of Mechanical Engineering, Purdue University
West Lafayette, IN 47907-1077, USA
daviesp@ecn.purdue.edu

Abstract The effect of noise on least-squares based estimation of nonlinear system parameters is described. While both bias and variance are affected by noise, bias is the more difficult problem to deal with. Variance can be reduced by using more data, but bias can, in some cases, become greater when more data is used. A parameter estimation technique, that utilizes the steady-state response of a fractional derivative model of the uni-directional dynamic response of a polyurethane foam block, is analyzed to illustrate why the estimation bias occurs. Also described is the influence of unmodeled dynamics on the estimates of parameters in a model of a mode of a cantilever beam.

1. Introduction

Experimental noise, model imperfections and environmental variations ensure that estimates differ for each set of experimental data used in parameter estimation. So a case can be made that the parameters of models of real systems should have statistical descriptions. Described below are the errors that arise when nonlinear system parameters are estimated by minimizing a sum of squares cost function and the measurements are corrupted by noise. While other approaches to parameter estimation are possible [7], and may lead to less bias in theory, they are often difficult to apply in practise due to the lack of data available for estimation, or the method requires knowledge of noise distributions which will, in general, not be known.

The case when the estimation equations are linear, but the system is nonlinear, is considered. Two systems are used to illustrate the discussion: a foam-mass system [4], and the modal response of a cantilever beam [6]. Ex-

N. Sri Namachchivaya and Y.K. Lin (eds.),
IUTAM Symposium on Nonlinear Stochastic Dynamics, 217-225.

pressions are derived for the asymptotic bias of the parameter estimates of the foam-mass system and are compared to the results from multiple simulations.

2. System Models

Foam Model and Identification. A fractional derivative model is utilized to model foam viscoelasticity [1] and the uni-directional motion of a foam-block system undergoing base excitation is modeled by [4]:

$$m\ddot{x} + c\dot{x} + kx + aD^{\alpha}(x) + k_3x^3 + k_5x^5 \quad = \quad f(t), \tag{1}$$

where m is the mass of the block, c the viscous damping coefficient, k, k_3 and k_5 are stiffness coefficients, and a and α are the viscoelastic parameters. x is the displacement of the mass relative to the base and $f(t)$ is the base excitation force. Here $D^{\alpha}(\cdot)$ is the fractional derivative operator [9] and the Laplace transform of this operator is $s^{\alpha}\mathcal{L}(\cdot)$.

When $f(t) = -mG\cos(\omega_f t)$, it can be assumed that the steady-state response is periodic. The harmonic balance method is used to derive an approximation to this steady-state response of the system. A one term harmonic balance approximation is derived by assuming that:

$$x(t) = Xe^{j\omega_f t} + \bar{X}e^{-j\omega_f t}; \quad X = A + Bj; \quad \bar{X} = A - Bj, \tag{2}$$

and substituting the assumed solution into the equation of motion. Equating like terms results in:

$$\begin{aligned}
F_1(\omega_f) &= PP|X|^2 + mGA/2 = 0, & (3)\\
F_2(\omega_f) &= QQ|X|^2 - mGB/2 = 0, & (4)\\
PP &= -m\omega_f^2 + k + 3k_3|X|^2 + 10k_5|X|^4 + a\omega_f^{\alpha}\cos(\pi/2\alpha), &\\
QQ &= c\omega_f + a\omega_f^{\alpha}\sin(\pi/2\alpha). &
\end{aligned}$$

The estimation of model parameters is based on (3)-(4) and it is assumed that A and B are known for a set of values of the excitation frequency ω_f. The estimation for the parameters (c, k, k_3, k_5, a, α) can be performed by minimizing the error function:

$$\mathcal{E} \quad = \quad \sum_{i=1}^{M} \left(|F_1(\omega_{fi})|^2 + |F_2(\omega_{fi})|^2 \right),$$

where the ω_{fi} are the frequencies at which the steady-state response is measured. This minimization results in a set of nonlinear equations that can be solved to obtain the system parameters. If α is known, the estimation of the remaining material parameters in (3)-(4) is a simple linear optimization problem. If α is not known, a sub-optimal method of estimation can be adopted [10].

All experimental data are corrupted with noise. It was observed in previous work that the bias increases as the frequency resolution increases although the variance decreases [4]. The variance effect is attributable to the use of more data sets at higher resolution. Both bias and variance increase with noise level. Due to the nonlinearity in the parameter estimation equations caused by the fractional derivative term, it is difficult to analytically predict the bias and variance of these estimates. However, in the suboptimal method some asymptotic results for the bias can be derived.

When α is fixed, the estimation equations are based on:

$$X\mathbf{z} = \left[R_1, R_2, \ldots, R_{\frac{N}{2}}, S_1, S_2, \ldots, S_{\frac{N}{2}}\right]^T + \mathbf{e}, \tag{5}$$

where elements of X are functions of A_j and B_j measured at the frequencies ω_{fj}, and $\mathbf{z}$ is the vector containing the material parameters, $R_j = m\omega_j^2 - mGA_j/2$ and $S_j = mGB_j/2$, and $\mathbf{e}$ represents model noise.

Beam Model and Identification. In this study, a horizontally mounted cantilevered beam is subjected to a harmonic external excitation in the form of motion of the clamped base. Only in-plane motions are considered. It is assumed that the transverse response of the beam is dominated by the second and third modes of vibration, the spatial dependence being characterized by the linear mode shapes, $\phi_2(s)$ and $\phi_3(s)$. Thus, the response can be written as: $v(s,t) = a_2(t)\phi_2(s) + a_3(t)\phi_3(s)$, where the amplitudes of the two modes are governed by [5]:

$$\begin{aligned}
&\ddot{a_2} + c\dot{a_2} + \omega_2^2 a_2 + d\dot{a_2}|\dot{a_2}| = -\left(4\alpha_1 a_2^3 + 3\alpha_2 a_2^2 a_3 + 2\alpha_3 a_2 a_3^2 + \alpha_4 a_3^3\right)\\
&-\left(\delta_1 a_2^2 + 2\delta_2 a_2 a_3 + \delta_3 a_3^2\right)\ddot{a_2} - \left(\delta_7 a_2^2 + \delta_8 a_2 a_3 + \delta_9 a_3^2\right)\ddot{a_3} - \left(\delta_1 a_2 + \delta_2 a_3\right)\\
&\dot{a_2}^2 - 2\left(\delta_2 a_2 + \delta_3 a_3\right)\dot{a_2}\dot{a_3} - \left[(\delta_8 - \delta_4)a_2 + (2\delta_9 - \delta_5)a_3\right]\dot{a_3}^2 - F_2 cos(\Omega t)\\
\\
&\ddot{a_3} + c\dot{a_3} + \omega_3^2 a_3 + d\dot{a_3}|\dot{a_3}| = -\left(\alpha_2 a_2^3 + 2\alpha_3 a_2^2 a_3 + 3\alpha_4 a_2 a_3^2 + 4\alpha_5 a_3^3\right)\\
&-\left(\delta_7 a_2^2 + \delta_8 a_2 a_3 + \delta_9 a_3^2\right)\ddot{a_2} - \left(\delta_4 a_2^2 + 2\delta_5 a_2 a_3 + \delta_6 a_3^2\right)\ddot{a_3}\\
&-\left[(2\delta_7 - \delta_2)a_2 + (\delta_8 - \delta_3)a_3\right]\dot{a_2}^2 - 2\left(\delta_4 a_2 + \delta_5 a_3\right)\dot{a_2}\dot{a_3}\\
&-\left(\delta_5 a_2 + \delta_6\right)a_3\dot{a_3}^2 - F_3 cos(\Omega t).
\end{aligned} \tag{6}$$

and F_2 and F_3 are the modal forcing terms. When only the second mode is considered ($a_3 = 0$), (6) can be simplified to [6]:

$$\left(\ddot{a_2} + {\omega_0}^2 a_2\right) = \left[-\dot{a_2}, F_2, -a_2^3, (a_2\dot{a_2}^2 + a_2^2\ddot{a_2}), -\dot{a_2}|\dot{a_2}|\right][c, \alpha_0, \alpha_1, \alpha_2, d]^T, \tag{7}$$

The elements $[c, \alpha_0, \alpha_1, \alpha_2, d]^T$, are the unknown parameters. In the continuous time approach, the system's input F and response a are measured and the data is sampled at times, $t_i, i = 1, \ldots, N$. $\dot{a}_2(t_i)$ and $\ddot{a}_2(t_i)$ are also determined from $a_2(t_i)$. Data at each t_i results in an equation of the form of (7), and the complete set of equations can be solved in a least squares sense to estimate $\mathbf{z}$.

3. Asymptotic Bias Analysis

Equation (5) and the set of equations based on (7) can be written in a more general form as $X\mathbf{z} = \mathbf{y} + \mathbf{e}$, where the matrix X ($N \times K$) and the vector $\mathbf{y}$ ($N \times 1$) contain the experimental data and $\mathbf{z}$ ($K \times 1$) contains the unknown parameters. Usually $N > K$. If $(X^T X)$ is of full rank then:

$$\mathbf{z} = (X^T X)^{-1} X^T \mathbf{y} + (X^T X)^{-1} X^T \mathbf{e} = \hat{\mathbf{z}} + (X^T X)^{-1} X^T \mathbf{e}.$$

Here, $\hat{\mathbf{z}}$ is the least squares estimate of $\mathbf{z}$. Ideally, if the elements of X are uncorrelated with the elements of $\mathbf{e}$, the least squares solution is an unbiased estimate of the true solution [8].

Consider the case when the data is contaminated with noise. Let the X_m and $\mathbf{y}_m$ denote the measured X and $\mathbf{y}$. So $N_x = X_m - X$ and $\mathbf{n}_y = \mathbf{y}_m - \mathbf{y}$. Then, by setting $\mathbf{e} = 0$ for convenience, we have $(X_m - N_x)\mathbf{z} = \mathbf{y}_m - \mathbf{y}$. By multiplying the above equation by X_m^T and noting that $X_m^T \mathbf{y}_m = X_m^T X_m \hat{\mathbf{z}}$, some additional manipulation yields:

$$\begin{aligned} \tfrac{1}{N}(X^T X + N_x^T X + X^T N_x + N_x^T N_x) \quad & \hat{\mathbf{b}}_N = \\ \tfrac{1}{N}\left[-(X^T N_x + N_x^T N_x)\mathbf{z} + X^T \mathbf{n}_y + N_X^T \mathbf{n}_y\right]. & \end{aligned} \tag{8}$$

$\hat{\mathbf{b}}_N = \hat{\mathbf{z}}_N - \mathbf{z}$ denotes an estimate of the bias from 1 set of N data points. When $N \to \infty$, the elements in the matrices in the above equation become correlation terms [3], and the matrices become deterministic. The asymptotic bias ($\mathbf{b}_\infty$) is the solution of (8) when $N \to \infty$.

To illustrate how nonlinear terms affect the bias, consider a subset of the fractional derivative model. Let $\mathbf{y} = [-mGA_1/2, \ldots, -mGA_N/2]^T$ and $X = [X_1^T X_2^T]$, where $X_1 = [A_1, \ldots, A_N]$ and $X_2 = [3B_1^2, \ldots, 3B_N^2]$. Assume that the measured values of A_j and B_j are corrupted by the noise and are replaced by $A_j + n_{Aj}$ and $B_j + n_{Bj}$, respectively. The N data points are P repeats at Q frequencies, i.e., $N = PQ$. Here Q is determined by the frequency resolution in the experiments and P is the number of repetition of the measurements. The noise variables n_{Aj} and n_{Bj} are assumed to be identically distributed and independent random variables.

Now consider the first term on the left hand side of (8),

$$\frac{1}{N} X^T N_X \quad = \quad \frac{1}{Q}\begin{bmatrix} \sum_{q=1}^{Q} A_q E_1 & \sum_{q=1}^{Q}(6A_q B_q E_2 + 9B_q^2 E_3) \\ \sum_{q=1}^{Q} 3B_q^2 E_1 & \sum_{q=1}^{Q}(18B_q^3 E_2 + 9B_q^2 E_3) \end{bmatrix}, \tag{9}$$

where $E_1 = \frac{1}{P}\sum_{p=1}^{P} n_{Ap}$, $E_2 = \frac{1}{P}\sum_{p=1}^{P} n_{Bp}$, and $E_3 = \frac{1}{P}\sum_{p=1}^{P}(n_{Bp})^2$. Note that A_j and B_j are the response data and they are deterministic. As $P \to \infty$, it can be shown that the above matrix becomes:

$$\lim_{N\to\infty} \frac{1}{N} X^T N_x \quad = \quad \frac{1}{Q}\begin{bmatrix} \sum_{q=1}^{Q} A_q E_a & \sum_{q=1}^{Q}(6A_q B_q E_b + 9B_q^2 E_{b2}) \\ \sum_{q=1}^{Q} 3B_q^2 E_a & \sum_{q=1}^{Q}(18B_q^3 E_b + 9B_q^2 E_{b2}) \end{bmatrix}. \tag{10}$$

where $E_a = E[n_A]$, $E_b = E[n_B]$, and $E_{b2} = E[n_B^2]$ and $E[\cdot]$ denotes the expected value [2]. Similarly, every element in (8) can be determined and (8) becomes:

$$\frac{1}{Q} D\mathbf{b}_\infty \quad = \quad \frac{1}{Q}\mathbf{v}$$

$$\begin{aligned}
D_{11} &= \sum_{q=1}^{Q} A_q^2 + 2A_q E[n_A] + E[n_A^2], \\
D_{12} &= \sum_{q=1}^{Q} (3A_q B_q^2 + 3B_q^2 E[n_A] + 6A_q B_q E[n_B] \\
&\quad +6B_q E[n_A]E[n_B] + 3A_q E[n_B^2] + 3E[n_A]E[n_B^2]) = D_{21}, \\
D_{22} &= \sum_{q=1}^{Q} \left(9B_q^4 + 36B_q^3 E[n_B] + 54B_q^2 E[n_B^2] + 36B_q E[n_B^3] + 9E[n_B^4]\right), \\
v_1 &= \sum_{q=1}^{Q} (-\tfrac{1}{2} mGA_q E[n_A] - \tfrac{1}{2} mGE[n_A^2] - A_q E[n_A] z_1 \\
&\quad -E[n_A^2] z_1 - 6A_q B_q E[n_B] z_2 - 6B_q E[n_A]E[n_B] z_2 \\
&\quad -3A_q E[n_B^2] z_2 - 3E[n_A]E[n_B^2] z_2), \\
v_2 &= \sum_{q=1}^{Q} (-\tfrac{3}{2} mGB_q^2 E[n_A] - 3mGB_q E[n_A]E[n_B] - \tfrac{3}{2} mGE[n_A]E[n_B^2] \\
&\quad -3B_q^2 E[n_A] z_1 - 6B_q E[n_A]E[n_B] z_1 - 3E[n_A]E[n_B^2] z_1 \\
&\quad -18B_q^3 E[n_B] z_2 - 45B_q^2 E[n_B^2] z_2 - 36B_q E[n_B^3] z_2 - 9E[n_B^4] z_2).
\end{aligned} \tag{11}$$

The asymptotic bias $\mathbf{b}_\infty$ is the solution of this equation.

When the full fractional derivative model is used, although the expressions for the matrices become more complicated, the asymptotic bias can be calculated following the same procedure described above. We now use this approach to compute results for the foam model.

4. Results and Discussion

Foam Model. It is assumed that the measurement noise n_A and n_B are uniformly distributed random noise with zero mean. Therefore, the moments of these random variables are:

$$E[(n_A)^k] = r^k/k,\ k \text{ even}; \qquad E[(n_A)^k] = 0,\ k \text{ odd}, \tag{12}$$

where r is one-half of the noise width. The effect of noise on $\mathbf{b}_\infty$ was calculated from (8) with r ranging from 0.1% to 10% of the peak value of the response (SNR: 65dB-25dB). The parameters are: $m = 3.38$ kg, $k = 3100$ N/m, $\alpha = 0.9$, $a = 14.19$ N/m^α, $k_3 = -8.4 \times 10^6$ N/m^3, $k_5 = 1.35 \times 10^{10}$ N/m^5. Frequencies range from 2 Hz to 8 Hz at 0.125Hz resolution level. Estimates of $\mathbf{b}_\infty$, normalized with respect to the true values, are shown in Figure 1. The

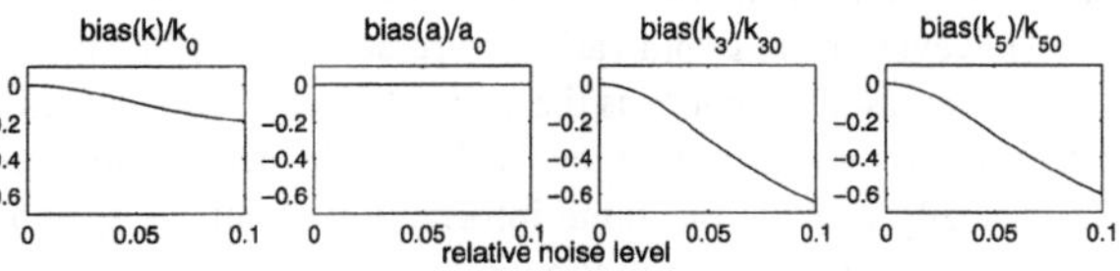

Figure 1. The changes in the asymptotic bias due to the noise levels.

asymptotic bias in $\hat{a}$ is very small, but the bias of other parameters increases as the noise level increases.

The parameter estimates ($\hat{\mathbf{z}}_N$) can be obtained by solving $X_m \hat{\mathbf{z}}_N = \mathbf{y}_m$ in a least squares sense with a finite number (N) of data points. The bias $E[\hat{\mathbf{z}}_N - \mathbf{z}]$ can therefore be estimated by running the simulations with many realizations of noise and averaging the results. The estimated biases from 1,000 simulations with a sample size of 56 ($P = 1$) to 7168 ($P = 128$) are shown in Figure 2. r was 5% of the peak response value. When $P > 64$, the estimated biases approach the asymptotic biases. However, when the number of data sets is small, the estimated bias is smaller than $\mathbf{b}_\infty$. Similar results have been found with nonlinear estimation of parameters when α is also an unknown parameter [4].

When only one set of data is used at each frequency resolution level, N increases with resolution level. Results from 10,000 simulations are shown in Figure 3 (a) where Level n

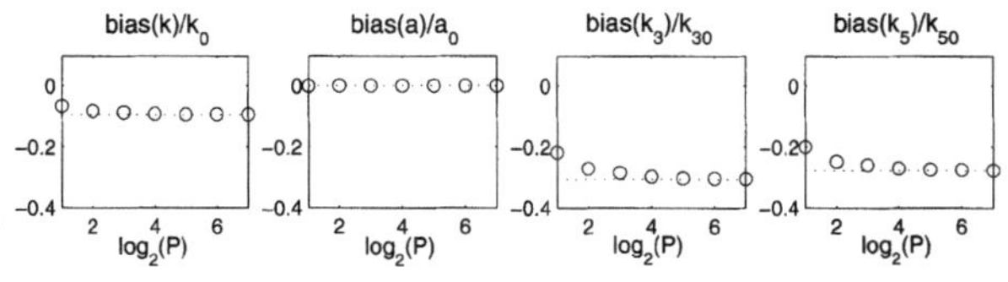

Figure 2. The change of the estimated bias with P, the number of repetitions. 'o' denotes the estimated bias from different sample sizes. '...' denotes the asymptotic bias.

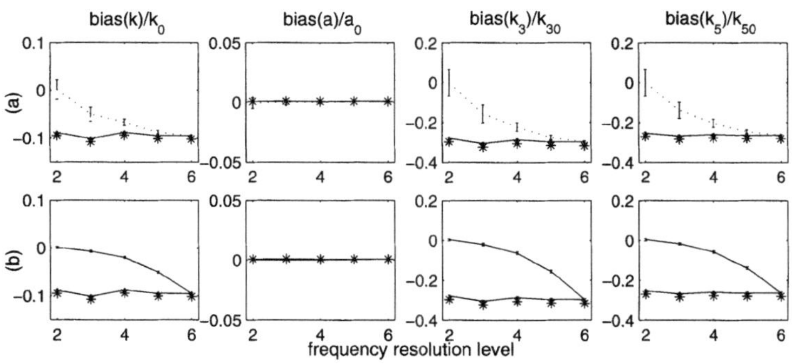

Figure 3. The normalized mean values of the bias and standard deviation of the estimates from different simulations based on 10,000 realizations. (a) '...' estimated bias from only one set of data. (b) '...' estimated bias from one set of averaged data. '—' estimated bias from multiple sets of data, $N = 224$. '$*$' asymptotic bias.

indicates a resolution of $\frac{1}{n}$ Hz. Bias increases and the variance decreases as the frequency resolution gets finer. However, the bias and variance are both large when the resolution is only 1 Hz (not shown). If P is increased so that N is the same at all resolution levels and $N = 224$, the results are almost the same for all resolution levels. This indicates that the frequency resolution levels above 0.5Hz basically provide the same information about the system and that using a larger number of data sets actually leads to more bias.

If several repeats of the data at low frequency resolution levels are available, the data can be averaged to reduce the noise level and then used to estimate the unknown parameters. Simulation results in Figure 3(b) show that both smaller bias and variance can be achieved if this is done. Thus, in an experiment it would be better to do many repeated measurements at a lower resolution level than to do fewer repetitions at a higher resolution level, so long as the resolution is ≥ 0.5Hz.

The response of the system was reconstructed from the estimated parameters (Figure 4). Consistent with the parameter estimates, the estimated responses from a lower frequency resolution level are less biased but have larger variance and at a higher resolution level, although the spread is narrower, the estimated response is more biased.

To investigate further the effect of the sample size on the estimates due to changes in frequency resolution, the probability density functions (pdf) were estimated from the results of 40,000 simulations. As seen from Figure 5, the pdfs become Gaussian as the number of data points used in the estimation increases. A Gaussian pdf is also plotted (using the estimated mean and variance) for reference.

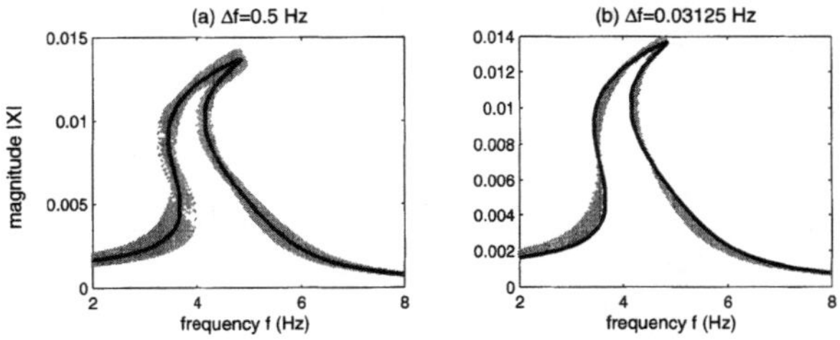

Figure 4. Reconstructed response of the system from the estimated parameters. (a) the estimates from $\Delta f = 0.5$Hz. (b) the estimates from $\Delta f = 0.0325$Hz. The dark line is the original system response and gray curves are the reconstructed response from 100 simulations.

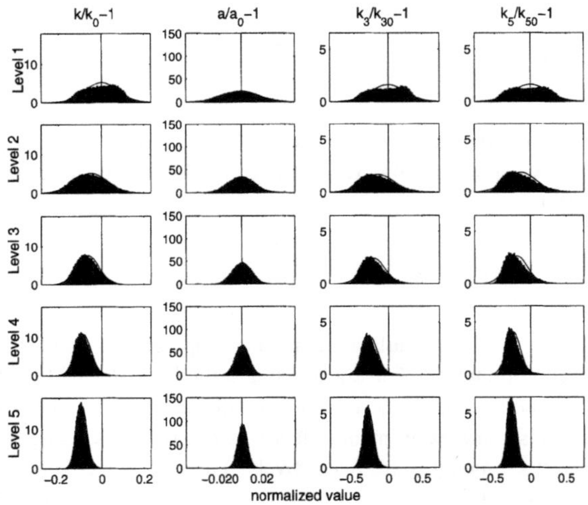

Figure 5. The estimated probability density functions of the estimated parameters for differen N based on 40,000 simulations. The results were normalized with respect to the true values. $\alpha_0 = 0.9$. Level L refers to $N = 56 \cdot 2^{L-1}$ points used in estimation.

The results of the least squares method are consistent with the results of the direct nonlinear estimation reported in [4]. However, the bias from the linear least-squares method is much less than that from nonlinear estimation, especially in the estimate of a. In the linear least-squares method, α is assumed known *a priori*, which is not the case in real experiments. The error in α will propagate into the estimates of the other parameters. To investigate this effect, the value of α was perturbed from -16% to 16% of its true value and the other parameters were estimated when no noise is present. As shown in Figure 6(a), the errors in parameter estimates increase as the error in α increases. The parameter most influenced by the error in α is a. The case when α is varied and its true value is known and noise is present ($r = 5\%$), is shown in Figure 6(b). The estimates of k, k_3 and k_5 are much more affected by the presence of noise and the value of α influences the result. When errors in α and noise are present, all the parameter estimates are affected shown in Figure 6(c). Therefore, it is expected that in the direct nonlinear estimation, the estimates of the parameters would tend to be more biased.

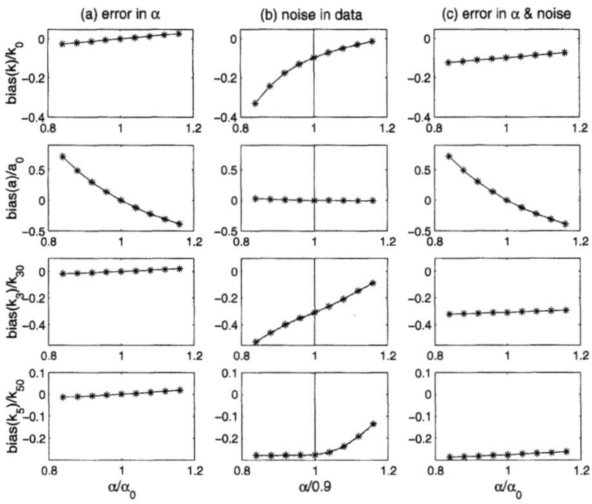

Figure 6. The normalized error in the estimates due to error in α and the presence of noise. (a) α varies but true value= α_0. (b) α varies and true value is used in estimation. (c) noise and error in α. $\alpha_0 = 0.9$.

Beam Model. Only simulation results are reported here and the specific issues addressed are: the appropriatenss of using the single mode model of (7) in modeling the experimental data, where another mode is being weakly excited; and the influence of noise on the resulting estimates. The experimental data was simulated by using the one-mode model in (7) and the two-mode model in (6) respectively. Parameters were chosen so that responses were close to reponses measured in an experiment [5]. Uniformly distributed random noise was added to the simulated response (a_2). The SNR varied from 10 dB to 50dB. NL, defined as $(-3\alpha_1 + 2\omega_0^2\alpha_2)$, was also calculated [6]. As shown in Figure 7, better estimates are achieved with higher SNR in both cases. With the one-mode model simulation, good estimates were obtained as seen from Figure 7. With the two-mode model simulation, larger bias is observed, especially in α_1 and α_2. However, the bias and variance in the estimates of NL are much smaller. This is consistent with a previous observation [5] that when the individual parameters α_1 and α_2 cannot be identified accurately, their combined effect, as represented by the NL term, may be estimated with much less error. This result indicates that the one-mode model may not be sufficient when another mode is excited, even if that excitation is weak.

5. Conclusion

It has been shown that the bias of estimates changes when different numbers of data sets are used in the parameter estimation; the bias only approaches the asymptotic prediction (within 1% relative error) when the number of data sets becom es very large ($N > 3584$). How unmodeled dynamics can affect parameter estimates was also illustrated. In the cases considered unmodeled dynamics had more influence on the bias than the noise. Based on these results, further analysis of the statistics of finite data set estimates is warranted.

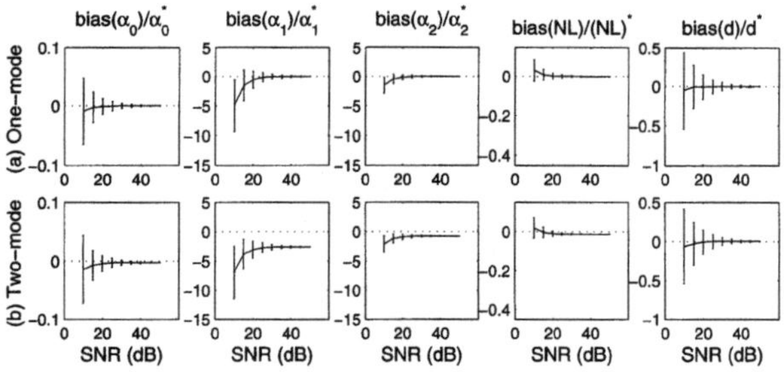

Figure 7. The normalized bias and standard deviation of the estimates at different signal-to-noise ratios (Beam Model).

References

[1] R. L. Bagley and P. J. Torvik, "A theoretical basis for the application of fractional calculus to viscoelasticity," *J. of Rheology*, vol. 27, no. 3, pp. 201–210, 1983.

[2] J. S. Bendat and A. G. Piersol, *Random Data: Analysis and Measurement Procedures*, John Wiley & Sons, 1986.

[3] S. Billings and Q. Zhu, "Rational model identification using an extended least-squares algorithm," *Int. J. of Control*, vol. 54, no. 3, pp. 529–546, 1991.

[4] R. Deng, P. Davies, and A. K. Bajaj, "Flexible polyurethane foam modeling and viscoelastic parameters identification for automotive seating applications," *J. of Sound and Vibration*, accepted for publication, 2002.

[5] T. Doughty, *System Identification of Modes in Nonlinear Structures*, Ph.D Thesis, Purdue University, West Lafayette IN, 2002.

[6] T. A. Doughty, P. Davies, and A. K. Bajaj, "A comparison of three techniques using steady-state data to identify nonlinear modal behavior of an externally excited cantilever beam," *J. of Sound and Vibration*, vol. 249, no. 4, pp. 785–813, 2002.

[7] C. Hsiao and Q. Wang, "Estimation of structural nonlinear errors-in-variables models by simulated least-squares method," *International Economic Review*, vol. 41, no. 2, pp. 523–542. 2000.

[8] L. Ljung, *System Identification: Theory for the User*, Prentice Hall PTR, Upper Saddle River, NJ, 1999.

[9] K. B. Oldham and J. Spanier, *An Introduction to Fractional Calculus*, Academic Press, 1974.

[10] R. Singh, P. Davies, and A. K. Bajaj, "Identification of nonlinear and viscoelastic properties of flexible polyurethane foam," *Nonlinear Dynamics*, accepted for publication, 2002.

STRUCTURAL VIBRATION MITIGATION USING DISSIPATIVE SMART DAMPING DEVICES

Erik A. Johnson and Baris Erkus
Department of Civil and Environmental Engineering
University of Southern California
Los Angeles, CA 90089-2531 USA
JohnsonE@usc.edu

Abstract An important problem in semiactive control of structures using "smart" damping devices for vibration mitigation is the development of an efficient control algorithm which accounts for the dissipative characteristics of the device. For smart dampers, the dissipative nature can be represented by a nonlinear inequality constraint that cannot be directly imposed on optimal control strategies using standard techniques. In previous research, the authors introduced a method using linear matrix inequality (LMI) methods to include a dissipativity constraint in a linear quadratic regulator (LQR) problem for semiactive control of structures. In this method, the LQR problem is defined as an eigenvalue problem (EVP) in terms of LMIs. Then, the dissipativity constraint is appended to the EVP in its weak expected value form. In this paper, this method is applied to realistic building and damper models. The structure is chosen to be a 9-story benchmark building and a recently-developed 20-ton magnetorheological (MR) fluid damper model as the control devices. It is found that the proposed method provides a more efficient controller, with reduced force levels and almost the same performance compared to a conventional LQR approach.

Keywords: structural control, smart damping technology, linear matrix inequalities, dissipative control

1. Introduction

In the field of structural control, semiactive control with "smart" damping devices has received special attention due to the attractive properties of these devices. From a stability point of view, which is an important issue in the seismic control of structures where the structure may exhibit nonlinear behavior, semiactive control provides better stability properties than a correspond-

N. Sri Namachchivaya and Y.K. Lin (eds.),
IUTAM Symposium on Nonlinear Stochastic Dynamics, 227-236.

ing active control strategy. In addition to stability concerns, large power requirements and issues about the reliability of active devices make semiactive control a sound alternative to active control.

There are various types of "smart" dampers and strategies for controlling structures using these devices (*e.g.*, [10]). Some examples of smart dampers are variable-orifice dampers, controllable friction dampers and controllable fluid dampers such as magnetorheological (MR) fluid dampers. Among these devices, the MR fluid damper is a recently developed device and has important advantages over its counterparts. When exposed to a magnetic field from the damper electromagnet, the MR fluid in the damper becomes a semi-solid, with a high yield strength. Various damper force levels may be achieved by varying the electromagnet current. Recently, a prototype 20-ton MR damper was built the by the Lord Corporation and tested at the University of Notre Dame [12]. A similar commercial device produced by the Sanwa Tekki Corporation has also been reported by [3]. As the behavior of these devices is highly nonlinear, mathematical damper modeling is governed by a set of nonlinear differential equations.

In practice, there is a common operational principle for all smart dampers: the force produced by the damper is changed by altering the damping properties associated with the devices. Therefore, smart dampers are controllable passive devices. From a theoretical viewpoint, this characteristic constitutes an important issue in the development of semiactive control strategies. While it improves the stability properties of the system, the damper can only produce a force which dissipates energy. In mathematical terms, dissipativity is a nonlinear inequality constraint that can be represented by $uv < 0$, where u is the damper force and v is damper velocity.

One control strategy that has been used to command smart dampers is a two-stage method known as *clipped-optimal* control (Figure 1). This strat-

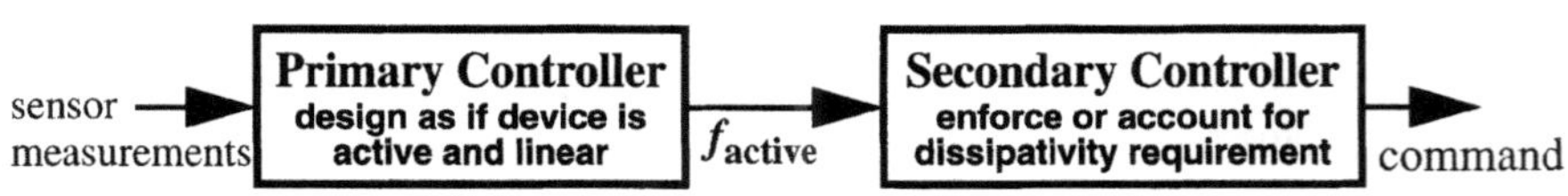

Figure 1. Semiactive control strategy.

egy has been implemented successfully in various types of problems (*e.g.*, [2, 8, 5]). The first stage of clipped-optimal control is a standard active control algorithm, which is generally chosen to be an H_2 optimal control such as a linear quadratic regulator (LQR). This primary controller assumes that the control device is fully active and linear. The second stage is a bang-bang type controller that switches the damper on (for maximum damping) or off (for minimum damping) comparing the primary controller and the force in the damper so that the damper tracks the primary control force history.

The assumptions used in the design of the primary controller play an important role in the performance of the clipped-optimal control. Clearly, the primary controller does not account for the dissipative nature of the damper. Therefore, the performance depends on the nature of the problem and the control design. If highly dissipative forces are required to achieve the design goal, clipped-optimal control can achieve the performance of the corresponding active system. On the other hand, if the primary controller produces highly nondissipative forces, the semiactive control will not be successful in mimicking the primary controller. To reduce this effect and to obtain maximal benefit from the damper, the nonlinear dissipativity property of the damper should be included in the design of the active primary controller. If the primary controller is an H_2-based optimal controller, the dissipativity can be approximately incorporated by representing the cross-terms as uv, though this approach is limited by the requirement that LQR optimization be well-posed and cannot be directly applied for multiple dampers. Another method is to impose the dissipativity as a constraint to the standard LQR problem. In this case, the solution of this problem is not trivial since the dissipativity is a nonlinear inequality constraint and special techniques should be employed.

In previous research, the authors developed a procedure for the solution of the LQR problem with a nonlinear dissipativity constraint [6] using linear matrix inequalities (LMIs). First, LQR was defined in terms of linear matrix inequalities with convex matrix inequality constraints and an LQR-equivalent eigenvalue problem (EVP) was defined. Then, the dissipativity constraint was attached to the EVP in its weak expected form, and the final problem was solved numerically. Two important benefits of this method were addressed. First, the LMI method eliminates the problem associated with LQR well-posedness. Second, it is more convenient for multi-damper systems. A two-degree-of-freedom structure with an ideal smart damper was considered as an example. It was found that dissipativity is considerably increased compared to the standard LQR method in the designs where the dissipativity is low.

In this paper, the proposed method that uses LMI techniques to append the dissipativity constraint to a standard LQR algorithm is tested for realistic smart damper and building models. First, brief definitions of LMIs, EVPs, LQR control problem are given. The derivation of the proposed method is reviewed. Then, a numerical example is considered to investigate the method. In this example, a realistic model of an MR damper is considered. A Kalman estimator is used to estimate states. The control is designed assuming a Kanai-Tajimi filtered white noise ground excitation. Several control designs are considered to investigate the dissipativity performance of the proposed method for a wide variety of control designs. The well-known 1940 El Centro earthquake ground acceleration record is used to excite the structure. Control design and analysis are performed using the Control Toolbox and LMI Control Toolbox of MATLAB® and simulations are carried out using SIMULINK®.

Numerical simulations show that the LMI method increases the dissipativity of the primary controller for the design parameters where the standard LQR yields low dissipativity levels. For these parameters, the proposed method shows an improvement in the control forces with response quantities similar to those in the conventional LQR approach. If, on the other hand, the LQR approach already gives highly dissipative forces, the LMI method provides only marginal improvements. It is also observed that as the LMI approach is a numerical method, the results are very sensitive to the numerical parameters of the solver. In the continuation of this research, these parameters should be investigated to improve the performance of the proposed method.

2. EVP Representation of an LQR Problem

2.1 Definitions

LMIs. Consider a matrix function given by

$$\mathbf{F}(\mathbf{x}) = \mathbf{F}_0 + x_1\mathbf{F}_1 + \ldots + x_n\mathbf{F}_n \tag{1}$$

where $\mathbf{x} = [x_1\ x_2\ \ldots\ x_n]^T$ is a real vector and $\mathbf{F}_0, \mathbf{F}_1, \ldots, \mathbf{F}_n$ are real symmetric matrices. A *linear matrix inequality* has the form

$$\mathbf{F}(\mathbf{x}) > 0 \tag{2}$$

Inequality (2) means that $\mathbf{F}(\mathbf{x})$ is a positive definite matrix. A vector $\mathbf{x}$ that satisfies inequality (2) is known as a *feasible solution* of the LMI.

The feasible solution set of (2) can be shown to be *convex* (*e.g.*, [1]). Convexity is an important property because powerful numerical techniques are available for the solution of problems involving convex constraints. In the control field, LMIs generally appear in *eigenvalue problems*. Eigenvalue problems have several representations, one of which is given by

$$\min_{\mathbf{x}} \mathbf{c}^T\mathbf{x} \quad \text{subject to } \mathbf{A}(\mathbf{x}) > 0 \tag{3}$$

where $\mathbf{A}(\mathbf{x})$ is of the form (1).

LQR Problem. Consider a linear time-invariant system

$$\dot{\mathbf{q}} = \mathbf{A}\mathbf{q} + \mathbf{B}\mathbf{u} + \mathbf{E}\mathbf{w} \qquad \mathbf{z} = \mathbf{C_z}\mathbf{q} + \mathbf{D_z}\mathbf{u} + \mathbf{F_z}\mathbf{w} \tag{4}$$

where $\mathbf{q}$ is the state vector, $\mathbf{u}$ is a vector of control forces, and $\mathbf{w}$ is a stationary zero-mean Gaussian white noise vector process disturbance with unit intensity.

The LQR problem is to find the control gain $\mathbf{K}$ that satisfies the optimization

$$\begin{array}{l} \min_{\mathbf{K}} E\left[\mathbf{z}^T\tilde{\mathbf{Q}}\mathbf{z} + \mathbf{u}^T\tilde{\mathbf{R}}\mathbf{u} + \mathbf{z}^T\tilde{\mathbf{N}}\mathbf{u} + \mathbf{u}^T\tilde{\mathbf{N}}^T\mathbf{z}\right] \\ \text{subject to (4) and } \mathbf{u} = -\mathbf{K}\mathbf{q} \end{array} \tag{5}$$

where $\tilde{\mathbf{Q}} \geq 0$ and $\tilde{\mathbf{R}} > 0$ are symmetric weighting matrices. To be well-posed [11], the parameters of LQR and H_2 control strategies should satisfy

$$\begin{bmatrix} \mathbf{Q} & \mathbf{N} \\ \mathbf{N}^T & \mathbf{R} \end{bmatrix} \geq 0 \qquad \begin{bmatrix} \tilde{\mathbf{Q}} & \tilde{\mathbf{N}} \\ \tilde{\mathbf{N}}^T & \tilde{\mathbf{R}} \end{bmatrix} \geq 0 \tag{6}$$

where $\mathbf{R} = \tilde{\mathbf{R}} + \mathbf{D}_\mathbf{z}^T\tilde{\mathbf{Q}}\mathbf{D}_\mathbf{z} + \mathbf{D}_\mathbf{z}^T\tilde{\mathbf{N}} + \tilde{\mathbf{N}}^T\mathbf{D}_\mathbf{z}$, $\mathbf{N} = \mathbf{C}_\mathbf{z}^T\tilde{\mathbf{Q}}\mathbf{D}_\mathbf{z} + \mathbf{C}_\mathbf{z}^T\tilde{\mathbf{N}}$ and $\mathbf{Q} = \mathbf{C}_\mathbf{z}^T\tilde{\mathbf{Q}}\mathbf{C}_\mathbf{z}$. It can be shown that LQR problem given above can be transformed into an equivalent EVP [6]

$$\begin{array}{ll} \min_{(\mathbf{X},\mathbf{S},\mathbf{Y})} & Tr(\mathbf{X}) + Tr(\mathbf{Q}\mathbf{S}) - Tr(\mathbf{Y}\mathbf{N}) - Tr(\mathbf{N}^T\mathbf{Y}^T) \\ \text{subject to} & \mathbf{A}\mathbf{S} - \mathbf{B}\mathbf{Y} + \mathbf{S}\mathbf{A}^T - \mathbf{Y}^T\mathbf{B}^T + \mathbf{E}\mathbf{E}^T \leq 0 \text{ and} \\ & \begin{bmatrix} \mathbf{X} & \mathbf{R}^{1/2}\mathbf{Y} \\ \mathbf{Y}^T\mathbf{R}^{1/2} & \mathbf{S} \end{bmatrix} > 0, \quad \mathbf{S} > 0 \end{array} \tag{7}$$

where the optimal control gain is given by $\mathbf{K} = \mathbf{Y}\mathbf{S}^{-1}$, matrix $\mathbf{S}$ is the optimal state covariance matrix when the problem is well-posed and there are no additional constraints, and $\mathbf{X} = \mathbf{R}^{1/2}\mathbf{K}\mathbf{S}\mathbf{K}^T\mathbf{R}^{1/2}$.

2.2 Dissipativity constraint

The LQR problem defined by (7) is in terms of inequalities, therefore another inequality constraint can be added to the problem. However, the dissipativity constraint $uv < 0$ for each smart damper cannot be directly added to the EVP since this constraint is deterministic while other variables of the EVP are expectations of system quantities. Therefore, the weaker constraint $E[uv] = E[(-\mathbf{K}\mathbf{q})(\mathbf{q}^T\mathbf{r}_v)] = -\mathbf{K}\mathbf{S}\mathbf{r}_v < 0$ is imposed, where $\mathbf{r}_v$ is a vector relating the state vector to the damper velocity. To force higher dissipativity levels, the constraint is strengthened to

$$-\mathbf{K}\mathbf{S}\mathbf{r}_v = -\mathbf{Y}\mathbf{r}_v < \gamma < 0 \tag{8}$$

where γ is a parameter to be determined by trial and error until a satisfactory solution is obtained.

3. A Numerical Example

In this section, the 9-story benchmark structure shown in Figure 2 is ana-

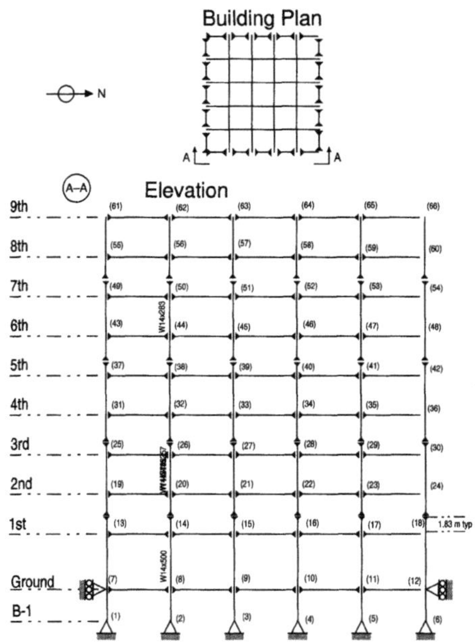

Figure 2. 9-story benchmark structure in plan and elevation views.

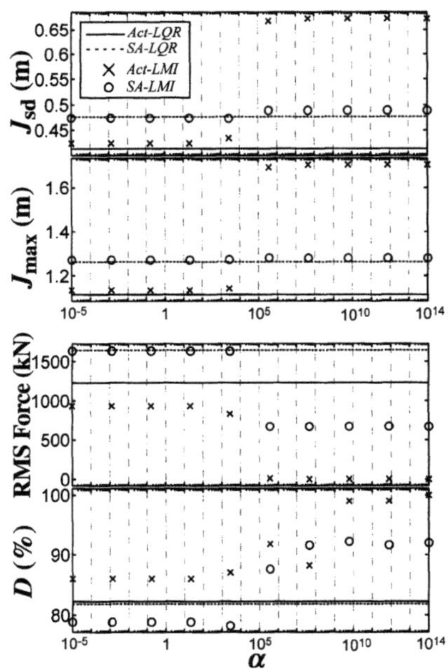

Figure 3. Performance indices, dissipativity index, RMS control force over range of response weights α ($\beta = 1$).

lyzed using the methods defined above. First, the validity of the EVP formulation of the LQR problem is verified. Then, for several values of the LQR design parameters, the structure is analyzed using standard LQR and using an LMI EVP with the added dissipativity constraint. Both control designs are applied (*i*) to a fully active system and (*ii*) as the primary controller in clipped optimal control of an MR damper. The systems investigated are summarized as follows:

- *Act-LQR:* This system is fully active and standard LQR is used as the control strategy.

- *Act-LMI:* This system is also fully active. The LMI method with the dissipativity constraint is used as the control strategy to observe how this constraint increases the dissipativity characteristics compared to the LQR.

- *SA-LQR:* This is a semiactive system where the smart damper is an MR damper. Control strategy *Act-LQR* is used as the primary controller.

- *SA-LMI:* This is also a semiactive system with an MR damper. Control strategy *Act-LMI* is used as the primary controller.

For the semiactive strategies, the secondary controller uses a bang-bang control law, following [2], to determine the voltage to send to the MR dampers: $v(t) = v_{\text{max}} H(f_{\text{measured}}[f_{\text{active}} - f_{\text{measured}}])$, where v_{max} is the maximum voltage, f_{measured} is the current force being exerted by the dampers, f_{active} is the force commanded by the primary controller and $H(\cdot)$ is the Heaviside unit step function.

3.1 9-story benchmark and EVP verification

The structure investigated is a 9-story benchmark model of a 9-story building designed for the Los Angeles area. This model, described by [7], is 150 ft square in plan and 122 ft tall. A nonlinear finite element model with 135 degrees of freedom (DOF) is the baseline model. A reduced-order 9DOF control-oriented input/output model is developed using a modal truncation approach [4]. The hysteretic model developed by [12] for the 20-ton MR damper is used herein. In this model, the time lag due to the fluid rheology (the transition to/from a semisolid state with changing magnetic field) and due to the RL circuit of the electromagnet coil is modeled as a second order filter. The damper is attached between the ground and the first floor. Since the 20-ton damper provides force levels too small for this structure, a 200-ton damper capacity is assumed; this could be generated with ten 20-ton devices each commanded identically. In the design of the controller and state estimator, earthquake ground motion is represented as a white noise filtered through a Kanai-Tajimi filter [9], where the filter parameters are estimated using several historical earthquake ground motion records. The sensor measurements are the ground acceleration, the absolute floor accelerations and the damper displacement. A Kalman estimator is used to estimate the full state of the 9DOF reduced model. Numerical simulations are carried out for a set of LQR parameters. The outputs to be minimized are chosen as absolute floor accelerations and interstory drifts.

To verify that the EVP given by (7) is equivalent to the LQR problem given by (5), the gains and of LQR and LMI approaches and covariance matrices of the states were computed using several values of weighting matrices $\mathbf{Q}$, $\mathbf{R}$ and $\mathbf{N}$. Both methods give identical results within numerical round-off accuracy if the LQR problem is well-posed.

3.2 The effects of the dissipativity constraint

The effects of the dissipativity constraint are investigated for the 9DOF system defined above. The output to be minimized is selected to be interstory drifts and absolute story accelerations. The LQR design parameters are taken as

$$\tilde{\mathbf{Q}} = \alpha \begin{bmatrix} \mathbf{I}_{9\times9} & \mathbf{0}_{9\times9} \\ \mathbf{0}_{9\times9} & \frac{\beta}{\omega_1^4}\mathbf{I}_{9\times9} \end{bmatrix} \qquad \tilde{\mathbf{R}} = \eta \qquad \tilde{\mathbf{N}} = \mathbf{0} \tag{9}$$

where ω_1 is the (uncontrolled) fundamental natural frequency. This form of $\tilde{\mathbf{Q}}$ allows one to select the relative importance of the drifts and absolute accelerations by adjusting β and the trade-off between performance and control effort by varying α. Herein, the 9DOF system is investigated for several values of α with $\beta = 1$ that satisfy inequality (6). 1940 El Centro earthquake ground acceleration data is used to excite the system.

The EVP solver of the LMI Control Toolbox has two important parameters. The first, which determines the accuracy, is chosen as 10^{-10}. The other parameter is the feasibility radius. It is observed that better results are achieved if γ is chosen close to the feasibility radius. After some trial simulations, the feasibility radius is set to 10^{30} and γ is -0.6×10^{29} N/m. η is taken as 10^{-20} $\mathrm{m}^2/\mathrm{N}^2$ for all simulations.

Two indices are introduced to evaluate the overall performances of the controlled systems: a cumulative root mean square standard deviation response index and a root mean square of peak responses. These indices are defined as follows:

$$J_{\mathrm{sd}}^2 = \sum \left[\sigma_{x_i}^2 + \sigma_{\ddot{x}_i^{\mathrm{a}}}^2/\omega_1^2\right], J_{\max}^2 = \sum \left[(x_i^{\max})^2 + (\ddot{x}_i^{\mathrm{a},\max})^2/\omega_1^2\right] \tag{10}$$

where σ is the standard deviation of the corresponding quantity, x_i is the interstory drift and $\ddot{x}_i^{\mathrm{a}}$ is the absolute acceleration of the i-th floor.

To determine how the dissipativity increases with the added dissipativity constraint, a dissipativity index is introduced as follows:

$$D = 1 - \frac{1}{N}\sum_{k=0}^{N-1} H\left[u(k\Delta t)v(k\Delta t)\right]. \tag{11}$$

This index is the fraction of time that the primary controller is commanding dissipative forces. Higher D shows an increased dissipativity.

Figure 3 shows the performance indices, dissipativity index and RMS control forces for several values of α when $\beta = 10$. For this set of control design parameters, it is immediately observed that the proposed method increases the dissipative nature of the primary controller considerably for some values of α, from about 82% dissipative to about 92% dissipative. The control force levels required for the proposed method are about half those of the LQR approach while having almost the same performance indices. For this range of design, the proposed method performs better than the conventional LQR-based clipped optimal control. In the light of the previous work, the main reason for the increased performance is associated with the increased dissipativity. However, for some control designs increased dissipativity may not always yield better performance as the control design itself may require highly nondissipative primary control forces.

This numerical example is more complex than the example implemented in the previous work by the authors. However, it is shown that proposed method has the capability of increasing dissipativity. The limits of this capability depend on many factors, including the control design goals, structural properties including the smart damper and estimators and other filters used. Also, the accuracy of numerical EVP solver effects the overall performance.

4. Conclusion

In this study, a nonlinear inequality constraint is introduced to a control design with a quadratic cost function to account for the dissipative nature of smart dampers in the semiactive control of structures for vibration mitigation. This is achieved by first representing an LQR problem as an EVP defined in terms of LMIs, and then appending the dissipativity constraint to the other constraints of the EVP. The proposed method is investigated by analyzing a 9DOF benchmark structure with ten 20-ton MR dampers for several control designs. Numerical simulations show that the additional constraint increases the dissipativity of the control forces even though its weak expected value form is used. Increased dissipativity resulted an increased overall performance.

Acknowledgments

The authors gratefully acknowledge the partial support of this research by the National Science Foundation under CAREER grant CMS 00-94030.

References

[1] G. E. Dullerud and F. Paganini, *A Course in Robust Control Theory: A Convex Approach*, Springer, New York, 2000.

[2] S. J. Dyke, B. F. Spencer, Jr., M. K. Sain, and J. D. Carlson, "Modeling and Control of Magnetorheological Dampers for Seismic Response Reduction," *Smart Materials & Struct.*, vol. 5, pp. 565–575, 1996.

[3] H. Fujitani, H. Sodeyama, K. Hata, N. Iwata, Y. Komatsu, K. Sunakoda, and S. Soda, "Dynamic Performance Evaluation of Magnetorheological Damper," *Proceedings of International Conference on Advances in Structural Dynamics*, Hong Kong, China, vol. 1, pp. 319–326, 2000.

[4] E. A. Johnson, "Nonlinear Seismic Benchmark Problem: Dissipativity and the 9-Story Benchmark Building." *2nd European Conf. Struct. Ctrl*, France, July 3–6, 2000.

[5] E. A. Johnson, G. A. Baker, B. F. Spencer, Jr., and F. Fujino, "Semiactive Damping of Stay Cables," *Journal of Engineering Mechanics*, ASCE, 2003, in press.

[6] E. A. Johnson and B. Erkus, "Structural Control with Dissipative Damping Devices," *2002 American Control Conference (ACC02)*, Anchorage, Alaska, May 8–10, 2002. Proceedings, pp. 2463–2468. CD-ROM Proceedings, paper ACC02-2463, file 441.pdf

[7] Y. Ohtori, R. E. Christenson, B. F. Spencer, Jr., and S. J. Dyke, "Benchmark Control Problems for Seismically Excited Nonlinear Buildings," *Proceedings of the 2nd European Conference on Structural Control*, 2000. An extended version of this paper is on web at http://www.nd.edu/ quake/papers/NLbench.pdf

[8] J. C. Ramallo, E. A. Johnson, and B. F. Spencer, Jr., "Smart' Base Isolation Systems," *Journal of Engineering Mechanics*, ASCE, vol. 128, no. 10, pp. 1088–1099, 2002.

[9] T. T. Soong and M. Grigoriu, "Random Vibration of Mechanical and Structural Systems," *Prentice Hall*, Englewood, Cliffs, NJ, 1993.

[10] B. F. Spencer, Jr. and M. K. Sain, "Controlling Buildings: A New Frontier in Feedback," *IEEE Control Systems Mag.*, vol. 17, no. 6, pp. 19–35, 1997.

[11] R. F. Stengel, *Optimal Control and Estimation*, Dover Publications, New York, 1994.

[12] G. Yang, B. F. Spencer, Jr., J. D. Carlson, and M. K. Sain, "Large-Scale MR Fluid Dampers: Modeling and Dynamic Performance Considerations," *Engineering Structures*, vol. 24, pp. 309–327, 2002.

ON PARAMETER ESTIMATION FOR LINEAR AND NONLINEAR STOCHASTIC SYSTEMS

Rafail Z. Khasminskii*

Department of Mathematics

Wayne State University

Detroit, MI, 48202, USA

rafail@math.wayne.edu

Abstract The overview of some results of [2], [3] on estimation of parameters in linear systems is presented. Some recent results of [4] concerning estimation of $f'(0)$ ($f(x)$ is a drift vector) for nonlinear systems are discussed too. It is shown that conditions for an existence of a consistent, and the more, asymptotically efficient estimator for $f'(0)$ are essentially less restrictive for the stable in probability systems. Some simple mechanical examples are presented and discussed.

Keywords: Asymptotic efficiency, stochastic differential equation, maximum likelihood estimator, stochastic stability

1. Linear Systems

Consider the system of linear Stochastic Differential Equations (SDE)

$$dX_t = A(\theta)X_t dt + \sum_{r=1}^{q} \sigma_r X_t dw_r(t). \tag{1}$$

Here $X_t \in \mathbf{R}^d$, $w_r(t)$, $r = 1, \ldots, q$, are independent standard scalar Wiener processes, $\sigma_1, \ldots, \sigma_q$ are real $d \times d$ matrices,

$$A(\theta) = A_0 + \sum_{j=1}^{k} A_j \theta_j, \tag{2}$$

where the $d \times d$ matrices A_0, A_j are known and the scalars θ_j, $j = 1, \ldots, k$, are unknown parameters. The estimation of parameters for a linear SDE of the

*Partially supported by NSF Grant DMS 9971608

N. Sri Namachchivaya and Y.K. Lin (eds.),
IUTAM Symposium on Nonlinear Stochastic Dynamics, 237-245.

form (1) was considered in [2], [3]. We suppose that the following conditions $(C_1) - (C_4)$ are fulfilled:

(C_1). The matrices $A_1, \ldots, A_k$ are linearly independent.

(C_2). $A_j \in \text{span}(\sigma_1, \ldots, \sigma_q), \quad j = 1, \ldots, k.$

(C_3). The weak Hörmander condition for the Markov diffusion process $\Lambda(t) = X(t)/|X(t)|$ with values on the unit sphere $\mathbf{S}^{d-1} \subset \mathbf{R}^d$ is fulfilled:

$$\dim LA\{h_0, h_1, ..., h_q\} = d - 1 \text{ for all } \lambda = x/|x| \in \mathbf{S}^{d-1},$$

where

$$h_0(\lambda) = \tilde{A}\lambda - (\tilde{A}\lambda, \lambda)\lambda, \quad \tilde{A} = A - \frac{1}{2}\sum_{r=1}^{q} \sigma_r^2,$$

$$h_r(\lambda) = \sigma_r\lambda - (\sigma_r\lambda, \lambda)\lambda, \quad r = 1, \ldots, q,$$

$LA\{\}$ denotes the Lie algebra generated by the vector fields which occur in the brackets (see [1]).

Introduce the diffusion matrix

$$B(x) = \sum_{r=1}^{q} \sigma_r x x^* \sigma_r^*$$

and denote by $B^+(x)$ the pseudoinverse of B.

(C_4). The matrix $B^+(\lambda)$, $|\lambda| = 1$, is continuous on $\mathbf{S}^{d-1}$.

Denote $P_\theta^{(T)}$ the probability measure corresponding to the process X_t, $0 \leq t \leq T$, generated by the system (1). It is easy to check (see [3, 4]) that the maximum likelihood estimator (MLE) $\hat{\theta}$ for θ based on the observation X_t, $0 \leq t \leq T$, is defined by the following system of linear algebraic equations

$$\frac{1}{2}\sum_{j=1}^{k} \int_0^T ((A_i^* B^+(X_t) A_j + A_j^* B^+(X_t) A_i) X_t, X_t) dt \cdot \theta_j \tag{3}$$

$$= \int_0^T (B^+(X_t) A_i X_t, dX_t - A_0 X_t dt), \quad i = 1, \ldots, k.$$

Denote by $H(X^T)$ the $k \times k$ matrix of the system (3) divided by T, and let $V(X^T)$ be a vector of right-hand side of (3). It is proven in [3] that under conditions $C_1 - C_4$ the matrix $H(X^T)$ is positive-definite a.s. and therefore

$$\hat{\theta} = \frac{1}{T} H^{-1}(X^T) V(X^T). \tag{4}$$

It was noticed in [3] that the likelihood ratio and, in particular, the vector $V(X^T)$ and the matrix $H(X^T)$ depend on the process $\Lambda(t)$ only. It means that the process $\Lambda(t)$ is the *sufficient statistics* for the estimation θ. In other words, the best estimation of θ can be obtained on the base observation the projection $\Lambda(t)$ of the process $X(t)$ on the unit sphere. We can conclude from this fact that the estimation performance for θ is the same for stable and unstable processes. We will see below that it is not the case for nonlinear systems.

It was shown in [2, 3] that under conditions $C_1 - C_4$ there exists (a.s.) a non-random limit

$$\lim_{T\to\infty} H(X^T) := I(\theta), \tag{5}$$

the matrix $TI(\theta)$ is the Fisher information matrix for the problem and the estimator (4) is asymptotically normal and asymptotically efficient for the wide class of a loss functions. In particular (here and below $\mathfrak{L}(.)$ is a distribution law of (.))

$$\mathfrak{L}(\sqrt{T}I^{1/2}(\theta)(\hat{\theta}-\theta)) \to \mathcal{N}(0, 1_k) \text{ as } T \to \infty, \tag{6}$$

where 1_k is the unit $k \times k$ matrix and $I(\theta)$ is defined by (5).

2. Nonlinear Systems

Another sort of problems arises when the observed process $X(t)$ is a solution of nonlinear SDE with the equilibrium point 0. The natural and important for applications problem here is a semiparametric estimation of the main part of a drift vector for the system at the vicinity of 0. So now we consider a system described by the nonlinear SDE

$$dX_t = f(X_t)dt + \sum_{r=1}^{q} b_r(X_t)dw_r(t), \tag{7}$$

where $X_t \in \mathbf{R}^d$, f and b_r, $r = 1, \ldots, q$, are d-dimensional vectors with $f(0) = b_r(0) = 0$, and, as before, $w_r(t)$ are independent standard Wiener processes. Assume that for $x \to 0$

$$f(x) = f'(0)x + \varphi(x), \quad \frac{|\varphi(x)|}{|x|} = O(|x|^\alpha), \alpha > 0, \tag{8}$$

$$b_r(x) = \sigma_r x + \psi_r(x), \quad \frac{|\psi_r(x)|}{|x|} = O(|x|^\alpha), \quad r = 1, \ldots, q, \tag{9}$$

where $f'(0)$ and $\sigma_r := b_r'(0)$ are $d \times d$ matrices. We suppose that all $b_r(x)$ are known vector-functions (see comments in [2]). The problem is to estimate the matrix $f'(0)$, or, a little more general, some unknown coefficients of this matrix. So it is convenient to write $f'(0)$ at the analogous to (2) form

$$f'(0) = A(\theta) = A_0 + \sum_{j=1}^{k} A_j\theta_j. \tag{10}$$

and consider the problem of estimation $\theta = (\theta_1, \ldots, \theta_k)$ on the base of the observation process

$$dX_t = A(\theta)X_t dt + \sum_{r=1}^{q} \sigma_r X_t dw_r(t) + \varphi(X_t)dt + \sum_{r=1}^{q} \psi_r(X_t)dw_r(t), \quad 0 \le t \le T. \tag{11}$$

It is clear that aforementioned estimation problem is inconsistent if the initial condition x_0 is arbitrary: the trajectory of X_t with positive probability will never visit a small neighborhood of the origin.

Consider, for instance, the one-dimensional system

$$dX_t = f(X_t)dt + b(X_t)dw_t, \quad X^x(0) = x. \tag{12}$$

and assume that $f(0) = b(0) = 0$ and

$$f(x) = f'(0)x + O(|x|^{1+\alpha}), \quad b(x) = b'(0)x + O(|x|^{1+\alpha}), \quad \alpha > 0, \tag{13}$$

as $x \to 0$. It was proven in [2] that for the linear system

$$d\bar{X}_t = \theta\bar{X}_t dt + \sigma\bar{X}_t dw_t, \quad \bar{X}_0 = x \neq 0, \tag{14}$$

the estimator

$$\hat{\theta}_T = \frac{1}{T}\int_0^T \frac{d\bar{X}_t^x}{\bar{X}_t^x} \left(\text{i.e.,}\hat{\theta}_T = \frac{1}{T}\ln\frac{\bar{X}_T^x}{x} + \frac{1}{2}\sigma^2\right) \tag{15}$$

is efficient. Moreover, it is easy to see that the distribution law of the normalized difference $\frac{\sqrt{T}}{\sigma}(\hat{\theta}_T - \theta)$ is exactly $\mathcal{N}(0,1)$, and there is no estimator with the quadratic risk less than σ^2/T uniformly in $\theta \in [\theta_0 - \delta, \theta_0 + \delta]$, $\theta_0 \in \mathcal{R}$, $\delta > 0$ is an arbitrary small constant.

It is natural to use the estimator (15) for the nonlinear equation (12). The following result concerning properties of this estimator was proven in [4].

Theorem 1 . *Suppose the coefficients of (12) satisfy the conditions (13) for some $\alpha > 0$ and*

$$f'(0) \le B \tag{16}$$

for a known constant B. Let $M := B - \frac{\sigma^2}{2} \ge 0$. Then the estimator

$$\hat{\theta}_T = \frac{1}{T}\int_0^T \frac{dX_t^{x_0}}{X_t^{x_0}} \tag{17}$$

with $x_0 = x_0(T)$ *satisfying the condition*

$$0 < |x_0| < e^{-(M+\varepsilon)T}, \tag{18}$$

$\varepsilon > 0$ *is a constant, has the property*

$$\mathcal{L}(\frac{\sqrt{T}}{\sigma}(\hat{\theta}_T - f'(0))) \to \mathcal{N}(0,1) \tag{19}$$

as $T \to \infty$.

The condition (18) is very restrictive (and impractical). But it is impossible to avoid it due to the fact that the system (18) can be unstable and it is impossible to create the asymptotically efficient estimator for unstable systems without exponential (with sufficiently large constant) in T closeness of the initial condition to the origin.

Situation is essentially better for the stable systems. Consider the linear approximation (variational equation) for (11):

$$d\bar{X}_t = A(\theta)\bar{X}_t dt + \sum_{r=1}^{q} \sigma_r \bar{X}_t dw_r(t). \tag{20}$$

Let the conditions $C_1 - C_4$ be fulfilled for the equation (20), and a solution of this equation be asymptotically stable in probability. Then it is proven in [4] that a small modification of the estimator (4) has an asymptotic efficiency arbitrary close to 1 if the initial condition x_0 for the process (11) satisfies the condition

$$x_0 = x_0(T) \to 0, \text{ as } T \to \infty. \tag{21}$$

The construction of this estimator consists from two steps. The first step is an observation for arbitrary $N > 1$ of $X^{x_0(T)}(T/N) = \tilde{x}(T)$. Then one constructs an estimator of the type (4) using the observation process $X^{\tilde{x}(T)}(t)$ on a time interval $[T/N, T]$ with the initial condition $\tilde{x}(T)$. More precisely, on the second step the estimator

$$\hat{\theta}_{[N,T]} = \frac{1}{T(1-1/N)} H^+(X^{\tilde{x}(T),[T/N,T]}) V(X^{\tilde{x}(T),[T/N,T]})$$

can be used. Here $H^+(X^{\tilde{x}(T),[T/N,T]})$ is the pseudo-inverse matrix (we cannot guarantee the existence of $H^{-1}(.)$ for nonlinear system) of the matrix $H(X^{\tilde{x}(T),[T/N,T]})$ with the elements $(i, j = 1, \ldots, k)$

$$H_{ij}(X^{\tilde{x}(T),[T/N,T]}) =$$
$$\tfrac{1}{2T(1-1/N)} \int_{T/N}^{T} ((A_i^* B^+(X^{\tilde{x}(T)}(t))A_j$$
$$+A_j^* B^+(X^{\tilde{x}(T)}(t))A_i) X^{\tilde{x}(T)}(t), X^{\tilde{x}(T)}(t))dt,$$

and $V(X^{\tilde{x}(T),[T/N,T]}) \in \mathbf{R}^k$ is the vector with coordinates

$$V_i(X^{\tilde{x}(T),[T/N,T]}) = \int_{T/N}^{T} (B^+(X^{\tilde{x}(T)}(t))A_i X^{\tilde{x}(T)}(t), \\ dX^{\tilde{x}(T)}(t) - A_0 X^{\tilde{x}(T)}(t)dt).$$

Theorem 2 *Let the conditions $C_1 - C_4$ be fulfilled for the equation (20), the solution of (20) be asymptotically stable in probability and the initial condition for the nonlinear SDE (7) satisfy the condition (21). Then for arbitrary $N > 1$*

$$\mathcal{L}(\sqrt{T(1-1/N)} I^{1/2}(\theta)(\hat{\theta}_{[N,T]} - \theta)) \to \mathcal{N}(0, 1_k) \text{ as } T \to \infty.$$

It follows from this theorem that the asymptotic efficiency of this estimator for any bounded loss function, as $T \to \infty$, and $x_0 = x_0(T) \to 0$ is equal to $(N-1)/N$. Thus one can see that for N large enough the asymptotic efficiency of the estimator $\hat{\theta}_{[N,T]}$ can be made an arbitrary close to 1.

The proof of this theorem uses essentially the following result.

Lemma 3 *Denote $\bar{\Lambda}_t^{(x_0)} = \bar{X}_t^{x_0}/|\bar{X}_t^{x_0}|$ and $\Lambda_t^{(x_0)} = X_t^{x_0}/|X_t^{x_0}|$ and assume that the trivial solution of (20) is asymptotically stable in probability and δ is an arbitrary small positive number. Then for x_0 with $|x_0| = |x_0(T)| \leq$* $\exp(-\delta T)$

$$P\left\{\lim_{T\to\infty} \sup_{0\leq t\leq T} |\Lambda_t^{(x_0)} - \bar{\Lambda}_t^{(x_0)}| \to 0\right\} = 1. \tag{22}$$

3. Examples

1. Consider a stochastic linear oscillator with unknown restoring force coefficient θ_1 and damping coefficient θ_2. We assume that these coefficients subject to the random perturbations by independent White Gaussian Noises (WGN) $\dot{w}_1$, $\dot{w}_2$ with the known intensities s_1, s_2 accordingly.

The equation of such system can be written formally as

$$\ddot{x} + (\theta_2 + s_2\dot{w}_2)\dot{x} + (\theta_1 + s_1\dot{w}_1)x = 0,$$

or, in a rigorous form as an Itô SDE of the form (1)

$$\begin{aligned} dX_1 &= X_2 dt, \\ dX_2 &= -\theta_1 X_1 dt - \theta_2 X_2 dt - s_1 X_1 dw_1 - s_2 X_2 dw_2. \end{aligned} \tag{23}$$

For this example

$$A(\theta) = \begin{pmatrix} 0 & 1 \\ -\theta_1 & -\theta_2 \end{pmatrix} = \theta_1 \begin{pmatrix} 0 & 0 \\ -1 & 0 \end{pmatrix} + \theta_2 \begin{pmatrix} 0 & 0 \\ 0 & -1 \end{pmatrix} + \begin{pmatrix} 0 & 1 \\ 0 & 0 \end{pmatrix};$$

$$\sigma_1 = \begin{pmatrix} 0 & 0 \\ -s_1 & 0 \end{pmatrix}; \sigma_2 = \begin{pmatrix} 0 & 0 \\ -s_2 & 0 \end{pmatrix}.$$

Thus the conditions $(C_1), (C_2)$ are valid for this example if $s_1 \neq 0;\ s_2 \neq 0$. It is easy to check (see [3]) that the condition (C_3) is also valid. The diffusion matrix for this model is

$$B(x) = \begin{pmatrix} 0 & 0 \\ 0 & s_1^2 x_1^2 + s_2^2 x_2^2 \end{pmatrix}.$$

Therefore

$$B^+(\lambda) = \begin{pmatrix} 0 & 0 \\ 0 & (s_1^2 \lambda_1^2 + s_2^2 \lambda_2^2)^{-1} \end{pmatrix}$$

and the condition (C_4) is also valid. For this example the estimator (4) has the form

$$\widehat{\theta} = -(TH_T(X^T))^{-1} \int_0^T \frac{X(t)dX_2}{\sigma}{}^2_1 X_1^2(t) + \sigma_2^2 X_2^2(t) \qquad (24)$$

with

$$H_T(X)^{(T)} = \frac{1}{T} \int_0^T \begin{pmatrix} X_1^2(t) & X_1(t)X_2(t) \\ X_1(t)X_2(t) & X_2^2(t) \end{pmatrix} (\sigma_1^2 X_1^2(t) + \sigma_2^2 X_2^2(t))^{-1} dt \qquad (25)$$

One can conclude from aforementioned results of [3] that this estimator is asymptotically efficient for a wide class of the loss functions.

2.. Consider now an analogous problem for the randomly perturbed non-linear oscillator

$$\ddot{x} + (\theta_2 + s_2 \dot{w}_2)\dot{x} + (\theta_1 + s_1 \dot{w}_1) \sin x = 0, \qquad (26)$$

We will consider this equation under the natural assumption

$$\theta_i > 0; i = 1, 2. \qquad (27)$$

The variational (linearized) equation for (26) coincides with (23). It is well known that the system (23) is asymptotically stable in probability if the condition (27) is valid and $|s_i|, i = 1, 2$ are small enough. Thus we can apply Theorem 2.2 and obtain the conclusion: the described in the Theorem 2.2 modification of the estimator (24),(25) has an asymptotic efficiency arbitrary close to 1 if $|s_i|$ are positive and small enough and the initial conditions satisfy (21).

It is instructive to consider the cases where one of the coefficients s_i is equal to 0. For $s_2 = 0$ the condition (C_2) is violated, and it implies the singularity of the θ_2 estimation problem. Informally, the reason of this singularity is the possibility to observe θ_2 almost without noise at the interval $[t_0 - \delta, t_0 + \delta]$, here t_0 is one of the roots of the equation $x_1(t_0) = 0$, and δ is small enough. More details for the linearized system can be found in [3], s.2.

3. Assume now that $s_1 = 0$ and consider a little more general problem estimation of $f'(0) = \theta$ for the system

$$\ddot{x} + a^2 \sin x + f(\dot{x}) + \sigma \dot{x} \dot{w} = 0. \tag{28}$$

Here a^2, σ are the known constants, and $f(0) = 0$ and $|f(x) - f'(0)x| \leq |x|^{1+\alpha}$ for some $\alpha > 0$. The variational equation for (28) has a form

$$\begin{aligned} d\bar{X}_1 &= \bar{X}_2 dt, \\ d\bar{X}_2 &= -a^2 \bar{X}_1 dt - \theta \bar{X}_2 dt - \sigma \bar{X}_2 dw. \end{aligned} \tag{29}$$

It is easy to check that the estimator

$$\hat{\theta} = -\frac{1}{T} \int_0^T \frac{a^2 \bar{X}_1(t)dt + d\bar{X}_2(t)}{\bar{X}_2(t)} \tag{30}$$

is asymptotically efficient for the system (29). So it looks natural to use this estimator for the equivalent to (28) nonlinear system

$$\begin{aligned} dX_1 &= X_2 dt, \\ dX_2 &= -a^2 \sin(X_1)dt - f(X_2)dt - \sigma X_2 dw. \end{aligned} \tag{31}$$

But the condition (C_4) is not valid for (29)(the matrix $B^+(\lambda)$ is discontinuous for $\lambda_2 = 0$), and as a result, the integral in (30) applied to the system (31) diverges with positive probability.

One can use however instead of it the estimator

$$\tilde{\theta} = -\frac{1}{T} \int_0^T \frac{a^2 \sin(X_1(t))dt + dX_2(t)}{X_2(t)} \tag{32}$$

It is not hard to check that (32) is asymptotically efficient estimator for $f'(0)$ if the initial condition in (31) is chosen close enough to the origin. It is possible also to weaken demands to the initial conditions by using an estimator, analogous to the one described in s.2.

4. Conclusion

1 One can see from the discussion above that properties of the best estimators for the matrix $A(\theta)$ in the linear system (1) and the matrix $f'(0)$ in nonlinear system (7) are very different. In the linear systems the estimation performance does not depend on the choice of initial conditions and the stability properties of the system under consideration: in all cases the same MLE (4) can be used, and this estimator is asymptotically efficient for a wide class of loss functions, for any choice of

(nonzero) initial conditions and as for stable so as for unstable systems.

2 In contrary to this, the properties of estimators for $f'(0)$ in *nonlinear* system (7) depend strongly on the choice of initial conditions and stability of the system. It is impossible to construct a consistent estimator for nonlinear systems if the initial condition is not close enough to the origin. The degree of closeness to the origin is essentially different for stable and unstable systems. The construction of asymptotically efficient estimator for unstable system requires the choice of initial condition exponentially (with respect to observation time T) close to the origin with sufficiently large exponent (see Theorem 2.1). This requirement is too restrictive and impractical: for instance, it follows from (18) that for $M = 1, T = 10$, $|x_0|$ has to be less than $\exp(-10)$. Thus we can conclude that in the most of applications it is impossible to estimate efficiently $f'(0)$ for unstable nonlinear systems.

   The situation is very different for the *asymptotically stable in probability* nonlinear systems. It follows from Theorem 2.2 that the essentially less restrictive condition $|x_0| = |x_0(T)| \to 0$, as $T \to \infty$ allows to construct the estimator for $f'(0)$ with asymptotic efficiency arbitrary close to 1.

3 It follows from the example **3** in s.3 that sometimes MLE for the linearized system does not work (and is not defined even) for a given nonlinear system in spite of the existence asymptotically efficient estimator for this nonlinear system. A construction of asymptotically efficient estimators in more general situations, when some of the conditions $(C_1) - (C_4)$ are not fulfilled, is an open problem.

References

[1] L. Arnold, E. Oeljeklaus, and E. Pardoux, "Almost sure and moment stability for linear Itô equations," in: Lyapunov exponents (L. Arnold and V. Wihstutz, eds.), Lecture Notes in Math., vol. 1186, Springer, Berlin, pp. 129–159, 1986.

[2] A. Jankunas and R. Khasminskii, "Estimation of parameters of linear stochastic differential equations," *Stoch. Proc. Applic.*, vol. 72, (1997), pp. 205–219, 1997.

[3] R. Khasminskii, N. Krylov, N. Moshchuk, "On the estimation of parameters for linear stochastic differential equations," *Probab. Theory Relat. Fields*, vol. 113, pp. 443–472, 1999.

[4] R. Khasminskii and G. Milstein, "On estimation of the linearized drift for nonlinear stochastic differential equations," *Stochastic and Dynamics*, vol. 1, no. 1, pp. 23–43, 2001.

CONTROL AGAINST LARGE DEVIATION FOR OSCILLATORY SYSTEMS

Agnessa Kovaleva
Russian Academy of Sciences
Mechanical Engineering Research Institute
Moscow 101990, Russia
a.kovaleva@ru.net

Abstract The problem of controlling a near-Hamiltonian noisy system so as to keep it within a domain of bounded oscillations has been studied intensively in the last decade. This paper considers a new class of problems associated with control against large deviation in a weakly perturbed system. An exponential risk-sensitive residence time criterion is introduced as a performance measure, and a related HJB equation is constructed. An averaging procedure is developed for deriving an approximate solution of the risk-sensitive control problem in the small noise limit. It is shown that the averaged HJB equation is reduced to a first order PDE with the coefficients dependent on the noise intensity in the leading order term, though this intensity tends to zero in the original system. Near optimal control is constructed as a nonlinear time-independent feedback with parameters dependent on the noise intensity in the small noise limit. An example illustrates an application of this method to a system with resonance dynamics and with non-white noise perturbations.

Keywords: Key words: large deviation, stochastic control, asymptotic methods

1. Introduction

In this paper, we discuss a problem of control against escape for a noisy system. We suppose that, in the absence of noise, the system has an asymptotically stable point. The domain of attraction of the stable point is usually considered as a reference region, and the control task is to prevent the system from escape from this domain. Noise-induced escape from the domain of attraction of the stable point is interpreted as a large deviation effect [13]. In particular, this implies that escape over the time interval $O(1)$ is a rare event, and the dominant contribution to the mean escape time and to the probability of escape comes from the paths with exponentially large time to escape [13]. Dupuis and Kushner [4] have proposed an exponential transfor-

N. Sri Namachchivaya and Y.K. Lin (eds.),
IUTAM Symposium on Nonlinear Stochastic Dynamics, 247-256.

mation of the escape criterion associated with large deviation. This transformation "biases" the criterion toward the paths with very large time to escape. Whittle [14] has brought into use the notion of "risk-sensitive control" and has demonstrated that the exponential transformation of standard criteria make them sensitive to small noise. Dupuis and McEneaney [5] and Fleming and Soner [7] have given a rigorous mathematical basis for this idea.

In this paper, we extend the risk-sensitive control approach to systems with averaging. We consider a problem of control against escape for a class of near-Hamiltonian systems in the plane. Extensions to systems in R^{2n} are quite obvious. A special case of a 3-dimensional system is discussed in Section 3.

Consider a process governed by the stochastic differential equations

$$\begin{aligned} \dot{q} &= X_p(q,p) \\ \dot{p} &= -X_q(q,p) + \varepsilon f(q,p,u) + \varepsilon\sigma(q,p)\dot{w}(t) \end{aligned} \tag{1}$$

where $\varepsilon > 0$ is a small parameter, $q, p \in G$, an admissible region G is an open domain in R^2 with boundary Γ. The term $f(q,p,u)$ incorporates additional nonconservative and control forces, u is a control defined on a compact set $U \subset R^1$. Coefficients of (1) are assumed to be sufficiently smooth for any admissible control u to guarantee necessary transformations.

We consider in details a pendulum-like system with the potential function $V(q)$ for which the reference mode of operation corresponds to "captured" motion (libration) within the potential well. The Hamiltonian of the system is

$$X(q,p) = \frac{1}{2}p^2 + V(q) \tag{2}$$

The potential $V(q)$ is assumed to have an unstable point q_u corresponding to the maximum of $V(q)$ and a stable point q_s corresponding to the minimum of $V(q)$ (Fig. 1). Escape trough the potential barrier $V(q_u)$ is considered as an event to be avoided. In the phase plane, the threshold Γ: $\{X(q_u, 0) = V(q_u) = X_\Gamma\}$ is associated with the unperturbed separatrix, energy in the domain inside the loop of the separatrix corresponds to "captured" oscillatory motion. In practice, the system should not enter a small neighborhood of the separatrix, in order to avoid chaotic transportation [8]. The admissible domain is defined as in Figure 1.

$$G : \{X(q,p) \leq X^* < X_\Gamma\} \tag{3}$$

For the further analysis, we need consideration of the uncontrolled counterpart of (1)

$$\begin{aligned} \dot{q} &= p \\ \dot{p} &= -V_q(q) + \varepsilon f^0(p,q) + \varepsilon\sigma(q,p)\dot{w}(t) \end{aligned} \tag{4}$$

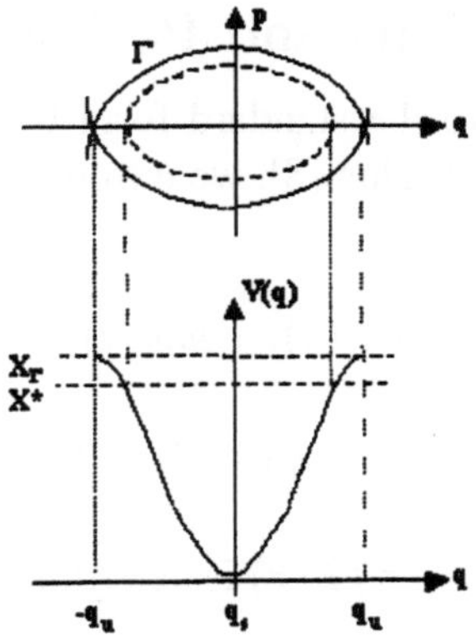

Figure 1. Phase portrait and potential.

where $f^0(q,p) = f(q,p,0)$. In the absence of perturbation ($\sigma = 0$), system (4) is assumed to have an asymptotically stable point with the domain of attraction $G_a \supset G$. In this case trajectories of system (4) converge to this point, with an exponentially small probability to leave the domain G [13].

Let $\tau = \varepsilon t$, $\tau_G = \inf\{\tau : q,p \notin G\}$ be the escape time, i.e., the first moment the system leaves the domain G. Under the assumption of asymptotic stability, the mean escape time $\mathbf{E}\tau_G \sim \exp(-C/\varepsilon)$ [13]. In order to take the exponentially large escape time into consideration, Dupuis and McEneaney [5] have introduced the exponential escape time criterion in the form

$$K^e(u) \quad = \quad -\varepsilon \ln \mathbf{E} \exp(\varepsilon^{-1}\tau_G) \tag{5}$$

For a deterministic system, criterion (5) coincides directly with escape time. We will show that the parameterization of the criterion with a small parameter ε allows construction of the feedback control sensitive to noise in the small noise limit, as $\varepsilon \to 0$ ("risk-sensitive control"). Criterion (5) is interpreted as a "risk-sensitive" counterpart of the standard mean escape time criterion in the small noise perspective.

Dupuis and McEneaney [5] have discussed risk-sensitive escape time control for a diffusion model with time-independent coefficients. This study demonstrates a similar approach to systems with averaging. Section 2 presents the equations of risk-sensitive control for a stochastic system in the plane. An example of a controlled resonance (Section 3) demonstrates the application of this approach to a system with three time-scales and with non-white noise perturbations.

2. Main Equations and Performance Criteria

We reduce system (1) to the standard form with slow and fast variables. Using the techniques of averaging [8], we introduce the new variables x and φ as

$$\begin{aligned} x(q,p) &= \frac{1}{2}p^2 + V(q), \quad p(x,q) = \pm\sqrt{2[x - V(q)]} \\ \frac{\partial\varphi}{\partial q} &= \frac{\omega(x)}{p(x,q)}, \quad \varphi(\text{mod}2\pi) \in Y \end{aligned} \tag{6}$$

where the frequency $\omega(x) = 2\pi/T(x)$ is determined through the period $T(x)$

$$T(x) = \oint\limits_{x\in G} \frac{dq}{p(x,q)} \tag{7}$$

The change of variables (6), with regard to the Itô formula [6], transforms equations (1) into the system with slow and fast variables [10]

$$\begin{aligned} dx &= F_1(\varphi,x,u,\varepsilon)d\tau + \varepsilon^{1/2}\sigma_1(\varphi,x)dw(\tau) \\ d\varphi &= [\varepsilon^{-1}\omega(x) + F_2(\varphi,x,u,\varepsilon)]d\tau + \varepsilon^{1/2}\sigma_2(\varphi,x)dw(\tau) \end{aligned} \tag{8}$$

In (8) we denote

$$F_1 = f(q,p,u)p + \frac{1}{2}\varepsilon\sigma^2(q,p), \quad \sigma_1 = \sigma(q,p)p \tag{9}$$

with the arguments $q = q(\varphi,x)$, $p = p(\varphi,x)$. Coefficients $F_2(\varphi,x,u,\varepsilon)$ and $\sigma_2(\varphi,x)$ require nontrivial derivation. We note that the resulting approximate equations do not involve these coefficients, and their precise form is insubstantial.

The control task is to prevent the system from leaving the admissible domain $G : \{x \le x^*\}$. We choose control u^ε maximizing the generalized escape time criterion

$$K^\varepsilon(u) = -\varepsilon \ln E \exp[-\varepsilon^{-1}\int_0^{\tau_G} L(x,u)d\tau] \tag{10}$$

It is straightforward to write the HJB equation related to system (8) with criterion (10) [7], [6]. Let $V^e(\varphi,x) = -\varepsilon \ln \Phi^e(\varphi,x)$ be the Bellman function associated with this problem. Starting with the PDE satisfied by Φ^e and taking the logarithm transformation, we formally obtain the equation

$$\begin{aligned} &\varepsilon^{-1}\omega(x)V^e_\varphi + H^e(\varphi,x,V^e_\varphi,V^e_x) - \tfrac{1}{2}(\sigma_1V^e_x + \sigma_2V^e_\varphi)^2 \\ &+\tfrac{\varepsilon}{2}(a_{11}V^e_{xx} + a_{22}V^e_\varphi + 2a_{12}V^\varepsilon_{x\varphi} + \sigma^2V^e_x) = 0, \quad \varphi,x \in Y\times G \\ &V^\varepsilon(\varphi,x) = 0, \quad \varphi\in Y,\ x = x^* \end{aligned} \tag{11}$$

where $a_{ij}(\varphi, x) = \sigma_i(\varphi, x)\sigma_j(\varphi, x)$ and

$$\begin{aligned} H^e(\varphi, x, V^e_\varphi, V^e_x) &= \max_{u \in U} h^e(\varphi, x, V^e_\varphi, V^e_x, u) \\ h^e(\varphi, x, V^e_\varphi, V^e_x, u) &= F_1(\varphi, x, u, \varepsilon)V^e_x + F_2(\varphi, x, u, \varepsilon)V^e_\varphi + L(x, u) \end{aligned} \quad (12)$$

Coefficients of (11) and (12) are assumed to be sufficiently smooth for a smooth solution $V^\varepsilon(x, \varphi)$ to exist in $Y \times \bar{G}$, uniformly in $\varepsilon > 0$, see [7] for details. As the function $V^\varepsilon(x, \varphi)$ is found, the solution of the associated optimal control problems is written in the form

$$u^\varepsilon(\varphi, x) = \arg\max_{u \in U} h^e(\varphi, x, V^e_\varphi, V^e_x, u) = U(\varphi, x, V^e_\varphi, V^e_x) \quad (13)$$

We need an asymptotic solution of (11) as $\varepsilon \to 0$. The asymptotic analysis of risk-sensitive escape time control for a system independent of the fast variable has been given by Dupuis and McEneaney [5]. The detailed asymptotic analysis of nonlinear PDEs associated with the singularly perturbed control problems has been presented by Alvarez and Bardi [1]. In both problems, the analysis is based on the viscosity solution method [3].

The fact that a continuous value function satisfies the appropriate HJB equation in the viscosity sense is a consequence of the dynamic programming principle [7]. Then, by combining the approaches from [5] and [1], we find that the second order derivatives are negligible in the asymptotic analysis [5], and the limit HJB equation can be obtained by averaging of the reduced first order PDE in the fast phase φ [1], [2].

The limit equation takes a form of the first order PDE with the coefficients and boundary conditions independent of φ. This allows construction of the asymptotic solution in the form $V^0(x)$ and reduction of the averaged HJB equation to the first order PDE

$$\begin{gathered} H^0(x, V^o_x) - \tfrac{1}{2}a^0(x)\,(V^o_x)^2 = 0, \quad x \in G \\ V^0(x) = 0, \quad x = x^* \end{gathered} \quad (14)$$

where

$$\begin{aligned} H^0(x, V^o_x) &= \tfrac{1}{2\pi} \int_{-\pi}^{\pi} H^e(\varphi, x, 0, V^e_x) d\varphi \\ a^0(x) &= \tfrac{1}{2\pi} \int_{-\pi}^{\pi} a_{11}(\varphi, x) d\varphi \end{aligned} \quad (15)$$

and

$$V^\varepsilon(x, \varphi) \to V^0(x), \qquad \varepsilon \to 0 \quad (16)$$

where the convergence is uniform in $Y \times \bar{G}$. From (13) and (14) we obtain the nearly-optimal control in the form

$$u^0(\varphi, x) = U^0(\varphi, x, V^0_x) = U(\varphi, x, 0, V^0_x) \quad (17)$$

Note that equation (14) involves the coefficient $a^0(x)$ depending on the noise intensity. This makes control (17) sensitive to noise in the small noise limit as $\varepsilon \to 0$.

3. Controlled Synchronization in the Presence of Noise

In this example, we demonstrate the extension of the asymptotic procedure to a resonance system with non-white noise perturbations. An extension of the risk-sensitive approach to non-resonance multi-frequency systems is obvious. In the non-resonance case, the multiphase averaging substitutes for the single phase averaging. However, an application of this approach to a resonance system requires a special procedure [10], [9]. The second problem to discuss is the application of the asymptotic method for a system with non-white noise when the HJB equation cannot be written directly. We will not consider these problems in a general setting, but take a simple example to demonstrate the substantial features of the asymptotic procedure.

Consider the equation of a Josephson junction with multiplicative noise

$$\ddot{\theta} + \sin\theta[k^2 + z(t)] = v \tag{18}$$

where v is a control force, $z(t) = a\cos t + \rho\xi(t)$, $\xi(t)$ is zero-mean Gaussian noise with spectral density $S(\omega)$ and variance $\mathbf{E}\xi^2(s) = 1$. We study the motion in the domain of fast rotation where $\dot{\theta} >> k$. In the absence of noise, fast rotation can be locked to weak periodic signal, and the resulting motion is close to rotation with angular velocity 1. The task of control v is to counteract the disrupting effect of weak noise and to sustain the synchronized motion.

Having regard to relationships between the parameters, we introduce the small parameter $\varepsilon = a/2$. A suitable parameterization transforms the equation of motion into the standard form with slow and fast variables

$$\begin{aligned} \dot{\theta} &= y \\ \dot{y} &= -\varepsilon\sin\theta[\varepsilon^{1/2}\gamma + 2\cos t + \varepsilon^{1/2}\Delta\xi(t)] + \varepsilon^{3/2}u \end{aligned} \tag{19}$$

where $k^2 = \varepsilon^{3/2}\gamma$, $\rho = \varepsilon^{3/2}\Delta$, $v = \varepsilon^{3/2}u$. The resonance condition is

$$y^* = 1 \tag{20}$$

We study small deviation of order $\varepsilon^{1/2}$ from frequency (20). Introduce the variables q and p by the formulas

$$y - 1 = \varepsilon^{1/2}p, \quad \theta - t = q \tag{21}$$

Thus we have

$$\begin{aligned} \dot{q} &= \varepsilon^{1/2}p \\ \dot{p} &= -\varepsilon^{1/2}[\sin q + \sin(2t+q)] - \varepsilon[\gamma + \Delta\xi It)]\sin(t+q) + \varepsilon u \end{aligned} \tag{22}$$

System (22) is similar to (1) and (2). Consider the Hamiltonian subsystem

$$\begin{aligned} \dot{q} &= \varepsilon^{1/2} p \\ \dot{p} &= -\varepsilon^{1/2} \sin q \end{aligned} \tag{23}$$

and introduce the variables x and φ by formulas (6), with the potential $U(q) = 1 - \cos q$. As a result, we obtain the equations

$$\begin{aligned} \dot{x} &= -\{\varepsilon^{1/2} \sin(2t+q) + \varepsilon[\gamma + \Delta\xi(t)] \sin(t+q) - \varepsilon u] p \\ \dot{\varphi} &= \varepsilon^{1/2}\omega(x) + \varepsilon k(x,\varphi)u + \varepsilon d(x,\varphi)\xi(t) + \varepsilon^{3/2} r(x,\varphi,t) \\ \dot{t} &= 1 \end{aligned} \tag{24}$$

where q, p are considered as functions of x, φ. As shown in [10], the terms $o(\varepsilon^{1/2})$ in the second equations are insubstantial for the further analysis. The time t is considered as the fast variable compared to the variables q and p. In order to make the time scales transparent, we employ the machinery of hierarchical averaging.

At the first stage, we use the partial stochastic averaging of system (24) in the fast time t. Taking into account the terms of order $\varepsilon^{1/2}$ and ε, we obtain the system for the slow variable x and the semi-fast variable φ [10]

$$\begin{aligned} dx &= p(x,\varphi)u d\tau + \varepsilon^{1/2}\sigma dw(\tau) \\ d\varphi &= \varepsilon^{-1/2}\omega(x) + k(x,\varphi)u + \varepsilon^{1/2}\delta(x,\varphi) dw(\tau) \end{aligned} \tag{25}$$

where $\tau = \varepsilon t$, $w(\tau)$ is standard Wiener process, the diffusion coefficient $\sigma^2 = \frac{1}{2}\Delta^2 S(1)$. As shown in Section 2, the precise form of the coefficients of the second equation in (25) is insubstantial for the further analysis. We see that, in the case of non-white noise, averaging in the fast time t removes the fast time from the equations and allows diffusion approximation (in the weak sense) of the perturbed process [11], [12]. The admissible domain G corresponds to the domain inside the separatrix of resonance [8] and is defined as $G : \{x \leq x^* < x_\Gamma = 2\}$.

The task is to maximize the mean time until escape from the admissible domain and minimize the control cost. In accordance with the problem, we introduce the criterion in the form

$$K^e(u) = -\varepsilon \ln \mathbf{E} \exp[-\varepsilon^{-1} \int_0^{\tau_G} (1 - u^2/2r) d\tau] \tag{26}$$

By (10), (26) we can write $L = 1 - u^2/2r$. The dynamic programming equation for the problem (25) and (26) is independent of t and is written as

$$\begin{gathered} \varepsilon^{-1/2}\omega(x) V^e_\varphi + H^e(\varphi, x, V^e_\varphi, V^e_x) - \tfrac{1}{2}(\sigma_1 V^e_x)^2 \\ +\varepsilon R(x, \varphi, V^e_{xx}, V^e_{\varphi\varphi}) = 0, \quad x < x^* \\ V^\varepsilon(x,\varphi) = 0, \quad x = x^* \end{gathered} \tag{27}$$

where

$$H^e(\varphi, x, V_x^e, V_\varphi^e) = \max_u\{u[p(x,\varphi)V_x^e + k(x,\varphi)V_\varphi^e] - \frac{u^2}{2r}\} + 1 \quad (28)$$

and the residual term R incorporates the second derivatives terms vanishing in the asymptotic analysis. This implies

$$u^e = r[p(x,\varphi)V_x^e + k(x,\varphi)V_\varphi^e] \quad (29)$$

and

$$H^e(\varphi, x, V_x^e, V_\varphi^e) = \frac{r}{2}[p(x,\varphi)V_x^e + k(x,\varphi)V_\varphi^e]^2 + 1 \quad (30)$$

Substitution of (30) into (25) and averaging in the semi-fast φ yield the reduced first order equation

$$\begin{aligned} \tfrac{\omega(x)}{2} I(x)(r-\sigma^2)(V_x^0)^2 = -1, \quad x < x^* \\ V^0(x) = 0, \quad x = x^* \end{aligned} \quad (31)$$

with the solution

$$V_x^0 = -\sqrt{\frac{2}{(\sigma^2 - r)\omega(x)I(x)}} = -g(x), \quad \sigma^2 > r \quad (32)$$

where $I(x)$ is the action integral corresponding to the energy level x [6], that is

$$I(x) = \frac{1}{2\pi}\oint_{x\in G} p(x,\varphi)d\varphi$$

From (16), (29), (32) we obtain

$$u^0 = -rg(x)p \quad (33)$$

Control (33) is organized as time-invariant feedback equivalent to additional viscous damping with the coefficient dependent on the noise intensity σ.

4. Conclusion

We have discussed a problem of control against noise-induced escape through the potential barrier. This process is considered as large deviation from an asymptotically stable point. To design a control sensitive to rare but large deviation, we have extended risk sensitive control theory to oscillatory systems. An exponential transformation of the mean escape time criterion "biases" the criterion towards the paths with rather long time to escape. Control designed on this criterion becomes sensitive to large deviation. The paper has developed an asymptotic procedure for the HJB equation of risk-sensitive control for a system with slow and fast variables. The limit equation has been obtained as a first order PDE independent of fast variables and much better behaved with regard to computational procedure than the full equation. A nearly optimal control maximizing the mean residence time is constructed as a time-invariant nonlinear feedback depending on the noise intensity in the small noise limit.

An example has been given to demonstrate the use of risk-sensitive control for a resonance system subjected to colored noise. The control task is to hold the frequency of nonlinear oscillations captured in the near resonance domain. The resonance system depends on three time-scales, and the hierarchical averaging techniques has been applied to separate the time-scales. Stochastic averaging in the fast time entails diffusion approximation of the original system by a diffusion with slow and semi-fast variables. This allows the application of the dynamic programming principle and the derivation of the HJB equation. At the second stage, the HJB equation is averaged in the semi-fast phase and the nearly-optimal control is found.

Acknowledgment

The work was supported in part by grant RFBR- 02-01-00011. Support provided by the U.S. National Institute of Standards and Technology is also gratefully acknowledged.

References

[1] O. Alvarez and M. Bardi, "Viscosity solutions methods for singular perturbations in deterministic and stochastic control." *SIAM J. Control Optim.*, vol. 40, pp. 1159–1188, 2001.

[2] A. Bensoussan, *Perturbation Methods in Optimal Control*, John Wiley, New York, 1988.

[3] M. G. Crandall, H. Ishii, and P. L. Lions, "User's guide to viscosity solutions of second order partial differential equations," *Bill. Amer. Math. Soc.*, vol. 27, pp. 1–67, 1992.

[4] P. Dupuis and H. J. Kushner, "Minimizing escape probabilities: A large deviation approach," *SIAM J. Contr. Optim.*, vol. 27, pp. 432–445, 1989.

[5] P. Dupuis and W. M. McEneaney, "Risk-sensitive and robust escape criteria." *SIAM J. Control Optim.*, vol. 35, pp. 2021–2049, 1997.

[6] W. H. Fleming and R. Rishel, *Deterministic and Stochastic Optimal Control*, Springer-Verlag, New York, 1975.

[7] W. H. Fleming and H. M. Soner, *Controlled Markov Processes and Viscosity Solutions*, Springer-Verlag, New York, 1992.

[8] J. Guckenheimer and P. Holmes, *Nonlinear Oscillations, Dynamical Systems, and Bifurcations of Vector Fields*, Springer-Verlag, New York, 1986.

[9] A. .S. Kovaleva, "Asymptotic solution of the optimal control problem for nonlinear oscillations in the neighborhood of a resonance," *J. Appl. Math. Mech.*, vol. 62, pp. 843–852, 1998.

[10] A. S. Kovaleva, "Near-resonance motion in systems with random perturbations," *J. Appl. Math. Mech*, vol. 62, pp. 43–49, 1998.

[11] A. S. Kovaleva, *Optimal Control of Mechanical Oscillations*, Springer-Verlag, Berlin, 1999.

[12] H. J. Kushner, *Weak Convergence Methods and Singularly Perturbed Stochastic Control and Filtering Problems*, Birkhauser. Boston, 1990.

[13] M. I. Freidlin and A. D. Wentzell, *Random Perturbations of Dynamical Systems*. Second Ed., Springer-Verlag. New York, 1998.

[14] P. Whittle, *Risk-Sensitive Optimal Control*, J. Wiley, New York, 1990.

ON PARAMETER ANALYSIS OF THE DIFFERENTIAL MODEL OF HYSTERESIS

F. Ma, H. Zhang
Department of Mechanical Engineering
University of California
Berkeley, California 94720, USA

A. Bockstedte
Mechanics and Ocean Engineering
Technical University of Hamburg-Harburg
D-21071 Hamburg, GERMANY

G. C. Foliente, and P. Paevere
CSIRO Infrastructure Systems Engineering
Highett, Victoria 3190, AUSTRALIA

Abstract The extended Bouc-Wen differential model is one of the most widely accepted phenomenological models of hysteresis in mechanics. It is routinely used in the characterization of nonlinear damping and in system identification. In this paper, the differential model of hysteresis is carefully re-examined and two significant issues are uncovered. First, it is found that the unspecified parameters of the model are functionally redundant. One of the parameters can be eliminated through suitable transformations in the parameter space. Second, local and global sensitivity analyses are conducted to assess the relative sensitivity of each model parameter. Through extensive Monte Carlo simulations, it is found that some parameters of the hysteretic model are rather insensitive. If the values of these insensitive parameters are fixed, a greatly simplified model is obtained.

Keywords: Hysteresis, differential model, parameter reduction, sensitivity analysis

N. Sri Namachchivaya and Y.K. Lin (eds.),
IUTAM Symposium on Nonlinear Stochastic Dynamics, 257-268.

1. Introduction

Structures exhibit inelastic behavior under severe cyclic loads associated with earthquakes, high winds, and recurrent waves. When the restoring force is plotted against the structural deformation, inelastic behavior often manifests itself in the form of hysteresis loops. The area enclosed by each loop is a measure of the energy dissipated over a complete cycle as the result of internal friction within the structure. In general terms, hysteresis refers to the hereditary and the memory nature of an inelastic system, in which the restoring force depends not only on the instantaneous deformation but also on the past history of deformation. Mechanical and structural systems capable of dissipating appreciable energy tend to possess large hysteresis loops. Hysteresis is thus particularly important in depicting the damping characteristics of inelastic systems. Yet, a fundamental and comprehensive theory of hysteresis has not been developed. In order to address practical problems in the contemporary design and analysis of structures, phenomenological models of hysteresis are often used.

In the past few decades, various phenomenological models of hysteresis have been proposed. One of the most widely accepted models is a differential model originally proposed by [3] and subsequently generalized by [15] and other researchers. In this model the restoring force and deformation are connected through a nonlinear differential equation containing unspecified parameters. By choosing the parameters suitably, it is possible to generate a large variety of different shapes of the hysteresis loops. The classical Bouc-Wen model contains only five loop parameters. The generalized differential model in its present form contains thirteen parameters; it can account for strength degradation, stiffness degradation, and even pinching characteristics of an inelastic structure. As the differential model becomes increasingly popular, questions concerning the complexity and consistency of the model are brought up more frequently.

The extended Bouc-Wen model of hysteresis is carefully re-examined in this paper. Two significant issues are uncovered. First, it is found that the unspecified parameters of the model are functionally redundant. In either the classical or contemporary version, one of the model parameters can be eliminated through suitable transformations in the parameter space. The number of model parameters can thus be reduced without affecting the system response. Second, local and global sensitivity analyses are carried out on the differential model of hysteresis. The global analysis employed is a probabilistic method recently expounded by Sobol and it can account for the mutual interactions of the parameters. Through extensive Monte Carlo simulations, the relative sensitivity of each parameter is assessed. It is found that some parameters are rather insensitive. These insensitive parameters should perhaps be set to constant values, thereby resulting in a greatly simplify hysteretic model.

2. Differential Hysteresis

To describe the differential model of hysteresis, consider an inelastic system possessing a finite number of degrees of freedom. Suppose the equation of motion can be decoupled and, along the direction of the generalized coordinate u, the system is governed by

$$m\ddot{u} + c\dot{u} + R_T(u,z) \quad = \quad F(t) \tag{1}$$

where u is the system displacement, z is an imaginary hysteretic displacement, and m, c are respectively the mass and damping coefficients. It is assumed that the system is subjected to a cyclic excitation $F(t)$. In the development of the differential model, the total restoring force $R_T(u,z)$ is separated into an elastic and a hysteretic component by

$$R_T(u,z) \quad = \quad \alpha k u + (1-\alpha) k z \tag{2}$$

where k is the stiffness coefficient and $0 \le \alpha \le 1$ is a weighting parameter. Obviously, the restoring force is purely hysteretic if $\alpha = 0$; it is purely elastic if $\alpha = 1$. Basic hystersis loops may be generated if the hysteretic displacement z and the total displacement u are connected by the nonlinear differential equation [15]

$$\dot{z} \quad = \quad A\dot{u} - \beta|\dot{u}||z|^{n-1}z - \gamma\dot{u}|z|^n. \tag{3}$$

There are five unspecified loop parameters A, α, β, γ, n in (1) and (3), which represent the classical Bouc-Wen model. Over the years, the original Bouc-Wen model has been extended several times. Many new parameters have been added to fit a variety of hysteretic shapes which include degradation and pinching. The result is a hysteresis model with thirteen loop parameters given by

$$\dot{z} \quad = \quad h(z)\left\{\frac{A\dot{u} - \nu(\beta|\dot{u}||z|^{n-1}z + \gamma\dot{u}|z|^n)}{\eta}\right\}. \tag{4}$$

In the above expression, ν and η are degradation shape functions [2] and $h(z)$ is a pinching shape function [1]. In general, degradation depends on the response duration and severity. A convenient measure of the combined effect of duration and severity is the energy

$$E(t) \quad = \quad \int_0^t (1-\alpha) k z \dot{u} dt \tag{5}$$

dissipated through hysteresis from initial time $t = 0$ to present time t. Since

$$\varepsilon(t) = \int_0^t z\dot{u}dt \tag{6}$$

is proportional to $E(t)$, it may also be used as a measure of duration and severity. Many functional relations between ν, η, and ε are possible. From practical considerations, both ν and η are assumed to depend linearly on ε as the system evolves:

$$\nu(\varepsilon) = 1 + \delta_\nu \varepsilon \tag{7}$$

$$\eta(\varepsilon) = 1 + \delta_\eta \varepsilon. \tag{8}$$

Two unspecified degradation parameters δ_ν and δ_η are thereby introduced. The pinching function $h(z)$ takes the form [4]

$$h(z) = 1 - \zeta_1 e^{-(z\,sgn(\dot{u}) - qz_u)^2/\zeta_2^2} \tag{9}$$

where $\mathrm{sgn}(\dot{u})$ is the signum function of $\dot{u}$ and z_u is the ultimate value of z given by

$$z_u = \left[\frac{A}{\nu(\beta + \gamma)}\right]^{1/n} \tag{10}$$

The two functions $\zeta_1(\varepsilon)$ and $\zeta_2(\varepsilon)$ are given by

$$\zeta_1(\varepsilon) = \zeta_s[1 - e^{(-p\varepsilon)}] \tag{11}$$

$$\zeta_2(\varepsilon) = (\psi + \delta_\psi \varepsilon)(\lambda + \zeta_1). \tag{12}$$

There is a total of six unspecified pinching parameters ζ_s, q, p, ψ, δ_ψ, and λ. Altogether there are thirteen loop parameters of hysteresis A, α, β, γ, n, δ_ν, δ_η, ζ_s, q, p, ψ, δ_ψ, and λ. This generalized model of hysteresis possesses all the important features observed in real structures, which include strength degradation, stiffness degradation, and pinching of the successive hysteresis loops.

3. Elimination of Redundant Parameter

Consider a nonlinear system governed by (1) and (4). From a strict mathematical standpoint, either $[u, \dot{u}, z]^T$ or $[u, \dot{u}, R_T(u, z)]^T$ can be taken as a solution vector. Note that z is an imaginary displacement while $R_T(u, z)$ is a measurable force. The system response is usually chosen as $[u, \dot{u}, R_T(u, z)]^T$ since it is directly connected to the hysteresis loops. Upon solution of (1) and (4), a graph of $R_T(u, z)$ against u is often compared to experimental data from cyclic performance tests. A good fit would indicate that the loop parameters are specified properly. It must be remembered, however, that the thirteen loop parameters are phenomenological parameters; they are not derived from fundamental equations of mechanics. It has never been claimed that all thirteen parameters are essential in producing the common features of hysteresis. After all, the physical meanings of many of these parameters have not been adequately clarified. It would be very useful, particularly in system identification, if the number of unspecified parameters can be reduced [18].

It will be shown that a transformation in the parameter space can be devised to freeze one of the loop parameters without affecting the system response. Since the hysteretic force $(1-\alpha)kz$ is proportional to k, the stiffness coefficient is used as an additional parameter to define the parameter set

$$P = (A, k, \alpha, \beta, \gamma, n, \delta_\nu, \delta_\eta, \zeta_s, q, p, \psi, \delta_\psi, \lambda). \tag{13}$$

Define a transformation from P into

$$\overline{P} = (\overline{A}, \overline{k}, \overline{\alpha}, \overline{\beta}, \overline{\gamma}, \overline{n}, \overline{\delta}_\nu, \overline{\delta}_\eta, \overline{\zeta}_s, \overline{q}, \overline{p}, \overline{\psi}, \overline{\delta}_\psi, \overline{\lambda}) \tag{14}$$

in such a way that:

$$\begin{aligned}
\overline{A} &= 1 && (15)\\
\overline{k} &= (\alpha + A - \alpha A)k && (16)\\
\overline{\alpha} &= \frac{\alpha}{\alpha + A - \alpha A} && (17)\\
\overline{\beta} &= A^{n-1}\beta && (18)\\
\overline{\gamma} &= A^{n-1}\gamma && (19)\\
\overline{n} &= n && (20)\\
\overline{\delta}_\nu &= A\delta_\nu && (21)\\
\overline{\delta}_\eta &= A\delta_\eta && (22)\\
\overline{\zeta}_s &= \zeta_s && (23)\\
\overline{q} &= q && (24)\\
\overline{p} &= Ap && (25)\\
\overline{\psi} &= \frac{1}{A}\psi && (26)\\
\overline{\delta}_\psi &= \delta_\psi && (27)\\
\overline{\lambda} &= \lambda. && (28)
\end{aligned}$$

It can be proved that when the parameters in $\overline{P}$ are used in (1) and (4), the same system response $[u, \dot{u}, R_T(u, z)]^T$ is obtained. The hysteretic traces under P and $\overline{P}$ are identical. The parameter A is redundant and can be eliminated from the differential model of hysteresis using transformation (15–28). The same can be said about the classical Bouc-Wen model (1) and (3), for which the number of parameters can be reduced from five to four.

4. Parameter Sensitivity Analysis

Assume that transformation (15–28) has been applied to fix the value of A at 1. There remain twelve loop parameters in the extended Bouc-Wen differential model. Does every parameter contribute equally to the system response? Will variations in some parameters combine to annul the effect of each other? Attempts were made in the past to understand the influence of each parameter on the system response. For example, [1] investigated the effect of variations of A, n, η, ν, ζ_1, and ζ_2 on the ratio between hysteretic and total displacement. Wong [17] studied the steady-state response of structures with different values of the loop parameters β, γ, and n. Despite numerous applications of the extended Bouc-Wen model [12, 16], there is hardly any systematic study of the influence of loop parameters on the system response. One basic objective of the current paper is to address this overlooked issue by conducting parameter sensitivity analysis. There is a large class of techniques for parameter sensitivity analysis. A review of some of the popular methods is given by [5, 8]. In this paper, the one-factor-at-a-time method and the Sobol sensitivity indices are used.

4.1 Local sensitivity analysis

The simplest way to conduct sensitivity analysis is to repeatedly vary one parameter at a time while holding the others fixed at chosen nominal values. This is referred to as one-factor-at-a-time method. It is easy to implement, often computationally inexpensive, and useful in providing a graphical representation of sensitivity ranking. However, it is a local method in the sense that it only addresses sensitivity relative to the chosen base values and not for the entire parameter space. How the base values should be chosen is often an issue itself. In addition, interactions of parameters cannot be addressed by this local method.

As a typical example of local sensitivity analysis, a shear wall of the type described in [18] is used. The shear wall has a natural frequency $f_n = 3.6$Hz and damping factor $\zeta = 0.02$. It is driven by the scaled E-W component of the 1940 El Centro earthquake over a duration of 10 seconds. Set as base values $\alpha = 0.05$, $\beta = 1$, $\gamma = 0.1$, $n = 1.5$, $\delta_\nu = 0.025$, $\delta_\eta = 0.15$, $\zeta_s = 0.85$, $q = 0.15$, $p = 2.5$, $\psi = 0.1$, $\delta_\psi = 0.005$, and $\lambda = 0.5$. Each of these twelve parameters is then varied, one at a time, by up to 50% from its base value while holding all other parameters at the base position. The choice of the base values is such that the resulting range spanned by each parameter is well within the usual range for fitting the experimental hysteresis loops of wood structures. Denote the system response by $[x_1, x_2, x_3]^T = [u, \dot{u}, R_T(u, z)]^T$ at the base values and let the response be $[y_1, y_2, y_3]^T$ when one parameter, say w, is varied. Define the root-mean-square error

$$e = \left[\sum_{i=1}^{M}(x_{1i} - y_{1i})^2 + (x_{2i} - y_{2i})^2 + (x_{3i} - y_{3i})^2\right]^{1/2} \tag{29}$$

where M is the number of sampling points. Put $M = 500$ in this example. A spider diagram is obtained when e_w is plotted against the varied parameter w, as shown in Figure 1. As each w is varied over its range, the maximum error

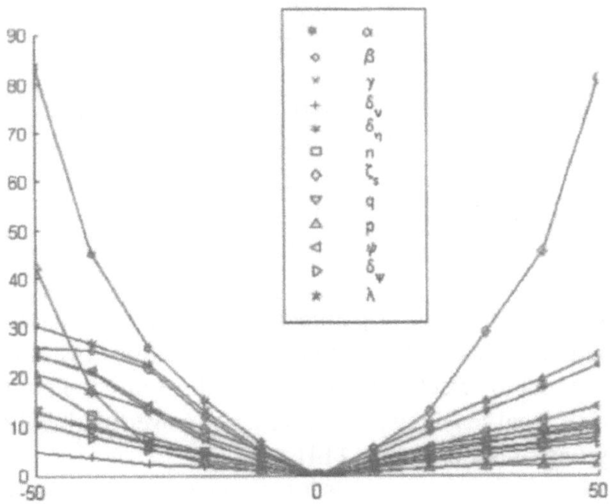

Figure 1. Spider diagram generated by the one-factor-at-a-time method

$$\|e_w\| = \max(e_w) \tag{30}$$

is recorded. These maximum values may be used as a measure of sensitivity. Based upon $\|e_w\|$, the twelve loop parameters of hysteresis are ranked in Table 1. Using two slightly different sets of base values, it is found that the sensitivity ranking given in Table 1 does not change.

4.2 Global sensitivity analysis

Most techniques for global sensitivity analysis were developed fairly recently. In global sensitivity analysis, it is not necessary to choose any base values. Sensitivity with respect to a parameter is assessed when all model parameters vary simultaneously, instead of one at a time. Oftentimes the sensi-

Table 1. Parameter sensitivity ranking

Parameter	Local analysis		Global analysis			
w	$\|e_w\|$	Rank	Sobol index		Total effect index	
			S_w	Rank	S_w^T	Rank
α	83.69	1	0.0894	2	0.2806	2
β	25.82	5	0.0410	4	0.1451	5
γ	24.31	6	0.0379	5	0.0962	7
n	19.10	8	0.0339	7	0.0930	8
δ_ν	4.78	12	0.0288	9	0.0151	12
δ_η	30.35	4	0.0308	8	0.2781	3
ζ_s	80.91	2	0.2803	1	0.6486	1
q	13.16	9	0.0377	6	0.0583	10
p	42.54	3	0.0255	10	0.0891	6
ψ	24.27	7	0.0449	3	0.1928	4
δ_ψ	10.49	11	0.0234	12	0.0574	11
λ	12.73	10	0.0238	11	0.0742	9

tivity with respect to a group of parameters can also be assessed. A sensitivity index is a number that gives quantitative information about the relative sensitivity with respect to selected parameters. In this sense the number $\|e_w\|$ defined in (30) is a local sensitivity index. A set of powerful global sensitivity indices is the group of Sobol indices [14].

Suppose $f(\mathbf{x}) = f(x_1, \ldots, x_n)$ is a real integrable function defined on the n-dimensional unit cube $I^n = \{\mathbf{x} | 0 \leq x_i \leq 1, i = 1, \ldots, n\}$. Let

$$f_0 = \int_{I^n} f(\mathbf{x}) d\mathbf{x} \tag{31}$$

$$f_i(x_i) = \int_{I^{n-1}} f(\mathbf{x}) d\mathbf{x}/dx_i - f_0 \tag{32}$$

where $d\mathbf{x}/dx_i$ denotes the product of all the dx_k except dx_i, and

$$f_{ij}(x_i, x_j) = \int_{I^{n-2}} f(\mathbf{x}) d\mathbf{x}/dx_i dx_j - f_0 - f_i(x_i) - f_j(x_j). \tag{33}$$

In this way one can define inductively all $f_{i_1,\ldots,i_s}(x_{i_1}, \ldots, x_{i_s})$ for any subset of variables $x_{i_1}, \ldots, x_{i_s}$ with indices $1 \leq i_1 < \cdots < i_s \leq n$. Furthermore, let

$$D = \int_{I^n} f(\mathbf{x}) d\mathbf{x} - f_0^2 \tag{34}$$

$$D_{i_1,\ldots,i_s} = \int_{I^n} \cdots \int f_{i_1,\ldots,i_s}^2 (x_{i_1}, \ldots, x_{i_s}) dx_{i_1}, \ldots, dx_{i_s}. \tag{35}$$

It has been established by Sobol that, under the assumption that $\mathbf{x}$ is uniformly distributed in I^n, the quantity

$$S_{i_1,\ldots,i_s} = D_{i_1,\ldots,i_s}/D \tag{36}$$

is a measure of the global sensitivity of the function $f(\mathbf{x})$ with respect to the group of variables $x_{i_1}, \ldots, x_{i_s}$. In fact, the quantity $S_{i_1,\ldots,i_s}$ is termed a Sobol sensitivity index. It can be proved that

$$\sum S_{i_1,\ldots,i_s} = \sum_{i=1}^{n} S_i + \sum \sum_{1 \le i < j \le n} S_{ij} + \ldots + S_{1,2,\ldots,n} = 1. \tag{37}$$

In other words, the sum over all the different groups of indices is one. There is a total $2^n - 1$ Sobol indices for $f(\mathbf{x})$. All the integrals in the Sobol indices can be evaluated by Monte Carlo simulations [11]. The amount of computational effort can be substantial if n is large. To speed up the convergence of Monte Carlo integration, probably the most effective way is to use quasi-random numbers [13] instead of pseudo-random numbers for choosing the samples. Other variance reduction schemes such as antithetic variates [6] may also be employed. In this paper, Latin hypercube sampling [10] is used instead of the crude Monte Carlo sampling. As a result, the sampling size is generally reduced to about one third of the original size.

Based upon the development of [13, 7] have introduced another set of sensitivity indices to measure the mutual interactions of parameters. A given parameter x_j can interact with other parameters in groups of two, three, or more members. In view of (37), the "total effect" index with respect to x_j is defined as

$$S_j^T = 1 - \sum_{i_k \neq j} S_{i_1,\ldots,i_s} \tag{38}$$

where the sum is over all the different groups of indices that do not include j. The index S_j^T provides a measure of the total sensitivity with respect to x_j, which includes its interactions with other model parameters. A scheme has been developed [7] for the calculation of the "total effect" indices that economizes on core memory and computing time. Both the Sobol index S_j and the total sensitivity index S_j^T will be computed in this paper.

In reporting the results of global sensitivity analysis, the same shear wall described earlier is again used. Recall that the shear wall has a natural frequency $f_n = 3.6$Hz and damping factor $\zeta = 0.02$, and is driven by the scaled E-W component of the 1940 El Centro earthquake over a duration of 10 seconds. Recall that the value of A is fixed at 1 by transformation (15-28). After an examination of the large amount of experimental data on wood structures, it is decided that acceptable ranges for the remaining twelve parameters are $0 \le \alpha \le 0.1$, $0.5 \le \beta \le 1.5$, $-0.3 \le \gamma \le 0.5$, $0 \le n \le 3$,

$0 \leq \delta_\nu \leq 0.05$, $0 \leq \delta_\eta \leq 0.3$, $0.7 \leq \zeta_s \leq 1$, $0 \leq q \leq 0.3$, $0 \leq p \leq 5$, $0 \leq \psi \leq 0.2$, $0 \leq \delta_\psi \leq 0.01$, and $0 \leq \lambda \leq 1$. The Sobol sensitivity index of the system response with respect to each of the parameter is evaluated through Monte Carlo simulations. In contrast to local analysis, no base values need to be specified. The "total effect" index associated with each parameter is then calculated. Even with a streamlined program, the computation of the total sensitivity indices often requires long run-time. The various indices are collected together in Table 1.

It is observed that the ranking generated by the local sensitivity analysis largely agrees with that by global analysis. Both the local and global methods indicate that α and ζ_s are the most sensitive parameters. On the other hand, δ_ν, δ_ψ, and λ are probably the least sensitive parameters. None of the unspecified parameters in the classical Bouc-Wen model are insensitive.

Information on parameter sensitivity is particularly valuable in system identification and system optimization. In these areas certain error functional is minimized. A sensitive parameter far away from its sought-after value will generally cause an appreciable increase in the error functional. As a result, convergence can be accelerated with various numerical schemes. On the other hand, an insensitive parameter tends to drift around its sought-after value since changes in its value contribute little to the error functional. This could cause many problems in convergence. Clearly, great simplification results if the insensitive parameters can be fixed. It must be emphasized that global sensitivity ranking relies appreciably on the range of variation of each parameter. Over the range specified, δ_ν, δ_ψ, and λ are rather insensitive parameters. This observation should be re-assessed for a different range of variation.

5. Conclusion

Although a limited set of data is presented, extensive numerical calculations have been performed by the authors to support any qualitative results herein. Additional results in simulations and in the construction of parameter-reducing transformations will be given by [9]. Major findings are summarized in the following statements.

1. The unspecified parameters of the differential model are functionally redundant. One of the parameters can be eliminated through suitable transformations in the parameter space. This is true for both the classical and the extended differential models of hysteresis.

2. In the extended differential model, there are three or four unspecified parameters that have been consistently ranked at or near the bottom of the sensitivity scale. If the values of these insensitive parameters can be fixed, a greatly simplified model is obtained.

3. Over a practical range, it appears that interactions among various unspecified parameters in the extended differential model are not significant.

The cyclic excitation $F(t)$ chosen for sensitivity analysis possesses a rich spectrum. Thus the above statements on sensitivity are applicable for a large class of cyclic excitation. However, these statements on sensitivity may not be valid for an excitation whose spectrum has little overlap with that of the chosen excitation. Among other things, it is hoped that the present paper would point to directions along which further research efforts to refine the differential model can be profitably made. As the differential model of hysteresis becomes increasingly popular in theory and applications, the update reported in this paper will hopefully allow it to be used with added confidence in the years to come.

Acknowledgments

We would like to thank Mr. Chinghang Ng for assistance in the preparation of the manuscript. This research has been supported in part by the National Science Foundation under Grant No. CMS-9908194. Opinions, findings, and conclusions expressed in this paper are those of the authors and do not necessarily reflect the views of the sponsor.

References

[1] T. T. Baber and M. N. Noori, "Modeling General Hysteresis Behavior and Random Vibration Application," *ASME Journal of Vibration, Acoustics, Stress and Reliability in Design*, vol. 108, pp. 411–420, 1986.

[2] T. T. Baber and Y. K. Wen, "Random Vibration of Hysteretic Degrading Systems," *ASCE Journal of the Engineering Mechanics*, vol. 107, pp. 1069–1087, 1981.

[3] R. Bouc, "Forced Vibration of Mechanical Systems with Hysteresis," *Proceedings of the 4th International Conference on Nonlinear Oscillations*, Prague, Czechoslovakia, 1967.

[4] G. C. Foliente, "Hysteretic Modeling of Wood Joints and Structural Systems," *ASCE Journal of Structural Engineering*, vol. 121, pp. 1013–1022, 1995.

[5] D. M. Hamby, "A Review of Techniques for Parameter Sensitivity Analysis of Environmental Models," *Environmental Monitoring and Assessment*, vol. 32, pp. 135–154, 1994.

[6] J. M. Hammersley and K. W. Morton, "A New Monte Carlo Technique: Antithetic Variates," *Proceedings of Cambridge Philosophical Society*, vol. 52, pp. 449–475, 1956.

[7] T. Homma and A. Saltelli, "Importance Measures in Global Sensitivity Analysis of Nonlinear Models," *Reliability Engineering and System Safety*, vol. 52, pp. 1–17, 1996.

[8] R. L. Iman and J. C. Helton, "The Repeatability of Uncertainty and Sensitivity Analyses for Complex Probabilistic Risk Assessments," *Risk Analysis*, vol. 11, pp. 591–606, 1991.

[9] F. Ma, H. Zhang, A. Bockstedte, G. C. Foliente, and P. Paevere, "Parameter Analysis of the Differential Model of Hysteresis," *ASME Journal of Applied Mechanics*, submitted for publication.

[10] M. D. McKay, R. J. Beckman, and W. J. Conover, "A Comparison of Three Methods for Selecting Values of Input Variables in the Analysis of Output from a Computer Code," *Technometrics*, vol. 21, pp. 239–245, 1979.

[11] W. H. Press, B. P. Flannery, S. A. Teukolsky, and W. T. Vetterling, *Numerical Recipes in C: The Art of Scientific Computing*, Cambridge University Press, United Kingdom, pp. 237–241, 1988.

[12] J. B. Roberts and P. D. Spanos, *Random Vibration and Statistical Linearization*, Wiley, New York, pp. 257–284, 1990.

[13] I. M. Sobol, "Quasi-Monte-Carlo Methods," *Progress in Nuclear Energy*, vol. 24, pp. 55-61, 1990.

[14] I. M. Sobol, "Sensitivity Estimates for Nonlinear Mathematical Model," *Mathematical Modeling and Computational Experiment*, vol. 1, pp. 407–414, 1993.

[15] Y. K. Wen, "Method for Random Vibration of Hysteretic Systems," *ASCE Journal of Engineering Mechanics*, vol. 102, pp. 249–263, 1976.

[16] Y. K. Wen, "Equivalent Linearization for Hysteretic Systems under Random Excitation," *ASME Journal of Applied Mechanics*, vol. 47, pp. 150–154, 1980.

[17] C. W. Wong, Y. Q. Ni, and S. L. Lau, "Steady-State Oscillation of Hysteretic Differential Model II: Performance Analysis," *ASCE Journal of Engineering Mechanics*, vol. 120, pp. 2299–2325, 1992.

[18] H. Zhang, P. Paevere, Y. Yang, G. C. Foliente, and F. Ma, "System Identification of Hysteretic Structures," in *Nonlinear and Stochastic Structural Dynamics*, 1999 IUTAM Symposium in Madras, India, S. Narayanan and R. N. Iyenger, eds., Kluwer, Dordrecht, The Netherlands, pp. 289–306, 2001.

STOCHASTIC PARAMETER ESTIMATION OF NON-LINEAR SYSTEMS

Marcello Vasta
DISAT, Faculty of Engineering
University of L'Aquila
*67040 Monteluco di Roio, L'Aquila, Italy**
m.vasta@ing.univaq.it

Abstract Methods to estimate unknown parameters in non-linear, randomly excited dynamic systems are presented. Attention is focused for the case where only the response is measurable and known only in terms of a stochastic process model. Three technique are discussed, based on response higher order spectral analysis, stochastic averaging and spectral representation processes. The approaches are illustrated through application to linear and non-linear oscillators under Gaussian and non-Gaussian excitation. The methods have applications in studies of the response of structures to random environmental loads, such as wind and ocean wave forces.

Keywords: stochastic parameter estimation, unknown random excitation, non-linear systems

1. Introduction

To validate design methods for predicting the dynamic response of structures to random environmental loads it is often necessary to compare predicted responses with those actually measured in practice, on full scale structures. Examples includes ships rolling in irregular waves and the vibration of tall, flexible buildings in a turbulent wind. Fitting dynamic models to measured data falls within the general field of system identification, on which there is an extensive literature (see e.g. [11]). Whilst mass and stiffness parameters in the governing equations of motion can be usually be computed with some accuracy, damping parameters are normally not quantifiable by theoretical means because of a lack of precise, quantitative models.

* *This work is devoted to the memory of Brian Roberts*

N. Sri Namachchivaya and Y.K. Lin (eds.),
IUTAM Symposium on Nonlinear Stochastic Dynamics, 269-278.

When simultaneous measurement of the system's excitation and response are obtainable a wide variety of appropriate system identification procedures are available (see e.g. [3]).

However, many cases occur where it is impractical, or actually impossible, to measure the excitation processes. As an example, for a flexible building responding to wind turbulence, although the oncoming velocity field is measurable, the wind loading is actually unmeasurable. In such situations, where the excitation is of an environmental nature and is essentially random in nature, it is necessary to adopt a stochastic model. The estimation problem is then to estimate both system and excitation model parameters, from measurements of the system response alone.

A technique, based on second order spectral analysis [5] has shown to produce reliable results and its applicability to MDOF systems has been demonstrated. These studies have revealed however that the second order spectral method is incapable of separating out the linear and non-linear damping contributions. Higher order spectral quantities, or polyspectra (see e.g. [10]), are appropriate for this purpose. This was shown in developing an identification technique, based on the use of fourth order spectral quantities (trispectrum), able to identify the parameters in highly non-linear dynamic oscillators to both Gaussian [12] and non-Gaussian [6] excitation.

An alternative energy based stochastic estimation method for nonlinear oscillators responding to stationary random excitation was presented in [7]. This method relies on the fact that, when the damping is light, the energy envelope of the response is approximately a one-dimensional Markov process, even if the excitation is nonwhite (see e.g. [4]). The proposed identification technique is an extension of an estimation scheme specific to ship roll motion in random seas [8]. The technique, originally developed to non linear oscillators under external excitation, has been recently extended to include the case of parametric excitation [9].

After reviewing the spectral and the energy based stochastic identification techniques, a new parameter estimation scheme is proposed, based on the assumption that the excitation belong to the class of spectral representation processes. The technique has been verified on linear systems under colored noise, leading to a quite accurate estimates of system parameters and input spectra.

2. The Estimation Problem

The dynamic analysis of structural and mechanical M degree of freedom structures subjected to random loads is usually performed by solving a system of non-linear differential equations of the general form

$$g_i[\mathbf{z}(t);\alpha] = f_i(t;\beta) \qquad i = 1,\ldots,M \tag{1}$$

where $\mathbf{z}(t) = [\mathbf{q}(t)^T\dot{\mathbf{q}}(t)^T\ddot{\mathbf{q}}(t)^T]^T$. In (1), $\mathbf{q}(t)$ represents an M-vector of random processes, f_i are excitation processes, g_i are non linear functions and α and β are vectors of parameters. In order to simplify the subsequent discussion it is assumed henceforth that $f_i(t;\beta)$ are zero-mean, symmetric stationary and ergodic processes. It is assumed here that simultaneous sample functions of the stationary response processes, $\mathbf{q}(t)$, can be measured over an interval of time $0 \leq t \leq T$. In contrast the excitation processes, $f_i(t;\beta)$, are assumed to be unmeasurable and known only in terms of a stochastic model. The elements of the parameter vector $\theta = (\alpha^T, \beta^T)^T$ are taken to be unknown and are to be estimated from experimental data. The estimation problem therefore consists in estimating the parameter vector θ by processing only finite duration sample functions of the response.

3. The Spectral Method

As the excitation and the response processes are assumed to be stationary and ergodic only one sample function is needed, for each random process, to obtain spectral information. The data may be divided into N blocks ($n = 1, 2, \ldots, N$) each of duration T, and, for each block, a Fourier transform may then be applied to both sides of (1), to obtain

$$X_{i;n}(\omega;\alpha) = Y_{i;n}(\omega;\beta) \tag{2}$$

where

$$X_{i;n} = \int_0^T g_{i;n}e^{-i\omega t}dt, \quad Y_{i;n} = \int_0^T f_{i;n}e^{-i\omega t}dt. \tag{3}$$

Suppose that the X and Y quantities are multiplied by their complex conjugates and divided by $2\pi T$. Thus

$$V_{ij;n} = \frac{1}{2\pi T} \cdot X_{i;n}X^*_{j;n}, \quad W_{ij;n} = \frac{1}{2\pi T} \cdot Y_{i;n}Y^*_{j;n} \tag{4}$$

For both linear and non-linear systems it is always possible to calculate $V_{ij;n}(\omega;\alpha)$ from measured response data. It can be easily shown that the expectation of $W_{ij;n}(\omega;\beta)$, here denoted as $\bar{W}_{ij}(\omega;\beta)$, coincide with the second order spectral excitation function $S_{ij}(\omega;\beta)$. The expectation of $V_{ij;n}(\omega;\alpha)$, $\bar{V}_{ij}(\omega;\alpha)$, can be estimated by averaging over the N blocks. The parameter vector θ can be estimated, for relatively large block lengths, by minimising the least-square cost function

$$J_2 = \sum_{ij}\sum_{k}[\bar{V}_{ij}(\omega_k;\alpha) - S_{ij}(\omega_k;\beta)]^2. \tag{5}$$

The J_2 functional can be minimised using a standard non-linear optimisation algorithm. Clearly the number of discrete frequencies, ω_k, at which the cost function J_2 is evaluated must be greater than the number of parameters to be estimated. The chosen frequencies should, moreover, concentrate around the dominant part of the response spectrum.

The second order spectral estimation scheme enables one to successfully estimates the absolute values of both linear and non-linear stiffness parameters, together with the excitation strength [5]. Unfortunately, however, does not provide a means of separating out the linear and non-linear contributions to the damping and only an equivalent linear damping can be estimated. In order to estimate the absolute levels of both linear and non-linear damping parameters it is necessary to consider higher order spectral quantities, or polyspectra, such as bispectra and trispectra relating to the memory of the response process (see e.g. [1]). They are also appropriate for estimating the presence of non-Gaussian excitation. Suppose that the following fourth order quantity of the g_i process, as given in (1), is defined by

$$\begin{aligned} V_{ijkl;n}(\omega_1,\omega_2,\omega_3;\alpha) &= \frac{1}{8\pi T^3} X_{i;n}(\omega_1)X_{j;n}(\omega_2)X_{k;n}(\omega_3)X_{l;n}(\bar{\omega}) \\ &- V_{ij;n}(\omega_1)V_{kl;n}(\omega_3)\Delta_T(\omega_1+\omega_2) \\ &- V_{ik;n}(\omega_3)V_{jl;n}(\omega_2)\Delta_T(\omega_1+\omega_3) \\ &- V_{il;n}(\omega_1)V_{jk;n}(\omega_2)\Delta_T(\omega_2+\omega_3) \end{aligned} \tag{6}$$

where $\bar{\omega} = \omega_1 + \omega_2 + \omega_3$ and the X and V quantities are defined by (3) and (4), respectively, and

$$\Delta_T(\omega) = \frac{1}{2\pi}\left[\frac{\sin(\omega T/2)}{(\omega T/2)}\right]^2. \tag{7}$$

A similar quantity, $W_{ijkl}(\omega_1,\omega_2,\omega_3;\beta)$ may be defined by replacing the X with Y quantities and the V with W quantities (see (3) and (4)). It can be easily shown that the expectation of $W_{ijkl;n}(\omega;\beta)$, here denoted as $\bar{W}_{ijkl}(\omega;\beta)$, coincide with the fourth order spectral excitation function $S_{ijkl}(\omega:\beta)$. The expectation of $V_{ijkl;n}$, $\bar{V}_{ijkl}$, can be estimated by averaging over the N blocks. It follows that, for relatively large blocks, the parameter vector θ can be estimated by minimising the least square cost function

$$J_4 = \sum_{ijkl}\sum_{rsu}[\bar{V}_{ijkl}(\omega_r,\omega_s,\omega_u;\alpha) - S_{ijkl}(\omega_r,\omega_s,\omega_u;\beta)]. \tag{8}$$

A considerable reduction of the computation effort may be achieved by restricting the frequencies to the principal domain, defined by [2]

$$\begin{cases} \omega_1 \geq \omega_2 \geq \omega_3 \\ \omega_1 + \omega_2 + \omega_3 \geq 0 \\ \omega_1 + \omega_2 + 2\omega_3 \geq 0 \\ \text{exactly one frequencies from} \quad \omega_1, \omega_2, \omega_3 \leq 0. \end{cases} \tag{9}$$

The higher order spectral method has been successfully applied to a single degree of freedom system with non-linear damping and stiffness components, and Gaussian as well as non Gaussian excitation [12],[6]. As an example, a Duffing oscillator with cubic non linear damping driven by a non Gaussian colored noise is herein considered. The excitation is analytically modelled as $f(t) = k_1 u + k_2 |u| u$, where $u(t)$ is a stationary Gaussian process, with a non-white spectrum, k_1 and k_2 being two input parameters. Figure 1 show the excitation estimates using second and second plus fourth order spectral estimates. Figure 2 show the estimates of the linear and non linear damping parameters for the same case.

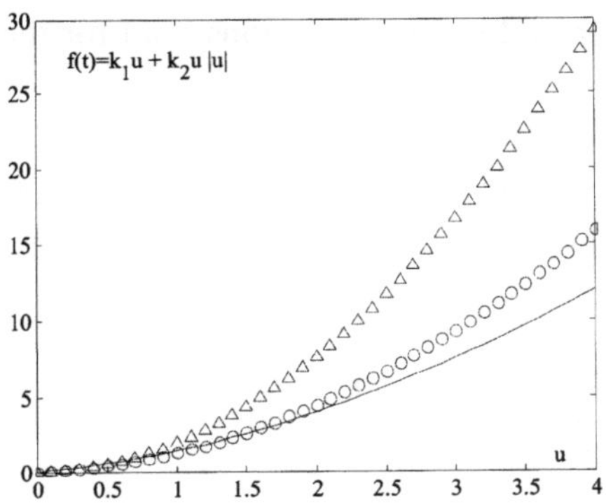

Figure 1. Variation of the excitation force $f(t)$ with velocity $u(t)$; $\triangle$ second order estimates; $\bigcirc$ second plus fourth order estimates

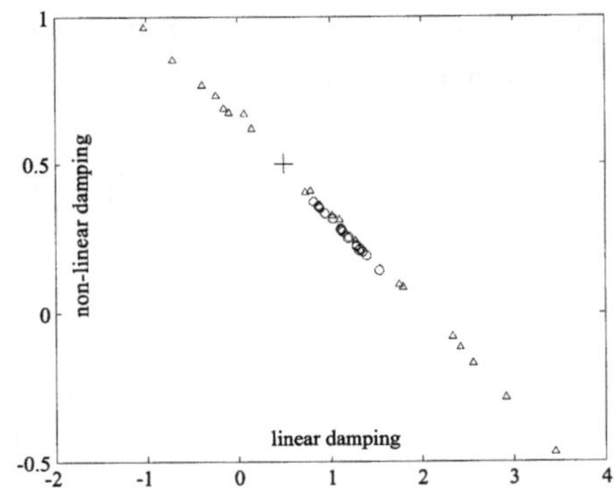

Figure 2. Variation of estimates of linear damping versus non-linear damping parameters; $\triangle$ second order estimates; $\bigcirc$ second plus fourth order estimates; $+$: exact value

4. Energy Based Stochastic Estimation Method

It is well known that when the damping is light, the energy envelope of the response is approximately a one-dimensional Markov process. An equation of motion of a SDOF system of the following general form is herein considered

$$\ddot{x} + \epsilon^2 \dot{x} h(x, \dot{x}) + g(x) = \epsilon y(t) \quad = \quad z(t) \tag{10}$$

where $\dot{x}h(x,\dot{x})$ is the nonlinear damping force and $g(x)$ is a nonlinear restoring force. The excitation $y(t)$ will be here modelled as a stationary random process, with zero mean and power spectrum $S_y(\omega)$. The total energy envelope process $E(t)$ associated with the response is the sum of kinetic and potential energies

$$E(t) = \frac{\dot{x}^2}{2} + \int_0^x g(\xi)d\xi. \tag{11}$$

The energy envelope is governed by the following stochastic Itô equation

$$dE = m(E)dt + D(E)^{1/2}dW \tag{12}$$

where $m(E)$ and $D(E)$ are the drift and diffusion coefficients defined, respectively, by

$$m(E) = \lim_{\Delta t \to 0} \frac{\Delta E}{\Delta t}, \qquad D(E) = \lim_{\Delta t \to 0} \frac{\Delta E^2}{\Delta t}. \tag{13}$$

In [8] was shown that the following averaged series expansions hold for the diffusion and drift coefficients

$$m(E) = -d(E) + \frac{\pi}{2} \sum_{n=1,3,\ldots}^{\infty} (s_n^2 + c_n^2) S_z[n\omega(E)] \tag{14}$$

$$D(E) = 2\pi E \sum_{n=1,3,\ldots}^{\infty} s_n^2 S_z[n\omega(E)] \tag{15}$$

where $d(E)$ is the average of the actual damping force and s_n, c_n are Fourier coefficients of the related free undamped phase angle. Numerical studies have shown that, in nearly all cases, the first terms in the summations in (14)-(15) are completely dominant. An estimates of the energy envelope $E(t)$, as well as for the potential energy $V(x(t))$, can be easily obtained if the response and velocity are known [7]. The stiffness parameters can then be estimated by linear regression, fitting the parametric form for $V(x(t))$. Suppose that an interval of time $\Delta t = m\delta t > \tau_{cor}$ is chosen and the increments $\Delta E_i = E(t_i) - E(t_{i-m})$ for $i = m+1, \ldots, N$ are computed. Estimates of the drift and diffusion coefficients may now be generated by averaging the ΔE_i values, and their square, within each slot of width δE_i

$$\hat{m}(E_j) = \frac{1}{n_j \Delta t} \sum_{n_j} \Delta E_i, \qquad \hat{D}(E_j) = \frac{1}{n_j \Delta t} \sum_{n_j} (\Delta E_i)^2 \tag{16}$$

where n_j stands for the number of δE_i values obtained for the j-th slot. Using (15) it is possible to generate an estimate of the power spectrum of the excitation process $y(t)$ directly from $\hat{D}(E_j)$. Thus

$$\hat{S}_y(\omega_j) = \frac{\hat{D}(E_j)}{2\pi E_j s_1^2}, \qquad \omega_j = \omega(E_j). \tag{17}$$

Once $\hat{S}_y(\omega_j)$ is determined then the averaged damping function can be estimated by using (14)

$$\hat{d}(E_j) = -\hat{m}(E_j) + \frac{\pi}{2}(s_1^2 + c_1^2)\hat{S}_y(\omega_j). \tag{18}$$

If a linear-in-the-parameters form of the damping function $\epsilon^2 \dot{x} h(x, \dot{x})$ is assumed, i.e.,

$$\epsilon^2 \dot{x} h(x, \dot{x}) = \sum_{i=1}^{p} \beta_i \dot{x} h_i(x, \dot{x}) \tag{19}$$

then

$$d(E) = \sum_{i=1}^{p} \beta_i H_i(E) \tag{20}$$

where

$$H_i(E) = \frac{2E}{T(E)} \oint h_i(E, \Phi_0) \sin^2 \Phi_0 dt. \tag{21}$$

The damping parameters can then be estimated by fitting the parametric form for $d(E)$ to the estimate of this function, using simple linear regression. As an example, a Duffing oscillator with linear-plus quadratic type of damping, driven by a Gaussian coloured noise is herein considered. Figure 3 show the variation of estimates of linear versus non-linear damping parameters for different excitation levels, Figure 4 show the estimates of the input power spectral density for the same case.

5. Identification to Spectral Representation Input Processes

The high computational effort of both spectral and energy methods is a limit for their applicability to large systems. Here a new method, based on the assumption that the excitation force belong to the class of spectral representation processes is presented. The method do not require arbitrary assumptions on the parametric form of the input spectra. Without loss of general-

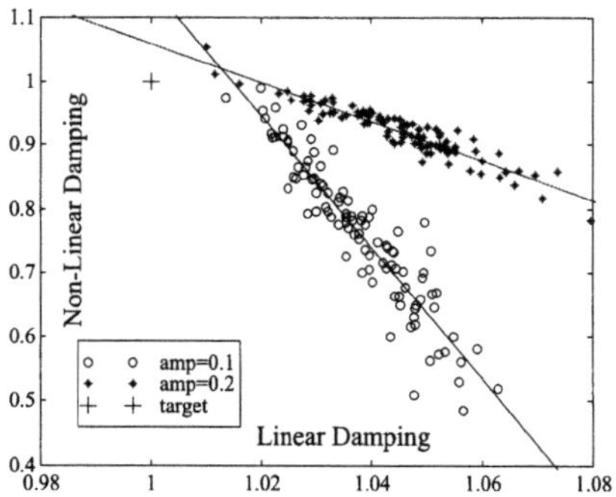

Figure 3. Variation of estimates of linear versus non-linear damping parameters for different excitation levels

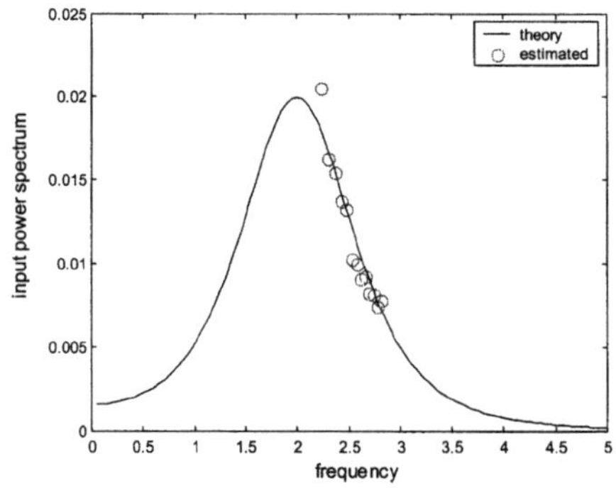

Figure 4. Estimates of the input power spectral density

ity, the identification technique is here introduced for SDOF non linear systems linear in the parameters, whose equation of motion takes the form

$$\ddot{x} + \sum_{j=1}^{n} a_j h_j(x, \dot{x}) \quad = \quad f(t) \tag{22}$$

where a_j are real coefficients and $h_j(x, \dot{x})$ are assigned non-linear functions. It will be assumed that the excitation process belong to the class of spectral representation processes defined by

$$f(t) \quad = \quad \sum_{k=1}^{N} \sqrt{2S_f(\omega_k)\Delta\omega} \cos(\omega_k t + \phi_k) \tag{23}$$

where $\omega_k = k\Delta\omega$, $(k = 1, \ldots, N)$, while ϕ_k are independent random phase angles uniformly distributed in $(0, 2\pi)$. In the frequency domain, by applying the Fourier operator (3), the equation of motion (22) becomes

$$-X(\omega) + \sum_{j=1}^{n} a_j H_j(\omega) \quad = \quad F(\omega) \tag{24}$$

Particularizing (24) for $\omega = \omega_k$, $(k = 1, \ldots, N)$, the following set of equations is obtained

$$-X(\omega_k) + \sum_{j=1}^{n} a_j H_j(\omega_k) = \Psi(\omega_k)\varphi(\omega_k, \phi_k, T) \qquad k = 1, 2, \ldots, N \tag{25}$$

where

$$\Psi(\omega) \quad = \quad \sqrt{2S_f(\omega)\Delta\omega} \tag{26}$$

and

$$\begin{aligned}\varphi(\omega,\phi,T) &= \frac{1}{2}\left[\cos\phi\left(1+\frac{\sin\Omega\cos\Omega}{\Omega}\right)+\sin\phi\left(\frac{\cos^2\Omega-1}{\Omega}\right)\right] \\ &+ \frac{i}{2}\left[\sin\phi\left(1-\frac{\sin\Omega\cos\Omega}{\Omega}\right)+\cos\phi\left(\frac{\cos^2\Omega-1}{\Omega}\right)\right] \quad (27)\end{aligned}$$

having set $\Omega = \omega T$. Equations (25) can be separate in their real and imaginary parts, yielding the following set of $2N$ equations

$$\begin{aligned}-X_{,r}\,\omega_k^2 + \sum_{j=1}^{n} a_j\, H_{j,r} &= \Psi_k\,\varphi_{,r} \\ -X_{,i}\,\omega_k^2 + \sum_{j=1}^{n} a_j\, H_{j,i} &= \Psi_k\,\varphi_{,i} \quad (28)\end{aligned}$$

where $\Psi_k = \Psi(\omega_k)$ and the subscripts i, r stand for the imaginary and real part of the related quantities. In matrix form, (28) becomes

$$\mathbf{A}_{2N\times n}[X(\omega)]\mathbf{a}_{n\times 1} + \mathbf{\Phi}_{2N\times N}[\phi]\mathbf{\Psi}_{N\times 1} + \mathbf{b}_{2N\times 1}[X(\omega)] = \mathbf{0} \quad (29)$$

having set with $\mathbf{a} = [a_1 \cdots a_n]^T$ and $\mathbf{\Psi} = [\Psi_1 \cdots \Psi_N]^T$ the system parameters and input spectrum unknown. For identification purposes the phase set angles ϕ is known only in distribution and if a different set $\hat{\phi}$ is adopted, the frequency equations of motion (29)lead to an error $\mathbf{e}$ defined as

$$\mathbf{A}[X(\omega)]\mathbf{a} + \mathbf{\Phi}[\hat{\phi}]\mathbf{\Psi} + \mathbf{b}[X(\omega)] = \mathbf{e}. \quad (30)$$

The parameter vectors $\mathbf{a}$ and $\mathbf{\Psi}$ can be obtained by the least square method, minimizing the variance of the error $E[\mathbf{e}^T\mathbf{e}]$ over the distribution of the phase angle. As an example, a linear oscillator driven by a Gaussian colored noise is herein considered. Figure 5 show the variation of estimates of natural frequency ω versus damping ratio ζ, Figure 6 show the estimates of the input power spectral density.

6. Conclusion

Methods to estimate unknown parameters in non-linear, randomly excited dynamic systems, have been presented. Three techniques have been discussed, based on response higher order spectra analysis, stochastic averaging and spectral representation processes. The approaches have been illustrated through applications to linear and non-linear oscillators under Gaussian and non-Gaussian excitation.

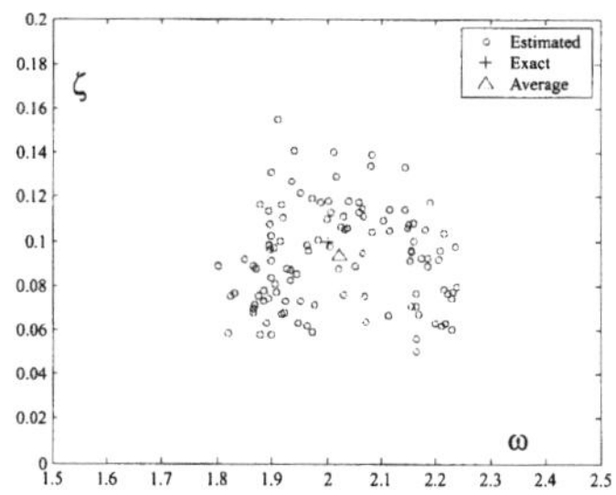

Figure 5. Estimates of natural frequency ω versus damping ratio ζ ; $\triangle$ average; $+$: exact value

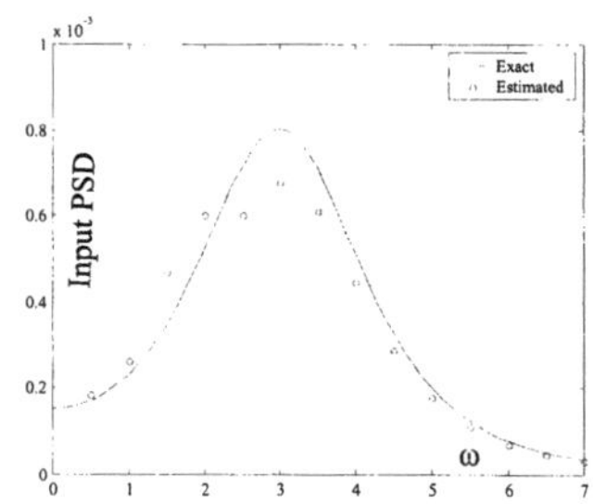

Figure 6. Estimates of the input power spectral density

References

[1] A. Blanc-Lapierre and R. Fortet, *Theories des Function Aleatories*, Paris: Masson, 1953.

[2] V. Chandran and S. Elgar, "A general procedure for the derivation of principal domains in of higher order spectra," *IEEE Transactions of Signal Processing*, vol. 42, pp. 229-233, 1994.

[3] F. Kozin and H. G. Natke, "System Identification Techniques," *Structural Safety*, vol. 3, pp. 269–316, 1986.

[4] S. Krenk and J. B. Roberts, "Local Similarity in Nonlinear Random Vibration," *Journal of Applied Mechanics*, vol. 66, pp. 225–235, 1999.

[5] J. B. Roberts, J. F. Dunne, and A. Debonos, "A Spectral Method for Estimation of Nonlinear System Parameters for Measured Response," *Probabilistic Engineering Mechanics*, vol. 10, pp. 199–207, 1995.

[6] J. B. Roberts and M. Vasta, "Parametric identification of systems with non-Gaussian excitation using measured response spectra," *Probabilistic Engineering Mechanics*, vol. 15, pp. 59-71, 2000.

[7] J. B. Roberts and M. Vasta, "Energy-Based Stochastic Estimation for Nonlinear Oscillators with Random Excitation," *Journal of Applied Mechanics*, vol. 67, pp. 763–771, 2000.

[8] J. B. Roberts and M. Vasta, "Markov modelling and stochastic identification for nonlinear ship rolling in random waves," *Phil. Trans. R. Soc. London A*, vol. 358, pp. 1917–1941, 2000.

[9] J. B. Roberts and M. Vasta, "Energy-Based Stochastic Estimation for Nonlinear Oscillators with Parametric Random Excitation," to appear on *Meccanica*, 2002.

[10] G. I. Schueller, (editor), *Structural Dynamics*, Berlin: Springer Verlag, 1991.

[11] H. Unbehauen and G. P. Rao, "Continuous Time Approaches to System Identification - A Survey," *Automatica*, vol. 26, pp. 23-35, 1990.

[12] M. Vasta and J. B. Roberts, "Stochastic Parameter Estimation of Non-linear Systems Using Higher Order Spectra," *Journal of Sound and Vibration*, vol. 213, pp. 201-221, 1998.

LYAPUNOV EXPONENT AND STABILITY OF OPTIMALLY CONTROLLED QUASI PARTIALLY INTEGRABLE HAMILTONIAN SYSTEMS

W. Q. Zhu and Z. L. Huang
Department of Mechanics, Zhejiang University
Hangzhou, Zhejiang 310027, P. R. China
wqzhu@yahoo.com, huangzhilong@yahoo.com

Abstract A controlled quasi partially integrable Hamiltonian system with pure parametric stochastic excitations is first reduced to averaged *Itô* equations by using the stochastic averaging method for quasi-Hamiltonian systems. A dynamical programming equation is then established for the ergodic control problem and solved to yield the optimal control law. The expression for the largest Lyapunov exponent is derived and the necessary and sufficient condition for the asymptotically Lyapunov stability with probability one of optimally controlled system is obtained. The system is stabilized by feedback control with proper choosing of the cost function and weighting matrix in the performance index for ergodic control problem.

Keywords: quasi-Hamiltonian system, stochastic excitation, stochastic optimal control, Lyapunov exponent, stochastic stabilization.

1. Introduction

Although the basic formulation and fundamental equations of stochastic stabilization have been known since 1960's, the only results of any significance for a long period were those pertaining to linear stochastic systems and employing quadratic performance index [4]. Only in recent years, the stabilization of nonlinear stochastic systems has received much attention and several interesting results have been obtained [1, 5].

In recent years, a nonlinear stochastic optimal control strategy and a procedure for evaluating the largest Lyapunov exponent of quasi Hamiltonian systems have been proposed [11, 10, 9]. In the present paper, the strategy and the procedure are combined to obtain the necessary and sufficient condition for

N. Sri Namachchivaya and Y.K. Lin (eds.),
IUTAM Symposium on Nonlinear Stochastic Dynamics, 279-288.

the asymptotic Lyapunov stability with probability one of controlled quasi partially integrable Hamiltonian systems, and to stabilize the systems.

2. Stochastic Averaging

Consider an n degree-of-freedom (DOF) weakly controlled quasi Hamiltonian system governed by the following equations of motion

$$\begin{aligned} \dot{Q}_i &= \frac{\partial H'}{\partial P_i} \\ \dot{P}_i &= -\frac{\partial H'}{\partial Q_i} - \varepsilon c_{ij}\frac{\partial H'}{\partial P_j} + \varepsilon u_i + \varepsilon^{1/2} f_{ik}\xi_k(t) \\ & i,j = 1,2,\ldots,n; \quad k = 1,2,\ldots,m \end{aligned} \tag{1}$$

where Q_i and P_i are generalized displacements and momenta, respectively; $H' = H'(\mathbf{Q},\mathbf{P})$ is twice differentiable Hamiltonian; ε is a small parameter; $\varepsilon c_{ij} = \varepsilon c_{ij}(\mathbf{Q},\mathbf{P})$, $\varepsilon u_i = \varepsilon u_i(\mathbf{Q},\mathbf{P})$ and $\varepsilon^{1/2} f_{ik} = \varepsilon^{1/2} f_{ik}(\mathbf{P},\mathbf{Q})$ represent, respectively, coefficients of quasi-linear dampings, unknown feedback control forces and amplitudes of stochastic excitations; $\xi_k(t)$ are Gaussian white noises in the sense of Stratonovich with correlation functions $E[\xi_k(t)\xi_l(t+\tau)] = 2D_{kl}\delta(\tau)$.

Equation (1) may be modeled as Stratonovich stochastic differential equations and then converted into Itô stochastic differential equations by adding the Wong-Zakai correction terms. Splitting these terms into conservative part and dissipative part and combining them with $-\partial H'/\partial Q_i$ and $-\varepsilon c_{ij}\partial H'/\partial P_i$, respectively, the resulting Itô equations are of the form

$$\begin{aligned} dQ_i &= \frac{\partial H}{\partial P_i} dt \\ dP_i &= -(\frac{\partial H}{\partial Q_i} + \varepsilon m_{ij}\frac{\partial H}{\partial P_i} - \varepsilon u_i)dt + \varepsilon^{1/2}\sigma_{ik} dB_k(t) \\ & i,j = 1,2,\ldots,n; \quad k = 1,2,\ldots,m \end{aligned} \tag{2}$$

where H and εm_{ij} are the modified Hamiltonian and modified damping coefficients, respectively; $\boldsymbol{\sigma}\boldsymbol{\sigma}^T = 2\mathbf{fDf}^T$; $B_k(t)$ are standard Wiener processes. Assume that the Hamiltonian system with Hamiltonian H is partially integrable and non-resonant and has $r(1 < r < n)$ independent first integrals $H_\eta, \eta = 1,\ldots,r$, which are in involution. The Hamiltonian is of the form

$$H = \sum_{\eta=1}^{r-1} H_\eta(q_\eta, p_\eta) + H_r(\mathbf{q}_2, \mathbf{p}_2) \tag{3}$$

where $\mathbf{q}_2 = [q_r,\ldots,q_n]^T, \mathbf{p}_2 = [p_r,\ldots,p_n]^T$. In the weakly controlled quasi partially integrable Hamiltonian system (2), H_η, H_r are slowly varying processes while Q_i, P_i are rapidly varying processes. Based on a theorem due to [3], H_η, H_r converge weakly to an r-dimensional diffusion process as $\varepsilon \to 0$ in a time interval of order ε^{-1}, which is governed by the following averaged Itô equations [10]:

$$dH_\alpha = \left[\bar{m}_\alpha(\mathbf{H}) + \varepsilon < u_i \frac{\partial H_\alpha}{\partial P_i} >\right] dt + \bar{\sigma}_{\alpha k}(\mathbf{H}) d\bar{B}_k(t) \tag{4}$$
$$\alpha = 1,2,\ldots,r; \quad i = 1,2,\ldots,n; \quad k = 1,2,\ldots,m$$

where $\mathbf{H} = [H_1, \ldots, H_r]^T$, $\bar{B}_k(t)$ are standard Wiener processes,

$$\begin{aligned}
\bar{m}_\alpha(\mathbf{H}) &= \epsilon < -m_{ij}\frac{\partial H}{\partial P_j}\frac{\partial H_\alpha}{\partial P_i} + \frac{1}{2}\sigma_{ik}\sigma_{jk}\frac{\partial^2 H_\alpha}{\partial P_i \partial P_j} > \\
\bar{b}_{\alpha\beta}(\mathbf{H}) &= \bar{\sigma}_{\alpha k}\bar{\sigma}_{\beta k}(\mathbf{H}) = \epsilon < \sigma_{ik}\sigma_{jk}\frac{\partial H_\alpha}{\partial P_i}\frac{\partial H_\beta}{\partial P_j} > \\
< \cdot > &= \frac{1}{(2\pi)^{r-1}T(H_r)}\int_\Omega \int_0^{2\pi} \left[\cdot / \frac{\partial H_r}{\partial p_r}\right] \\
&\quad d\boldsymbol{\theta} dq_r \cdots dq_n dp_{r+1} \cdots dp_n \qquad (5) \\
T(H_r) &= \int_\Omega \left(1/\frac{\partial H_r}{\partial p_r}\right) dq_r \cdots dq_n dp_{r+1} \cdots dp_n
\end{aligned}$$

in which $\boldsymbol{\theta} = [\theta_1, \theta_2, \cdots, \theta_{r-1}]^T$,

$$\Omega = (q_r, \ldots, q_n; p_{r+1}, \ldots, p_n) | H_r(q_r, \ldots, q_n, 0, p_{r+1}, \ldots, p_n) \leq H_r. \tag{6}$$

Note that u_i are unknown so far, the averaging of terms $u_i \partial H_\alpha / \partial P_i$ will be performed later.

3. Ergodic Control

Assume that there exists a stationary ergodic solution of (4) for a given control. Then the following ergodic control problem can be formulated: finding Markov control $u_i = u_i(\mathbf{Q}, P)$ to minimize performance index

$$J = \lim_{T\to\infty} \frac{1}{T}\int_0^T (f(\mathbf{H}(s)) + < \mathbf{u}^T(s)\mathbf{R}\mathbf{u}(s) >) ds \tag{7}$$

where $\mathbf{u} = [u_1, \ldots, u_n]^T$, $f(\mathbf{H})$ and $\mathbf{R}$ are cost function and positive definite weighting matrix, respectively. A stochastic dynamical programming equation for this ergodic control problem can be established based on the dynamical programming principle [6] as follows:

$$\min_{\mathbf{u}} \left\{ \frac{1}{2}\bar{b}_{\alpha\beta}(\mathbf{H})\frac{\partial^2 V}{\partial H_\alpha \partial H_\beta} + \left[\bar{m}_\alpha(\mathbf{H}) + \varepsilon < u_i \frac{\partial H_\alpha}{\partial P_i} >\right] \frac{\partial V}{\partial H_\alpha} + f(\mathbf{H}) + < \mathbf{u}^T \mathbf{R} \mathbf{u} > \right\} = \gamma \tag{8}$$

where V is called value function,

$$\gamma = \lim_{T\to\infty} \frac{1}{T}\int_0^T [f(\mathbf{H}(s))+ < \mathbf{u}^{*T}(s)\mathbf{R}\mathbf{u}^*(s) >]ds \tag{9}$$

is the optimal average cost. $\mathbf{u}^*$ is the optimal control and obtained from the necessary conditions for minimizing the left hand side of (8)

$$\begin{aligned} u_i^* &= -\frac{1}{2}(\mathbf{R}^{-1})_{ij}\frac{\partial V}{\partial H_\alpha}\frac{\partial H_\alpha}{\partial P_j} \\ &\quad i,j = 1,2,\ldots,n; \quad \alpha = 1,2,\ldots,r. \end{aligned} \tag{10}$$

Substituting u_i^* into (8) for replacing u_i and completing the averaging of $u_i^*\frac{\partial H_\alpha}{\partial P_i}$ lead to the final dynamical programming equation. In the case of diagonal $\mathbf{R}$, it is

$$\begin{aligned} f(\mathbf{H}) + \bar{\bar{m}}_\alpha(\mathbf{H})\frac{\partial V}{\partial H_\alpha} - \frac{1}{4R_i} < \left(\frac{\partial H_\alpha}{\partial P_i}\right)^2 > \left(\frac{\partial V}{\partial H_\alpha}\right)^2 + \\ \frac{1}{2}\bar{b}_{\alpha\beta}(\mathbf{H})\frac{\partial^2 V}{\partial H_\alpha \partial H_\beta} = \gamma \\ i = 1,2,\ldots,n; \quad \alpha,\beta = 1,2,\ldots,r. \end{aligned} \tag{11}$$

Solving (11) for $\frac{\partial V}{\partial H_\alpha}$ and then substituting them into (10) yield the optimal control forces u_i^*.

4. The Largest Lyapunov Exponent

Substituting u_i^* into (4) for replacing u_i and completing the averaging of $u_i^*\frac{\partial H_\alpha}{\partial P_i}$ lead to the following fully averaged Itô equations for optimally controlled system:

$$\begin{aligned} dH_\alpha &= \bar{\bar{m}}(\mathbf{H}_\alpha dt + \bar{\sigma}_{\alpha k}(\mathbf{H})d\bar{B}_k(t) \\ &\quad \alpha = 1,2,\ldots,r; \quad k = 1,2,\ldots,m \end{aligned} \tag{12}$$

where

$$\bar{\bar{m}}_\alpha(\mathbf{H}) = \bar{m}_\alpha(\mathbf{H}) - \frac{1}{2R_i} < \left(\frac{\partial H_\alpha}{\partial P_i}\right)^2 > \frac{\partial V}{\partial H_\alpha}. \tag{13}$$

Suppose that $\mathbf{p} = \mathbf{q} = 0$ is the unique elliptic fixed point of the associated Hamiltonian system and the stochastic excitations in (1) are pure parametric (multiplicative). Then $\mathbf{H} = 0$ is the trivial solution of (12). Furthermore, assume that the drift and diffusion coefficients are homogeneous in H_α of degree one as $|\mathbf{H}| \to 0$. Otherwise, (12) is linearized at $\mathbf{H} = 0$. Introducing new variables

$$\rho = lnH^{1/2} = \frac{1}{2}lnH = \frac{1}{2}ln\left(\sum_{\alpha=1}^{r} H_\alpha\right) \tag{14}$$

$$\Lambda_\alpha = H_\alpha/H, \quad \alpha = 1,2,\ldots,r. \tag{15}$$

The Itô equations for ρ and Λ_α are obtained from (12) by using Itô differential rule as follows:

$$d\rho = Q(\mathbf{\Lambda})dt + \Sigma_k(\mathbf{\Lambda})d\bar{\bar{B}}_k(t) \tag{16}$$

$$d\Lambda_\alpha = F_\alpha(\mathbf{\Lambda})dt + G_{\alpha k}(\mathbf{\Lambda})d\bar{\bar{B}}_k(t) \tag{17}$$

where $\mathbf{\Lambda} = [\Lambda_1, \Lambda_2, \ldots, \Lambda_r]^T$, $\bar{\bar{B}}_k(t)$ are standard Wiener processes,

$$\begin{aligned}
Q(\mathbf{\Lambda}) &= \frac{1}{2}\sum_{\alpha=1}^{r} \bar{\bar{m}}_\alpha(\mathbf{\Lambda}) - \frac{1}{4}\sum_{\alpha,\beta=1}^{r}\sum_{k=1}^{m} \bar{\sigma}_{\alpha k}\bar{\sigma}_{\beta k}(\mathbf{\Lambda}) \\
F_\alpha(\mathbf{\Lambda}) &= -\Lambda_\alpha \sum_{\beta=1}^{r} \bar{\bar{m}}_\beta(\mathbf{\Lambda}) + \bar{\bar{m}}_\alpha(\mathbf{\Lambda}) + \frac{1}{2}\Lambda_\alpha \sum_{\beta,\beta'=1}^{r}\sum_{k=1}^{m} \bar{\sigma}_{\beta k}\bar{\sigma}_{\beta' k}(\mathbf{\Lambda}) \\
&\quad -\frac{1}{2}\sum_{\beta=1}^{r}\sum_{k=1}^{m} \bar{\sigma}_{\alpha k}\bar{\sigma}_{\beta k}(\mathbf{\Lambda}) \qquad (18) \\
G_{\alpha k}(\mathbf{\Lambda}) &= \bar{\sigma}_{\alpha k}(\mathbf{\Lambda}) - \Lambda_\alpha \sum_{\beta=1}^{r} \bar{\sigma}_{\beta k}(\mathbf{\Lambda}).
\end{aligned}$$

Define the Lyapunov exponent of fully averaged Itô (12) as the asymptotic rate of the exponential growth of the square-root of the sum of r first integrals, i.e.,

$$\lambda = \lim_{T\to\infty} \frac{1}{T} \ln \left(\sum_{\alpha=1}^{r} H_\alpha(T)\right)^{1/2} = \lim_{T\to\infty} \frac{1}{2T} \ln H(T). \tag{19}$$

By using a procedure similar to that due to [9], the following expression for the largest Lyapunov exponent for optimally controlled system, λ_1^c, can be obtained:

$$\lambda_1^c = \int \bar{Q}(\mathbf{\Lambda}')p(\mathbf{\Lambda}')d\mathbf{\Lambda}' \tag{20}$$

where $\mathbf{\Lambda}' = [\Lambda_1, \Lambda_2, \ldots, \Lambda_{r-1}]^T$; $\bar{Q}(\mathbf{\Lambda}')$ is obtained from $Q(\mathbf{\Lambda})$ with Λ_r replaced by $1 - \sum_{\eta=1}^{r-1} \Lambda_\eta$; $p(\mathbf{\Lambda}')$ is the stationary probability density of $\mathbf{\Lambda}'$ obtained from solving the reduced Fokker-Planck equation associated with (17) with Λ_r replaced by $1 - \sum_{\eta=1}^{r-1} \Lambda_\eta$. The largest Lyapunov exponent of uncontrolled system, λ_1^u, is obtained similarly by letting $u_i^* = 0$, i.e., $\bar{\bar{m}}_\alpha$ replaced by $\bar{m}_\alpha$. Averaging is a nonlinear mixing operation. Based on the Oseledec multiplicative ergodic theorem [7], λ_1^u and λ_1^c should be the first approximation of the largest Lyapunov exponent of uncontrolled and controlled quasi partially integrable Hamiltonian system (1).

Define the Lyapunov stability and asymptotic Lyapunov stability with probability one using the norm $(\Sigma_{\alpha=1}^r H_\alpha)^{1/2}$. Then, based on the Oseledec multiplicative ergodic theorem [7], the necessary and sufficient condition for the asymptotic Lyapunov stability with probability one of the trivial solution of the optimally controlled system (1) is obtained approximately by letting λ_1^c to be negative. For the problem of feedback stabilization, let the cost function $f(\mathbf{H})$ and weighting matrix $\mathbf{R}$ in performance index (7) have some undetermined parameters. λ_1^c will depend on these undetermined parameters and the undetermined parameters are determined by minimizing λ_1^c.

5. Example

Consider a 3DOF system governed by the following equations of motion:

$$\begin{aligned}
\dot{Q}_1 &= P_1 \\
\dot{P}_1 &= -\omega_1^2 Q_2 - \beta_{11}P_1 - \beta_{12}P_2 - \beta_{13}P_3 + u_1 + f_{1k}P_k\xi_k(t) \\
\dot{Q}_2 &= P_2 \\
\dot{P}_2 &= -\frac{\partial U}{\partial Q_2} - \beta_{21}P_1 - \beta_{22}P_2 - \beta_{23}P_3 + u_2 + f_{2k}P_k\xi_k(t) \\
\dot{Q}_3 &= P_3 \\
\dot{P}_3 &= -\frac{\partial U}{\partial Q_3} - \beta_{31}P_1 - \beta_{32}P_2 - \beta_{33}P_3 + u_3 + f_{3k}P_k\xi_k(t) \\
& \quad k = 1,2,3
\end{aligned} \tag{21}$$

where β_{ij} and f_{ij} are constants; $\xi_k(t)$ are independent Gaussian white noises in the sense of Stratonovitch with intensities $2D_k$;

$$U(Q_2, Q_3) = k(\omega_2^2 Q_2^2 + \omega_3^2 Q_3^2)^\delta, k, \delta > 0, \quad \delta \neq 1. \tag{22}$$

System (21) can be modeled as Itô equations with the Hamiltonian

$$H = H_1 + H_2 \tag{23}$$

where

$$H_1 = (P_1^2 + \omega_1^2 Q_1^2)/2; \quad H_2 = (P_2^2 + P_3^2)/2 + U(Q_2, Q_3) \tag{24}$$

$U(Q_2, Q_3)$ is inseparable. So, the Hamiltonian system with Hamiltonian H is partially integrable with two independent first integrals H_1, H_2 in involution. Assume that β_{ij}, D_k, u_i are of the same order of ε. Applying the stochastic averaging method for quasi partially integrable Hamiltonian systems [10], the following averaged equations are obtained:

$$\begin{aligned} dH_1 &= [\bar{m}_1(H_1, H_2) + < u_1 P_1 >]dt + \bar{\sigma}_{1k} dB_k(t) \\ dH_2 &= [\bar{m}_2(H_1, H_2) + < u_2 P_2 + u_3 P_3 >]dt + \bar{\sigma}_{2k} dB_k(t) \\ & k = 1, 2, 3 \end{aligned} \tag{25}$$

where $\bar{m}_\alpha, \bar{\sigma}_{\alpha k}$ can be obtained by using formulas in (5).

For the ergodic control problem with performance index (7), the optimal control law is of the form of (10) with $i, j = 1, 2, 3$ and $\alpha = 1, 2$. In the case of diagonal **R**, it is

$$u_1^* = -\frac{1}{2R_1}\frac{\partial V}{\partial H_1}P_1, u_2^* = -\frac{1}{2R_2}\frac{\partial V}{\partial H_2}P_2, u_3^* = -\frac{1}{2R_3}\frac{\partial V}{\partial H_2}P_3 \tag{26}$$

and the final dynamical programming equation is

$$\begin{aligned} &\frac{1}{2}\bar{b}_{11}\frac{\partial^2 V}{\partial H_1^2} + \frac{1}{2}\bar{b}_{22}\frac{\partial^2 V}{\partial H_2^2} + \bar{m}_1\frac{\partial V}{\partial H_1} + \bar{m}_2\frac{\partial V}{\partial H_2} - \frac{1}{4R_1} < P_1^2 > \left(\frac{\partial V}{\partial H_1}\right)^2 \\ &\qquad - \left[\frac{1}{4R_2} < P_2^2 > + \frac{1}{4R_3} < P_3^2 >\right]\left(\frac{\partial V}{\partial H_2}\right)^2 + f(\mathbf{H}) = \gamma. \end{aligned} \tag{27}$$

For the purpose of stability analysis or feedback stabilization, only the asymptotic solution of (27) near $\mathbf{H} = 0$ is necessary. In order that $\bar{\bar{m}}_\alpha$ are homogeneous in H_1, H_2 of degree one as $|\mathbf{H}| \to 0$, $f(\mathbf{H})$ should be of the form

$$f(\mathbf{H}) - \gamma = k_1 H_1 + k_2 H_2 + o(H_1, H_2) \tag{28}$$

where k_1 and k_2 are constants. Then, the solution to (27) is of the form

$$V(\mathbf{H}) = c_1 H_1 + c_2 H_2 + o(H_1, H_2) \tag{29}$$

where c_1, c_2 are constants. Substituting (28) and (29) into (27) leads to the equation for unknown c_1 and c_2 for the problem of stability analysis of controlled system, or to the equation for unknown k_1, k_2, R_1, R_2, R_3 for the problem of feedback stabilization. Substituting u_1^* into (25) for replacing u_i and completing the averaging yield the fully averaged Itô equations for controlled system

$$\begin{aligned} dH_1 &= \left[\bar{m}_1(H_1, H_2) - \frac{c_1}{2R_1}H_1\right]dt + \bar{\sigma}_{1k} dB_k(t) \\ dH_2 &= \left[\bar{m}_2(H_1, H_2) - \left(\frac{1}{2R_2} + \frac{1}{2R_3}\right)c_2 H_2\right]dt + \bar{\sigma}_{2k} dB_k(t) \\ & k = 1, 2, 3. \end{aligned} \tag{30}$$

For the stability analysis of optimally controlled system (21), k_1, k_2, R_1, R_2, R_3 are given and c_1, c_2 are obtained from solving final dynamical programming (27). The largest Lyapunov exponent λ_1^c is obtained by using (20). The necessary and sufficient condition for the asymptotic Lyapunov stability with probability one is then obtained by letting $\lambda_1^c < 0$. For feedback stabilization of system (21), c_1, c_2 and λ_1^c are functions of undetermined parameters k_1, k_2, R_1, R_2, R_3, which are determined by minimizing λ_1^c.

It is seen from (26) and (29) that the optimal control forces are actually linear damping forces with damping coefficients depending on the parameters of the system, excitations and performance index. These control forces reduce the largest Lyapunov exponent and stabilize the system. To obtain a general idea as how the feedback control affects on the stability, a set of numerical results are obtained. The largest Lyapunov exponents of uncontrolled and controlled systems (21), λ_1^u and λ_1^c, as function of stochastic excitation intensity, are shown in Figure 1. It is seen that the system is indeed stabi-

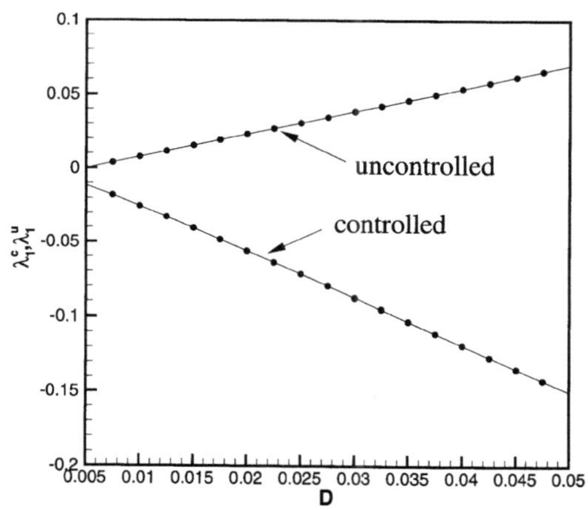

Figure 1. The largest Lyapunov exponent λ_1^c and λ_1^u of system (18) versus intensity of stochastic excitations. $\beta_{11} = 0.02$, $\beta_{22} + \beta_{33} = 0.02$, $f_{ij} = 1(i,j = 1,2,3), \delta = 2.0, k_1 = k_2 = 0.0001$, $R_1 = R_2 = R_3 = 1.0$, $D_1 = D_2 = D_3 = D$. – analytical results; · from digital simulation

lized by feedback control and the stability margin is larger for stronger stochastic excitation as long as the feedback control law is unchanged.

6. Conclusion

The procedure proposed in the present paper for analysing the asymptotic Lyapunov stability with probability one of controlled quasi partially integrable Hamiltonian systems or for designing a feedback control to stabilize the systems can be applied to quasi non-integrable Hamiltonian system by letting $r = 1$ and to quasi integrable Hamiltonian system by letting $r = n$. Compared with control Lyapunov function method, the proposed procedure is much simpler and, the necessary and sufficient stability condition rather than sufficient stability condition is obtained by using the proposed procedure.

In this paper, the local asymptotic Lyapunov stability with probability one of the trivial solution and, correspondingly, the largest Lyapunov exponent near $\mathbf{H} = 0$ are considered. In the proposed procedure, instead of Enclidean norm, a new norm $(\Sigma_{\alpha=1}^{r} H_\alpha)^{1/2}$ is introduced in the definitions of stability and Lyapunov exponent, and, instead of directly linearizing the original system, the averaged Itô equations are linearized if they are not homogeneous in H_α of order one. The advantages of this new procedure are as follows: the dimension of equations involved is lower and the largest Lyapunov exponent can be evaluated easier; the new procedure can be applied directly to nonlinear stochastic systems and to the case with vanishing linear stiffness term, etc. For Duffing-van der Pol oscillator subject to multiplicative excitation of wide-band random process, it has been shown [8] that the largest Lyapunov exponent evaluated by using such a procedure is the same in the first approximation as that obtained by using the method of small noise expansion.

In this paper, only non-resonant case has been considered. In resonant case, e.g., there are β resonant relations in the integrable sub-Hamiltonian system, there will be β averaged Itô equations for the combinations of angle variables added to (4), and the dimension of dynamical programming (8) will be $r + \beta$. Similarly, (17) will be expanded to include β averaged Itô equations for the combinations of angle variables. The largest Lyapunov exponent can be evaluated by using a similar procedure. Of course, it will be more difficult to evaluate the largest Lyapunov exponent in this case since a higher-dimensional reduced Fokker-Planck associated with expanded (17) has to be solved.

Acknowledgments

The work reported in this paper was supported by the National Natural Science Foundation of China under Grant No. 19972059 and 10002015, and the Special Fund for Doctor Programs in Institutions of Higher Learning of China under Grant No. 20020335092.

References

[1] D. Florchinger, "Feedback stabilization of affine in the control stochastic differential systems by the control Lyapunov function method," *SIAM J.Control and Optimization* vol. 35, pp. 500–511, 1997.

[2] R. Z. Khasminskii, "Necessary and sufficient conditions for the asymptotic stability of linear stochastic systems," *Theory of Probability and Application*, vol. 12, pp. 144–147, 1967.

[3] R. Z. Khasminskii, "On averaging principle for stochastic Itô equation," *Kibernetika*, vol. 4, pp. 260–279, 1968. (In Russian).

[4] R. Z. Khasminskii, *Stochastic Stability of Differential Equation*, Sijthoff and Noordhoff, Rockville, MD, 1980.

[5] M. Krstic and H. Deng, *Stabilization of Nonlinear Uncertain Systems*, London: Springer, 1998.

[6] J. Kushner, "Optimality conditions for the average cost per unit time problem with a diffusion model,"*SIAM J. Control and Optimization* vol. 16, pp. 330–346, 1978.

[7] V. I. Oseledec, "A multiplicative ergodic theorem: Lyapunov charecteristic number for dynamical systems," *Trans. Moscow Math. Society*, vol. 19, pp. 197–231, 1968.

[8] W. Q. Zhu, "Lyapunov exponent and stochastic stability of quasi- non-integrable Hamiltonian systems," *Int. J. Non-linear Mech.* In press.

[9] W. Q. Zhu and Z. L. Huang, "Lyapunov exponents and stochastic stability of quasi-integrable-Hamiltonian systems," *ASME J. Appl. Mech.*, vol. 66, pp. 211–217, 1999.

[10] W. Q. Zhu, Z. L. Huang, and Y. Suzuki, "Stochastic averaging and Lyapunov exponent of quasi partially integrable Hamiltonian systems," *Int. J. Non-linear Mech.*, vol. 37, pp. 419–437, 2002.

[11] W. Q. Zhu, Z. G. Ying, and T. T. Soong, "An optimal nonlinear feedback control strategy for randomly excited structural systems," *Nonlinear Dynamics*, vol. 24, pp. 31–51, 2001.

IV

PART 4: MODELING

Due to environmental fluctuations, temperature variations and other unknown effects, any engineering system is, inherently, subjected to random excitation. Choosing a model for the input or the environmental forcing, such as wave, wind or earthquake excitation, depends on the random nature and duration of the forcing. The *fourth* part of the proceedings deals with modeling of such stochastic uncertainties, which for many systems can either be internal or external, or a combination of both. This part also deals with various approximate analytical methods (e.g. equivalent linearization techniques) for nonlinear multi-degree-of-freedom systems. The fourth *Lead Lecture* of the Symposium, given by Kurt Wiesenfeld of the Georgia Institute of Technology, USA, is the lead paper of this part. The rest of the papers presented during the Symposium, on the general topic of *modeling and random vibrations* are compiled in alphabetical order of the last name of the first author.

Kurt Wiesenfeld provided a brief overview of stochastic resonance and discussed modeling issues related to human sensory systems, where Brownian motion of hair bundles improves the mechanoelectrical transduction of extremely weak signals. Some of the recent theoretical results on stochastic resonance can also be found in Part-I of this proceedings.

Mircea Grigoriu and Radoslaw Iwankiewicz considered the system response to non-Gaussian excitations. Mircea Grigoriu presented results on the dynamics and stability of some linear and nonlinear systems driven by Poisson and Lévy white noise. A method was developed for finding the tail properties of the distribution of the stationary states. Monte Carlo simulation and Hill estimates were used to confirm the analytical results. Radoslaw Iwankiewicz considered dynamical systems with multiplicative excitation of random impulses. It was found, using a generalized Itô differential rule, that the diffusion term should be corrected in this case rather than the drift term. Raouf Ibrahim studied the dynamic behavior of a disc brake subjected to a follower frictional force governed by random friction coefficient. Both transverse and circumferential displacements are found to be essentially non-Gaussian, nonstationary, and narrow-band processes. Werner Schiehlen investigated the impact processes using different time scales. The elastodynamic effects were computed on a fast time scale using analytical and numerical models.

Guoqiang Cai considered first the case of a quasi-linear system, namely, a linear system under both additive and multiplicative (parametric) excitations, for which exact solutions for the stationary moments and response spectral densities were obtained from sets of linear algebraic equations. Mikhail Dimentberg considered the generalized Lotka-Volterra model with white-noise random variations of the preys' reproduction factor which simulate the environmental variations. For continuously acting random processes, it may be reasonable to assume stationarity. The random forces due to earthquake require non-stationary excitation models which is the topic of the following two papers. Sami Masri studied a model structure subjected to an ensemble of ground-motion recordings from the 1994 Northridge (California) Earthquake (nonstationary excitations) using equivalent linearization techniques. Y. K. Wen proposed a new method for characterization and simulation of a nonstationary random process based on samples of the process. He also demonstrated its potential for engineering applications when dealing with nonstationary, nonlinear random processes.

STOCHASTIC RESONANCE: FROM EARTH'S ICE AGES TO HUMAN SENSORY SYSTEMS

Kurt Wiesenfeld
School of Physics, Georgia Institute of Technology,
Atlanta, GA 30332, USA
kurt.wiesenfeld@physics.gatech.edu

Fernán Jaramillo
Department of Biology, Carleton College
Northfield, MN 55057, USA
fjaramil@carleton.edu

Abstract Stochastic resonance has been studied intensively by physicists over the past decade or so. The phenomenon is seen in an extraordinarily wide variety of physical systems. More recently, researchers have asked whether stochastic resonance maybe relevant to problems in sensory biology. We give a brief overview of the subject and discuss our work on single hair cells in the auditory system, where Brownian motion of hair bundles improves the mechanoelectrical transduction of extremely weak signals.

Keywords: Stochastic Resonance, hair cells, noise

1. Introduction

Noise is usually considered a nuisance, degrading the performance of dynamical systems. But in nonlinear systems that display stochastic resonance, the presence of noise enhances the detection of weak signals. This puts stochastic resonance in the more general class of phenomena where random fluctuations play a constructive role. Stochastic resonance is now firmly established as a common phenomenon evidenced in a wide variety of physical systems.

Today, one can find in the physics literature several hundreds of research papers on the subject. The range of experiments which show SR encompasses electrical, optical, magnetic, chemical, and mechanical systems. More

N. Sri Namachchivaya and Y.K. Lin (eds.),
IUTAM Symposium on Nonlinear Stochastic Dynamics, 291-306.

recently, the subject has crossed disciplinary boundaries, as researchers ask whether it may be relevant to problems in sensory biology. The interested reader can find a comprehensive technical review [9], as well as more leisurely minireviews [17, 3, 16] and a popular review [12].

2. Stochastic Resonance in Physics

Stochastic resonance (SR) was originally conceived in 1981 as a mechanism that might explain why the Earths ice ages occur more or less periodically every 100,000 years. The suggestion was that a minor periodic oscillation in the shape of the Earths orbit, though on its own too small an effect to cause large climatic changes, could nevertheless entrain the energy available in large scale meterological fluctuations, resulting in the observed regularity. The principle did not rest on the specifics of atmospheric and geophysical dynamics. Instead, the idea was that a typical bistable system could show a kind of nonlinear cooperative phenomenon that coupled large random and small periodic influences [2, 14, 15, 1].

The first laboratory study of SR soon followed using a Schmitt trigger, a simple two-state electronic switch. The input to the trigger was the sum of two independent sources, one random and the other periodic. The amplitude of the periodic input was kept well below the threshold necessary to induce deterministic switching between the two states. The presence of the noisy input enabled switching events, and the strength of the noise was systematically varied. The experimental data were staggering: not only could an increase in the input noise lead to greater power at the signal frequency, but near the optimal operating point the broadband part of the output power practically vanished! This work introduced what has become the most frequently used definition of SR: output showing a maximum in the signal-to-noise ratio at a nonzero value of input noise. This led to a new and provocative formulation of the effect. Suppose one has a detector which is trying to extract a weak periodic signal embedded in a large amount of background noise. Then under the right circumstances, increasing the amount of input noise can increase the output signal-to-noise ratio [8] as shown in figure 1.

These striking experimental results achieved instant oblivion. Then in 1988, SR was reported in experiments on a laser system. The type of laser used was a ring laser, in which the light path forms a closed ring. The ring laser can operate in either of two modes, with the beam travelling around the ring in either the clockwise or counterclockwise sense but not both. Ordinarily, which of these modes is observed depends on fluctuations in the initial conditions. If the laser is repeatedly turned on and off, then on average each mode is observed half the time. However, it is possible to manipulate the laser to differentially favor one mode over the other using an optically active crystal. In the ex-

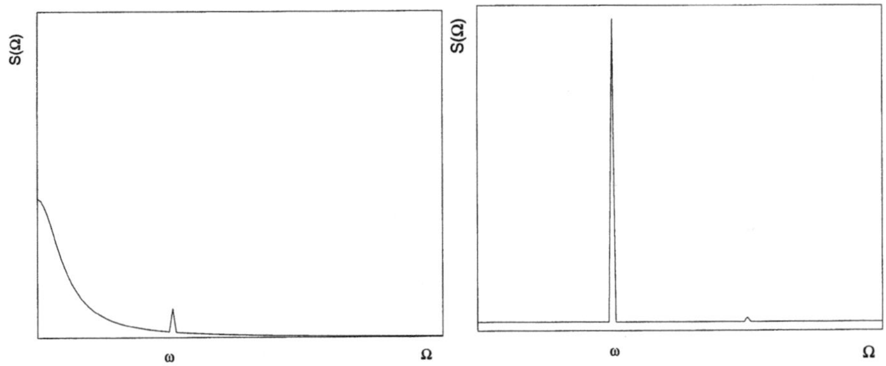

Figure 1. Output power spectra in the Schmitt trigger experiment showed an improvement in signal to noise ratio when some noise was added at the input. The lower specturm has the larger level of input noise. (After [8])

periment, the crystal acoustically excited with the sum of a small periodic signal and a large noise component. The periodic perturbation was too weak to induce on its own a change in the active laser mode, but the noise could. Taking the output as the intensity of the clockwise laser mode, a time series looked like random on-off switching with a regular component which became evident only at a non-zero level of input noise. As shown in the figure 2, the signal to noise ratio first increased and then decreased as the input noise level was systematically raised.

As it happened, the ring laser experiment ignited a period of intense investigation among researchers which continues even today. Stochastic resonance has turned out to be surprisingly commonplace and robust, occurring in experiments spanning the range of electrical, optical, magnetic, biological, chemical, and mechanical systems.

During this same period, a large body of theoretical work was developed. The interested reader can find a detailed overview of the main theoretical advances in the review article in [9]. In hindsight, the reason why SR is so widely and robustly observed is clear: there is no single set of specific ingredients which are essential for the effect. Instead, several mechanisms can lead to SR. For example, the attractors of the noise-free system can be equilibria or limit cycles or chaotic attractors; the input noise can be white or colored or even deterministic chaos rather than random; the signal can be periodic or quasiperiodic or still more complicated (anything that carries information will do). Theories have been developed using a variety of frameworks, including Langevin equations, Fokker-Planck equations, rate equations, time de-

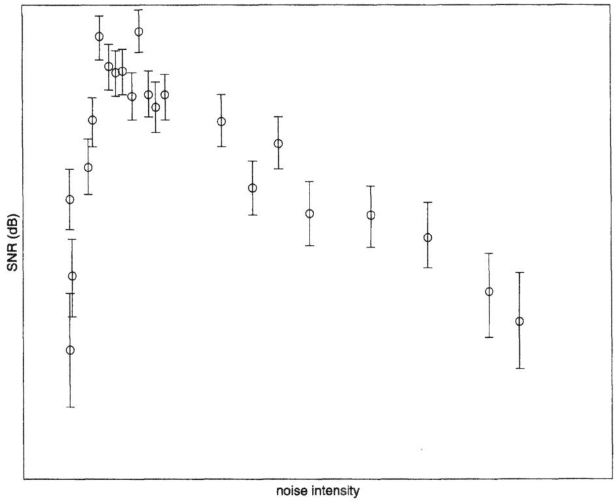

Figure 2. Experiments on a ring laser showed that the largest output signal to noise ratio occurs for a non-zero level of input noise. (After [13])

pendent Poisson processes, fluctuating rate processes, and linear response theory. It remains an open question whether a unified theory exists which can encompass the entire range.

Does SR have any practical applications? This remains an open question. There have been proposals for technological uses of SR; a couple of existing patents are based on the effect [10, 4]. And researchers are pursuing strategies to use SR in biomedical contexts, where one hopes to retrofit an existing system rather than engineer and insert an entirely new device. The analogy is that it is a useful option to correct a person's eyesight with prescription glasses (or contact lenses) rather than surgery, despite advances in laser surgery. One very nice example of such an application is a simple device to help people who have a tendency to lose their balance even while standing still. Though you are unaware of it, when standing still you are constantly righting yourself as you tilt this way and that under the influence of gravity – you are essentially an inverted pendulum! – with the information of your current tilt angle provided by sensors in your ankle. With age the sensory feedback system tends to degrade so that the critical tilt angle at which you right yourself grows larger. If this angle is too large, the feedback comes too late and you fall down. The solution proposed by researchers at Boston University is a stocking which provides low level random stimulation. Recent experiments on healthy individuals shows an improvement in maintaining "verticality" [6].

3. Brownian motion and hair bundles

Our current research joins a growing effort to explore whether SR plays a useful role in sensory biology. We have settled on hair cells which play a key role in the auditory system of many animals (including humans). Specifically, we are studying the effect of noise on the ability of the hair bundles embedded to detect faint stimuli. In the traditional view of the auditory system, noise in the form of Brownian motion of cochlear components is thought to limit the sensitivity of the ear. With the advent of stochastic resonance, this assumption needs to be reconsidered. And indeed, experiments show that noisy motion of a hair cell's bundle improves mechanoelectrical transduction via SR. (See Figures 3 and 3).

In the experiment, hair cells were obtained from the maculae sacculi of leopard frogs (*Rana pipiens*). Mechanical stimuli were conveyed by a glass rod attached to a bundle tip. The rod motion was the sum of a 2 nm sinusoidal component and a random component (effectively white noise). The resulting mechanoelectrical transduction current was recorded with a patch-clamp whole-cell recording electrode, and from this time series we constructed the power spectrum and signal to noise ratio.

Interestingly, in these experiments the optimal signal to noise ratio occurred with a noise level of a couple of nanometers, which agrees rather well with estimates of the equilibrium Brownian motion of hair bundles in living organisms. This has encouraged us to take seriously the idea that noise may play an important functional role in the auditory system in the detection of very weak acoustic signals. See Figure 3.

From a theoretical point of view, the hair cell is a particularly good candidate for detailed study because the transduction is a mechanical process and the physiology of the bundle is well characterized. We are therefore able to write down the governing equations of motion incorporating the known structure of the hair bundle and its response to stimuli. Random forces occur naturally as a result of Brownian motion, and we understand at a fundamental level how to include these random forces. We can also move beyond the *in vitro* situation, adding further deterministic and random stimuli to account for external manipulation used in certain experiments. This allows us to probe both equilibrium and non-equilibrium behaviors. Though conceptually straightforward, the resulting Langevin model is fairly elaborate insofar as a hairbundle comprises many individual stereocilia (some several tens of them) plus a comparable number of mechanically sensitive ion channels. To date our work has focussed on simpler versions of the full problem. The results thus far are encouraging, with the model exhibiting the typical transduction responses observed in hair cell experiments, including SR.

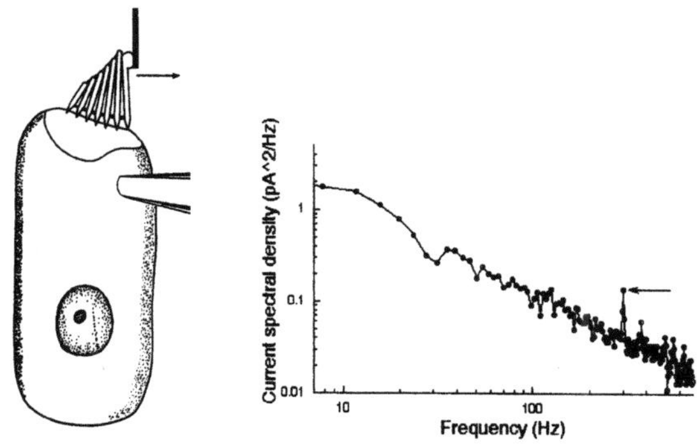

Figure 3. (top) Schematic of the hair cell experiment in which one bundle on a single hair cell is stimulated by an external probe whose position is controlled. (bottom) Resulting power spectrum when the probe position is comprised of random noise and a weak period component at 300 Hz.

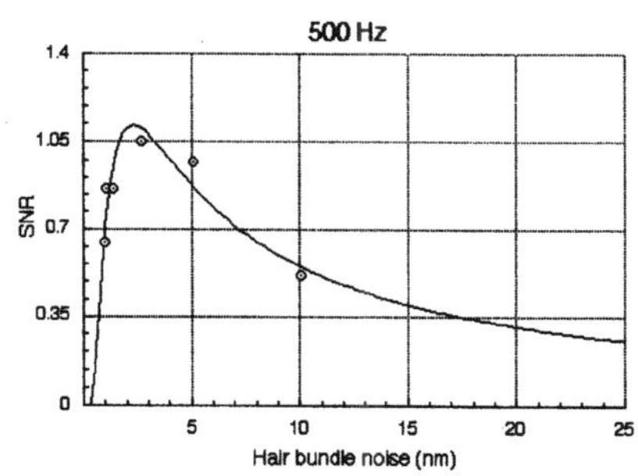

Figure 4. Output signal to noise ratio versus input noise strength from experiments on a single hair cell. The noise strength is taken to be the root mean squared amplitude of the random part of the bundle position. The solid line represents a best fit to the theoretically derived expression SNR = (a/x) exp(-b/x) where x is the noise strength and a and b are constants [11].

Our study of noisy hair bundle dynamics builds on the earlier work of physiologists who developed the so-called gating spring picture and the associated two state rate model [7]. Including the role of random stimuli and fluctuations is a natural and important step in the study of mechanotransduction in hair cells. One way to do this is to introduce stochastic components directly into the transition rates, resulting in a stochastic rate theory. Such a theory was developed to augment the rate theory model by including a stochastic component to the transition rates. In the stochastic resonance literature, such a stochastic rate theory was introduced by [5]. That theory highlights possible universal features of diverse systems which exhibit stochastic resonance, but leaves open certain conceptual problems insofar as it introduces randomness in two distinct ways: an underlying event rate which characterizes a random process, and random variations of a parameter on which this rate depends.

Treating the problem at the Langevin level moves us beyond the rate theory and provides us a way of elucidating the fundamentals which underpin the doubly stochastic description. By considering a model which is based directly and explicitly on the physical properties of hair cells, we can keep clear the distinct roles of the two coupled simultaneous random processes, channel transitions and hair bundle Brownian fluctuations.

Another intriguing aspect of the hair cell problem that these investigations may resolve involves an outstanding evolutionary puzzle: why is it that hair cells with free-standing hair bundles exist at all? Most hair bundles are not free-standing but rather are attached to overlying structures which deliver mechanical stimuli. In the cochlea both types exist: why hasn't evolution selected one type over the other? The new hypothesis is that noise is needed to detect the very weakest signals, and this is accomplished better in the free-standing bundles which undergo significant Brownian motion; conversely, the transduction of strong signals is degraded by noise, and detecting these signals is the job of the cells with attached hair bundles.

Acknowledgments

We thank Joe Neff for his participation in this project. This work is supported by grants from the National Institutes of Health (DC02235-08) and the Office of Naval Research (N00014-99-1-0592).

References

[1] R. Benzi, G. Parisi, S. Sutera, and A. Vulpiani, *Tellus*, vol. 34, pp. 10–18, 1982.

[2] R. Benzi, S. Sutera, and A. Vulpiani, *J. Phys. A*, vol. 14, pp. L453–L457, 1981.

[3] A. R. Bulsara and L. Gammaitoni, *Physics Today*, vol. 49, pp. 39–45, 1996.

[4] A. R. Bulsara, *et al*, "Controlled stochastic resonance circuit", U.S. Patent No. 6285249, 2001.

[5] S. M. Bezrukov and I. Vodyanov, *Nature*, London, vol. 378, p. 362, 1995.

[6] J. J. Collins, *et al*, *Phys. Rev. Lett.*, in press.

[7] D. P. Corey and A. J. Hudspeth, *J. Neurosci.*, vol. 3, pp. 962–976, 1983; J. Howard, W. M. Roberts, and A. J. Hudspeth, *Ann. Rev. Biophys. and Biophys. Chem.*, vol. 17, pp. 99–124, 1988.

[8] S. Fauve and F. Heslot, *Phys. Lett. A*, vol. 97, pp. 5–7, 1983.

[9] L. Gammaitoni, P. Hanggi, P. Jung, and F. Marchesoni, *Rev. Mod. Phys.*, vol. 70, pp. 223–288, 1998.

[10] A. D. Hibbs, "Detection and communications device employing stochastic resonance", U.S. Patent No. 5574369, 1996.

[11] F. Jaramillo and K. Wiesenfeld, *Nature Neuroscience*, vol. 1, pp. 384–388, 1998.

[12] F. Moss and K. Wiesenfeld, *Sci. Am.*, 1995.

[13] B. McNamara, K. Wiesenfeld, and R. Roy, *Phys. Rev. Lett.*, vol. 60, pp. 2626–2629, 1988.

[14] C. Nicolis, *Sol. Phys.*, vol. 74, p. 473, 1981.

[15] C. Nicolis, *Tellus*, vol. 34, pp. 1–9, 1982.

[16] K. Wiesenfeld and F. Jaramillo, *Chaos*, vol. 8, pp. 539–548, 1998.

[17] K. Wiesenfeld and F. Moss, *Nature*, London, vol. 373, pp. 33–36, 1995.

RESPONSE SPECTRAL DENSITIES OF SYSTEMS UNDER BOTH ADDITIVE AND MULTIPLICATIVE EXCITATIONS

G. Q. Cai
Center for Applied Stochastics Research
Florida Atlantic University
777 Glades Rd., Boca Raton, FL 33431
USA
caig@fau.edu

Abstract Solution techniques are developed to obtain the response spectral densities for systems under both additive and multiplicative random excitations, where the multiplicative excitations appear in the linear terms. For the case of a quasi-linear system, in which the damping and stiffness forces are linear, exact solutions for the stationary moments and response spectral densities are obtained from sets of linear algebraic equations. When the damping forces and/or the stiffness forces are nonlinear, the system can be replaced by an equivalent quasi-linear system for which exact solutions are available. A numerical example shows that the procedure yields quite accurate results when compared with those obtained from Monte Carlo simulations even for the case of strong nonlinearity.

Keywords: Random vibration, nonlinear system, spectral density, parametric excitation

1. Introduction

Many engineering systems are excited by dynamic loads which can be modeled as random processes, such as a building under earthquake or wind forces, a ship subjected to sea wave forces, etc. The response of such a system, for example, displacement, velocity, stress or strain, is also a random process. Solution for the system response in terms of the response spectral density is desirable since it gives the energy distribution in the frequency domain. For linear systems under additive excitations, the solutions are well known, which can be expressed in terms of the spectral densities of the excitations and the frequency response functions of the system. When a nonlinear system is under additive excitations, the concept of frequency response is no longer valid,

N. Sri Namachchivaya and Y.K. Lin (eds.),
IUTAM Symposium on Nonlinear Stochastic Dynamics, 299-306.

and various methods have been developed to obtain response spectral densities [11, 12, 3, 14, 1, 2, 15, 16, 13, 17, 4]. The frequency response concept is also not applicable for a linear system if it is subjected to both additive and multiplicative excitations. [5] proposed a version of linearization to replace the multiplicative excitation by an additive excitation. [6] introduced a "measuring filter" to obtained an exact spectral density for a linear oscillator under both additive and multiplicative excitations. [9] applied the stochastic averaging method to obtain the response spectral density on the condition of a certain energy level, and then averaged it over the entire energy range. As it turned out, the result was the same as that of [6] for the case of a linear oscillator.

In this paper, a simpler procedure for obtaining response spectral densities is proposed for systems under both additive and multiplicative random excitations. The stochastic differential equations of Itô type and an integral transformation are employed. If the system is linear, exact solutions are obtainable by solving a set of complex linear algebraic equations. In the case of a nonlinear system, a linearization scheme is applied to replace the nonlinear damping and stiffness forces while keeping the excitation terms unchanged. In such a way, the characteristics of multiplicative excitations are preserved. A numerical example is given to show the accuracy of the present procedure.

2. Analysis

Consider a stochastically excited system governed by the following equations

$$\ddot{q}_j + h_j(\mathbf{q}, \dot{\mathbf{q}}) = \sum_{i=1}^{n} [a_{ji}\dot{q}_i\eta_i(t) + b_{ji}q_i\gamma_i(t)] + \xi_j(t) \qquad (j = 1, 2, \ldots, n) \tag{1}$$

where $h_j(\mathbf{q}, \dot{\mathbf{q}})$ include both damping and stiffness forces, $\eta_i(t)$,$\gamma_i(t)$ and $\xi_j(t)$ are Gaussian white noises. As indicated in (1), the multiplicative excitations appear in the linear terms. For simplicity, it is assumed that the additive excitations $\xi_j(t)$ are not correlated with the multiplicative excitations.

The case of linear damping and stiffness is considered first, namely, $h_j(\mathbf{q}, \dot{\mathbf{q}}) = \sum_{i=1}^{n}(\alpha_{ji}q_i + \beta_{ji}\dot{q}_i)$. Such a system is said to be quasi-linear since multiplicative excitations are present, and the principle of superposition is no longer applicable. Letting $q_j = X_j$ and $\dot{q}_j = X_{j+n}$, a set of Itô stochastic differential equations [7] is derived from (1) as follows

$$
\begin{aligned}
dX_j &= X_{j+n}dt \\
dX_{j+n} &= \sum_{i=1}^{2n} C_{ji}X_i dt + \sigma_j(\mathbf{X})dB_j(t) \quad (j = 1, 2, \dots, n)
\end{aligned} \tag{2}
$$

where $B_j(t)$ are independent unit Wiener processes, and C_{ji} and $\sigma_j(\mathbf{X})$ can be derived from (1) (see e.g., [10]). Note that the Wong-Zakai correction terms are linear in the present case, and have been incorporated in (2) in the coefficients C_{ji}. Denoting $m_{ij} = E[X_i X_j]$, and using the Itô stochastic differential rule [8], a set of ordinary differential equations for the second-order statistical moments m_{ij} can be obtained. These equations are closed in the second order, and they can be solved exactly to obtain either transient or stationary second-order moments. Similarly, higher-order moments can be obtained exactly.

Multiplying $X_k(t_0)$, $t_0 = t - \tau$, on both sides of (2) and taking the ensemble average, we obtain a set of equations for the correlation functions $R_{ij}(\tau) = E[X_i(t)X_j(t_0)]$ as follows

$$
\begin{aligned}
\frac{dR_{jk}(\tau)}{d\tau} &= R_{j+n,k}(\tau) \\
\frac{dR_{j+n,k}(\tau)}{d\tau} &= \sum_{i=1}^{2n} C_{ji}R_{ik}(\tau) \; (j = 1, 2, \dots, n; k = 1, 2, \dots, 2n)
\end{aligned} \tag{3}
$$

Equations (3) can be solved with initial conditions $R_{ij}(0) = m_{ij}$, which have been obtained from the second-order moment equations. A direct procedure to obtain the spectral densities is given below, although they can also be obtained as the Fourier transform of the correlation functions.

Define the following integral transformation

$$
\bar{\Phi}_{ij}(\omega) = \Im[R_{ij}(\tau)] = \frac{1}{\pi}\int_0^\infty R_{ij}(\tau)e^{-i\omega\tau}d\tau \tag{4}
$$

It can be shown that

$$
\Im[\frac{dR_{ij}(\tau)}{d\tau}] = i\omega\bar{\Phi}_{ij}(\omega) - \frac{1}{\pi}m_{ij} \tag{5}
$$

Using (4) and (5), (3) can be transformed to

$$
\begin{aligned}
i\omega\bar{\Phi}_{jk}(\omega) - \frac{1}{\pi}m_{jk} &= \bar{\Phi}_{j+n,k}(\omega) \\
i\omega\bar{\Phi}_{j+n,k}(\omega) - \frac{1}{\pi}m_{j+n,k} &= \sum_{i=1}^{2n} C_{ji}\bar{\Phi}_{ik}(\omega) \\
(j = 1, 2, \dots, n; k &= 1, 2, \dots, 2n)
\end{aligned} \tag{6}
$$

The set of $\bar{\Phi}_{ij}(\omega)$ can be solved from (6). The spectral density functions are then obtained from

$$\Phi_{ii}(\omega) = Re[\bar{\Phi}_{ii}(\omega)], \quad \Phi_{ij}(\omega) = \frac{1}{2}[\bar{\Phi}_{ij}(\omega) + \bar{\Phi}^*_{ji}(\omega)] \tag{7}$$

It is noticed that (6) is a set of complex linear algebraic equations, and analytical solutions can be obtained for low dimensional systems. For high dimensional cases, analytical solutions may be tedious, but numerical solutions can be carried out quite simply.

For the case of nonlinear damping and stiffness forces, we replace these forces by linear ones, namely replacing system (1) by a quasi-linear one

$$\ddot{q}_j + \sum_{i=1}^{n}(\alpha_{ji}q_i + \beta_{ji}\dot{q}_i) = \sum_{i=1}^{n}[a_{ji}\dot{q}_i\eta_i(t) + b_{ji}q_i\gamma_i(t)] + \xi_j(t)$$
$$(j = 1, 2, \ldots, n) \tag{8}$$

The equivalent linear coefficients α_{ji} and β_{ji} are chosen to minimize the following mean-square differences

$$E[h_j(\mathbf{q}, \dot{\mathbf{q}}) - \sum_{i=1}^{n}(\alpha_{ji}q_i + \beta_{ji}\dot{q}_i)]^2 \quad (j = 1, 2, \ldots, n) \tag{9}$$

to yield

$$E[\mathbf{X}\mathbf{X}'] \begin{bmatrix} \boldsymbol{\alpha}' \\ \boldsymbol{\beta}' \end{bmatrix} = E[\mathbf{X}\mathbf{h}'] \tag{10}$$

where a prime denotes a matrix transposition, $\mathbf{X} = \{X_1 X_2 \ldots\ldots X_{2n}\}'$, $\boldsymbol{\alpha}= [\alpha_{ij}]$, $\boldsymbol{\beta}= [\beta_{ij}]$, and $\mathbf{h} = \{h_1 h_2 \ldots\ldots h_n\}'$. In many cases, functions $h_j(\mathbf{q}, \dot{\mathbf{q}})$ are polynomials of q_j and $\dot{q}_j$; thus the right hand sides of (10) can be calculated exactly for the replacing quasi-linear system (8). Iteration is then applied to equation set (10) to solve for the unknown coefficients α_{ij} and β_{ij}. The assumption of Gaussian distribution for the responses in the conventional linearization procedure is avoided here. The preservation of the parametric excitations in the quasi-linear system (8) is obviously an important feature of the approximation procedure.

3. Example

Consider a single-degree-of-freedom system governed by the following equation

$$\ddot{q} + h(q, \dot{q}) = \dot{q}\eta(t) + q\gamma(t) + \xi(t) \tag{11}$$

The spectral densities of the Gaussian white noises $\eta(t)$, $\gamma(t)$ and $\xi(t)$ are $K_{\eta\eta}$, $K_{\gamma\gamma}$ and $K_{\xi\xi}$, respectively, and it is assumed that only the two parametric excitations are correlated with a cross-spectral density $K_{\eta\gamma}$.

The case of a quasi-linear system is considered first, namely, $h(q,\dot{q}) = 2\alpha\dot{q} + \Omega^2 q$. Letting $X_1 = q$ and $X_2 = \dot{q}$, a set of Itô stochastic equations are derived from (11) as follows

$$\begin{aligned} dX_1 &= X_2 dt \\ dX_2 &= (-2\alpha X_2 + \pi K_{\eta\eta} X_2 - \Omega^2 X_1 + \pi K_{\eta\gamma} X_1)dt \\ &\quad + [2\pi(K_{\gamma\gamma} X_1^2 + K_{\eta\eta} X_2^2 + K_{\xi\xi} + 2K_{\eta\gamma} X_1 X_2)]^{1/2} dB(t) \end{aligned} \tag{12}$$

where $B(t)$ is a unit Wiener process. Using the Itô stochastic differential rule [8], equations for the second-order moments are obtained as follows

$$\begin{aligned} \frac{dm_{11}}{dt} &= 2m_{12} \\ \frac{dm_{12}}{dt} &= m_{22} - (\Omega^2 - \pi K_{\eta\gamma})m_{11} - (2\alpha - \pi K_{\eta\eta})m_{12} \\ \frac{dm_{22}}{dt} &= -2(\Omega^2 - \pi K_{\eta\gamma})m_{12} - 2(2\alpha - \pi K_{\eta\eta})m_{22} \\ &\quad + 2\pi(K_{\gamma\gamma} m_{11} + K_{\eta\eta} m_{22} + 2K_{\eta\gamma} m_{12} + K_{\xi\xi}) \end{aligned} \tag{13}$$

The stationary solutions of (13) are

$$\begin{aligned} m_{11} &= \frac{\pi K_{\xi\xi}}{2(\alpha - \pi K_{\eta\eta})(\Omega^2 - \pi K_{\eta\gamma}) - \pi K_{\gamma\gamma}}, \\ m_{22} &= m_{11}(\Omega^2 - \pi K_{\eta\gamma}), \quad m_{12} = 0 \end{aligned} \tag{14}$$

The validity of (14) requires

$$\alpha > \pi K_{\eta\eta}, \quad \Omega^2 > \pi K_{\eta\gamma}, \quad 2(\alpha - \pi K_{\eta\eta})(\Omega^2 - \pi K_{\eta\gamma}) > \pi K_{\gamma\gamma} \tag{15}$$

Now multiplying $X_1(t_0)(t_0 = t-\tau)$ on both sides of (12) and taking ensemble averaging, we obtain a set of equations for the correlation functions as follows

$$\begin{aligned} \frac{dR_{11}(\tau)}{d\tau} &= R_{21}(\tau) \\ \frac{dR_{21}(\tau)}{d\tau} &= -(\Omega^2 - \pi K_{\eta\gamma})R_{11}(\tau) - (2\alpha - \pi K_{\eta\eta})R_{21}(\tau) \end{aligned} \tag{16}$$

(16) can be solved with initial conditions $R_{ij}(0) = m_{ij}$. Using (4) and (5), equations in (16) are transformed to

$$\begin{aligned} &i\omega\bar{\Phi}_{11}(\omega) - \bar{\Phi}_{21}(\omega) = \frac{1}{\pi} m_{11} \\ &(\Omega^2 - \pi K_{\eta\gamma})\bar{\Phi}_{11}(\omega) + (2\alpha - \pi K_{\eta\eta} - i\omega)\bar{\Phi}_{21}(\omega) = 0 \end{aligned} \tag{17}$$

Similarly, we have

$$
\begin{aligned}
&i\omega\bar{\Phi}_{12}(\omega) - \bar{\Phi}_{22}(\omega) = 0 \\
&(\Omega^2 - \pi K_{\eta\gamma})\bar{\Phi}_{12}(\omega) + (i\omega + 2\alpha - \pi K_{\eta\eta})\bar{\Phi}_{22}(\omega) = \frac{1}{\pi} m_{22}
\end{aligned}
\tag{18}
$$

Spectral density functions can then be solved from (17) and (18) as follows

$$
\begin{aligned}
\Phi_{11}(\omega) &= \frac{m_{11}}{\pi\Delta}(2\alpha - \pi K_{\eta\eta})(\Omega^2 - \pi K_{\eta\gamma}), \quad \Phi_{22}(\omega) = \omega^2 \Phi_{11} \\
\Phi_{12}(\omega) &= -\frac{m_{11}}{\pi\Delta}(\Omega^2 - \omega^2 - \pi K_{\eta\gamma})(\Omega^2 - \pi K_{\eta\gamma})
\end{aligned}
\tag{19}
$$

where

$$
\Delta = (\Omega^2 - \omega^2 - \pi K_{\eta\gamma})^2 + \omega^2(2\alpha - \pi K_{\eta\eta})^2 \tag{20}
$$

Both the stationary moments of (14) and the spectral densities of (19) are exact solutions. The spectral density $\Phi_{11}(\omega)$ in (19) is in agreement with those obtained by [6] and [9]. It can be seen from (19) that conditions in (15) are also required for meaningful spectral densities.

Now consider a nonlinear system with cubic nonlinearity in both the damping and stiffness forces, namely,

$$
\ddot{q} + 2\alpha\dot{q} + \lambda\dot{q}^3 + \Omega^2 q + \delta q^3 = \dot{q}\eta(t) + q\gamma(t) + \xi(t) \tag{21}
$$

which can be replaced by an equivalent quasi-linear system as follows

$$
\ddot{q} + 2\alpha_{eq}\dot{q} + \Omega_{eq}^2 q = \dot{q}\eta(t) + q\gamma(t) + \xi(t) \tag{22}
$$

The equivalent linearization coefficients α_{eq} and Ω_{eq} can be calculated, according to (10), from

$$
\alpha_{eq} = \alpha + \frac{\lambda E[\dot{q}^4]}{2m_{22}}, \quad \Omega_{eq}^2 = \Omega^2 + \frac{\delta E[q^4]}{m_{11}} \tag{23}
$$

In (23), the second-order moments m_{11} and m_{22} are given in (14) with α and Ω replaced by α_{eq} and Ω_{eq} respectively, and the fourth-order moments $E[q^4]$ and $E[\dot{q}^4]$ can be obtained from the fourth-order moment equations derived from (22) by using Itô differential rule, similarly to the second-order moments. Since the fourth-order moment equations of the replacing quasi-linear system (22) are closed, namely, they only contain fourth-order and lower-order moments, they can be solved exactly.

Numerical calculations are carried out for system (21) with $\alpha = 0.4$, $\Omega = 6$, and two different sets of nonlinear coefficients λ and δ. One set of $\lambda = 0.1$ and $\delta = 5$ corresponds to a weak nonlinearity, while the other of $\lambda = 1$ and $\delta = 50$ represents a quite strong nonlinearity. The three white noises are assumed to be uncorrelated, and their spectral densities are $K_{\gamma\gamma} = 0.5$, $K_{\eta\eta} = 0.05$ and $K_{\xi\xi} = 1$, respectively. Figure 1 shows the analytical results calculated from the present procedure, as well as those obtained from Monte Carlo simulations. It is seen that they agree quite well.

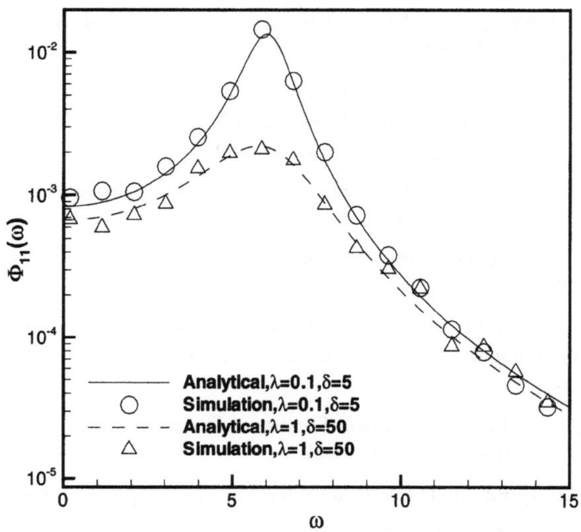

Figure 1. Spectral density of displacement response $X(t)$

4. Conclusion

For a randomly excited quasi-linear system, namely a linear system with multiplicative random excitations appearing in linear terms, exact statistical moments and spectral densities of responses can be obtained either analytically or numerically using Itô differential rule and an integral transformation, as presented in the paper. For the case of a nonlinear system, equivalent linearization procedure is applicable to replace the original system by a quasi-linear one, for which the exact solutions are available. The procedure is applicable to multi-degree-of-freedom systems.

References

[1] S. Bellizzi and R. Bouc, "Spectral Response of Asymmetrical Random Oscillators," *Probabilistic Engineering Mechanics*, vol. 11, pp. 51–59, 1996.

[2] S. Bellizzi and R. Bouc, "Analysis of Multi-Degree of Freedom Strongly Non-Linear Mechanical Systems With Random Input," *Probabilistic Engineering Mechanics*, vol. 14,

pp, 229–256, 1999.

[3] R. Bouc, "The Power Spectral Density of Response for a Strongly Non-Linear Random Oscillator," *Journal of Sound and Vibration*, vol. 175, pp. 317–331, 1994.

[4] G. Q. Cai and Y. K. Lin, "Response Spectral Densities of Strongly Nonlinear Systems Under Random Excitation," *Probabilistic Engineering Mechanics*, vol. 12, pp. 41–47, 1997.

[5] R. J. Chang, "A Practical Technique for Spectral Analysis of Nonlinear Systems Under Stochastic Parametric and External Excitations," *Journal of Vibration and Acoustics*, vol. 113, pp. 516-522, 1991.

[6] M. Dimentberg and Y. K. Lin, "An Exact Solution for Response Spectral Density of a Single-Degree-of-Freedom System Under Both Parametric and Additive White Noise Excitations," *Journal of Applied Mechanics*, vol. 69, pp, 399–400, 2002.

[7] K. Itô, "On Stochastic Differential Equations," *Memoirs of the American Mathematical Society*, vol. 4, pp. 289–302, 1951.

[8] K. Itô, "On a Formula Concerning Stochastic Differentials," *Nagoya Mathematical Journal*, vol. 3, pp. 55–65, 1951.

[9] S. Krenk, Y. K. Lin, and F. Rudinger, "Effective System Properties and Spectral Density in Random Vibration with Parametric Excitation," *Journal of Applied Mechanics*, vol. 69, pp. 161–170, 2002.

[10] Y. K. Lin and G. Q. Cai, *Probabilistic Structural Dynamics, Advanced Theory and Applications*, McGraw-Hill, New York, 1995.

[11] R. N. Miles, "An Approximate Solution for the Spectral Response of Duffing's Oscillator with Random Input," *Journal of Sound and Vibration*, vol. 132, pp. 43–49, 1989.

[12] R. N. Miles, "Spectral Response of a Bilinear Oscillator," *Journal of Sound and Vibration*, vol. 163, pp. 319–326, 1993.

[13] R. V. Roy and P. D. Spanos, Power Spectral Density of Nonlinear System Response: The Recursive Method, *Journal of Applied Mechanics*, vol. 60, pp. 358–365, 1993.

[14] C. Soize, Stochastic Linearization Method with Random Parameters for SDOF Nonlinear Dynamical Systems: Prediction and Identification Procedure, *Probabilistic Engineering Mechanics*, vol. 10, pp. 143–152, 1995.

[15] P. D. Spanos and M. G. Donley, Equivalent Statistical Quadratization for Nonlinear Systems, *Journal of Engineering Mechanics*, vol. 117, pp. 1289–1310, 1991.

[16] M. A. Tognarelli, J. Zhao, K. B. Rao, and A. Kareem, Equivalent Statistical Quadratization and Cubicization for Nonlinear Systems, *Journal of Engineering Mechanics*, vol. 123, pp. 512–523, 1997.

[17] W. Wedig, "Power Spectra of Nonlinear Dynamic Systems Analysis Via Generalized Hermite Polynomials, *Probabilistic Methods in Applied Physics, Lecture Notes in Physics*, eds. P. Kree and W. Wedig, Springer-Verlag, Berlin, pp. 310–326, 1995.

STOCHASTIC LOTKA-VOLTERRA SYSTEM

Mikhail F. Dimentberg
Mechanical Engineering Department
Worcester Polytechnic Institute
Worcester, MA, USA
diment@WPI.EDU

Abstract Generalized Lotka-Volterra LV model is considered, with the equation for preys reproduction containing quadratic term that accounts for interspecies competition and term with white-noise random variations of the preys reproduction factor which simulate the environmental variations. An exact solution is obtained for the corresponding Fokker-Planck-Kolmogorov equation for joint stationary probability density (p.d.f.) of the population sizes. Both population sizes are shown to be independent gamma-distributed stationary random processes. Increasing level of the environmental variations and/or approaching bifurcation point, which corresponds to extinction of predators, may lead to an intermittent behavior, whereby one or both population sizes experience very rare and violent short pulses or outbreaks while remaining on a very low level most of the time. This intermittency is described analytically in terms of the derived p.d.f. as well as by applying theory of excursions of random functions and by predicting p.d.f. of peaks and troughs in the predators population size.

Keywords: No keywords submitted.

1. Stochastic L-V Model in Population Dynamics and its Analysis

Random oscillations are considered for two interacting populations of the predator-prey or parasite-host type. The phenomenon is described by the following stochastic Lotka-Volterra (L-V) model [10, 12, 2, 3, 4]

$$\begin{aligned}\dot{u} &= -mu + k\beta uv, V\dot{v} = \alpha v[1+\xi(t)]\,[1+\xi(t)] - \beta uv - \gamma v^2,\\ &\langle \xi(t)\rangle = 0, \langle \xi(t)\xi(t+\tau)\rangle = D\delta(\tau)\end{aligned} \quad (1)$$

where $u(t)$ and $v(t)$ are population sizes of predators (or parasites) and preys (or hosts), respectively, whereas $\xi(t)$ is a zero-mean Gaussian random white noise in the Stratonovich sense with intensity D and ξ is Dirac delta-function;

N. Sri Namachchivaya and Y.K. Lin (eds.),
IUTAM Symposium on Nonlinear Stochastic Dynamics, 307-317.

all coefficients in (1) are positive. Special case $\gamma = 0$, $\xi(t) \equiv 0$ corresponds to the classical L-V model [10, 12], with its singular points being $u = \alpha/\beta$, $v = m/k\beta$. This classical model, however, is conservative and cannot "withstand" random environmental variations [5]. Thus, to study its steady-state random response some kind of "damping" should be accounted for. Such a damping is provided by the interspecies competition term γv^2 in the second (1). Upon addition of this term the stable equilibrium point shifts to $u_0 = (\alpha - \gamma m/k\beta)/\beta; v_0 = m/k\beta$ and becomes asymptotically stable (either focus or node), whereas an additional equilibrium state appears at $u^* = 0$, $v^* = \alpha/\gamma$. The latter is unstable if $v_0 < v^*$ or $\gamma < \gamma^* = \alpha k\beta/m$. At the bifurcation point $\gamma = \gamma^*$ these two equilibrium states merge, with the state $u^* = 0$, $v^* = \alpha/\gamma$ becoming stable for $\gamma > \gamma^*$. Physical meaning of these transformations is clear: beyond this transcritical bifurcation point growth of preys population is bounded by their interspecies competition rather than by activity of predators, who become extinct because of food shortage.

The random variations in the preys reproduction rate α simulate temporal variations of the environmental conditions (actually temporal random variations in the death rate of predators m were also accounted for in [10, 12]). These variations lead to random oscillations in the sizes of both populations, which can be analyzed by using transformations $x = lnu$, $y = lnv$. The SDEs (1) are reduced then to the form

$$\begin{aligned} \dot{x} &= \partial H/\partial y, \\ \dot{y} &= -\partial H/\partial x - g \cdot (\partial H/\partial y) + \alpha\xi(t) \end{aligned} \tag{2}$$

where

$$H(x, y) = k\beta \exp y - my + \beta \exp x - (\alpha - \gamma m/k\beta)x \tag{3}$$

and $g = \gamma/k\beta$. Note that H is not a Hamiltonian of the classical L-V system; it is reduced to the latter only when $\gamma m/k\beta << \alpha$.

Joint p.d.f. $p(x, y, t)$ of the transformed random state variables as governed by the SDE (2) satisfies the well-known Fokker-Planck-Kolmogorov (FPK) equation (3,4,5,6)

$$\frac{\partial p}{\partial t} = -\frac{\partial}{\partial x}\left(\frac{\partial H}{\partial y} \cdot p\right) + \frac{\partial}{\partial y}\left[\frac{\partial H}{\partial x} \cdot p + g \cdot \frac{\partial H}{\partial y} \cdot p\right] + \frac{\alpha^2 D}{2} \cdot \frac{\partial^2 p}{\partial y^2} . \tag{4}$$

Direct substitution shows, that the PDE (4) has the following exact stationary (independent of time) solution (3, 4, 5)

$$p(x, y) = C \exp\lfloor -\left(2\gamma/k\beta D\alpha^2\right) H(x, y)\rfloor . \tag{5}$$

Here C is a normalization constant, which may be found after returning to the original state variables u, v in the solution (5). The resulting joint stationary p.d.f. $w(u, v)$, as defined within the first quadrant of the (u, v)-plane, is then found to be a product of the individual one-dimensional p.d.f.s of the population sizes of $u(t)$ and $v(t)$ [2, 3, 4]

$$w(u,v) = p(\delta u/k) \cdot p(\delta v),$$

where

$$p(z) = z^{z_0-1}e^{-z}/\Gamma(z_0) \text{ and } \delta = 2\gamma/D\alpha^2. \tag{6}$$

Here the universal expression for the p.d.f. $p(z)$ of gamma-distribution is used where Γ is the gamma-function: whereas $z = \delta u/k$ and $z = \delta v$ for the stationary p.d.f.s of $u(t)$ and $v(t)$, respectively, with $z_0 = \delta u_0/k$ and $z_0 = \delta v_0$. Thus, both steady-state population sizes are independent gamma-distributed stationary random processes with mean values $u_0 = (\alpha - \gamma m/k\beta)/\beta = (\alpha/\beta)(1 - v_0/v^*)$; $v_0 = m/k\beta$ and standard deviations $\sigma_u = \sqrt{u_0 k/\delta}$, $\sigma_v = \sqrt{v_0/\delta}$. They do exist irrespectively of the intensity level D of the environmental variations provided that $\gamma < \gamma^*$. Thus, stationary environmental variations cannot by themselves lead to the extinction of the population(s) (of course, this conclusion is true only as long as the present ODE model of the population dynamics is valid, that is provided that both population sizes remain large compared with unity). On the other hand, if $\gamma > \gamma^*$ then the p.d.f. of predators or parasites population degenerates into the Dirac delta-function, implying extinction of predators, or parasites (whereas that of the preys or hosts remain on a relatively low level because of high self-limitation).

2. Intermittency in Population Sizes and its Analytical Description

However, an important change with the population(s) may appear indeed for sufficiently large Ds. Namely, if $\delta v_0 < 1(\delta u_0/k < 1)$, then $w_v(v)(w_u(u))$ has a singularity at $v = 0(u = 0)$. The singularity is integrable however, so that the corresponding stationary p.d.f. still does exist. The p.d.f. of predators is particularly prone to such a change in the vicinity of the bifurcation point, i.e., for small $1 - v_0/v^*$. Moreover, the intermittent behavior should be expected of $v(t)(u(t))$ whenever $\delta v_0 << 1(\delta u_0/k << 1)$, so that $\sigma_v >> v_0(\sigma_u >> u_0)$, with rare and short pulse-like intensive outbreaks in $v(t)(u(t))$ and low-level oscillations between the pulses. Whilst this expectation may be generally relevant in view of expressions for the first- and second-order moments of any stationary ergodic random process, for the present case it is also based on the following expression for a ratio of stay times above and below mean level z_0 for a gamma-distributed random process $z(t)$ obtained by integrating the p.d.f. (6):

$$\lambda = \frac{\text{Prob}(z > z_0)}{\text{Prob}(z < z_0)} = \frac{\int_{z_0}^{\infty} p(z)dz}{\int_0^{z_0} p(z)dz} = \frac{\Gamma(z_0, z_0)}{\Gamma(z_0) - \Gamma(z_0, z_0)}. \tag{7}$$

Here the function Γ that depends on two arguments is the incomplete gamma-function. As can be seen from the Fig. 1, this ratio of stay times approaches

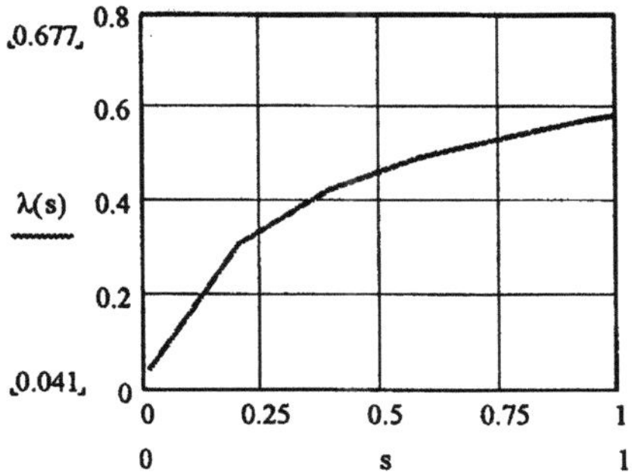

Figure 1. Ratio of stay times of a gamma-distributed process $z(t)$ above and below its mean level $s = z_0$.

zero with $z_0 \to 0$ and thus may be regarded as an index for the on-off intermittency. It may be added that this index of intermittency may also be applicable to the case of intermittency in a nonlinearly damped single-degree-of-freedom system with parametric excitation, as studied in [7, 9]. In that case however, the gamma-distribution governs not the response itself but its amplitude (according to the asymptotic stochastic averaging analysis); thus, the intermittency in the present case of the L-V system was called "quasistatic" [4].

Yet another index for the on-off intermittency in the predators population is an expected number $n_+(u)$ of upcrossings per unit time of a given arbitrary level u by $u(t)$ (crossings of the given level with positive derivative $\dot{u}$). Substituting (6), together with the first (1) into the basic relation [1] for upcrossings yields [2, 3, 4]

$$\begin{aligned} n_+(u) &= \int_0^\infty \dot{u} w(u, \dot{u}) d\dot{u} = \int_{v_0}^\infty k\beta u(v - v_0) w(u, v) dv = \\ &= (k\beta/\delta)(\delta v_0)^{\delta v_0} (\delta u/k)^{\delta u_0/k} v \\ &\quad \exp(-\delta v_0 - \delta u/k)[\Gamma(\delta v_0)\Gamma(\delta u_0/k)]^{-1} \end{aligned} \tag{8}$$

The latter formula provides just the expected frequency of oscillations if upcrossings of the mean or expected level ($u = u_0$) are considered:

$$\begin{aligned} n_+(u_0) &= n_\infty f(\delta v_0) f(\delta u_0/k), \\ n_\infty &= \lim_{\delta \to \infty} n_+(u_0) = \Omega/2\pi, \\ \Omega &= \beta(k u_0 v_0)^{1/2} \\ f(z) &= \left[(2\pi)^{\frac{1}{2}} z^{z-\frac{1}{2}} \exp(-z)\right] / \Gamma(z) \end{aligned} \tag{9}$$

(the quantity Ω can be clearly identified here as the systems natural frequency of small oscillations, that is, oscillations with small deviations of $u(t)$, $v(t)$ from their expected values). It may also be shown that the expected number $n_-(u)$ of downcrossings, or crossings of the given level with negative $\dot{u}$ is governed by the same expressions (8), (9). The expression (9) clearly indicates in particular, that the outbreaks in $u(t)$ may become very rare indeed when u_0 is small – particularly in the vicinity of the bifurcation point where the predators are close to extinction. The relations for excursions may be important for evaluation of the systems parameters from observed response data, which may be highly uncertain in case of field observations rather than that of the laboratory tests. In case of the intermittent behavior it would be particularly important to use the excursion data at sufficiently high levels, where adequate accuracy may be expected.

There is also another important potential application of the expressions for crossings. Namely, they may be used to predict certain first-passage probabilities according to the Poisson-pulse-train approximation for large deviations from mean: $P(T) = \exp(-nT)$ [1]. Here $P(T)$ is probability for zero crossings of a given level by $u(t)$ during time interval T, whereas $n = n_+(u) = n_-(u)$ is the expected number of crossings. Thus, so-called quasiextinction probability [6] can be predicted if a certain low level of u is selected. Probability for reaching a given large level of u during outbreak, or epidemic can also be estimated in this way. (In both applications Poisson approximation should be adequate as long as the events of interest are rare indeed). An example of potential specific applications is prediction of outbreaks in budworms in forests of Quebec, Canada, where seven outbreaks were reported since 1710 [11].

It may also be of interest to study peaks and/or troughs of $u(t)$, so as to predict (probabilistically) level of the intermittent outbreaks in the parasites population size together with their frequency. In view of the first (1) peak or trough of $u(t)$ corresponds to a crossing of the level v_0 by $v(t)$. Denoting the corresponding values of $u(t)$ by $U(t)$, we may then use (2) for $H(U, v_0)$ as an implicit definition of U in terms of H. Thus, returning in (2) to the original variables u, v and imposing relations $u = U$, $v = v_0$ yields then

$$H = k\beta v_0 - \min \ln v_0 + \beta U - (\alpha - \gamma m/k\beta) \ln U. \qquad (10)$$

Consider now the stationary solution (5) to the FPK (4), which may be regarded as a probability density $p(H)$ of H. Introducing nonlinear transformation (10) we may obtain p.d.f. of peaks/troughs $p(U)$ in terms of $p(H)$ using basic formula for such a transformation

$$\begin{aligned} p(U) &= p(H(U)) \mid dH/dU \mid \\ &= C\beta \mid 1 - u_0/U \mid v_0^{\delta v_0} \exp(-\delta v_0) U^{\delta u_0/k} \exp(-\delta U/k). \quad (11) \end{aligned}$$

Here peaks and troughs correspond to ranges $U > u_0$, V, and $U < u_0$, respectively. Applying normalization condition for this p.d.f. within each of these ranges and using relation for the incomplete gamma-function $\Gamma(z_0+1, z_0) = z_0\Gamma(z_0, z_0) + z_0^{z_0}\exp(-z_0)$ yields finally

$$p(U) = (\delta/k) \mid 1 - u_0/U \mid (U/u_0)^{\delta u_0/k} \exp\left[-(\delta/k)(U - u_0)\right]. \quad (12)$$

It seems appropriate now to present results of direct numerical simulations of the SDEs (1) so as to illustrate qualitative changes in the populations behavior with increasing γ up to its bifurcation value γ^*. This is done in Figures 2-5 for the following values of parameters of the "classical" L-V model:

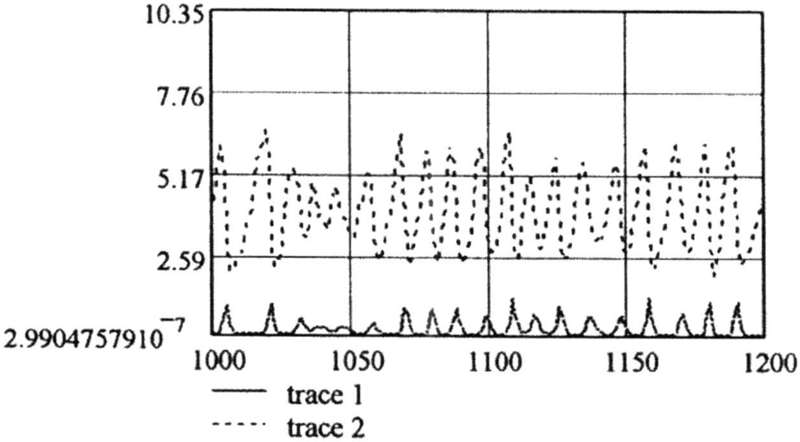

Figure 2. Samples of $u(t)$ (trace 1) and $v(t)$ (trace 2) for $m = 4$, $\alpha = 0.25$, $k = \beta = 1$, $\gamma = 0.005$.

$m = 4$, $\alpha = 0.25$, $k = \beta = 1$ so that $\gamma^* = 0.0625$. The case in Fig. 2 may be regarded as a "quasihamiltonian" one since $v_0/v^* = 0.08 << 1$ for $\gamma = 0.005$ and $H(x, y)$ is rather close to the Hamiltonian of the classical L-V system as defined by (2) with $\gamma = 0$; in particular, shift of the expected predators population size from its "classical" value is just 8%.

The case represented in Figure 3, where $v_0/v^* = 0.4$ for $\gamma = 0.025$, may be regarded as an intermediate one. On the other hand, the on-off intermittency in predators population size $u(t)$ is clearly seen in Figure 4 where $v_0/v^* = 0.8$ for $\gamma = 0.05$ and especially in Figure 5 where $v_0/v^* = 0.96$ for $\gamma = 0.06$ and the outbreaks in $u(t)$ are seen to be less frequent than in Figure 4 — as it should be expected in view of the relations (8), (9) for excursions.

These examples relate to a system with relatively high death rate of predators and low reproduction rate of preys. Therefore this system is prone to intermittency mainly in the predators populations. Samples of $v(t)$ in the Figures 4 and 5 look rather common indeed, although sharp drops can be clearly

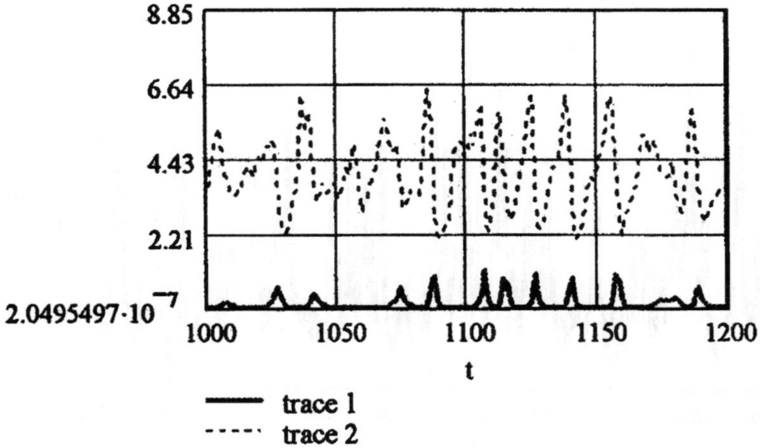

Figure 3. Samples of $u(t)$ (trace 1) and $v(t)$ (trace 2) for $m = 4$, $\alpha = 0.25$, $k = \beta = 1$, $\gamma = 0.005$.

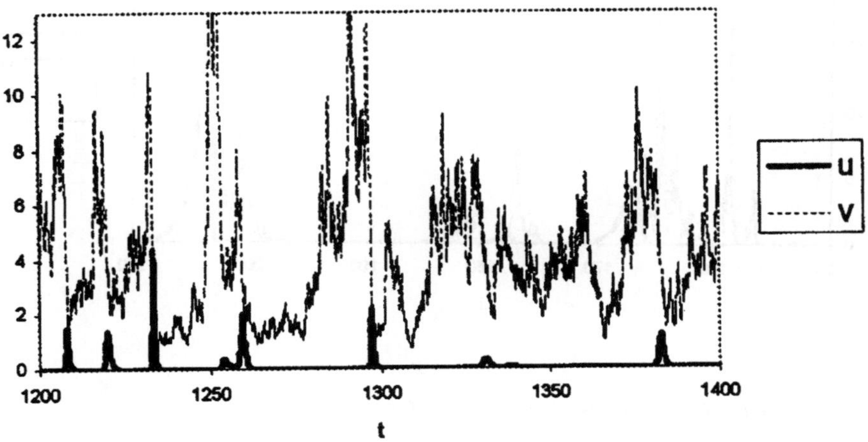

Figure 4. Samples of $u(t)$ and $v(t)$ for $m = 4$, $\alpha = 0.25$, $k = \beta = 1$, $\gamma = 0.05$.

seen which are coherent with outbreaks in $u(t)$. On the other hand, intermittency in both predators and preys populations is possible as well, as predicted by the present theory and illustrated by simulation results in Figure 6 for the case $m = \alpha = k = \beta = 1$, $\gamma = 0.5$.

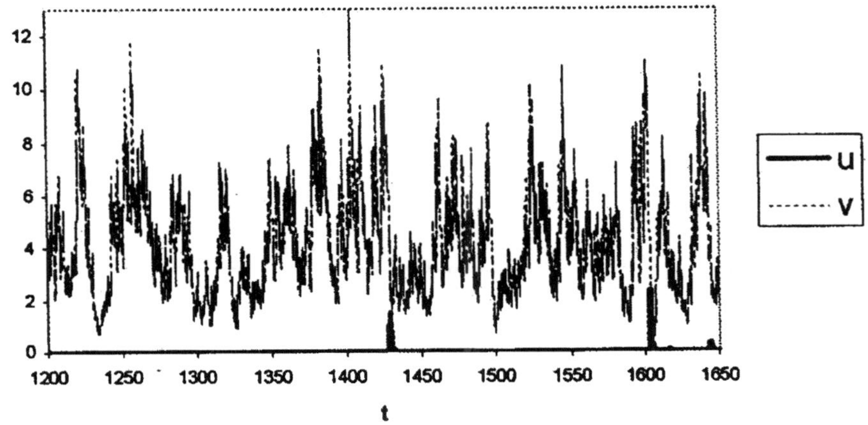

Figure 5. Samples of $u(t)$ and $v(t)$ for $m = 4$, $\alpha = 0.25$, $k = \beta$, $V = 1$, $\gamma = 0.6$.

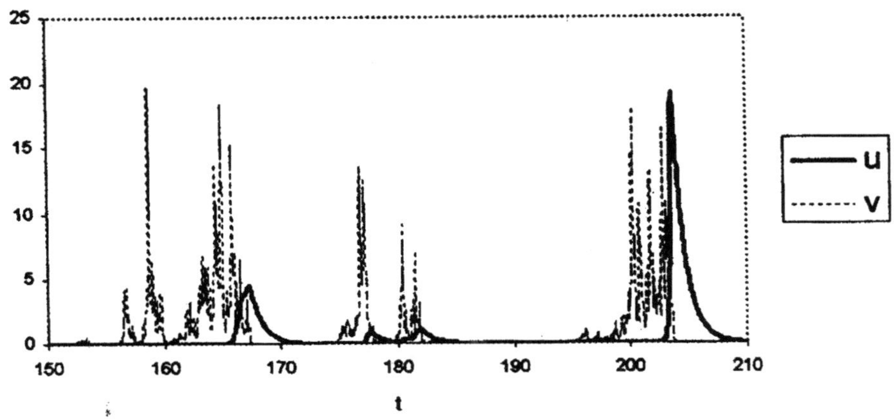

Figure 6. Samples of $u(t)$ and $v(t)$ for $m = \alpha = k = \beta = 1$, $\gamma = 0.5$.

3. Expected Time for Quasiextinction: First-Passage Problem

Yet another analytical study can be performed of the first-passage problem for the system (1) for the quasihamiltonian case, where γ and D are proportional to a small parameter (and the function $H(t)$ is close to the Hamiltonian of the classical L-V system). The assumption of independent crossings of a given level, or Poisson approximation for quasiextinction probability may become inaccurate in this case except for very high excursion levels. The reason is that the excursions may appear in clusters, which is a well-known trend for a lightly damped SDOF linear oscillator [2]. An alternative approach is to apply the asymptotic method of quasiconservative averaging as described originally in [8] to derive a single first-order SDE for the slowly varying $H(t)$ and solve the first-passage problem for the reduced system. To this end consider $H(t)$ as a state variable and use the SDE (1) to obtain

$$\dot{H} = (\partial H/\partial x)\dot{x} + (\partial H/\partial y)\dot{y} = (\partial H/\partial y)[-\gamma/k\beta + \alpha \cdot \xi(t)]. \quad (13)$$

Applying averaging over the natural period $T_0(H)$ of the corresponding Hamiltonian system, i.e., one with $\gamma = 0$, $D = 0$ [2, 8] yields the approximate single "shortened" SDE

$$\begin{aligned} \dot{H} &= (\gamma/k\beta)S(H)/T_0(H) + \left[D\alpha^2 S(H)/T_0(H)\right]^{\frac{1}{2}} \varsigma(t) \\ T_0(H) &= \oint (\partial H/\partial y)^{-1} dx, S(H) = \oint (\partial H/\partial y) dx. \end{aligned} \quad (14)$$

Here $\varsigma(t)$ is a zero-mean white noise of unit intensity and the integrals are calculated along the closed-loop trajectories of the corresponding Hamiltonian system.

The process $H(t)$ as governed by the SDE (14) has an exponential stationary p.d.f. $p(H)$ which may be obtained as the solution to the corresponding FPK equation; this p.d.f. coincides with one obtained through appropriate change of variables in the exact solution (5) for $p(x, y)$. The SDE (14) may also be used to solve the following first-passage problem: find an expected time $T(H)$ for reaching a given level H^* selected as that of quasiextinction of $u(t)$ or $v(t)$. Here H is a value of $H(t)$ at the initial time instant, where $H_0 \leq H \leq H^*$ and $H_0 = H(x_0, y_0)$ is value of H at the systems equilibrium state $x_0 = lnu_0$, $y_0 = lnv_0$. The expected quasiextinction time $T(H)$ satisfies a second-order inhomogeneous ODE which has a well-known quadrature solution [2]. This solution for the specific case of the SDE (14) may be written as [2, 3]

$$\begin{aligned} T(H) &= \frac{2}{D\alpha^2} \int_H^{H^*} \frac{\exp(\delta H'/k\beta)}{S(H')} \\ &\quad \left\{ \int_{H_0}^{H'} T_0(H'') \exp(-\delta H''/k\beta) dH'' \right\} \cdot dH' \\ &\cong \frac{2}{D\alpha^2} \cdot \int_H^{H^*} T_0(H'') \exp(-\delta H''/k\beta) dH'' \\ &\quad \cdot \int_{H_0}^{H^*} \frac{\exp(\delta H'/k\beta)}{S(H')} dH' \\ &\cong (D\alpha^2 k\beta^2)[T_0(H_0)/S(H^*)] \cdot \exp[(\delta/k\beta)(H^* - H_0)] \end{aligned} \quad (15)$$

Here $T_0(H_0) = 2\pi/\Omega$, $\Omega = \sqrt{am}$ is a natural period of small-amplitude oscillations around mean values of population sizes. The approximate versions of (15) are valid for large values of $\delta/k\beta$, in which case $T_0(H'')$ and $S(H')$ are slowly varying compared with $\exp(-\delta H''/k\beta)$ and $\exp(\delta H'/k\beta)$, respectively, these exponent having sharp peaks at $H'' = H_0$ and $H' = H^*$. The final approximate expression for $T(H)$ is seen to be independent of H, i.e., of the initial value of $H(t)$; this is usually the case whenever stationary p.d.f. of the response is attained within the given domain long before of the expected time of the systems escape.

It may be noted in conclusion that the above asymptotic quasiconservative analysis may be extended to the cases of more complicated interactions between predators and preys than in the SDEs (1). Actually the product-type nonlinearities represented by terms $k\beta uv$ and βuv account just for the expected number of encounters of predators with preys as the main factor of their interaction. Whilst this may very well be the case indeed for parasite-host pairs, more complicated pursuit/evasion activity of predators and preys during their encounters may be of importance in some cases. This can be accounted for by replacing the above interaction terms by $kuV(v)$ and $uV(v)$, respectively, where $V(v)$ is a so-called functional response function which approaches a certain finite limit with $v \to \infty$ [12]. The approximate quasiconservative stochastic averaging analysis for the generalized L-V system can be performed for such cases provided that $V(v)$ is close to a linear function (difference is proportional to small parameter). The corresponding expressions for $p(H)$ and $T(H)$ can be found in [2, 3].

References

[1] H. Cramer, M. R. Leadbetter, *Stationary and Related Stochastic Processes* (John Wiley & Sons, New York, 1967)

[2] M. F. Dimentberg, *Statistical Dynamics of Nonlinear and Time-Varying Systems* (Research Studies Press, Taunton, 1988).

[3] M. F. Dimentberg, *Random Processes in Dynamic Systems with Variable Parameters* (Nauka, Moscow, 1989)(in Russian).

[4] M. F. Dimentberg, Lotka-Volterra system in a random environment, *Physical Review E*, vol. 65, 036204 (2002).

[5] C. W. Gardiner, *Handbook of Stochastic Methods* (Springer-Verlag, Berlin, 1985).

[6] L. R. Ginzburg, L. B. Slobodkin, K. Johnson, and A. G. Bindman, Quasiextinction probabilities as a measure of impact on population growth. *Risk Analysis* vol. 2, pp. 171–181, 1982.

[7] R. Ibrahim, Stabilization and stochastic bifurcation, with application to nonlinear ocean structures. In: *Stochastically Excited Nonlinear Ocean Structures* (Editors M.F.Shlesinger, T.Swean, World Scientific, Singapore, 1998), pp. 1–52.

[8] R. Z. Khasminskii, On the behavior of a conservative system with small friction and small random noise (in Russian). *Prikladnaya Matematika i Mechanika*, vol. 28, no. 5, 1964.

[9] P. S. Landa, et al., On-off intermittency in a pendulum with a randomly vibrating suspension axis. *Chaos, Solitons & Fractals* **9**, pp. 157–159, 1998.

[10] R. M. May, *Stability and Complexity in Model Ecosystems* (Princeton University Press, Princeton, 1973).

[11] T. Royama, *Analytical Population Dynamics*, (Chapman & Hall, London, UK, 1992).

[12] Ju. M. Svirejev, D. O. Logofet, *Stability of Biological Communities* (Nauka, Moscow, 1978)(in Russian).

DYNAMIC SYSTEMS DRIVEN BY POISSON/LÉVY WHITE NOISE

Mircea Grigoriu and Gennady Samorodnitsky
Cornell University, Ithaca, NY 14853–3501
mdg12@cornell.edu

Abstract Properties are determined for the state of some linear and nonlinear dynamic systems driven by Poisson and Lévy white noise. The state of dynamic systems with Poisson or Lévy white noise is a semimartingale with samples that are right continuous with left limits so that classical methods of random vibration developed for continuous semimartingales do not apply. It is shown that the Itô formula for semimartingales can be used to develop equations for the characteristic function of the state of a broad class of systems driven by non-Gaussian noise and assess the stability of the stationary states of these systems. However, it is not possible to obtain equations for the characteristic function of the state of some nonlinear systems with additive Lévy noise. For these systems a method is developed for finding the tail properties of the distribution of the stationary state. Monte Carlo simulation and Hill estimates are used to confirm analytical results.

Keywords: Itô formula, Lévy processes, Lyapunov exponent, non-Gaussian white noise, semimartingales, stochastic processes, stochastic stability

1. Introduction

A main objective of the classical methods for random vibration is the determination of properties of the state $\boldsymbol{X}(t)$, $t \geq 0$, of linear and nonlinear dynamic systems driven by Gaussian white noise. The probability law of the state $\boldsymbol{X}$ of a linear system is completely defined by its second moment properties since $\boldsymbol{X}$ is a Gaussian process. However, the second moment properties provide only a partial characterization for the state $\boldsymbol{X}$ of nonlinear systems, and a broad range of methods and heuristic approximations have been developed for characterizing $\boldsymbol{X}$ [13]. Methods are also available for assessing the stability of the stationary solutions of nonlinear systems with multiplicative Gaussian noise [2, 8].

This paper develops methods for (1) finding properties of the state $\boldsymbol{X}$ of linear and nonlinear systems driven by Poisson and Lévy white noise and (2) as-

N. Sri Namachchivaya and Y.K. Lin (eds.),
IUTAM Symposium on Nonlinear Stochastic Dynamics, 319-330.

sessing the stability of the stationary states of these systems. The driving noise processes are defined as the formal derivatives of compound Poisson and Lévy processes. The analysis is based on the Itô formula for semimartingales, properties of Lévy processes, Monte Carlo simulation, and Hill estimates for the tail of the distribution of $\boldsymbol{X}$. This version of Itô's formula simplifies to a more familiar form for systems driven by Poisson white noise ([5], Example 4.51, p. 274). Numerical results are used to demonstrate the application of the proposed methods and illustrate properties of the state $\boldsymbol{X}$ for some dynamic systems with additive and multiplicative non-Gaussian noise.

2. Why Non-Gaussian White Noise?

Suppose that $\boldsymbol{X}(t) \in \mathbb{R}^d$, $t \geq 0$, is the solution of the stochastic differential equation

$$d\boldsymbol{X}(t) = \boldsymbol{a}(\boldsymbol{X}(t))dt + \boldsymbol{b}(\boldsymbol{X}(t))d\boldsymbol{B}(t), \quad t \geq 0, \tag{1}$$

where $\boldsymbol{a}$ and $\boldsymbol{b}$ are $(d, 1)$ and (d, d') matrices with entries depending on the state $\boldsymbol{X}$ and $\boldsymbol{B}(t)$, $t \geq 0$, denotes an $\mathbb{R}^{d'}$-valued Brownian motion. If the entries of $\boldsymbol{a}$ and $\boldsymbol{b}$ are, for example, uniform Lipschitz functions, the solution $\boldsymbol{X}$ of (1) exists and is unique, and $\boldsymbol{X}$ is an $\mathbb{R}^d$-valued continuous semimartingale called diffusion process ([5], Section 4.7.1.1).

The diffusion process $\boldsymbol{X}$ defined by (1) has been used to model the state of a broad range of mechanical, biological, physical, and other systems. However, there are phenomena that cannot be modeled by diffusion processes, but can be described by the solution of (1) in which the Gaussian noise is replaced by a Lévy noise or by the sum of a Gaussian process and a compound Poisson process, for example, the time series in Figure 1. The jumps of the Lévy and

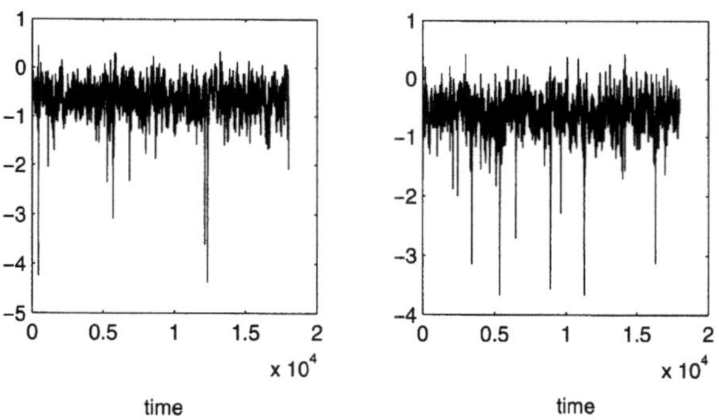

Figure 1. Wind pressures recorded at corners and eaves

compound Poisson processes produce jumps in the state $\boldsymbol{X}$ of the type seen in Figure 1. Other examples can be found in [3] and [5] (Section 7.4.1).

3. Linear Systems

Let X be the solution of the stochastic differential equation

$$dX(t) = -\rho X(t-)dt + dS(t), \quad t \geq 0, \quad X(0) = x \in \mathbb{R}, \tag{2}$$

driven by $S(t) = \sigma C(t)$ or $S(t) = \sigma L_\alpha(t)$, where $C(t) = \sum_{k=1}^{N(t)} Y_k$, $t \geq 0$, is a compound Poisson process, N denotes a Poisson process with intensity $\lambda > 0$, Y_k, $k = 1, 2, \ldots$, are independent identically distributed random variables, $L_\alpha(t)$, $t \geq 0$, is an α-stable process with independent increments $dL_\alpha(t)$ following the distributions $dL_\alpha(t) \sim S_\alpha((dt)^{1/\alpha}, 0, 0)$, $0 < \alpha \leq 2$, ([4], Section 3.5.3, [12], Section 1.1), and σ is a constant.

Suppose first that the input in (2) is $S(t) = \sigma C(t)$. Figure 2 shows samples of X for $\rho = 1$, $\sigma = 1$, several values of λ, and $Y_k \sim N(0, \gamma)$ such that $\lambda\gamma =$

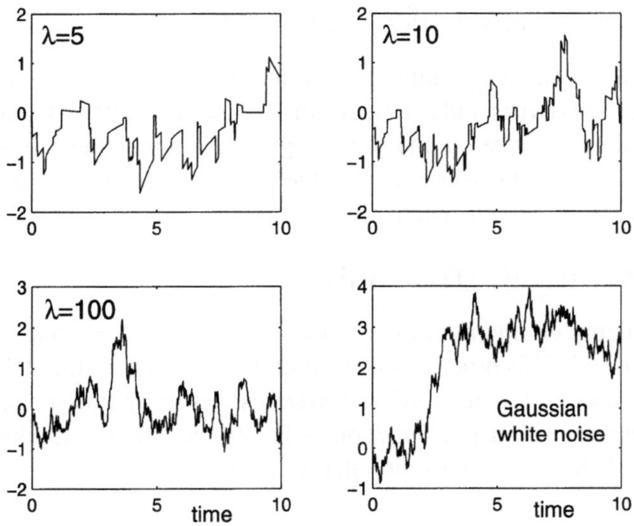

Figure 2. Samples of X for $\rho = \sigma = 1$, $Y_k \sim N(0, \gamma)$, and $\lambda\gamma = 1$.

1. Although the driving noise process $\sigma C(t)$ has the same second moment properties as a Brownian motion for each λ, the samples of X depend strongly on the value of λ. The process $S(t) = \sigma C(t)$ experiences large infrequent jumps for small values of λ. For larger values of λ the jumps of S are more frequent and smaller.

The characteristic function $\varphi(u;t) = E[\exp(iuX(t))]$ of $X(t)$ satisfies the equation $\partial\varphi/\partial t = -\rho u \partial\varphi/\partial u + \lambda\varphi[\varphi_y(u\sigma) - 1]$ with the initial and boundary conditions $\varphi(u;0) = e^{iux}$, $\varphi(0;t) = 1$, and $\varphi(u;t) \to 0$ as $|u| \to \infty$, where $\varphi_y(v) = E[\exp(ivY_1)]$ denotes the characteristic function of Y_1. That φ satisfied the above partial differential equation follows from the complex-valued version of the Itô formula for semimartingales ([5], Section 7.2.2).

Suppose now that $S = \sigma L_\alpha$. A partial differential equation can be developed for the characteristic function of X by using Itô's formula for semimartingales. Alternatively, properties of α-stable random variables show that $X(t)$ is an α-stable random variable with skewness 0, location 0, and scale $\left[(1-e^{-\rho\alpha t})/(\rho\alpha)\right]^{1/\alpha}$ for $X(0) = 0$ and $\sigma = 1$ ([4], Section 3.5.4). Hence, the input L_α and the output X have the same heavy tails characterized by ([12], Property 1.2.15, p. 16)

$$P(X(t) > x), P(L_\alpha(t) > x) \sim x^{-\alpha}, \quad \text{as } x \to \infty. \tag{3}$$

4. Nonlinear Systems

Let X be the solution of the stochastic differential equation

$$dX(t) = a(X(t-))dt + b(X(t-))dS(t), \quad t \geq 0, \tag{4}$$

where S is a real-valued semimartingale and the functions a, b are such that the solution of (4) exists and is unique. In all cases the drift a is assumed to be a polynomial in the state. In the following section b is a linear function of X and S is a compound Poisson process. Then b and S are taken to be a constant and an α-stable process, respectively.

4.1 Multiplicative noise

The geometric compound Poisson process is defined and some of its properties established. This process is used to find conditions for the stability of the stationary solution of some nonlinear systems driven by Poisson white noise. Then a phase transition phenomenon is illustrated for two randomized versions of the Verhulst model for population growth.

Example 1. Let X be the solution of

$$dX(t) = cX(t-)dt + \sigma X(t-)dC(t), \quad t \geq 0, \tag{5}$$

where $X(0) = x \in \mathbb{R}$ denotes the initial state and $1 + \sigma Y_1 > 0$ a.s. Figure 3 shows samples of X for $X(0) = x = 1$, two values of λ, and random variables Y_k uniformly distributed in $(-1/2, 1/2)$. The sample with $\lambda = 5$ is characterized by an increasing trend while the sample for $\lambda = 30$ approaches 0 as time increases.

The Itô formula for $\mathbb{R}^d$-valued semimartingales applied to $\exp(cX_1(t) + X_2(t))$ shows that $X(t) = X(0)\exp(tZ(t))$ is the solution of (5), where $Z(t) = c + C^*(t)/t$, $C^*(t) = \sum_{k=1}^{N(t)} \ln(1+\sigma Y_k)$, $dX_1(t) = dt$, and $dX_2(t) = dC^*(t)$. Since $1+\sigma Y_1 > 0$ a.s., we have $\lim_{t\to\infty} Z(t) = c + \lambda E[\ln(1+\sigma Y_1)]$

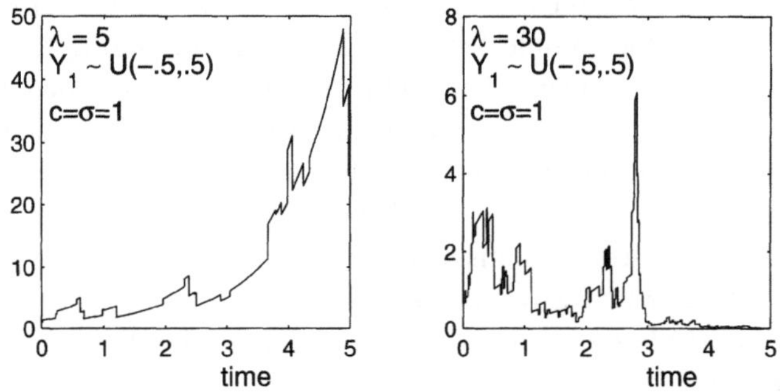

Figure 3. Two samples of X for $\lambda = 5$ and $\lambda = 30$ starting at $X(0) = 1$.

so that $X(t) \to 0$ a.s. as $t \to \infty$ if $c + \lambda E[\ln(1 + \sigma Y_1)] < 0$. If $c + \lambda E[\ln(1 + \sigma Y_1)] > 0$, the state converges to infinity as time increases. These observations are consistent with the samples in Figure 3. The values of $c + \lambda E[\ln(1 + \sigma Y_1)]$ with $\sigma = c = 1$ are 0.79 and -0.24 for $\lambda = 5$ and $\lambda = 30$, respectively.

Example 2. Let X be defined by the stochastic differential equation

$$dX(t) = (\beta X(t-) - X(t-)^m)dt + X(t-)dC(t), \quad t \geq 0, \tag{6}$$

where $m \geq 2$, $1 + Y_1 > 0$ a.s., and $\alpha, \sigma, \beta = \alpha + \sigma^2/2$ are some constants.

Figure 4 shows two samples of X corresponding to $\lambda = 30$, $\sigma = 1$, $X(0) =$

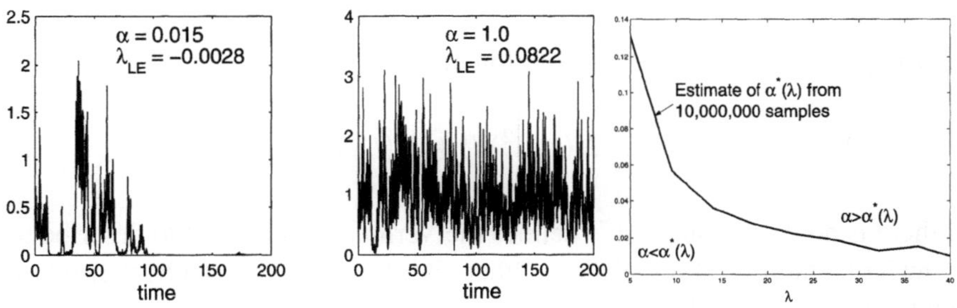

Figure 4. Two samples of X for $X(0) = 0.5$ and two values of α and boundary of the stability region

0.5, $Y_1 \sim U(-a, a)$, $a = \sigma\sqrt{3/\lambda}$, and two values of α. Since $Y_1 \sim U(-a, a)$ and $a = \sigma\sqrt{3/\lambda}$, the driving noise has the same second moment properties as a Brownian motion scaled by σ for each value of λ. The Poisson driving noise is scaled such that it has the same second moment properties as the standard Brownian motion. The sample of X for $\alpha = 0.015$ approaches 0 as time increases suggesting that the trivial solution is stable asymptotically as $t \to \infty$. On the other hand, the sample for $\alpha = 1.0$ seems to belong to a stationary process with non-zero mean. The right plot in Figure 4 shows the regions $\alpha < \alpha^*(\lambda)$ and $\alpha > \alpha^*(\lambda)$ in which the trivial and non-trivial stationary solutions are stable for values of λ in the range $[5, 40]$. This plot suggests that $\alpha^*(\lambda)$ approaches 0 as $\lambda \to \infty$, that is, when C approaches a Brownian motion. This observation is consistent with a result in [1] for a system with state defined by (6) with $\sigma B(t)$ in place of $C(t)$.

Suppose that (6) has a stationary solution X_s and let $\tilde{X} = X - X_s$. Then $\tilde{X}$ is defined approximately by $d\tilde{X}(t) = \zeta(t-)\tilde{X}(t-)dt + \tilde{X}(t-)dC(t)$, where $\zeta(t) = \beta - mX_s(t-)^{m-1}$. According to Example 1 we have $\tilde{X}(t) = \tilde{X}(0)\exp(t\tilde{Z}(t))$, where $C^*(t) = \sum_{k=1}^{N(t)} \ln(1 + Y_k)$ and $\tilde{Z}(t) = \left(\int_0^t \zeta(s-)ds + C^*(t)\right)/t$. If $\lim_{t\to\infty} \tilde{Z}(t) < 0$ a.s., X_s is stable as $t \to \infty$. If $X_s = 0$, then $\tilde{Z}(t) = \beta + C^*(t)/t$ so that $\lim_{t\to\infty} \tilde{Z}(t) = \beta + \lambda E[\ln(1 + Y_1)]$ since $C^*(t)/t \to \lambda E[\ln(1 + Y_1)]$ a.s. as $t \to \infty$. Hence, the trivial stationary solution is stable a.s. for $\alpha < \alpha^*(\lambda) = -(\sigma^2/2 + \lambda E[\ln(1 + Y_1)])$. If a non-trivial stationary solution X_s is an ergodic process, then $\tilde{Z}(t) = \beta - m\frac{1}{t}\int_0^t X_s(u-)^{m-1}du + C^*(t)/t$ converges a.s. to $\beta - mE[X_s(t-)^{m-1}] + \lambda E[\ln(1 + Y_1)]$ as $t \to \infty$. This solution is stable a.s. for $\alpha > \alpha^*(\lambda)$ since $E[X_s(t-)^{m-1}] = \beta + \lambda E[\ln(1+Y_1)]$. The Itô formula and considerations as in [1] give the expression $\lambda_{LE} = \lim_{t\to\infty} \frac{1}{t} \ln \| \tilde{X}(t) \| = \alpha + \frac{\sigma^2}{2} + \lambda E[\ln(1 + Y_1)]$ for the Lyapunov exponent ([5], Example 8.56, p. 660). The values of λ_{LE} in Figure 4 are consistent with the behavior of the samples in this figure. For example, a negative value of λ_{LE} implies that the trivial stationary solution is stable a.s.

Example 3. Let X be the solution of

$$dX(t) = (\rho X(t-) - X(t-)^2)dt + X(t-)dS(t), \quad t \geq 0. \tag{7}$$

If there is no driving noise, (7) defines the Verhulst model used to describe the growth of a biological population. Otherwise, this equation can be viewed as a randomized version of the Verhulst model.

The solution of (7) is $x(t) = x(0)e^{\rho t}\left[1 + x(0)(e^{\rho t} - 1)/\rho\right]^{-1}$ for $S = 0$, and can be interpreted as a measure of the population size, for example, the average population per unit area. Positive values of ρ describe a favorable environment while $\rho < 0$ corresponds to a hostile environment ([7], Section 6.4). The steady-state state $x_s(t) = 0$ for $\rho < 0$ becomes unstable at $\rho = 0$ and a new steady-state state $x_s(t) = \rho$ emerges. The corresponding degenerate densities of x_s are $f_s(x) = \delta(x)$ and $f_s(x) = \delta(x - \rho)$ for $\rho < 0$ and $\rho > 0$, respectively. The system undergoes a phase transition at $\rho = 0$.

The process X defined by (7) with $S = \sigma B$ is ([9], Section 4.4)

$$X(t) = \frac{X(0)\exp[(\rho - \sigma^2/2)t + \sigma B(t)]}{1 + X(0)\int_0^t \exp[(\rho - \sigma^2/2)s + \sigma B(s)]ds}. \tag{8}$$

The density f_s of the stationary solution X_s can be obtained from the Fokker-Planck equation corresponding to (7) with $S(t) = \sigma B(t)$. If $\rho > \sigma^2/2$, the density f_s exists and is $f_s(x) = c x^{2(\rho/\sigma^2 - 1)} \exp(-2x/\sigma^2)$, $x > 0$, where $c > 0$ is a constant. If $\rho < \sigma^2/2$, then $f_s(x) = \delta(x)$. The mean and the variance of X_s for $\rho > \sigma^2/2$ are $E[X_s(t)] = \rho$ and $\text{Var}[X_s(t)] = \rho\sigma^2/2$, so that the expectation of the non-trivial stationary solution coincides with the steady-state solution x_s of the deterministic Verhulst model. Also, $X_s = x_s = 0$ for $\rho < \sigma^2/2$. Figures 5 and 6 show samples of X and the marginal density of the stationary solution X_s for $\sigma = 1$ and $\rho = 0$, 3/4, and 2. The samples are consistent with the functional form of the stationary density f_s. Note also that the functional form of the stationary density f_s experiences abrupt changes, referred to as transitions, at $\rho = \sigma^2/2$ and $\rho = \sigma^2$. In contrast, the deterministic Verhulst model has a single transition at $\rho = 0$. The transitions of f_s represent qualitative changes in the properties of the state of the randomized Verhulst model. They are caused by random fluctuations in the environment since

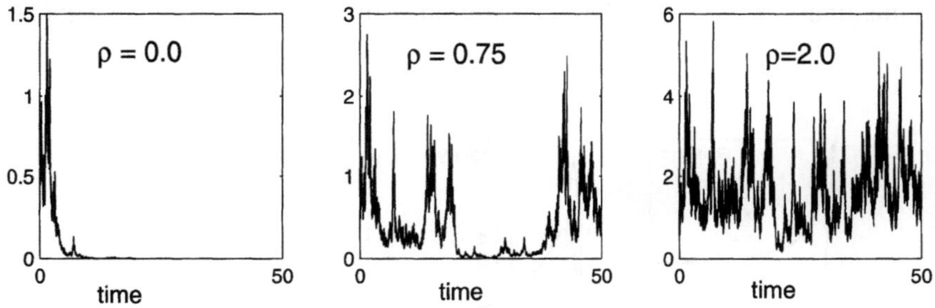

Figure 5. Samples of X for $\sigma = 1$ and $\rho = 0$, 3/4, and 2

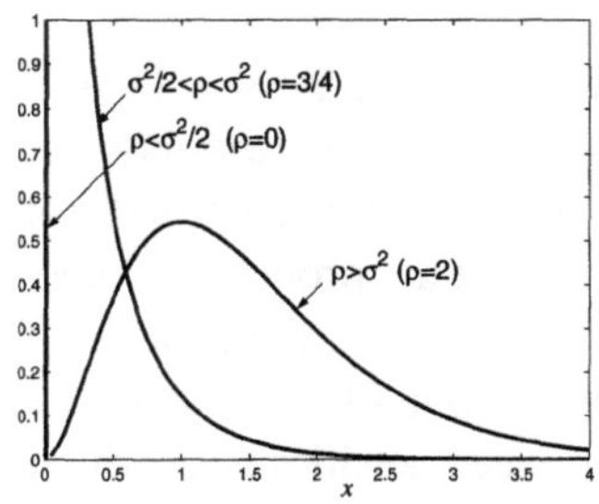

Figure 6. The density of X for $\sigma = 1$ and $\rho = 0$, 3/4, and 2

$E[\rho dt + \sigma dB(t)]/dt = \rho$ has the same value as in the deterministic Verhulst model. For this reason the qualitative changes exhibited by f_s are referred to as noise-induced transitions ([7], Section 6.4).

Suppose now that $S(t) = C(t)$ in (7). The characteristic function $\varphi(u;t) = E[\exp(iuX(t))]$ of $X(t)$ satisfies the equation

$$\frac{\partial \varphi}{\partial t} = \rho u \frac{\partial \varphi}{\partial u} + i \frac{\partial^2 \varphi}{\partial u^2} + \lambda \left[\int_{\mathbb{R}} \varphi(u(1+y);t) dF_y(y) - \varphi(u;t) \right],$$

where F_y denotes the distribution of the jumps Y_k of the compound Poisson process C. This equation results from the Itô formula for semimartingales applied to $\exp(iuX(t))$. A finite difference approximation has been used to find the solution of this equation. Figure 7 shows in the left two plots the real and imaginary parts of the characteristic function $\varphi_s(u) = \lim_{t\to\infty} \varphi(u;t)$ of

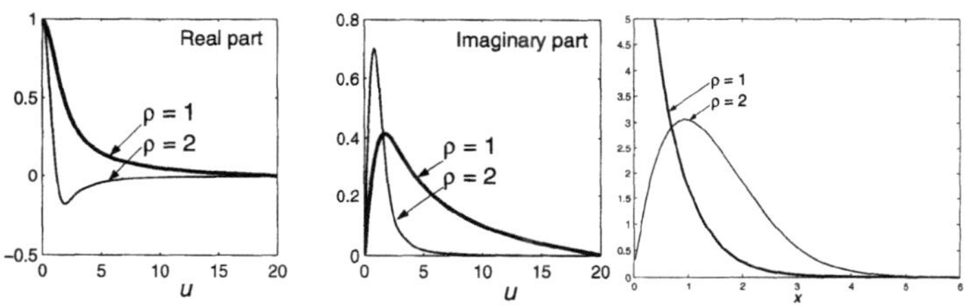

Figure 7. Real and imaginary parts of characteristic function of X_s and its density

the stationary solution X_s for $\lambda = 1$, $Y_1 \sim U(-a, a)$, $a = 0.5$, and $\rho = 1, 2$. The right plot in the figure gives the corresponding densities f_s of X_s. There is a qualitative difference between the functional form of the densities f_s for $\rho = 1$ and $\rho = 2$. The stationary process X_s spends most time in a small vicinity of zero for $\rho = 1$. For $\rho = 2$ the process X_s oscillates about a non-zero value in a vicinity of the mode of f_s. The precise values of ρ at which the stationary density f_s experience transitions is difficult to find in this case since the expression of f_s is not known so that both φ_s and f_s have to be obtained numerically.

4.2 Additive noise

Suppose that X is the solution of the stochastic differential equation

$$dX(t) = -(X(t-) + X(t-)^3)dt + \sigma dL_1(t), \quad t \geq 0, \tag{9}$$

where L_1 is the α-stable process L_α with $\alpha = 1$. The characteristic function of the increments $dL_1(t)$ of L_1 is $E[\exp(iu)dL_1(t)] = \exp(-|u|dt)$ so that (9)

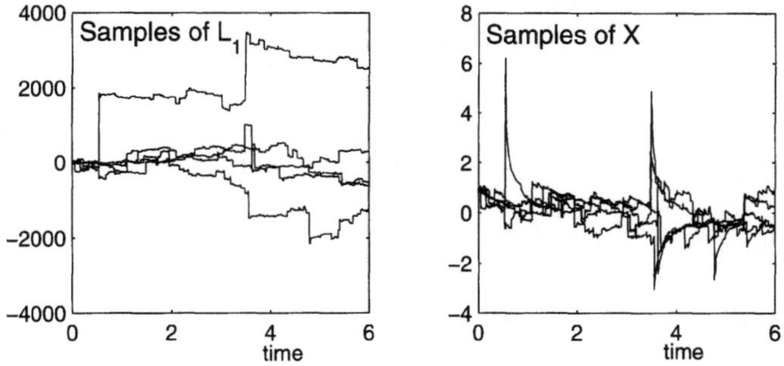

Figure 8. Five samples of L_1 and X

is driven by a Cauchy noise. Figure 8 shows five samples of L_1 and the corresponding samples of X for $X(0) = 1$ and $\sigma = 1$. The samples of both processes exhibit frequent small jumps and infrequent large jumps. This observation is used later to establish an asymptotic approximation for the tail of the distribution of X. But first it is attempted to find an equation for the characteristic function of X.

The Itô formula for semimartingales applied to $\exp(iuX(t))$ yields the partial differential equation $\partial\varphi/\partial t = u\left(-\partial\varphi/\partial u + \partial^3\varphi/\partial u^3\right) - \sigma|u|\varphi$ for the characteristic function $\varphi(u;t)$ of $X(t)$ if the expectation is calculated term by term. Accordingly, $\varphi_s(u) = \lim_{t\to\infty}\varphi(u;t) = c_1 e^{-\alpha_1|u|} + c_2 e^{-\alpha_2|u|}$, where $\alpha_k, c_k \geq 0$, $k = 1, 2$, are some constants and $c_1 + c_2 = 1$. The density of the stationary solution X_s is $f_s(x) = \sum_{k=1}^{2} c_k\alpha_k/[\pi(\alpha_k^2 + x^2)]$ so that the tail of the distribution F_s of X_s is $1 - F_s(x) \sim x^{-1}$ as $x \to \infty$. This result is incorrect since $X(s-)^3\exp(iuX(s-))$ is not integrable so that the expectation of the Itô formula cannot be calculated term by term. The actual tail of the distribution of X_s is $1 - F_s(x) \sim x^{-3}$ as $x \to \infty$ [11]. The main idea of the proof of this result is only presented here.

The jumps $\Delta X(t) = X(t) - X(t-)$ of X are equal to the jumps $\Delta L_1(t) = L_1(t) - L_1(t-)$ of L_1 scaled by σ so that only the large jumps of L_1 need to be considered to assess the behavior of the tail of the distribution of X_s. Since the collection of the large jumps of a Lévy process defines a compound Poisson process ([10], Theorem 42, p. 32), it is sufficient to find the properties of $P(X_s(t) > x)$ as $x \to \infty$ by considering the solution of (9) with L_1 replaced by a compound Poisson process $\tilde{C}$ with adequate properties.

Let $\tilde{C}$ be a compound Poisson process with jumps that follows the distribution $P(\tilde{Y}_1 > u) \sim u^{-1}$, $u \geq u_0 > 0$, and arrive at an average rate $\tilde{\lambda} > 0$. We continue to denote by X the state of (9) with L_1 replaced by $\tilde{C}$. Let $T_2 < T_1 < 0$ be the times of the last two jumps of X prior to 0. Between the consecutive jumps of $\tilde{C}$ the state satisfies the differential equation $\dot{X}(t) = -X(t) - X(t-)^3$. If $X(T_1) > 0$, we have $\dot{X}(t) \leq -X(t)^3$ in $[T_1, 0]$ so that $\{X(0) > u\} \subseteq \{|T_1| \leq u^{-2}\}$. We have $P(X(0) > u) = P(A \cup B) \leq P(A) + P(B)$, where $A = \{X(0) > u, X(T_1-) \leq \varepsilon u\}$, $B = \{X(0) > u, X(T_1-) > \varepsilon u\}$, and $\varepsilon \in (0, 1/2)$. Relatively simple but long calculations give $P(A) \leq (2/3)(1 + 2\varepsilon)\tilde{\lambda} u^{-3}$. We also have

$$P(B) \leq P(C) \leq \left(1 - e^{\tilde{\lambda}/u^2}\right)\left(1 - e^{\tilde{\lambda}/(\varepsilon u)^2}\right) \leq \tilde{\lambda}^2 \varepsilon^{-2} u^{-4}$$

so that $\limsup_{u\to\infty} u^3 P(X(0) > u) \leq (2/3)(1 + 2\varepsilon)\tilde{\lambda}$ holds for any $\varepsilon \in (0, 1/2)$ and implies $\limsup_{u\to\infty} u^3 P(X(0) > u) \leq (2/3)\tilde{\lambda}$ by letting $\varepsilon \to 0$, where $C = \{|T_1| \leq u^{-2}, |T_2 - T_1| \leq (\varepsilon u)^{-2}\}$. Since it can also be shown that $\liminf_{u\to\infty} u^3 P(X(0) > u) \geq (2/3)\tilde{\lambda}$, we have the asymptotic property $\lim_{u\to\infty} u^3 P(X(0) > u) = (2/3)\tilde{\lambda}$.

Hence, the tail of the distribution of the stationary state has the property $1 - F_s(x) \sim x^{-3}$ as $x \to \infty$. The complete proof and additional results are in [11].

That the tail of the distribution of X_s has the above property is confirmed by a Monte Carlo experiment. Let $X_i = X(i\Delta t)$, $i = 1, 2, \ldots, n$, be a sample of X in (9), where $\Delta t > 0$ is a time step. Denote by $X_{(1)} \geq \cdots \geq X_{(n)} \geq \cdots$ the corresponding order statistics and let $H_{k,n} = (1/k)\sum_{i=1}^{k} \log\left(X_{(i)}/X_{(i+1)}\right)$ be the Hill estimator based on the $k+1$ upper order statistics of the sample of X. The tail of the stationary state X_s has the property $1 - F_s(x) \sim x^{-\rho}$ as $x \to \infty$ if $1/H_{k,n}$ converges in probability to a constant ρ as $k, n \to \infty$ such that $k/n \to 0$ [6].

Figure 9 shows Hill estimates $H_{k,n}$ and their inverse $1/H_{k,n}$ for several ra-

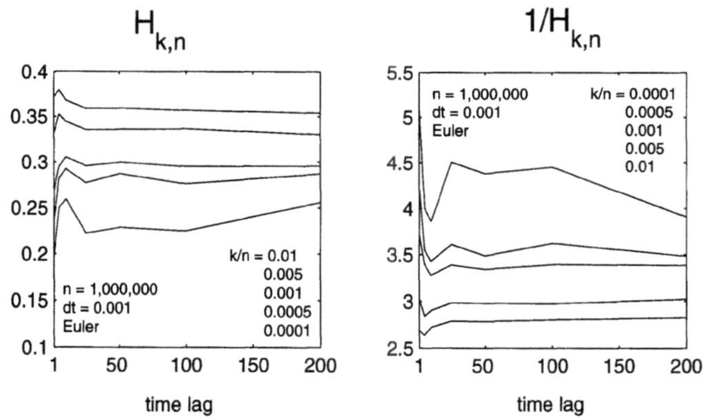

Figure 9. Hill estimates for $X(0) = 1$ and $\sigma = 1$

tios k/n. An Euler finite difference approximation of (9) with time step $\Delta t = 0.001$ was used to generate samples of X. The time lag on the horizontal axes indicates the number of data skipped in the calculation of Hill estimates. For example, a time lag of 50 means that the time series $X(i\Delta t), X((i + 50)\Delta t)X((i + 100)\Delta t), \ldots$ was used to calculate Hill estimates. All Hill estimates are based on records of length $n = 1,000,000$ irrespective of the time lag. Although the Hill estimates show some variability, they cluster consistently around the value $\rho = 3$ obtained by theoretical arguments.

5. Conclusion

Probability law and other properties have been determined for the state of a broad range of linear and nonlinear systems driven by Poisson and Lévy white noise. The Poisson and Lévy non-Gaussian white noise processes are defined as the formal derivative of compound Poisson and Lévy processes, respectively. Examples have been presented to demonstrate the need in science and engineering for models defined by stochastic differential equations driven by Poisson and Lévy white noise processes.

The analysis is based on the Itô formula for semimartingales, properties of Lévy processes, Monte Carlo simulation, and Hill estimates. It was shown that the Itô formula can be used to develop equations for the characteristic function of the state of a broad range of systems with additive and multiplicative non-Gaussian white noise and to assess the stochastic stability of the stationary state of these systems. It was found that the Itô formula cannot deliver equation for the characteristic function of, for example, some nonlinear systems with additive Lévy white noise. A method was developed for these systems to find properties of the tail of the distribution of their stationary state. Analytical findings have been confirmed by Monte Carlo simulation and Hill estimates.

References

[1] S. T. Ariaratnam. Some illustrative examples of stochastic bifurcation. In J. M. T. Thompson and S. R. Bishop, editors, *Nonlinearity and Chaos in Engineering Dynamics*, pages 267–274. John Wiley & Sons, Ldt., New York, 1994.

[2] S. T. Ariaratnam and H. N. Pi. On the first-passage time for envelope crossing for a linear oscillator. *International Journal of Control*, vol. 18, no. 1, pp. 89–96, 1973.

[3] P. Ditlevsen. Observations of α–stable noise induced millennial climate changes from an ice–core record. *Geophysical Research Letters*, vol. 26, no. 10, pp. 1441–1444, 1999.

[4] M. Grigoriu. *Applied Non-Gaussian Processes: Examples, Theory, Simulation, Linear Random Vibration, and MATLAB Solutions*. Prentice Hall, Englewoods Cliffs, NJ, 1995.

[5] M. Grigoriu. *Stochastic Calculus. Applications in Science and Engineering*. Birkhäuser, Boston, 2002.

[6] B. M. Hill. A simple general approach to inference about the tail of a distribution. *The Annals of Statistics*, vol. 3, no. 5, pp. 1163–1174, 1975.

[7] W. Horsthemke and R. Lefever. Phase transitions induced by external noise. *Physics Letters*, vol. 64A, no. 1, pp. 19–21, 1977.

[8] R. Z. Khasminskii. Necessary and sufficient conditions for the asymptotic stability of linear systems. *Theory of Probability and Its Applications*, vol. 12, no. 1, pp. 144–147, 1967.

[9] P. E. Kloeden and Platen E. *Numerical Solutions of Stochastic Differential Equations*. Springer–Verlag, New York, 1992.

[10] P. Protter. *Stochastic Integration and Differential Equations*. Springer–Verlag, New York, 1990.

[11] G. Samorodnitsky and M. Grigoriu. Tails of solutions of certain nonlinear stochastic differential equations driven by heavy tailed Lévy motions. *Stochastic Processes and Their Applications*, 2002. in press.

[12] G. Samorodnitsky and M. S. Taqqu. *Stable Non–Gaussian Random Processes. Stochastic Models with Infinite Variance*, Birkhäuser, New York, 1994.

[13] T. T. Soong and M. Grigoriu. *Random Vibration of Mechanical and Structural Systems*. Prentice Hall, Englewood Cliffs, NJ, 1993.

STOCHASTIC DYNAMICS OF FRICTION-INDUCED VIBRATION IN DISC BRAKES

S. Qiao and R. A. Ibrahim
Wayne State University
Department of Mechanical Engineering
Detroit, MI
ibrahim@eng.wayne.edu

Abstract The dynamic behavior of a disc brake subjected to a follower frictional force governed by either a constant or random friction coefficient is studied. Under constant frictional load, the circumferential motion is characterized by a quasi-periodic vibration while the transverse deflection is mainly governed by a periodic motion. The numerical simulation indicates that the response amplitude of the circumferential vibration is larger than that of the transverse vibration. Under random frictional load, both transverse and circumferential displacements are found to be essentially non-Gaussian, non-stationary, and narrow-band processes. The energy of the transverse motion considerably concentrates at a dominant frequency. On the other hand, the spectral density of circumferential motion is mainly distributed at two different frequencies.

1. Stochasticity of Friction

The value of the static friction coefficient, at which macroscopic sliding occurs, depends on the rate of increase of the tangential force. It is known [3] that the static friction coefficient increases with time due to the creep rate of compression of the asperities, resulting in an increase in the junction areas, and the shear strength of the junctions due to the existence of the cold-welding effect. Ibrahim, *et al.* [5] conducted a series of experimental tests to characterize the interfacial forces in metal-to-metal contact under harmonic excitation, and found that both the static and kinetic friction coefficients exhibit random time variations.

Some early work suggested that self-sustained oscillations leading to noise were the result of time variation of friction coefficient, either unequal static and dynamic values or reducing friction with increasing rubbing speed. Rabinowicz [8] and Rabinowicz, *et al.* [9] studied the statistical nature of friction in a copper-copper system with a sliding speed of $0.1 cm/s$. Rozeanu [10,

N. Sri Namachchivaya and Y.K. Lin (eds.),
IUTAM Symposium on Nonlinear Stochastic Dynamics, 331-342.

11] presented a statistical formulation of friction system and the unsteady nature of friction. Kilburn [6] conducted an experimental investigation to measure the frequency content (spectrum) and the statistical distribution of the amplitudes of the friction force signal. Aronov, *et al.* [1, 2] conducted experimental measurements using a pin-on-disc model and found that the measured friction force and velocity of the pin along the friction force change randomly with time. Recently, Qiao and Ibrahim [7] and [4] conducted a series of experimental tests on a pin-on-disc system. Typical time history records of contact forces exhibited continuous random fluctuations. The time variation of the natural frequency was attributed to the time variation of the boundary conditions (contact forces). The probability density function (pdf) of the friction force was essentially non-Gaussian.

2. Analytical Modeling

Consider a uniform, axisymmetric, moderately thick annular disc of thickness h, clamped at the inner radius $r = a$ and free at the outer radius b. The disc is subjected to a distributed, sector-shaped, in-plane tangentially frictional load over a sector of an angle ψ as shown in Figure 1. The mate-

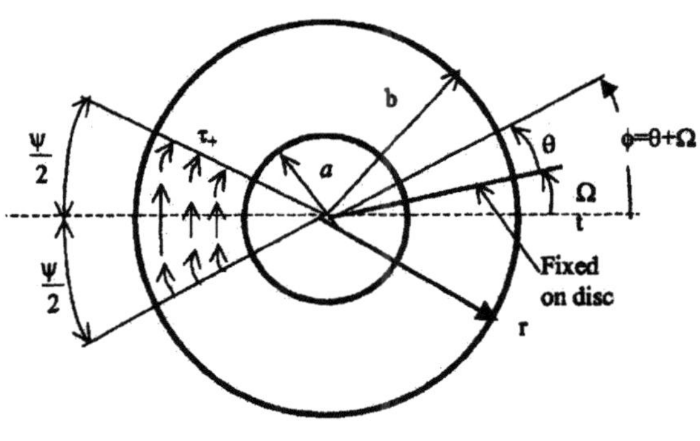

Figure 1. Schematic diagram of disc brake

rial of the disc is assumed isotropic and elastic of density ρ, Young's modulus E, and Poisson's ratio ν. It rotates in space with angular velocity Ω, while observation is made from a stationary coordinate frame r, ϕ, z. The moving frame r, θ, z is attached to the disc such that θ and ϕ are related through the relationship $\phi = \theta + \Omega t$. The traction τ_+ acts on one face of the disc and τ_- on the other face. Generally, $\tau_+ \neq \tau_-$, and by definition, they are directed tangent to the disc's instantaneous transverse profile w. For small w, the traction τ_+ (or τ_-) decomposes into the disc's equilibrium plane with components very close to τ_+ (or τ_-), and $\tau_{+w,\theta/r}$, (or $\tau_{-w,\theta/r}$) normal to the disc surface. In order to simplify the analytical model and yet capture the fundamental dynamical features of the disc, we will assume the disc does not have initial static disc deflection and the rotating speed is relatively low so that gyroscopic and centrifugal forces can be neglected. Mindlin's thick plate theory, takes into account the influence of transverse shear deformation and rotary inertia effects, will be used. The equations of motion and boundary conditions of the disc are derived using Hamilton's principle, i.e.,

$$\delta \int_{t_1}^{t_2} (T - U + W_{nc})\, dt = 0 \tag{1}$$

where T and U are the kinetic and potential energies, respectively, and W_{nc} is the work done by non-conservative forces. These are given by the following expressions

$$\begin{aligned} T &= \frac{1}{2}\iiint_V \rho \left[(u_{r,t})^2 + (u_{\theta,t})^2 + (u_{z,t})^2\right] dV \\ &= \frac{1}{2}\rho \iint_A \left(\frac{h^3}{12}\dot{\varphi}_r^2 + \frac{h^3}{12}\dot{\varphi}_\theta^2 + hwm\right) r d\theta dr \end{aligned} \tag{2}$$

$$\begin{aligned} U &= \frac{1}{2}\int_0^{2\pi}\int_a^b \left\{ D\left[\left(\frac{\partial \varphi_r}{\partial r}\right)^2 + \frac{1}{r^2}\left(\frac{\partial \varphi_\theta}{\partial \theta} + \varphi_r\right)^2 \right.\right. \\ &+ \frac{2}{r}\nu \frac{\partial \varphi_r}{\partial r}\left(\frac{\partial \varphi_\theta}{\partial \theta} + \varphi_r\right) \\ &+ \frac{1-\nu}{2r^2}\left(\varphi_\theta - r\frac{\partial \varphi_\theta}{\partial r} - \frac{(\partial \varphi_r)^2}{\partial \theta}\right) \\ &+ kGh\left[\left(\frac{\partial w}{\partial r} + \varphi_r\right)^2 + \left(\frac{1}{r}\frac{\partial w}{\partial \theta} + \varphi_\theta\right)^2\right]\right\} r dr d\theta \end{aligned} \tag{3}$$

To determine the external work done, W_{nc}, we define

$$\tau(r,\theta,z,t) = \left[\tau_+(t)\delta\left(z - \frac{h}{2}\right) + \tau_-(t)\delta\left(z + \frac{h}{2}\right)\right] H(\theta)$$

where $\delta()$ is here is the Dirac delta function and

$$H(\theta) = \begin{cases} 1 & -\psi/2 \le \theta \le \psi/2 \\ 0 & \text{elsewhere} \end{cases}$$

The work done, W_{nc} is

$$\begin{aligned} W_{nc} &= \iiint_V \left(\frac{\tau}{r} \frac{\partial w}{\partial \theta} w - \tau z \varphi_\theta \right) r dr d\theta dz \\ &= \iint_A \left(\frac{\tau_z}{r} \frac{\partial w}{\partial \theta} w - \tau_\theta \frac{h}{2} \varphi_\theta \right) r dr d\theta \end{aligned} \tag{4}$$

where $\tau_z = (\tau_+ + \tau_-)H(\theta) = \hat{\tau}(t)H(\theta)$, $\tau_\theta = (\tau_+ - \tau_-)H(\theta) = \tilde{\tau}(t)H(\theta)$. u_r, u_θ, and u_z are the displacements of a material point in the disc in the r, θ and $z-$directions, respectively, while w is the transverse deflection of a point on the middle surface of the disc. φ_r and φ_θ are the slopes in the respective radial and circumferential planes. $D = Eh^3/[12(1-\nu^2)]$, $k = \pi^2/12$, and $G = E/[2(1+\nu)]$. Carrying out the variation term by term and expressing the coordinate with reference to the fixed frame, r, ϕ, z,

$$w(r,\theta,t) \;=\; w(r,\phi - \Omega t, t) \;=\; w_z(r,\phi,t) \tag{5}$$

such that the corresponding derivatives take the form

$$\frac{\partial^m w}{\partial r^m} = \frac{\partial^m w_z}{\partial r^m}, \quad \frac{\partial^m w}{\partial \theta^m} = \frac{\partial^m w_z}{\partial \phi^m}, \quad \frac{\partial^m w}{\partial t^m} = \left(\frac{\partial}{\partial t} + \Omega \frac{\partial}{\partial \phi} \right)^m w_z \tag{6}$$

gives three coupled equations, which are not easy to handle due to their complexity. Instead, attention will be given only to transverse and circumferential motions. Assume that the slope in the radial plane, φ_r, is very small comparing with w_z and φ_ϕ such that φ_r could be neglected, we obtain the following two-coupled equations

$$\begin{aligned} \left[\rho h \left(\frac{\partial^2}{\partial t^2} + 2\Omega \frac{\partial^2}{\partial t \partial \phi} + \Omega^2 \frac{\partial^2}{\partial \phi^2} \right) - kGh\nabla^2 - \frac{2}{r}\tau_z \frac{\partial}{\partial \phi} \right] w_z & \\ -2kGh \frac{\partial \varphi_\phi}{r \partial \phi} &= 0 \end{aligned} \tag{7}$$

$$\begin{aligned} -\tau_\theta \frac{h}{2} = kGh \frac{\partial w_z}{r \partial \phi} + \Bigg\{ & \frac{\rho h^3}{12} \left(\frac{\partial^2}{\partial t^2} + 2\Omega \frac{\partial^2}{\partial t \partial \phi} + \Omega^2 \frac{\partial^2}{\partial \phi^2} \right) \\ & -D \left[\nabla^2 - \frac{1+\nu}{2r} \left(\frac{\partial}{\partial r} + r \frac{\partial^2}{\partial r^2} \right) - \frac{1-\nu}{2r^2} \right] + kGh \Bigg\} \varphi_\phi \end{aligned} \tag{8}$$

where $\nabla^2 = \frac{\partial^2}{\partial r^2} + \frac{1}{r}\frac{\partial}{\partial r} + \frac{1}{r^2}\frac{\partial^2}{\partial \theta^2}$. The corresponding boundary conditions are

$$w_z = \varphi_\phi = 0 \text{ at } r = a, \text{ and } w_{z,r} = 0, \quad b\varphi_{\phi,r} - \varphi_\phi = 0, \text{ at } r = b. \tag{9}$$

Since the natural frequencies for all modes with one or more nodal circles are greater than the natural frequencies of modes with zero nodal circles and up to three nodal diameters, the current analysis will include only modes with zero nodal circles, $i = 0$. One may write the solution of (7) and (8) in terms of Fourier sine and cosine series expansion:

$$W_z(r,\phi,t) = \sum_{j=0}^{J} W_j(r)\left[F_j(t)\cos j\phi + H_j(t)\sin j\phi\right] \tag{10}$$

$$\varphi_\phi(r,\phi,t) = \sum_{j=0}^{J} \Phi_j(r)\left[S_j(t)\cos j\phi + T_j(t)\sin j\phi\right] \tag{11}$$

where the index j gives the number of circumferential nodal diameters of the disc with zero nodal circles. The weighting functions related to the assumed solutions, in the form of a single mode with j nodal diameters, are

$$\bar{w}_z(r,\phi,t) = W_j(r)\left[\bar{F}_j(t)\cos j\phi + \bar{H}_j(t)\sin j\phi\right] \tag{12}$$

$$\bar{\varphi}_\phi(r,\phi,t) = \Phi_j(r)\left[\bar{S}_j(t)\cos j\phi + \bar{T}_j(t)\sin j\phi\right] \tag{13}$$

where $F_j(t)$, $H_j(t)$, $S_j(t)$, and $T_j(t)$ are unknown functions of time to be determined. The same with over bar are arbitrary functions of time. The partial differential equations of motion of the disc are discretized using Galerkin's method. The method gives the following set of ordinary differential equations

$$\ddot{F}_j + 2j\Omega\dot{H}_j + \left[\frac{K_W}{M_W} - j^2\Omega^2 - \frac{Q_W}{M_W}\tilde{\tau} f_F(t)\right]F_j - \frac{Q_W}{M_W}\tilde{\tau} f_{HF}(t)H_j - \frac{K_{W\Phi}}{M_W}T_j = 0 \tag{14}$$

$$\ddot{H}_j + 2j\Omega\dot{F}_j - \left[\frac{K_W}{M_W} - j^2\Omega^2 - \frac{Q_W}{M_W}\tilde{\tau} f_H(t)\right]H_j + \frac{Q_W}{M_W}\tilde{\tau} f_{FH}(t)F_j + \frac{K_{W\Phi}}{M_W}S_j = 0 \tag{15}$$

$$\ddot{S}_j + 2j\Omega\dot{T}_j + \frac{K_{W\Phi}}{M_\Phi}H_j + \left[\frac{K_\Phi}{M_\Phi} - j^2\Omega^2\right]S_j = -\frac{Q_\Phi}{M_\Phi}\tilde{\tau} f_S(t) \tag{16}$$

$$\ddot{T}_j - 2j\Omega\dot{S}_j + \frac{K_{W\Phi}}{M_\Phi}F_j + \left[\frac{K_\Phi}{M_\Phi} - j^2\Omega^2\right]T_j = -\frac{Q_\Phi}{M_\Phi}\tilde{\tau} f_T(t) \tag{17}$$

where

$$\begin{aligned} f_F(t) &= \cos(2j\Omega t)\sin j\psi + j\psi, \\ f_H(t) &= \cos(2j\Omega t)\sin j\psi - j\psi \\ f_{FH}(t) &= f_{HF}(t) = \sin(2j\Omega t)\sin j\psi, \\ f_S(t) &= \cos(2j\Omega t)\sin(j\psi/2) \\ f_T(t) &= \sin(2j\Omega t)\sin(j\psi/2) \end{aligned} \tag{18}$$

It is clear that the follower-type frictional stresses result in parametric excitation terms in transverse motion. For the circumferential direction, vibration may directly result from external excitations also caused by frictional stresses. To a great extent, the dynamical behavior of the disc depends on the nature of the excitations that will be harmonic if $\tilde{\tau}$ and $\hat{\tau}$ are deterministic under given conditions or stochastic if $\tilde{\tau}$ and $\hat{\tau}$ are random. Theoretically, (14)–(17) can be solved for any single mode with j nodal diameters if the radial mode shapes are known.

The radial mode can be chosen as either a linear combination of Bessel functions of the first and second kinds or some kinds of high order polynomials in r. To reasonably simplify the analysis, $W_j(r)$ and $\Phi(r)$ are assumed polynomials in r defined by

$$\begin{aligned} W_j(r) &= (r-a)^{j+2}(a_j^W + b_j^W r), \\ \Phi_j(r) &= (r-a)^{j+1}(a_j^\phi + b_j^\phi r) \end{aligned} \tag{19}$$

where a_j^W, b_j^W, a_j^ϕ and b_j^ϕ are coefficients to be determined from the boundary conditions. Another two equations needed to determine the coefficients a_j^W, b_j^W, a_j^ϕ and b_j^ϕ of (19) are the normalizing conditions given by

$$W_j(b) = \Phi_j(b) = 1. \tag{20}$$

Substituting (20) into (19) and solving for a_j^W, b_j^W, a_j^ϕ, and b_j^ϕ gives

$$\begin{aligned} a_j^W &= \frac{(j+3)b-a}{(b-a)^{J+3}}, \quad b_j^W = \frac{j+2}{(b-a)^{J+3}}, \\ a_j^\phi &= \frac{(j+1)b}{(b-a)^{J+2}}, \quad b_j^\phi = \frac{bj+a}{b(b-a)^{J+2}} \end{aligned} \tag{21}$$

Although equations (14)–(17), (19), and (21) constitute a complete vibration model for describing the dynamical behavior of the disc in transverse and circumferential directions, they are still too complicated for analytical or numerical solutions for F_j, H_j, S_j, and T_j. At this stage, we focus on the case of single mode based on some further simplifications of the circumferential modes.

Some simpler mode shapes are possible. For instance, the transverse deflection and circumferential rotation may be written as

$$W(r,\phi) = \sum_{j=0}^{n} W_j(r)\cos j\phi, \quad \Phi(r,\phi) = \sum_{j=0}^{n} \Phi_j(r)\sin j\phi \tag{22}$$

where

$$\begin{aligned} w_z(r,\phi,t) &= \sum_{j=0}^{n} W_j(r)F_j(t)\cos j\phi, \\ \varphi_\phi(r,\phi,t) &= \sum_{j=0}^{n} \Phi_j(r)T_j(t)\sin j\phi \end{aligned} \tag{23}$$

Note that $\Phi(r,\phi) = 0$ for $j = 0$ in this case. Equations (23) have been verified to be proper assumed forms for mode shapes with enough computational accuracy in numerical analysis. Consequently, the following simplified equations of motion are obtained for the $j - th$ mode

$$\ddot{F}_j + \left[\frac{K_W}{M_W} - j^2\Omega^2 - \frac{Q_W}{M_W}\tilde{\tau} f_F(t)\right] F_j - \frac{K_{W\Phi}}{M_W} T_j = 0 \tag{24}$$

$$\ddot{T}_j + \frac{K_{W\Phi}}{M_\Phi} F_j + \left[\frac{K_\Phi}{M_\Phi} - j^2\Omega^2\right] T_j = -\frac{Q_\Phi}{M_\Phi}\tilde{\tau} f_T(t) \tag{25}$$

These equations can be written in the non-dimensional form

$$w'' + 2\zeta_w w' + \left[1 - j^2\bar{\Omega}^2 - q_w\hat{\tau}(\tau) f_F(\tau)\right] w - C_w v = 0 \tag{26}$$

$$v'' + 2\zeta_v \varpi v' + \left[\varpi^2 - j^2\bar{\Omega}^2\right] v + C_v w = -q_v\tilde{\tau}(\tau) f_T(\tau) \tag{27}$$

where a prime denotes differentiation with respect to $\tau = \omega_z t$. The following non-dimensional parameters have been introduced

$$\begin{aligned} \omega_z &= \sqrt{K_W/M_W}, \quad w = F_j/b, \quad v = T_j h/2b, \\ \varpi^2 &= K_\Phi/(M_\Phi\omega_z^2) \quad C_v = \frac{hK_{w\Phi}}{2M_\Phi\omega_z^2}, \quad q_w = \frac{Q_W}{M_W\omega_z^2}, \\ C_w &= \frac{2K_{w\Phi}}{hM_W\omega_z^2}, \quad q_v = \frac{hQ_\Phi}{2bM_\Phi\omega_z^2}, \quad \bar{\Omega} = \Omega/\omega_z, \end{aligned} \tag{28}$$

where ζ_w and ζ_v are linear damping ratios for transverse and circumferential displacements, respectively, and are introduced to account for energy dissipation. v stands for the non-dimensional displacement in tangential direction of the upper surface of the disc.

2.1 Dynamics under Constant Friction Coefficient

Consider constant contact forces in (26, 27), a case in which the relationship between the frictional force and normal force obeys the classical Coulomb law. The numerical solutions of (26, 27) are obtained using Runge-Kutta method with a non-dimensional time increment of $\Delta\tau = 0.001$, and zero initial conditions. A typical time history record and its FFT at disc rotation speed $\bar{\Omega} = 0.006$ is shown in Figure 2 for $C_w = 0.07$, $C_v = 0.15$, $\varpi = 1.6$, $q_w\hat{\tau} =$

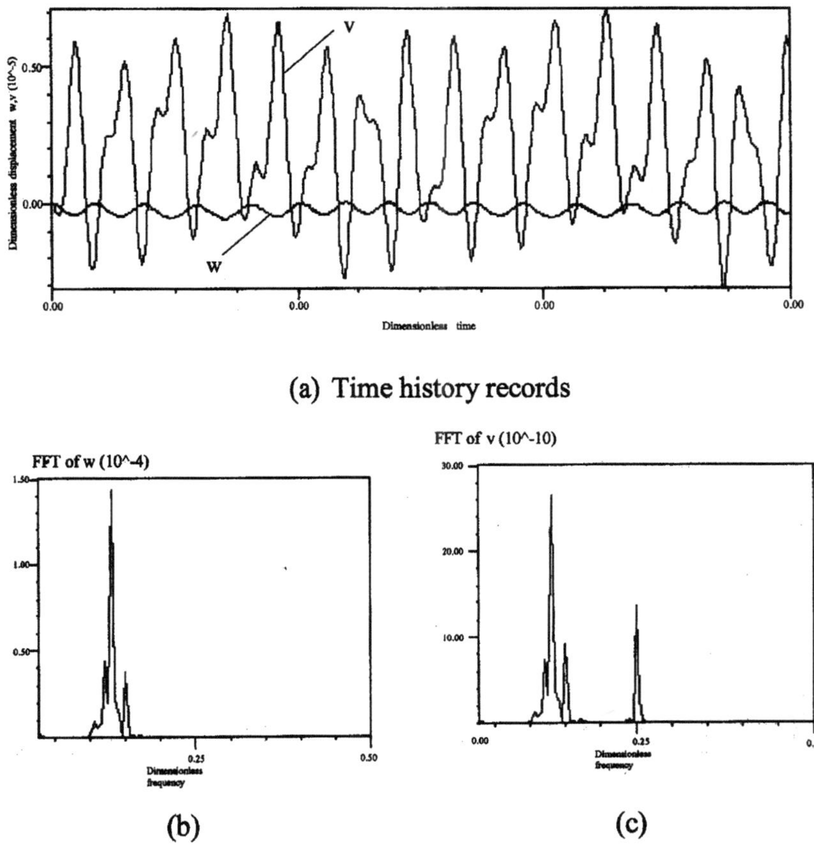

Figure 2. FFT of transverse (b), and circumferential (c) motions

1.3, $q_v\tilde{\tau} = 0.23$, $\zeta_w = 0.02$, and $\zeta_v = 0.02$. The transverse, $w(\tau)$, and circumferential, $v(\tau)$, displacements are plotted in the same time history record for the purpose of comparison, indicating the importance of the circumferential displacement of the disc. It is seen that both $w(\tau)$ and $v(\tau)$ time histories are always asymmetric about their undeformed positions. The FFT

diagrams, shown in Figures 2(b) and 2(c), exhibit more frequency components, which may be attributed to the fact that $f_F(\tau)$ takes a non-periodic form $(f_F(\tau) = \cos(2j\bar{\Omega}\tau)\sin j\psi + j\psi)$. However, only one dominant peak in the FFT of transverse deflection $w(\tau)$ appears at about $0.129(2\pi) = 0.81$, while two dominant peaks in the FFT of circumferential displacement $v(\tau)$ occur at approximately $0.129(2\pi) = 0.81$ and $0.25(2\pi) = 1.57$, respectively. In other words, the transverse deflection of the disc behaves much like periodic motion governed mainly by the first mode associated with the first damped frequency and some trivial frequency components, while the circumferential motion seems to be a quasi-harmonic motion at least containing two distinct dominant frequencies.

2.2 Dynamics under Random Friction Coefficient

As indicated in Section 1, measured contact forces between a rotating disc and stationary pin/pads displayed an essential nature of wide band random process. Thus, it is reasonable to approximately regard frictional stresses $\hat{\tau}$ and $\tilde{\tau}$ as white noise random processes. The equations of motion (28) are written in terms of the state vector form through the coordinate transformation $(X_1, X_2, X_3, X_4) = (w, w', v, v)'$. In this case, $\hat{\tau}$ and $\tilde{\tau}$ are assumed Gaussian, stationary white noises with zero means, and correlation time that is much smaller than any characteristic period of oscillation of the system. It is also assumed that $\hat{\tau}$ and $\tilde{\tau}$ are independent and uncorrelated. It should be emphasized that $f_F(\tau)$ is a non-periodic function, which results in non-stationarity of the excitation.

The dynamic characteristics of the disc as described by (26, 27) are estimated using Monte Carlo simulation. The simulation was conducted for different values of the non-dimensional rotating speed $\bar{\Omega}$, radii ratio a/b, and thickness ratio h/b, to examine their effects on the disc dynamic behavior. The mean square time histories are plotted (not shown for space limitation), and as expected, both processes $P_w(\tau) = q_w W_w(\tau) f_F(\tau)$ and $P_v(\tau) = q_v W_v(\tau) f_T(\tau)$ are non-stationary since the fluctuations of their mean squares do not vanish within the given duration. The numerical results are further analyzed and the associated statistical parameters such as mean square, autocorrelation function and power spectra density function for both transverse deflection and circumferential displacement are evaluated.

Time history records, mean squares, autocorrelation and power spectral density functions of both transverse deflection and circumferential displacement for different rotational speeds revealed that both transverse and circumferential displacements, are narrow band. The mean square responses are non-stationary mainly because the excitation is non-stationary. Another source responsible for the fluctuations is partially due to the very short simulation duration within which the mean square is not reaching a stationary state. It is this short duration that mimics the practical braking process, which can be actually rendered as a kind of transition.

Figures 3(a) and 3(b) show pdf plots of transverse and circumferential dis-

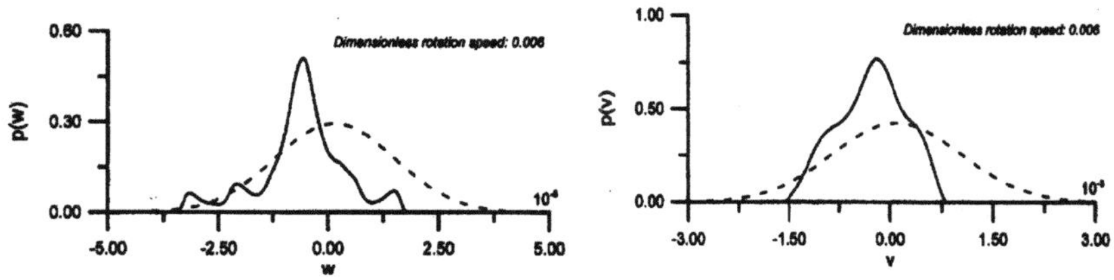

Figure 3. Pdf of transverse and circumferential displacements

placements, respectively, where the dashed curves correspond to the pure Gaussian distributions. The estimated pdf diagrams of the transverse deflection are found to deviate significantly from normality, with multi-peaks and asymmetry.

Other simulations were performed for different values of radii ratio, a/b, and it was found that an increase of the radii ratio a/b results in an almost linear increase of non-dimensional central frequencies of the response. Physically, this means that the stiffness of the disc is increasing, and meanwhile the mass of the disc is decreasing. Figures 4(a) and 4(b) describe the pdf di-

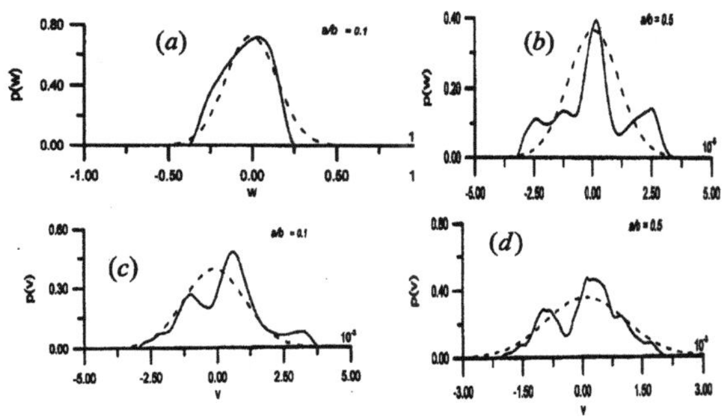

Figure 4. Pdfs of (a,b) transverse; (c,d) circumferential motions

agrams for the cases of $a/b = 0.1$, and 0.5. The estimated probability density functions of the transverse deflection are found to have a single peak when $a/b = 0.1$, which becomes three or four peaks when a/b increases. The change of the number of peaks on the pdf curve is an indicator that stochastic bifurcation or instability might occur. On the other hand, the probability density function curves of the circumferential motion exhibit relatively similar shape containing three peaks. Both cases still deviate significantly from normality with an asymmetric feature.

3. Conclusion

The general equations of motion for a rotating disc under stationary distributed in-plane frictional loads has been developed using Hamilton's principle on the basis of Mindlin theory. By applying the generalized Galerkin's method, the three coupled partial differential equations of motion are reduced to two coupled ordinary differential equations for transverse and circumferential motions. The numerical analyses for damped vibration of the disc with constant frictional loads reveal that the circumferential motion is characterized by a quasiperiodic vibration while the transverse deflection is mainly governed by a periodic vibration. Both transverse and circumferential displacements are not symmetric about their underformed positions. With random contact forces, Monte Carlo simulations revealed that both transverse and circumferential displacements are essentially non-Gaussian, non-stationary, and narrow-band processes. Characterized by a sharp peak of the power spectral density curve, the vibration energy of the transverse motion considerably concentrates at a dominant frequency, whereas that of the circumferential motion possesses two frequency components. The increase of inner radius or thickness of the disc results in a corresponding increase of the response frequency. On the other hand, increasing rotation speed of the disc within the low speed range under consideration can hardly change the response frequency of the system. Probability distribution functions of the response are irregular and asymmetric, whose number of the peak and amplitude values are changed by different disc speed and inner radius, implying possible bifurcations and instability in a stochastic sense.

References

[1] V. Aronove, A. F. D'Souza, S. Kalpakjian, and I. Shareef, "Experimental investigation of the effect of system rigidity on wear and friction-induced vibrations," *J Lub Tech*, vol. 105, pp. 206–211, 1983.

[2] V. Aronove, A. F. D'Souza, S. Kalpakjian, and I. Shareef, "Interaction among friction, wear, and system stiffness, Part 2: Vibrations induced by dry friction," *ASME J Tribology*, vol. 106, pp. 59–62, 1984.

[3] R. A. Ibrahim, "Friction-induced vibration, chatter, squeal, and chaos, Part I: Mechanics of contact and friction, Part II: Dynamics and modeling," *ASME Appl Mech Rev*, vol. 47, no. 7, pp. 209–253, 1994.

[4] R. A. Ibrahim, S. Madhaven, S. L. Qiao, and W. K. Chang, "Experimental investigation of friction-induced noise in disc brake systems," *Int J Vehicle Design*, vol. 23, no. (3/4), pp. 365–377, 2000.

[5] R. A. Ibrahim, S. A. Zielke, and K. Popp, "Characterization of interfacial forces in metal-to-metal contact under harmonic excitation," *J Sound Vib*, vol. 220, no. 2, pp. 365–377, 1999.

[6] R. F. Kilburn, "Friction viewed as a random process," *J Lubrication Tech*, vol. 96, pp. 291–299, 1974.

[7] S. L. Qiao and R. A. Ibrahim, "Stochastic dynamics of systems with friction-induced vibration," *J Sound Vib*, vol. 223, no. 1, pp. 115–140, 1999.

[8] E. Rabinowicz, "Autocorrelation analysis of sliding process," *J Appl Phys*, vol. 27, pp. 131–135, 1955.

[9] E. Rabinowicz, B. G. Rightmire, C. E. Tedholm, and R. E. Williams, "The statistical nature of friction," *Trans ASME*, vol. 77, pp. 931–934, 1955.

[10] L. Rozeanu, "Statistical formulation of friction systems," *Wear*, vol. 25, pp. 29–37, 1973.

[11] L. Rozeanu, "Friction transients," *J Amer Soc Lubrication*, vol. 19, pp. 257–266, 1976.

DYNAMICAL SYSTEMS WITH MULTIPLICATIVE RANDOM IMPULSE PROCESS EXCITATION

Radoslaw Iwankiewicz
University of the Witwatersrand, Johannesburg,
Private Bag 3, WITS 2050, SOUTH AFRICA
radiwank@mech.wits.ac.za

Abstract Dynamical systems with multiplicative (parametric) random impulse process excitation are considered. A technique is developed to convert the usual differential equation of motion (physical equation) into an Itô's type stochastic differential equation for random impulse process excitation multiplicative to the velocity term of the equation of motion. The use is made of the generalized Itô's differential rule and of the properties of the increments of the underlying regular random counting process. It is found that the diffusion term rather than the drift term has to be corrected. For linear and quadratic velocity terms the corrected diffusion terms are given in a closed form.

Keywords: Dynamical systems, random impulses, stochastic point process, Itô's stochastic differential equation, Itô's differential rule

1. Introduction

As is well known when a dynamical system is subjected to a Gaussian white noise process the differential equation of motion cannot be interpreted in the usual sense; it has to be converted into its meaningful stochastic counterparts: Stratonovich or Itô's type equations. If the Gaussian white noise in the first order dynamical system is multiplicative (excitation is parametric), or in the second order dynamical system it is multiplicative to the velocity-dependent term, the drift term of the Itô's type stochastic differential equation is not the same as that of the physical equation; it has to be completed by an extra term called Wong and Zakai [14] or Stratonovich [9] correction term.

If a dynamical system is driven by a Poisson, or a more general random, impulse process, the usual differential equation of motion is not mathematically meaningful because the dynamic response process is not continuously differentiable. The velocity process reveals finite jumps at random occur-

N. Sri Namachchivaya and Y.K. Lin (eds.),
IUTAM Symposium on Nonlinear Stochastic Dynamics, 343-352.

rence times of impulses, so its paths are only piece-wise continuously differentiable. In other words, in every infinitely short time interval there may be, with non-zero probability, a jump. It is then natural to seek an Itô's type stochastic differential equation as a valid counterpart of the usual differential equation. Then the generalized Itô's differential rule [3, 4, 8] and other tools of the theory of non-diffusive Markov processes may be used for analysis for dynamical systems [6, 5].

The question arises whether or not for dynamical systems with multiplicative random (e.g., Poisson) impulse process excitation, the Itô's type stochastic differential equation has the same drift and "diffusion" terms as the original differential equation, i.e., if for this class of problems the Itô's and Stratonovich type equations are the same or not. There exist two opposed answers to this question. In [1] the dynamical systems under non-Gaussian process excitations with independent increments, called sth order delta-correlated processes, were considered. Using formal calculus the authors found that the "diffusion" term of the Itô's stochastic differential equation has another form than the term in the original differential equation. The corrected term was found to be given by a recursive expression, which for $s = 2$ reduces to the classical Wong and Zakai (Stratonovich) corrected term. Author of the paper [10] presented opposing finding, based on the theory of semimartingales and stochastic integrals. His finding is that Itô's and Stratonovich type interpretations are the same for stochastic differential equations driven by Poisson impulse process, and that the "diffusion" term is the same.

It is worth mentioning that for some problems driven by a Poisson impulse process [11, 12] the equations are directly formulated in terms of stochastic differentials, so there is no need for any converting.

In this paper the first- and second-order dynamical systems with multiplicative (parametric) random impulse process excitation are considered. In the second-order dynamical system the impulse process excitation is multiplicative to the velocity term. A technique is developed to convert the usual differential equation of motion (physical equation) into an Itô's type stochastic differential equation for general, not necessarily Poisson, regular stochastic point processes. Using an expansion similar to the Taylor expansion in the impulse magnitude (tantamount to the generalized Itô's differential rule) and the properties of the increments of the underlying regular stochastic point process the expression for the corrected diffusion term is obtained. For diffusion terms linear and quadratic in velocity the closed-form expressions for the corrected term are found.

2. Dynamical Systems With Multiplicative Random Impulse Noise: Statement of the Problem

Consider the following deterministic first-order differential equation

$$\dot{Y}(t) = aY(t) + bY(t)f(t), \quad Y(0) = Y_0, \tag{1}$$

with continuous excitation $f(t)$, whose solution is

$$Y(t) = Y_0 \exp\left(at + b\int_0^t f(\tau)d\tau\right) = Y_0 \exp\left(at + bF(t)\right), \tag{2}$$

where $F(t)$ is a continuously differentiable function representing the excitation: $\dfrac{dF(t)}{dt} = f(t)$.

Consider now a first-order dynamical system with multiplicative random impulse process excitation

$$\dot{Y}(t) = aY(t) + bY(t)\xi(t), \quad Y(0) = Y_0, \tag{3}$$

where

$$\xi(t) = \sum_{i=1}^{\mathcal{N}(t)} P_i \delta(t - t_i), \tag{4}$$

where $\mathcal{N}(t)$ is a regular random counting (stochastic point) process giving the number of events (impulse arrivals) in the time interval $(0, t]$, including the one that possibly occurs at t. Thus the paths of $\mathcal{N}(t)$ are càdlàg (continue à droite, limite à gauche - right continuous with left limits). The impulse magnitudes P_i are assumed to be independent, identically distributed random variables.

The driving impulse process $\xi(t)$ is characterized by a marked (compound) stochastic point process $\Gamma(t)$, which may be conveniently represented in integral form as

$$\Gamma(t) = \sum_{i=1}^{\mathcal{N}(t)} P_i = \int_0^t d\Gamma(\tau) = \int_0^t P(\tau)d\mathcal{N}(\tau), \tag{5}$$

where $P(\tau)$ corresponds to the mark variable P_i, i.e., to the random magnitude of the impulse occurring in the time interval $(\tau, \tau + d\tau]$, $d\mathcal{N}(\tau) = \mathcal{N}(\tau + d\tau) - \mathcal{N}(\tau)$ and $d\Gamma(\tau) = P(\tau)d\mathcal{N}(\tau)$ denote, respectively, the increments of the random counting process and of the marked stochastic point process in the time interval $(\tau, \tau + d\tau]$. It must be emphasized that the notation $d\mathcal{N}(\tau)$ is purely conventional, $d\mathcal{N}(\tau)$ is not a differential in the usual sense, it is never infinitely small, as an increment of a regular counting process it may only assume values 0 or 1 [13]. Likewise the increment $d\Gamma(\tau) = P(\tau)d\mathcal{N}(\tau)$ equals a random jump given by a random impulse magnitude $P(\tau)$ or zero.

The solution of (3) reveals jumps at the times t_i of occurrence of the impulses, hence the process $Y(t)$ is only piecewise (between the impulses) continuously differentiable. The value $Y(t_i)$ is not defined, there are only its left and right limits, which of course differ. Hence the differential equation (3) is not mathematically meaningful. If we try to integrate it with respect to time in the usual way, making use of the properties of Dirac delta pseudofunction $\delta(t - t_i)$, we encounter the fundamental difficulty, as the Dirac delta coincides with the jumps in $Y(t)$, hence it cannot be arbitrarily assigned to the value $Y(t_{i-})$ immediately before the jump or to the value $Y(t_{i+})$ immediately after the jump.

The problem is to find a stochastic differential equation (in Itô's sense) counterpart of the "formal" equation (3), whose solution would be a stochastic counterpart of the result (2), i.e.,

$$Y(t) = Y_0 \exp\left(at + b\Gamma(t)\right). \tag{6}$$

Let us denote $X(t) = \Gamma(t)$, hence

$$dX(t) = d\Gamma(t) = P(t)d\mathcal{N}(t). \tag{7}$$

Consider the process $X(t)$ governed by the Itô's type stochastic differential equation

$$dX(t) = a\Big(t, X(t)\Big)dt + c\Big(t, P(t), X(t)\Big)d\mathcal{N}(t), \tag{8}$$

where $dX(t) = X(t + dt) - X(t)$. Consistently with the definition of $\mathcal{N}(t)$, the paths of $X(t)$ are càdlàg as well. Here again the notation $dX(t)$ is purely conventional, as it is not necessarily infinitely small. If no impulse occurs in $(t, t + dt]$ then $dX(t) = a(t, X(t))dt$, but if an impulse does occur, then $dX(t) = a(t, X(t))dt + c(t, P(t), X(t))$, where the second term is a finite jump change. In other words, if at the time $t = t_i$ an impulse of magnitude P_i occurs, the finite jump change in the state variable equals $\Delta X(t) = X(t_{i+}) - X(t_{i-}) = c(t_{i-}, P_i, X(t_{i-}))$.

The generalized Itô's differential rule pertinent to the process $X(t)$ governed by (8) has the form (cf. [4, 5])

$$\begin{aligned} dV(t, X(t)) &= \frac{\partial V(t, X(t))}{\partial t}dt + \frac{\partial V(t, X(t))}{\partial X}a\left(t, X(t)\right)dt \\ &+ \left[V\left(t, X(t) + c\Big(t, P(t), X(t)\Big)\right) - V\Big(t, X(t)\Big)\right]d\mathcal{N}(t), \end{aligned} \tag{9}$$

or an equivalent, Taylor expansion, form [5]

$$
\begin{aligned}
dV(t,X(t)) = \frac{\partial V(t,X(t))}{\partial t} dt + \frac{\partial V(t,X(t))}{\partial X} a\,(t,X(t))\,dt \\
+\sum_{k=1}^{\infty} \frac{1}{k!} \frac{\partial^k V(t,X(t))}{\partial X^k} c^k\Big(t,P(t),X(t)\Big)\Big(d\mathcal{N}(t)\Big)^k.
\end{aligned}
\tag{10}
$$

where $dV(t,X(t)) = V\Big(t+dt, X(t+dt)) - V(t,X(t)\Big)$, is the increment in the time interval $(t, t+dt]$, and due to the finite jumps $d\mathcal{N}(t)$, this increment may also be finite.

It is also worth noting that the above generalized Itô's differential rule holds for every regular stochastic point process $\mathcal{N}(t)$, which has a finite number of points in a finite time interval [8]. So it need not be a Poisson process. The difficulty is, however, that if the increment of the random counting process $\mathcal{N}(t)$ are not independent, then $V(t,X(t))$ is not statistically independent of the increments $d\mathcal{N}(t)$, which causes that the averaging of the differential rule cannot be effectively performed.

For the compound function $Y(t) = V(t,X(t)) = Y_0 \exp\,(at + bX(t))$ of the process $X(t)$ governed by (7) we obtain from (9)

$$
\begin{aligned}
& dY(t) = dV(t,X(t)) = aY_0 \exp\Big(at + bX(t)\Big)dt + \\
& Y_0\Big[\exp(at)\exp\Big(b(X(t)+P(t))\Big) - \exp(at)\exp\Big(bX(t)\Big)\Big]d\mathcal{N}(t) = \\
& aY_0 \exp\Big(at + bX(t)\Big)dt \\
& +Y_0 \exp\Big(at + bX(t)\Big)\Big(\exp(bP(t)) - 1\Big)d\mathcal{N}(t),
\end{aligned}
\tag{11}
$$

hence the stochastic differential equation is obtained as

$$
dY(t) = aY(t)dt + \Big(\exp(bP(t)) - 1\Big)Y(t)d\mathcal{N}(t), \quad Y(0) = Y_0. \tag{12}
$$

It may be verified that the solution of this equation is given by (6) indeed. In order to solve this equation we use the resolving function $V(Y(t)) = \ln Y(t)$ and apply the generalized Itô's differential rule (9).

3. Itô's Stochastic Differential Equation for Dynamical Systems with Multiplicative Random Impulse Noise

3.1 General procedure

Consider a second-order dynamical system with multiplicative random impulse process excitation $\xi(t)$, governed by the usual equation of motion

$$\begin{aligned} \ddot{Y}(t) &= a\Big(t, Y(t), \dot{Y}(t)\Big) + b\Big(t, Y(t), \dot{Y}(t)\Big)\xi(t), \\ \dot{Y}(0) &= \dot{Y}_0, \quad Y(0) = Y_0, \end{aligned} \tag{13}$$

which can be written in a standard form for the state vector $\mathbf{Y}(t) = [Y(t), \dot{Y}(t)]^T = [Y_1(t), Y_2(t)]^T$ as

$$\dot{\mathbf{Y}}(t) = \begin{bmatrix} \dot{Y}_1(t) \\ \dot{Y}_2(t) \end{bmatrix} \begin{aligned} &= Y_2(t), \\ &= a\,(t, \mathbf{Y}(t)) + b(t, \mathbf{Y}(t))\xi(t). \end{aligned} \tag{14}$$

The increment $dY_2(t) = Y_2(t+dt) - Y_2(t)$ during the time interval $(t, t+dt]$ consists of the continuous, smooth part due to the drift term $a\,(t, \mathbf{Y}(t))$ and of the discontinuous (jump) part due to an impulse occurring in the time interval $(t, t+dt]$. At the time of an impulse occurrence there is a jump in the velocity process. The process $Y_2(t)$ may be regarded as a compound function (strictly speaking as a functional) of the excitation process $\Gamma(t)$, i.e., $Y_2 = Y_2[t, \Gamma(t)]$. In order to evaluate $dY_2(t)$ we may use the Itô's differential rule (10) for $Y_2(t) = V(t, X(t))$, with $X(t)$ governed by (7), which gives the result similar to the Taylor expansion in the impulse magnitude (cf. [5, 7])

$$\begin{aligned} dY_2(t) &= Y_2(t+dt) - Y_2(t) = a\,(t, \mathbf{Y}(t))\,dt + \frac{\partial Y_2}{\partial \Gamma} d\Gamma(t) \\ &\quad + \frac{1}{2}\frac{\partial^2 Y_2}{\partial \Gamma^2}\Big(d\Gamma(t)\Big)^2 + \frac{1}{3!}\frac{\partial^3 Y_2}{\partial \Gamma^3}\Big(d\Gamma(t)\Big)^3 + \ldots, \end{aligned} \tag{15}$$

where $\dfrac{\partial Y_2}{\partial \Gamma} = b(t, \mathbf{Y}(t))$ is a gradient of the state variable $Y_2(t)$ with respect to the excitation process $\Gamma(t)$. For example if we use (2) or (6) then the gradient is obtained, respectively, as

$$\frac{\partial Y(t)}{\partial F} = bY(t), \quad \frac{\partial Y(t)}{\partial \Gamma} = bY(t). \tag{16}$$

Note that for a continuous excitation $f(t)$, of a bounded variation, the Taylor expansion (15) yields an obvious result

$$\begin{aligned} dY_2(t) &= Y_2(t+dt) - Y_2(t) = a\,(t, \mathbf{Y}(t))\,dt + \frac{\partial Y_2}{\partial F} dF \\ &= a\Big(t, \mathbf{Y}(t)\Big)dt + b\Big(t, \mathbf{Y}(t)\Big)dF(t), \end{aligned} \tag{17}$$

because all higher order powers $(dF)^k = f^k(t)(dt)^k$, for $k > 1$ are infinitely small of orders higher than dt.

In the present problem we first have to take into account that due to the regularity of the random counting process $\mathcal{N}(t)$, the following replacement holds with probability 1 (cf. [5, 13])

$$\begin{aligned}\Big(d\Gamma(t)\Big)^k &= \Big(P(t)d\mathcal{N}(t)\Big)^k = P^k(t)\Big(d\mathcal{N}(t)\Big)^k = P^k(t)d\mathcal{N}(t),\\ & k > 1.\end{aligned} \tag{18}$$

Hence the Taylor expansion (15) is infinite; all terms are of the same order $d\mathcal{N}(t)$.

Higher-order partial derivatives are found as (cf. [1])

$$\begin{aligned}\frac{\partial^2 Y_2}{\partial \Gamma^2} &= \frac{\partial}{\partial \Gamma}\left(\frac{\partial Y_2}{\partial \Gamma}\right) = \frac{\partial}{\partial Y_1}\left(\frac{\partial Y_2}{\partial \Gamma}\right)\frac{\partial Y_1}{\partial \Gamma} + \frac{\partial}{\partial Y_2}\left(\frac{\partial Y_2}{\partial \Gamma}\right)\frac{\partial Y_2}{\partial \Gamma}\\ &= \frac{\partial b\big(t,\mathbf{Y}\big)}{\partial Y_2} b\big(t,\mathbf{Y}\big),\end{aligned} \tag{19}$$

$$\begin{aligned}\frac{\partial^3 Y_2}{\partial \Gamma^3} &= \frac{\partial}{\partial \Gamma}\left(\frac{\partial^2 Y_2}{\partial \Gamma^2}\right) = \frac{\partial}{\partial Y_1}\left(\frac{\partial^2 Y_2}{\partial \Gamma^2}\right)\frac{\partial Y_1}{\partial \Gamma} + \frac{\partial}{\partial Y_2}\left(\frac{\partial^2 Y_2}{\partial \Gamma^2}\right)\frac{\partial Y_2}{\partial \Gamma}\\ &= \frac{\partial}{\partial Y_2}\left(\frac{\partial b\big(t,\mathbf{Y}\big)}{\partial Y_2} b\big(t,\mathbf{Y}\big)\right) b\big(t,\mathbf{Y}\big)\\ &= \frac{\partial^2 b\big(t,\mathbf{Y}\big)}{\partial Y_2^2} b^2\big(t,\mathbf{Y}\big) + \left(\frac{\partial b\big(t,\mathbf{Y}\big)}{\partial Y_2}\right)^2 b\big(t,\mathbf{Y}\big),\end{aligned} \tag{20}$$

where it has been used that the gradient $\dfrac{\partial Y_1}{\partial \Gamma} = 0$.

The above results are fully comparable to those in [1]. Here, however, the expansion (15) is infinite, which is consistent with the interpretation of the random impulse process as an infinite-order delta correlated process (cf. [9]).

If the term $b\big(t,\mathbf{Y}(t)\big)$ is only a function of the state variable $Y_1(t)$, all higher-order derivatives in (15) are equal to zero and the diffusion term is not corrected.

3.2 Dynamical systems with linear term $b(t, \mathbf{Y}(t))$

Consider a second-order dynamical system, under multiplicative random impulse process excitation. If the term $b(t, \mathbf{Y}(t))$ in the equation of motion

(13) equals $b(t, \mathbf{Y}(t)) = bY_2(t)$, the partial derivatives in (15) become:

$$\begin{aligned}\frac{\partial Y_2}{\partial \Gamma} &= bY_2,\\ \frac{\partial^2 Y_2}{\partial \Gamma^2} &= b^2 Y_2,\\ \frac{\partial^3 Y_2}{\partial \Gamma^3} &= b^3 Y_2,\\ \frac{\partial^4 Y_2}{\partial \Gamma^4} &= b^4 Y_2, \quad \text{e.t.c.}\end{aligned} \tag{21}$$

Obviously the infinite series in (15) may be summed up

$$\begin{aligned}&bP(t)Y_2 + \frac{1}{2}b^2P^2(t)Y_2 + \frac{1}{3!}b^3P^3(t)Y_2 + \frac{1}{4!}b^4P^4(t)Y_2 + \cdots\\ &= \Big(\exp\big(bP(t)\big) - 1\Big)\, Y_2,\end{aligned} \tag{22}$$

hence the stochastic equations are arrived at the form (cf. (12) and [2])

$$\begin{aligned}dY_1(t) &= Y_2(t)dt,\\ dY_2(t) &= a\left(t, \mathbf{Y}(t)\right)dt + \Big(\exp\big(bP(t)\big) - 1\Big)\, Y_2 d\mathcal{N}(t).\end{aligned} \tag{23}$$

3.3 Dynamical systems with quadratic term $b(t, \mathbf{Y}(t))$

Consider a first-order dynamical system, with quadratic term $b\big(t, Y(t)\big) = bY^2(t)$, governed by the differential equation

$$\dot{Y}(t) = aY^2(t) + bY^2(t)\xi(t) \quad Y(0) = Y_0, \tag{24}$$

The partial derivative terms in (15) are

$$\begin{aligned}\frac{\partial Y}{\partial \Gamma} &= bY^2,\\ \frac{1}{2}\frac{\partial^2 Y}{\partial \Gamma^2} &= b^2 Y^3,\\ \frac{1}{3!}\frac{\partial^3 Y}{\partial \Gamma^3} &= b^3 Y^4,\\ \frac{1}{4!}\frac{\partial^4 Y}{\partial \Gamma^4} &= b^4 Y^5, \quad \text{e.t.c.,}\end{aligned} \tag{25}$$

which forms the geometric progression, hence the infinite sum in (15) is

$$\begin{aligned}&bP(t)Y^2 + b^2P^2(t)Y^3 + b^3P^3(t)Y^4 + b^4P^4(t)Y^5 + \cdots\\ &\qquad = \frac{bP(t)Y^2}{1 - bP(t)Y},\end{aligned} \tag{26}$$

and the resulting stochastic equation is

$$dY(t) = aY^2(t)dt + \frac{bP(t)Y^2}{1 - bP(t)Y} d\mathcal{N}(t), \quad Y(0) = Y_0. \tag{27}$$

To solve this equation we use the resolving function $V(Y(t)) = \dfrac{1}{Y(t)}$ and apply the generalized Itô's differential rule (9), which gives

$$\begin{aligned} dV(Y) &= d\left(\frac{1}{Y}\right) = -\frac{1}{Y^2} aY^2 dt + \left(\frac{1}{Y + \frac{bPY^2}{1-bPY}} - \frac{1}{Y}\right) d\mathcal{N}(t) \\ &= -adt - bP(t) d\mathcal{N}(t). \end{aligned} \tag{28}$$

Integration over time from 0 to t yields

$$Y(t) = \frac{Y_0}{1 - Y_0\Big(at + b\Gamma(t)\Big)} \tag{29}$$

This is a stochastic analogue of the deterministic solution of (24) driven by a continuous excitation $f(t)$, in the same way as in the case of a linear diffusion term the stochastic solution (6) is an analogue of the solution (2) to the deterministic (1).

4. Conclusion

A technique is devised to convert the usual differential equation of motion into an Itô's type stochastic differential equation for random (not necessarily Poisson) impulse process excitations multiplicative to the velocity term of the equation of motion. The use is made of the generalized Itô's differential rule and of the properties of the increments of the underlying regular random counting process. It is found that the diffusion term rather than drift term has to be corrected. For linear and quadratic velocity terms the closed-form expressions for corrected diffusion terms are obtained.

References

[1] M. Di Paola and G. Falsone, "Stochastic Dynamics of Nonlinear Systems Driven by Non-normal Delta-correlated Processes," *Journal of Applied Mechanics, ASME*, vol. 60, pp. 141–148, 1993.

[2] M. Di Paola and M. Vasta, "Stochastic Integro-differential and Differential Equations for Non-linear Systems Excited by Parametric Poisson Pulses," *Int. J. Non-linear Mechanics*, vol. 32, no. 5, pp. 855–862, 1997.

[3] I. I. Gikhman and A. J. Dorogovtsev, "On stability of solutions of stochastic differential equations," Ukrain., Math. Zhurnal, vol. 17, pp. 229–250, 1965 (in Russian).

[4] I. I. Gikhman and A. W. Skorokhod, *Stochastic differential equations*, Berlin, Springer-Verlag, 1972.

[5] R. Iwankiewicz and S. R. K. Nielsen, "Advanced methods in stochastic dynamics of nonlinear systems," *Vibration Theory*, vol. 4., Aalborg University Press, Denmark, 1999.

[6] R. Iwankiewicz and S. R. K. Nielsen, "Solution techniques for pulse problems in nonlinear stochastic dynamics," *Probabilistic Engineering Mechanics*," vol. 15, no. 1, pp. 25–36, 2000.

[7] H. U. Köylüoğlu, S. R. K. Nielsen, and A. S. Çakmak A., "Fast Cell-to-Cell Mapping (Path Integration) with Probability Tails for the the Stochastic Response of Non-Linear White Noise and Poisson Driven Systems," *Structural Safety*, vol. 17, pp. 151–165, 1995.

[8] D. L. Snyder, *Random point processes*, New York, John Wiley, 1975.

[9] R. L. Stratonovich, "A new form of representing stochastic integrals and equations," *SIAM J. Control*, vol. 4, pp. 362–371, 1966.

[10] M. Grigoriu, "The Itô and Stratonovich Integrals for Stochastic Differential Equations with Poisson White Noise," *Probabilistic Engineering Mechanics*, vol. 13, no. 3, pp. 175–182, 1998.

[11] R. C. Merton, "Option pricing when underlying stock returns are discontinuous," *J. Financial Economics*, vol. 3, pp. 125–144, 1976.

[12] A. Tylikowski, "Vibration of a harmonic oscillator due to a sequence of random collisions," *Proc. of the Inst. of Machine Design Foundation*, no. 13, Warsaw University of Technlogy, 1982, (in Polish).

[13] S. K. Srinivasan, *Stochastic point processes and their applications*, Griffin, London, 1974.

[14] E. Wong and M. Zakai, "On the relation between ordinary and stochastic differential equations," *International Journal of Engineering Sciences*, vol. 3, pp. 213–229, 1965.

MULTISCALE IMPACT MODELS: MULTIBODY DYNAMICS AND WAVE PROPAGATION

W. Schiehlen and R. Seifried
Institute B of Mechanics
University of Stuttgart
Pfaffenwaldring 9
D-70550 Stuttgart, GERMANY
[wos,rs]@mechb.uni-stuttgart.de

Abstract For the investigation of impact processes different time scales are used. Using multibody dynamics the overall motion is computed on a slow time scale. The elastodynamic effects are computed on a fast time scale using analytical and numerical models. Then, the fast time scale results are used to model the impact on the slow time scale accurately. Experimental validation proves the accuracy of the impact simulations on multiple time scales for impacts on an aluminum rod, a brass rod, and a half-circular plate as well as an aluminum ball. While the rods produce clearly deterministic signals, the ball responds with random signals.

Keywords: Impact, multiscale problem, FEM, multibody dynamics, coefficient of restitution

1. Introduction

There is no doubt that the method of multibody systems is most efficient for the dynamical analysis of the overall motion of machines, vehicles and all kinds of mechanical devices. Collisions in multibody systems may result in impacts between the bodies involved. During impact kinetic energy is lost by plastic deformations, viscoelastic effects and wave propagation. For rigid body models this energy loss is measured macromechanically by the coefficient of restitution. This coefficient has to be estimated from experiments and experience but it cannot be computed within the multibody system approach. In this paper elastic impacts without any plastic deformations or viscoelastic effects are considered. However, the impacts generate waves in the bodies which propagate until they vanish due to material damping. Especially in slender bodies, such as rods, beams, plates and shells, a large amount of the ini-

N. Sri Namachchivaya and Y.K. Lin (eds.),
IUTAM Symposium on Nonlinear Stochastic Dynamics, 353-362.

tial kinetic energy is transformed into wave propagation. To investigate these high frequency phenomena, the equations of motion for elastodynamics are introduced and solved by D'Alembert's approach for wave propagation or discretization by finite elements, respectively. These very expensive computations on a fast time scale can be limited to approximately five cycles of the lowest frequencies of the response signal. Then, these results are used to compute the coefficient of restitution and is fed back to the multibody system equations and the solution continues on the related slow time scale.

The efficiency and accuracy of this approach is shown by numerical and experimental investigations on a slow and fast time scales for the impact of a steel sphere on different objects: a slender aluminum cylinder called aluminum rod in the following, a brass rod, a half-circular plate and a short aluminum cylinder called ball in the following due to its similar behavior to an aluminum ball of same mass. The three investigated aluminum objects have approximately the same mass. The material and geometric data of the colliding bodies are summarized in Table 1.

Table 1. Geometrical and material data

	steel sphere	alu. rod	brass rod	alu. plate	alu. ball
radius [mm]	15	10	10	200	36.48
thickness [mm]	–	–	–	5.1	–
length [mm]	–	1000	500	–	73.68
E-modulus [GPa]	210	70.5	100.6	72.55	70.5
poisson ratio	0.3	0.33	0.36	0.33	0.33
density [kg/m^3]	7780	2710	8403	2710	2710
mass [kg]	0.11	0.851	1.319	0.868	0.851

2. Multibody Systems with Impact

Before and after impact the multibody system motion is clearly divided into two periods by different sets of initial conditions. The equations of motion of both periods read as

$$\boldsymbol{M}(\boldsymbol{x})\ddot{\boldsymbol{x}} + \boldsymbol{k}(\boldsymbol{x}, \dot{\boldsymbol{x}}) = \boldsymbol{q}(\boldsymbol{x}) \qquad (1)$$

where $\boldsymbol{x}(t)$ is the global position vector featuring f degrees of freedom, $\boldsymbol{M}(\boldsymbol{x})$ the inertial matrix, $\boldsymbol{k}(\boldsymbol{x}, \dot{\boldsymbol{x}})$ the vector of Coriolis and gyroscopic forces and $\boldsymbol{q}(\boldsymbol{x})$ the vector of the applied forces. For more details see [10].

The steady motion of the first period of the mechanical system is interrupted by collision. The hypothesis of an infinitely large impact force leads to the force impulse

$$\triangle \boldsymbol{p} = \lim_{\varepsilon \to 0} \int_{t_1-\varepsilon}^{t_1+\varepsilon} \boldsymbol{f}(\tau)\, d\tau \tag{2}$$

changing the velocities of the motion. Due to the infinitely large impact force all other forces can be neglected during impact. This results in an unsteady system

$$\boldsymbol{M}(\boldsymbol{x})\ddot{\boldsymbol{x}} + \boldsymbol{k}(\boldsymbol{x}, \dot{\boldsymbol{x}}) = \boldsymbol{q}(\boldsymbol{x}), \quad 0 < t < t_1 - \varepsilon\,,\; t_1 + \varepsilon < t < \infty\,. \tag{3}$$

describing more precisely the motion before and after impact. At the instant of the impact $t = t_1$, the initial velocities of the second period have to be reset, however the displacements remain unchanged. For the instant of impact the impulse equations include the coefficient of restitution which represents the energy loss during impact. This approach results finally in linear complimentary problems (LCPs) as shown in [8].

3. Elastodynamic Contact Using Wave Propagation

Using the equation of motion for elastodynamics and solving them by D'Alembert's approach for wave propagation is a very time effective method [4, 5] to simulate impacts on the fast time scale. However, this approach is limited to geometrically simple bodies, such as the longitudinal impact of a sphere on a rod as shown in the following. The longitudinal waves in a rod [1] are governed by the partial differential equation

$$\frac{\partial^2 u(x,t)}{\partial t^2} = c^2 \frac{\partial^2 u(x,t)}{\partial x^2} \quad \text{with} \quad c = \sqrt{\frac{E}{\rho}} \tag{4}$$

where c represents the wave speed, E, ρ are the rod's Young's modulus and density, respectively. According to D'Alembert's approach the general solution of Eq. (4) reads as

$$u(x,t) = f(x - ct) + g(x + ct), \tag{5}$$

where f and g are real functions representing a forward and a backward traveling wave, respectively. Using Eq. (4) and Eq. (5) with the dynamic boundary condition

$$-EA\frac{\partial u(0,t)}{\partial x} = F(t) \tag{6}$$

for the struck end, the impact of a sphere on a rod is modeled. Thereby the contact force $F(t)$ between the sphere and the rod is described by a Hertzian elastostatic contact law [6]

$$F(t) = max[0, k\delta(t)^{3/2}]. \tag{7}$$

The contact stiffness k and the indentation $\delta(t)$ read as

$$k = \frac{4\sqrt{R_s}}{3(\frac{1-\nu_s^2}{E_s} + \frac{1-\nu^2}{E})} \quad \text{and} \quad \delta(t) = x_s(t) - u(0,t) \tag{8}$$

where R_s, E_s, E, ν_s, ν are radius, Young's modulus and Poisson's ratio of the sphere and the rod and $x_s(t)$ describes the sphere's position. The good agreement of this analytical model with experimental results is shown in [5].

4. Contact Simulation Using Finite Elements

For investigation of elastodynamic impact effects on the fast time scale involving bodies with complex contact and geometric shapes, numerical methods such as the FEM have to be used. The solution of a frictionless static contact problem is the nodal displacement vector $\boldsymbol{u}$ which minimizes the potential energy function $\Pi(\boldsymbol{u})$ of the FE system considering the impenetrability condition $\boldsymbol{g}(\boldsymbol{u}) \geq \boldsymbol{0}$ where the penetrations of the bodies are summarized in $\boldsymbol{g}(\boldsymbol{u})$. Due to its simplicity the penalty method is used in most commercial FE-codes for solving this restricted minimization problem. Other common approaches are the Lagrange multiplier method, the augmented Lagrangean method and hybrid methods.

Using the penalty method the potential energy function $\Pi(\boldsymbol{u})$ is extended by a penalty term of the form $\frac{1}{2}P\boldsymbol{g}^T\boldsymbol{g}$ where P is called the penalty factor. The minimization of the extended potential results in the nodal displacements $\boldsymbol{u}$. The contact force is computed by $\boldsymbol{F} = P\boldsymbol{g}$. For penalty factors P approximating infinity the penetration vanishes and the solution converges to physical reality. However, too high penalty factors result in numerical instabilities. An optimal choice of the penalty factor for static contact is described by Nour-Omid and Wriggers [7].

For impact problems the influence of the penalty factor and the discretization is investigated in [11] using the sphere to rod impact as a test example. It is shown that the choice of the penalty factor and the discretization of the contact area have a significant influence on the time response of the calculated impact force. Especially the independence of the results from the choice of the penalty factor has to be checked by additional simulations. Also great attention has to be given to the evaluation of the resulting wave propagation in the elastic bodies which requires a small element size.

5. Experimental Validation

For the experimental validation of the simulation results for the impact processes on a slow and a fast time scale the bodies are suspended in a frame with thin kevlar wires like pendulums [5]. At the beginning the steel sphere is released by an magnet from a predefined height. The rod, plate or ball, respectively are initially in rest. The longitudinal impact on the rods and the ball occurs along the central axis of the bodies and the radial impact on the half-circular plate occurs along the symmetry axis on the curved side of the plate. When rebounding after impact the sphere is set to rest again. Thus, no second impact occurs and the rod or plate is moving like a free pendulum.

For the measurement of displacements and velocities of the bodies, Laser-Doppler-Vibrometers of type OFV-3000/OFV-302 made by Polytec GmbH [9] are used. The vibrometer utilizes an interferometric technique to measure displacement signals. Velocity signals are measured based on the Doppler-effect and the direction of motion is obtained by using a Bragg-cell. The measured signals are sampled with a frequency of $200kHz$.

The left plot of Fig. 1 shows the measured velocities at the free end of

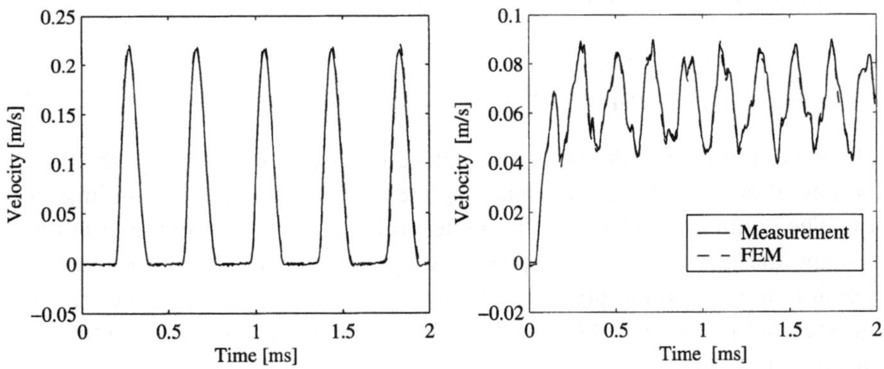

Figure 1. Velocity of the free end of the aluminum rod and the half-circular plate

the aluminum rod on the fast time scale. During impact kinetic energy is transformed into strain energy which propagates as wave away from the contact region. This wave is reflected at the free end of the rod. The investigated rods are so long that the contact is dissolved before the reflected wave reaches the struck end. Then, the wave is reflected again. After the wave passes a point of the rod, e.g., the free end, this point remains in rest again until the wave passes through it again. The right plot shows the measured velocity of the half-circular plate. In the half-circular plate the initiated waves are multiply reflected at the boundaries and then the motion is described by a superposition of a rigid body motion and an irregular oscillation. In both plots the results of the FE-simulation using ANSYS [2] are indicated, too. This shows the very good agreement of FE-simulations and experiments for the fast time scale. For more details see [11, 12].

6. Computation of the Coefficient of Restitution

For the simulation on the slow time scale using the multibody system approach the coefficient of restitution is required to describe the kinetic energy lost during impact. Therefore, the kinetic coefficient of restitution by Poisson [8, 13] is used. The coefficient of restitution by Poisson is defined as the ratio of linear momentum during the compression and restitution phases of the impact. The coefficient of restitution cannot be determined within the multibody system approach but must be estimated by costly experiments [3] or experience. However, using the results of the simulation on the fast time scale the coefficient of restitution can be determined computationally.

Following the assumptions of rigid body impact, the impact duration is infinitesimally small and all forces but the impact force are negligibly small. The linear momentum balance in the central impact line of two colliding bodies during the compression and restitution phases reads as

$$\triangle p_c = m_1(v_1^- - v), \qquad \triangle p_c = m_2(v - v_2^-), \tag{9}$$

$$\triangle p_r = m_1(v - v_1^+), \qquad \triangle p_r = m_2(v_2^+ - v), \tag{10}$$

where v_1^- and v_2^- are the velocities before impact and v_1^+ and v_2^+ are the velocities after impact and m_1 and m_2 are the masses of the colliding bodies. The common velocity of the bodies at the end of the compression phase is v and the linear momentum produced by the impact force during the compression and restitution phase is denoted by $\triangle p_c$ and $\triangle p_r$, respectively. Using the coefficient of restitution by Poisson, the total linear momentum produced by the impact reads as

$$\triangle p = \triangle p_c + \triangle p_r = \triangle p_c(1 + e). \tag{11}$$

Rearranging Eq. (9) to (11), the coefficient of restitution can be expressed in dependence of the total linear momentum during impact as,

$$e = \frac{(m_1 + m_2)\triangle p}{m_1 m_2(v_1^- - v_2^-)} - 1. \tag{12}$$

The total linear momentum $\triangle p$ is known from the previously performed simulations on the fast time scale as a time integral of the computed impact force. Therefore, the coefficient of restitution can be determined by the simulation results on the fast time scale.

With an initial velocity of $0.3\frac{m}{s}$ for the steel sphere, a coefficient of restitution of 0.625 is computed for the impact on the aluminum rod. Using this value, the rigid body motion of the bodies can be simulated efficiently using multibody dynamics. The simulation result is shown in Fig. 2 and compared with measurements. This shows the good agreement of the rigid body

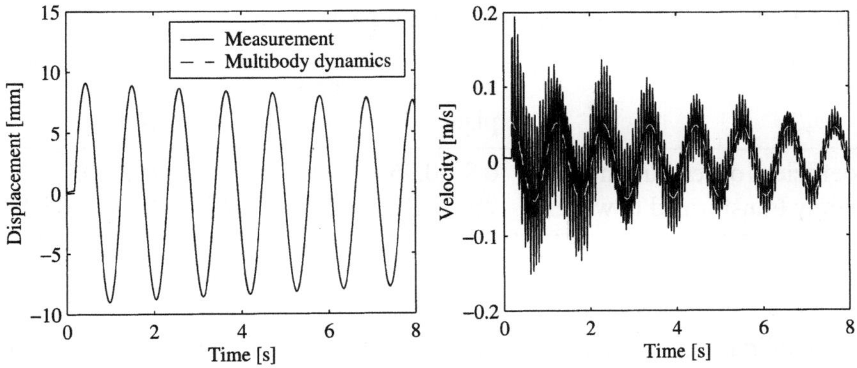

Figure 2. Slow time scale: aluminum rod

motion. The velocities differ at the beginning due to the strong wave phenomena which cannot be captured by the rigid body model.

The coefficient of restitution depends not only on the material parameters, the contact geometry and the body geometry but also on the initial velocity [3]. The influence of the velocity is shown in the right plot of Fig. 3 for the half-

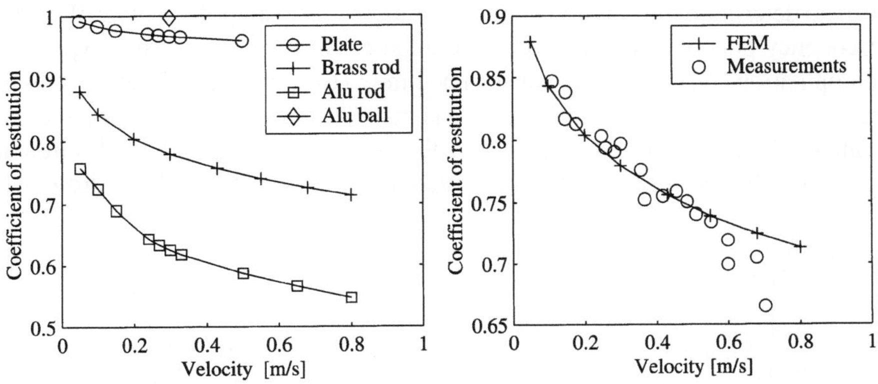

Figure 3. Computed coefficient of restitution and comparison with measurements for the brass rod

circular plate and both rods in a range of 0.05 to $0.8\frac{m}{s}$. The computed kinetic energy loss during impact is purely due to the initiation of waves in the bodies and is summarized in Table 2. The simulation of the impact on the aluminum ball shows that very little energy is transformed into wave propagation and the computed coefficient of restitution is close to 1. Due to the short dimensions of the ball, multiple reflected waves travel during impact with high frequency back and forth through the ball and, therefore, the ball behaves nearly rigidly.

Table 2. Kinetic energy transformed to vibrations during impact

	alu. plate	brass rod	alu. rod
coefficient of restitution	0.99-0.96	0.88-0.71	0.77-0.55
energy transformed to waves	2% – 7%	21% – 46%	36% – 62%

To validate the computed coefficients of restitution, first experiments are performed for the impact on the brass rod, shown in the left plot of Fig. 3. The measured coefficients are always the mean value of 5 measurements. The computed and measured coefficients agree very well for velocities below $0.55\frac{m}{s}$. For higher velocities the measured coefficient drops significantly below the computed value. This is due to plastic deformations which are not included in the simulations.

7. Signal Analysis

The simulation on the fast time scale does not only give valuable information to determine the coefficient of restitution but also helps to analyze the high frequency phenomena initiated by the impact. However, these very expensive computations on a fast time scale can be limited to approximately five cycles of the lowest frequencies of the response signal. The spectral density of the computed velocity signals (see Fig. 1) for the aluminum rod and the half-circular plate are shown in Fig. 4. The longitudinal waves in the rod show strong pe-

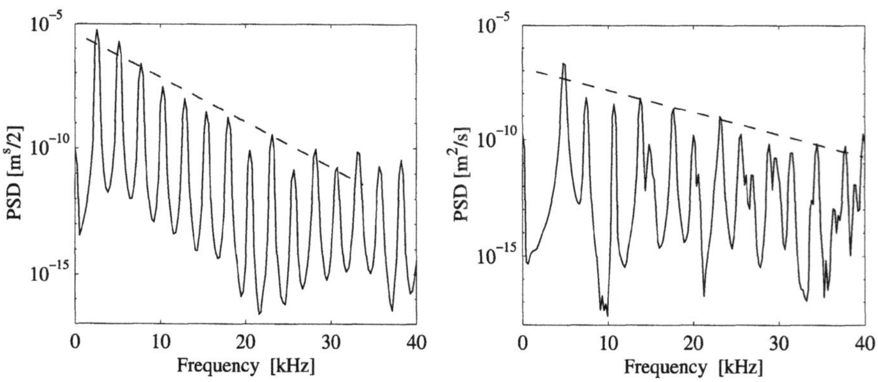

Figure 4. Spectral density for the aluminum rod and plate

riodicity where the strength of the excited modes decrease strongly with increasing frequency. In contrast the plane waves in the plate show a highly irregular pattern, even for the linear structural model. Thus, the decrease of the modes' strength with increasing frequency is less significant. The impact on the aluminum ball results in a rigid body motion of the ball which is superimposed by small vibrations as shown in Fig 5. The lowest eigenfre-

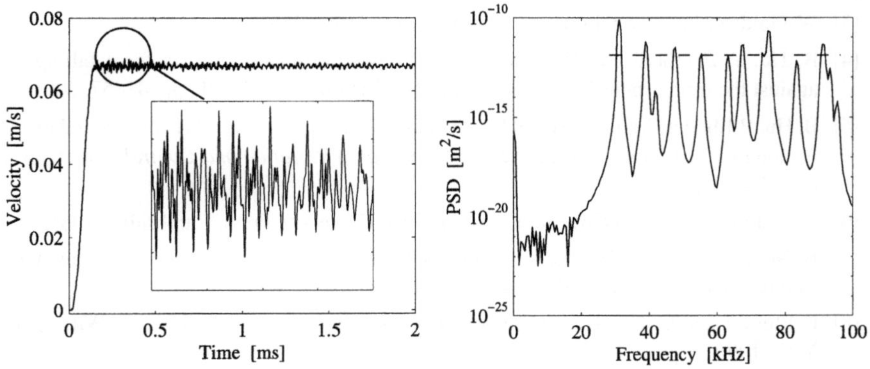

Figure 5. Impact on the ball

quency of the ball is in the range of $30kHz$. The signal analysis of the vibrations show that for a wide frequency range the modes are excited mostly equally, similar to white noise.

Signal analysis of the simulation results helps to investigate experimental observations such as the tone and noise produced. The impact on the rods produces strong periodic waves which propagate in the rod for a long time before they vanish due to material damping. This produces a clear, harmonical, enduring tone. Due to irregular mode excitation the impact on the plate produces a metallic sound. For the impact on the ball only a very short dull noise is observed, due to a relatively equal excitation of higher modes featuring white noise as found in stochastic processes, too.

8. Conclusion

The main results of the paper include impacts on rods with longitudinal waves, impacts on a plate with plane waves and impacts on a ball. While the longitudinal waves show strong periodicity, the plane and spatial waves may be highly irregular even for linear structural dynamics models. In conclusion, the interaction of multibody dynamics and wave propagation represents an efficient tool for the dynamical analysis of nonlinear systems with impacts. The required coefficient of restitution is determined from the fast time scale results. The approach allows the consideration of nonlinear material properties, too. Then, stochastic methods for the dynamical analysis will be most helpful.

References

[1] J. Achenbach, *Wave Propagation in Elastic Solids*, North Holland, Amsterdam, 1975.

[2] ANSYS Inc.: *ANSYS Theory Reference*, Release 5.4., Canonsburg, ANSYS Inc., 1997.

[3] W. Goldsmith, *Impact*, London: Edward Arnold Ltd., 1960.

[4] B. Hu, P. Eberhard and W. Schiehlen, "Symbolical impact analysis for a falling rod against the rigid ground," *Journal of Sound and Vibration*, vol. 240, pp. 41–57, 2001.

[5] B. Hu, P. Eberhard, and W. Schiehlen, "Comparison of analytical and experimental results for longitudinal impacts on elastic rods," *Journal of Vibration and Control*, vol. 9, no. 1, pp. 157–174, 2003.

[6] K. L. Johnson, *Contact Mechanics*, Cambridge University Press, Cambridge, 1985.

[7] B. Nour-Omid and P. Wriggers, "A note on the optimum choice for penalty parameters," *Communications in Applied Numerical Methods*, vol. 3, pp. 581–585, 1987.

[8] F. Pfeiffer and C. Glocker, *Multibody Dynamics with Unilateral Contacts*, Wiley, New York, 1996.

[9] Polytec GmbH: *Vibrometer's manual for Polytec Vibrometer Series OFV-3000/OFV-302, OFV501 and OFV502*. Manual No. VIB-MAN-9308-e04/01, Polytec GmbH, 1994.

[10] W. Schiehlen, "Multibody System Dynamics: Roots and Perspectives," *Multibody System Dynamics*, vol. 1, pp. 149–188, 1997.

[11] R, Seifried, B. Hu, and P. Eberhard, "Numerical and experimental investigation of radial impacts on a half-circular plate," *Multibody Systems Dynamics*, (accepted for publication).

[12] R. Seifried, "Experimentelle und numerische Untersuchung von gesto"senen Kreisscheiben." Diplomarbeit DIPL-85. Stuttgart: University of Stuttgart: Institut B of Mechanics, 2001.

[13] W. J. Stronge, *Impact Mechanics*, Cambridge: Cambridge University Press, 2000.

USE OF ORTHOGONAL DECOMPOSITION OF DATA-BASED EXCITATION PROCESSES FOR NONSTATIONARY RESPONSE OF NONLINEAR SYSTEMS

Andrew W. Smyth
Dept. of Civil Engineering & Engineering Mechanics
Columbia University
New York, New York, USA
smyth@civil.columbia.edu

Sami F. Masri
School of Engineering
University of Southern California
Los Angeles, California, USA
masri@usc.edu

Abstract The topic of stochastic response of dynamic systems has been an area of considerable interest for some time in the analysis of risk and structural reliability. A substantial reference list of work in this area can be found in [4]. The authors, in previous work [3, 6], have developed a method which can analyze the response of linear multi-degree-of-freedom systems to completely general data-based nonstationary excitations in a highly efficient and analytical form. This was demonstrated with successful application to a model structure subjected to an ensemble of ground-motion recordings from the 1994 Northridge (California) Earthquake. The authors have now extended this work to nonlinear system response by using equivalent linearization techniques [7]. This paper summarizes the extension to the analysis of nonlinear systems, and explores the range of application of the technique.

Keywords: Equivalent Linearization, Random Vibrations, Nonstationary, Karhunen Loéve

N. Sri Namachchivaya and Y.K. Lin (eds.),
IUTAM Symposium on Nonlinear Stochastic Dynamics, 363-373.

1. Overview of Nonstationary Excitation Compact Representation

The fundamental starting point for the development of the new technique is in the compact analytical data approximation used for the nonstationary excitation. After applying Karhunen-Loéve spectral decomposition to the measured excitation process covariance matrix $\mathfrak{C}_k(t_1, t_2)$, and least-squares fitting the eigenvectors with orthogonal polynomials, the excitation data is condensed into the following approximate form:

$$\hat{\mathfrak{C}}_k(t_1, t_2) = \sum_{i=1}^{k} \lambda_i \sum_{j=0}^{m_i-1} \sum_{\ell=0}^{m_i-1} H_{ij} H_{i\ell} T_j(t_1') T_\ell(t_2'), \tag{1}$$

where the λ_i's are the truncated series of eigenvalues, the $T_j(t_1')$'s are Chebyshev polynomials, and the H's are weighting coefficients for the Chebyshev polynomials in the least-squares fitting of the eigenvectors. In this equation, $t_i' = 2t_i/t_{max} - 1$. For additional details the reader is referred to [3]. In effect, the conversion from measured data to analytical representation involves the approximation of the discrete *eigenvectors* of the excitation covariance matrix by analytical *eigenfunctions* (given as a polynomial series).

This compact analytical expression is in such a form which allows the closed-form solution of the nonstationary response problem for general (i.e., without assuming proportional damping) linear systems of the form:

$$\dot{\mathbf{x}} = \mathbf{A}\mathbf{x} + \mathbf{B}\mathbf{u}. \tag{2}$$

2. Statistical Linearization and Nonstationary Excitation

Since in the case of nonlinear system response it is difficult to obtain probabilistic response solutions for nonstationary (and even stationary) excitations, researchers have turned to approximating the nonlinear system as an *equivalent linear* system. The reader may refer to *Random Vibration and Statistical Linearization* [4] for a comprehensive treatment of the subject.

Consider the following nonlinear *n* degree of freedom equation of motion

$$\mathbf{M}\ddot{\mathbf{q}}(t) + \mathbf{C}\dot{\mathbf{q}}(t) + \mathbf{K}\mathbf{q}(t) + \mathbf{g}(\mathbf{q}(t), \dot{\mathbf{q}}(t)) = \mathbf{F}(t) \tag{3}$$

where $\mathbf{g}$ is a nonlinearity dependent upon the system displacements and velocities, and $\mathbf{F}(t)$ is a random vector process of length n. After some manipulation, this nonlinear system can be written in its equivalent linear state-space form (see [4]):

$$\dot{\mathbf{z}} = \mathbf{G}(t)\mathbf{z} + \mathbf{f} \tag{4}$$

where

$$\mathbf{z}(t) = \begin{bmatrix} \mathbf{q}_0(t) \\ \dot{\mathbf{q}}_0(t) \end{bmatrix} \text{ and } \mathbf{q}_0(t) = \mathbf{q}(t) - \boldsymbol{\mu}_\mathbf{q}(t)$$

$$\mathbf{G}(t) = \begin{bmatrix} \mathbf{0} & \mathbf{I} \\ -\mathbf{M}^{-1}(\mathbf{K}+\mathbf{K_e}) & -\mathbf{M}^{-1}(\mathbf{C}+\mathbf{C_e}) \end{bmatrix} \tag{5}$$

$$\mathbf{f} = \begin{bmatrix} \mathbf{0} \\ -\mathbf{M}^{-1}\mathbf{F}_0 \end{bmatrix} \quad \text{and} \quad \mathbf{F}_0(t) = \mathbf{F}(t) - \boldsymbol{\mu}_\mathbf{F}(t)$$

and where $\mathbf{C_e}$ and $\mathbf{K_e}$ are *time-dependent* equivalent linear terms which account for the nonlinear $\mathbf{g}$; this contrasts with stationary statistical linearization problems where these matrices are constant.

The goal is to get a second order probabilistic description of the response state $\mathbf{z}$, i.e. $\mathbf{V}(t) \equiv E\{\mathbf{z}(t)\mathbf{z}^T(t)\}$. The differential equation which defines the evolution of $\mathbf{V}(t)$ is

$$\dot{\mathbf{V}} = \mathbf{G}(t)\mathbf{V} + \mathbf{V}\mathbf{G}^T(t) + \mathbf{U}(t) + \mathbf{U}^T(t) \qquad \text{where} \quad \mathbf{U}(t) = E\{\mathbf{z}\mathbf{f}^T\} \tag{6}$$

After incorporating the compact form of the excitation process, one may write $\mathbf{U}(t)$ as

$$\mathbf{U}(t) = \sum_{i=1}^{k} \lambda_i \sum_{j=0}^{m_i-1} \sum_{\ell=0}^{m_i-1} H_{ij} H_{i\ell} T_j(t') \mathbf{U}^{(\ell)}(t) \tag{7}$$

and after some further manipulation $\mathbf{U}^{(\ell)}(t)$ can be expressed as

$$\dot{\mathbf{U}}^{(\ell)}(t) = \mathbf{G}(t)\mathbf{U}^{(\ell)}(t) + \mathbf{R}T_\ell(t') \tag{8}$$

where $\mathbf{R}$ is a very simple constant matrix which defines the degrees of freedom to which the force is applied. For example, in the single input case

$$\mathbf{W_f}(t_1, t_2) \equiv E\{\mathbf{f}(t_1)\mathbf{f}^T(t_2)\} = \mathbf{R}\mathcal{C}_{ss}(t_1, t_2) \tag{9}$$

This new solution is highly efficient, contrasting with an analogous solution by [4] which is limited to a restrictive class of nonstationary excitations, and involves 3-index terms $\mathbf{U}^{(ijk)}(t)$ which increases the number of terms in the nested summations in (7). For a detailed comparison of the two approaches, the reader is referred to Smyth and Masri, (2002).

3. Example Application

In [7], several applications are presented to demonstrate the effectiveness of the simulation technique. The methodology was used to obtain the probabilistic response of a single-degree-of-freedom Duffing oscillator to several excitations. Firstly, to demonstrate that the methodology could at least perform well on relatively restrictive types of excitations to which other methods are well suited, a modulated stationary white noise process was used as excitation [4, 5]. The transient response obtained using the technique presented here agreed very well with that obtained through Monte Carlo simulation. As part of this demonstration, it was also shown that, as expected, the effectiveness of the proposed approach deteriorates with increased system nonlinearity.

In this paper, an additional application is presented which uses real-world measured nonstationary excitation data. Some additional discussion is provided on the convergence of the estimated dynamic response with that obtained through a more computationally intensive Monte Carlo simulation.

The input excitations were taken from a large data-set of ground motions available from a recent earthquake in Taiwan. In total 1191 records were chosen to represent the nonstationary excitation process. Because this example application is for demonstration purposes only, the units were assumed to be force rather than acceleration. The records were synchronized to all begin as they passed a 1% g trigger level. Then, the remaining 1501 points were recorded. The time scale was stretched to correspond with the system under consideration, and to allow comparison with results from [7]. In addition, the records were all normalized so that they each had the same peak magnitude of 1. The covariance matrix of the recorded ensemble is shown in Figure 1.

The convergence of the eigenvalues of the excitation matrix is shown in Figure 2. The eigenvectors of the matrix were fitted with various orders of Chebyshev polynomials, and an approximate analytical form of the excitation covariance matrix was obtained as in (1). The simple system considered is the same as that considered in [7]. It is a simple single-degree-of-freedom Duffing oscillator, whose dynamics can be described by the following equation of motion:

$$\ddot{q} + 2\zeta\dot{q} + q + \lambda q^3 = s(t) \tag{10}$$

In this single-degree-of-freedom system, the system matrices become scalar values, i.e., $\mathbf{M} = 1$, $\mathbf{C} = 2\zeta$, $\mathbf{K} = 1$. For this example, the value of $\zeta = 0.1$ is fixed. For the analysis of the response of nonlinear systems, the nonlinear restoring force term needs to be converted into a time varying equivalent linear term. Here, this is done by assuming that the response is Gaussian, and based on [1], one can take the partial of the nonlinear term with respect to the response quantity (say x) to determine the equivalent linear term.

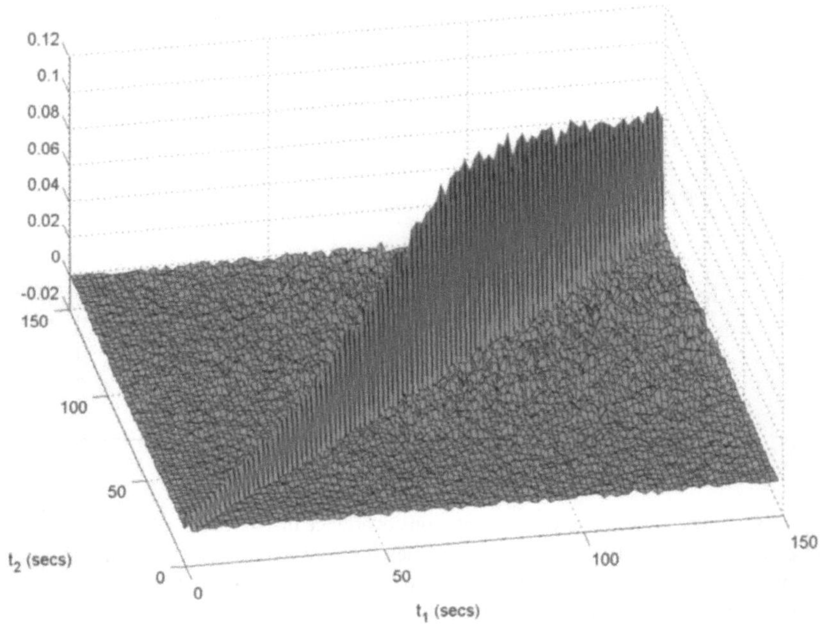

Figure 1. Covariance matrix $\mathfrak{C}_{ss}(t_1, t_2)$ of the excitation process derived from 1191 earthquake ground motion records measured from a single seismic event.

3.1 Wide-banded real world excitation data example

In [7] an example was provided where the estimated response from a Gaussian excitation process agreed very well with that obtained by directly using Monte Carlo simulation. Here, however we test the methodology for a clearly non-Gaussian excitation process to explore the limitations of the methodology. By first taking 100 eigenvectors and corresponding eigenvalues, and performing a 300-order Chebyshev polynomial fit, the system transient root-mean-square (RMS) response was obtained. This is shown in Figure 3. Interestingly, compared with the performance of the Gaussian input case presented in [7], the agreement is still quite good. Increasing the number of eigenvalues included in the series approximation of the excitation, and the order of polynomial fit of the eigenvectors to $k = 200$ and $m_i = 500$, and even $k = 300$ and $m_i = 750$, improves the performance of the method. In the case of linear systems, as was shown in [3], improving the fidelity of the analytically estimated excitation process was enough to ensure excellent agreement with that calculated through direct Monte Carlo simulation. In

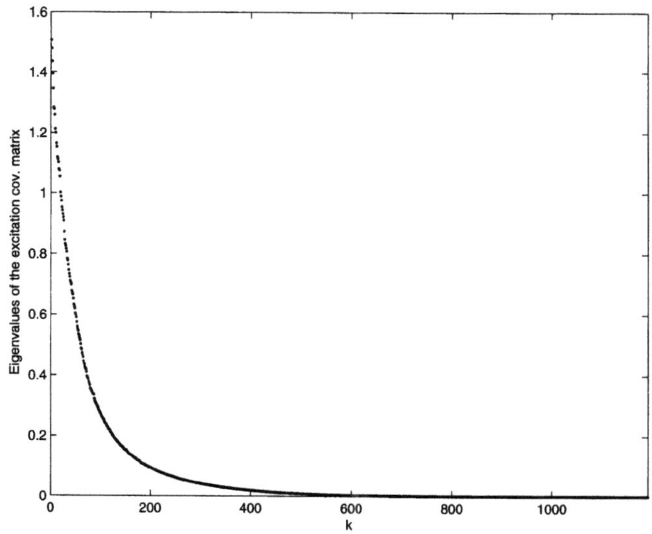

Figure 2. The eigenvalues of the excitation covariance matrix $\mathcal{C}_{ss}(t_1, t_2)$ shown in Figure 1. Notice that there are 1191 non-zero eigenvalues (i.e., the number of records in the ensemble), and that because of the rapid convergence most of the energy in the input process can be captured by taking only the first few hundred eigenvalues.

fact, in the linear case this would converge to the true response if the excitation was fitted to high enough accuracy. Clearly this tendency to converge towards the true solution is observed here for this example, however this conclusion will not hold true for all excitation and nonlinear model combinations. Figure (4) shows a typical sample of nonlinear behavior of the Duffing oscillator subjected to this excitation.

3.2 Narrow-banded excitation

In addition to the raw data being used as input excitation the data-set was also narrow-band filtered, with a pass-band from 0-0.5Hz. The natural frequency of the system falls within this range. It has often been observed that narrow-bandedness in the response greatly effects results obtained through equivalent linearization techniques [2]. A quick check comparison of the curtosis and the skewness of the un-filtered and filtered data indicates an equal level of non-Gaussian excitation. The skewness is close to zero, and the curtosis varies from about 15 to 3 (the true value for a Gaussian process). Interestingly, when the response of the nonlinear system from filtered input process is computed, the estimated response compares very well with that computed through the Monte Carlo simulation.

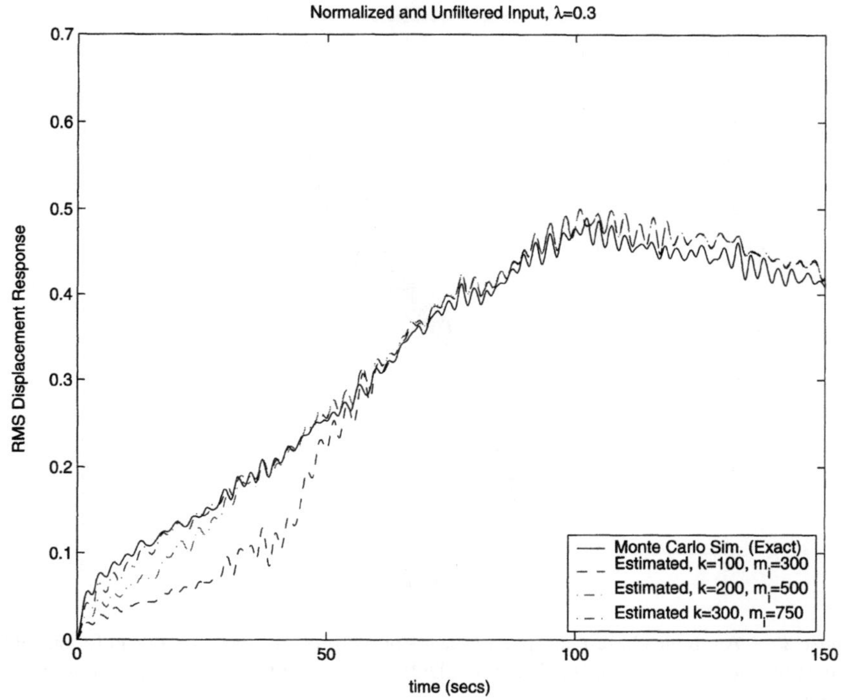

Figure 3. RMS transient displacement response due to the nonstationary earthquake data-set excitation. The solid line is the 'exact' response as calculated by direct Monte Carlo simulation. The dashe lines correspond to the estimated response for various level of input process approximation. ($\lambda = 0.3$)

In all cases the reduction in computer time for the proposed method, and for the 1191 input realizations, is on the order of 1–2 hours for the Monte Carlo simulation versus about 30–60 seconds, respectively, for the method presented here. Figure (5) shows the generally good agreement for the nonstationary response of the system, even for the case where $k = 100$ and $m_i = 300$ in the excitation approximation stage.

3.3 Response estimation with a highly non-Gaussian excitation process

The method has clearly demonstrated its capacity to estimate the response of this simple Duffing oscillator quite well for Gaussian and moderately non-Gaussian excitations. If the earthquake data records are not all "normalized" so that their peak values are all identical then the ensemble represents large variation in the excitation magnitude, and hence is a highly non-Gaussian input. This can be seen by plotting the skewness and curtosis of the excita-

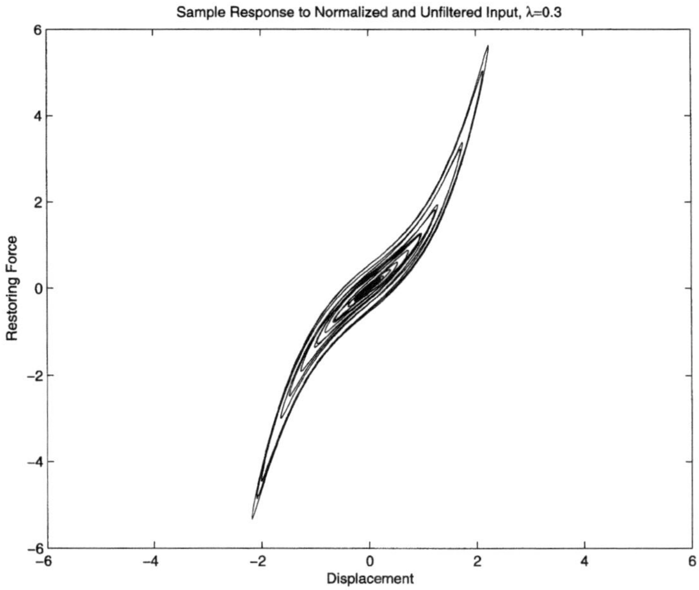

Figure 4. A sample phase-plane plot of the restoring force vs. the displacement for one of the realizations of the response of the hardening Duffing oscillator to the unfiltered but normalized excitation.

tion without the normalization step. In addition, so that one more or less compares the same levels of excitation and corresponding nonlinear behavior, the ensemble was multiplied by 12, which had the effect of making the average peak of all the records equal to 1. The curtosis and skewness of this data set is shown in Figure (6). Note that at no point is the curtosis below even 5, and more often than not, it is extremely high. For this data set, not surprisingly, as shown in Figure (7), regardless of the fidelity of the analytically approximated excitation process, the response does not converge to the exact response. It should be noted, that had the input not been multiplied by 12, that because of relatively lower nonlinear behavior, the equivalent linear based technique performs quite well.

4. Conclusion

The proposed method presents a highly efficient alternative to direct Monte Carlo simulation for computing the nonstationary response of nonlinear systems to nonstationary excitations. From the examples presented here with highly wide-banded input excitation, the results appear to deteriorate only slightly. Clearly from the final example, when presented with a highly non-Gaussian excitation process, the estimation process does not provide an ac-

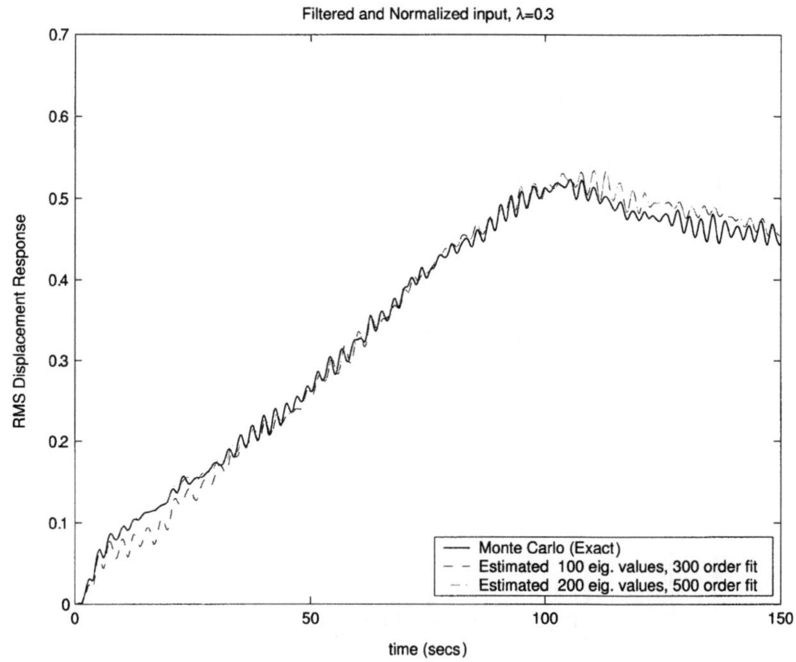

Figure 5. A comparison of the exact and estimated RMS transient displacement response due to the narrow-band filtered excitation process. Here, again, $\lambda = 0.3$.

curate response. Although potentially highly challenging for nonstationary excitation and responses, further work should be conducted to better quantify this relationship between non-Gaussianity and equivalent linearization error for the method.

Acknowledgments

This study was supported in part by the National Science Foundation and the Air Force Office of Scientific Research.

References

[1] I. E. Kazakov, "Generalization of the method of statistical linearization to multidimensional systems," *Auto. Remote Control*, vol. 26, pp. 1201–1206, 1965.

[2] L. D. Lutes, The gaussian assumption in equivalent linearization, In *Eight ASCE Speciality Conference on Probabilistic Mechanics and Structural Reliability, PMC2000*, number PMC2000-334. Univ. of Notre Dame, 2000.

[3] S. F. Masri, A. W. Smyth, and M. I. Traina, "Probabilistic representation and transmission of nonstationary processes in multi-degree-of-freedom systems," *Jo. of Applied Mechanics, ASME*, vol. 65, pp. 398–409, 1998.

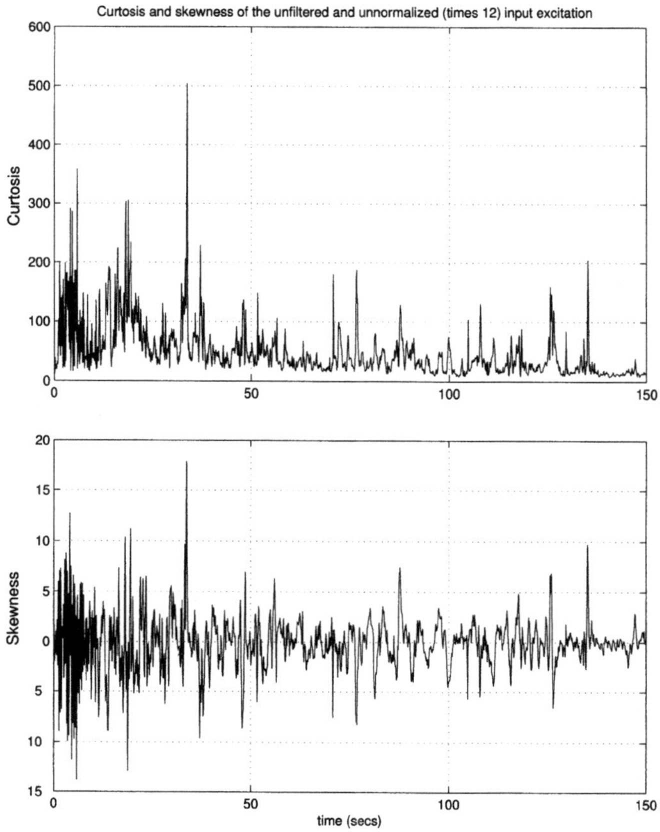

Figure 6. Curtosis and skewness of the data ensemble. This data ensemble is the raw recorded data multiplied by 12 so that the mean of the peaks of the realizations equals 1 (for comparison with the previously "normalized" case).

[4] J. B. Roberts and P. D. Spanos, *Random Vibration and Statistical Linearization*, John Wiley & Sons, New York, NY, 1990.

[5] M. Sakata and K. Kimura, "Calculation of the non-stationary mean square response of a non-linear system subjected to non-white excitation," *Jo. of Sound and Vibration*, vol. 73, no. 3, pp. 333–343, 1980.

[6] A. W. Smyth, *Experimental and Analytical Studies in Nonlinear System Identification and Modeling for Structural Control*, Ph.d. thesis, University of Southern California, Los Angeles, CA, 1998.

[7] A. W. Smyth and S. F. Masri, "Nonstationary response of nonlinear systems using equivalent linearization with a compact analytical form of the excitation process," *Probabilistic Engineering Mechanics*, vol. 17, no. 1, pp. 97–108, 2002.

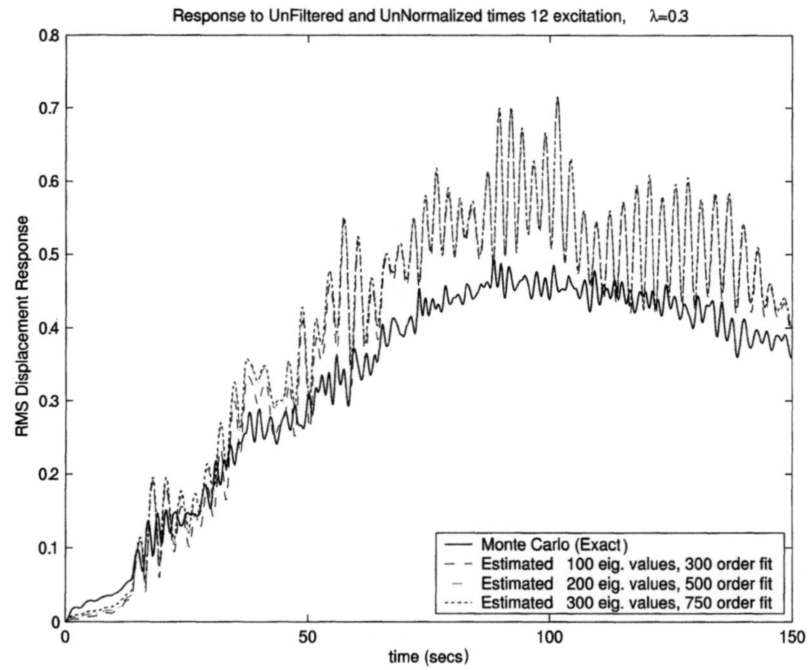

Figure 7. RMS of the transient displacement response due to the unfiltered and unnormalized data-set (multiplied by 12).

HILBERT SPECTRAL DESCRIPTION AND SIMULATION OF NONSTATIONARY RANDOM PROCESSES

Y. K. Wen and Ping Gu

Department of Civil Engineering

University of Illinois at Urbana-Champaign

Urbana, IL 61801, USA

wen@uiuc.edu

pinggu@uiuc.edu

Abstract A new method is proposed for characterization and simulation of a nonstationary random process based on samples of the process. The theoretical background is that of the Hilbert Huang Transform (HHT). Samples of a random process $X(t)$ can be decomposed into a summation of modal functions whose Hilbert transforms can be used to describe the amplitude and frequency changes with time. A Hilbert spectrum is then defined to describe the time-varying spectral content of the sample process. The method is also extended to characterization of vector processes. It can be conveniently applied to simulation of nonstationary random processes based on observed sample functions. The simulated processes approach Gaussian and will have the Hilbert spectra equal to the target spectra. Unlike current procedures such as those based on the evolutionary process, no assumptions of any functional forms for the spectra are necessary for parameter estimation, which are unknown a priori. Applications to spectral characterization and simulation of multivariate earthquake ground motions show that the Hilbert spectra give a clear description of the spectral content change with time. The simulated samples have the desired frequency and amplitude variation with time. The method has great potential for engineering applications when dealing with non-stationary, nonlinear random processes.

Keywords: Nonstationary random processes, simulation, Hilbert Transform, vector process

1. Introduction

Uncertainty modeling is an essential component in performance evaluation of engineering systems since most demands on the systems fluctuate in time and space in a random manner. For this purpose stationary random pro-

N. Sri Namachchivaya and Y.K. Lin (eds.),
IUTAM Symposium on Nonlinear Stochastic Dynamics, 375-384.

cesses have been widely used. Stationarity is of course a mathematical idealization as most real engineering problems are transient in nature. Theoretical treatment of nonstationary processes has not been as well developed as for stationary processes mainly due to the difficulty in frequency domain (spectral) analysis of such processes. The spectral analysis of random processes traditionally relies on the Fourier decomposition of the process, i.e., representing the processes by sine and cosine waves extending without bounds over time and space. It is inherently difficult when applied to nonstationary processes due to the "uncertainty principle" that one cannot simultaneously localize or resolve both time and frequency in a wave form to an arbitrarily high degree. For example, all currently available methods of spectral analysis such as double frequency spectrum, instantaneous spectrum and physical spectrum, have some drawbacks (e.g., [4]). As a result, in most engineering applications, an equivalent stationary process, piece-wise stationary process, or a slowly varying (evolutionary) process [5] is used as an approximation. They all have limitations to various degrees.

Recently, [1] introduced a new method of spectral analysis of nonstationary processes based on Hilbert transform and the concept of instantaneous frequency. Unlike Fourier transform, Hilbert transform emphasizes local behavior and does not decompose the process into sine and cosine waves of fixed frequencies. The Hilbert energy or amplitude spectrum as function of time and frequency can be then described in terms of the amplitude functions and instantaneous frequency. No assumption of slowly time-varying behavior of the processes is required and no sine or cosine waves of fixed frequencies are used. The Hilbert spectrum gives a much sharper description of the frequency content of the physical phenomena as function of time. It is a new approach to the challenging problem of frequency domain analysis of nonstationary processes.

The emphasis thus far has been on data analysis and interpretation of a given time series of a real physical phenomenon (e.g., Huang et. al, 1999, and [3]). Much less has been done on further development of the method as an analytical tool for spectral description of nonstationary uni-variate and multivariate (vector) processes. Such extension is essential since many important natural phenomena are nonstationary multivariate processes, such as the six components of the earthquake motions or the three velocity components of wind and wave at a given point. With the ever-increasing power of computer and capacity of computational mechanics, simulation has become an indispensable tool in engineering. The potential of this method in simulation of nonstationary/nonlinear processes has not been explored. In view of the possible applications in performance evaluation of complex and nonlinear systems, the impact of such a simulation method would be indeed very significant.

2. Hilbert-Huang Transform (HHT) and Empirical Mode Decomposition (EMD)

The background of the HHT+EMD method is briefly described. Details can be found in [1]. Any time series can be decomposed into a series of Intrinsic Mode Functions (IMF). An IMF is defined as a time series whose number of extrema and number of zero crossings must either equal or differ at most by one at any point. The mean value of the envelopes defined by the local maxima and minima is zero. Therefore, a given time series $X(t)$ can be given by:

$$X(t) = \sum_{j=1}^{n} C_j(t) + r_n(t) \tag{1}$$

in which $C_j(t)$ is the j-th IMF and $r_n(t)$ is the residue that can be the mean trend or a constant. The construction of C_j from $X(t)$ requires a repeated "sifting" process. The procedure is referred to as Empirical Mode Decomposition (EMD). Taking Hilbert transform of $C_j(t)$,

$$D_j(t) = \frac{1}{\pi} P \int \frac{C_j(\tau)}{t-\tau} d\tau \tag{2}$$

in which P indicates Cauchy principal value, one can form an analytical function

$$Z_j(t) = C_j(t) + iD_j(t) = a_j(t)e^{i\theta_j(t)}. \tag{3}$$

The *instantaneous frequency* of the IMF is given by

$$\omega_j(t) = \frac{d\theta_j(t)}{dt}. \tag{4}$$

One can then express $X(t)$ as follows

$$X(t) = \text{Re} \left\{ \sum_{j=1}^{n} a_j(t) e^{i\theta_j(t)} \right\} + r_n(t) \tag{5}$$

in which Re denotes the real part. $a_j(t)$, $j = 1$ to n therefore control the amplitude variation; each has an instantaneous frequency at a given time. Collectively, the two functions $a_j(t)$ and $\theta_j(t)$ define the Hilbert Amplitude Spectrum $H(\omega, t)$ or simply Hilbert Spectrum. Similarly, $H^2(\omega, t)$ in terms of $a_j^2(t)$ and $\theta_j(t)$, defines the energy distribution with respect to time and frequency. The marginal Hilbert energy spectrum is obtained by integration over the duration,

$$h(\omega) = \int_0^T H^2(\omega, t) dt. \tag{6}$$

Because of the local nature of the Hilbert transform and the concept of instantaneous frequency, the Hilbert marginal spectrum describes energy contribution corresponding to a given frequency at certain time but not the entire duration of the process, as would be the case for a Fourier spectrum. This is an important fundamental difference between these two approaches.

As an example, the method is applied to the well-known Newhall ground motion of 1994 Northridge earthquake in SAC-2 ground motion records [8]. The time history of the ground acceleration (in cm/sec2) is shown at the top of Figure 1. The Hilbert spectrum is calculated and shown in Figure 2. Note

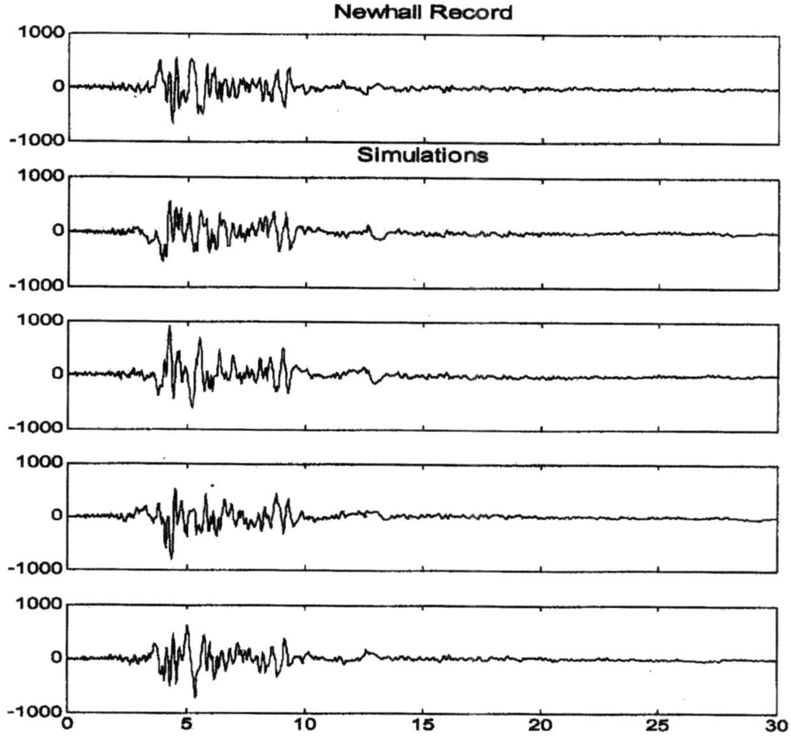

Figure 1. Recorded (top figure) and Simulated Newhall Ground Motions

that the Hilbert transform can be carried out efficiently via a FFT procedure. As can be seen, one distinct advantage of the Hilbert spectrum is that the energy distribution over time and frequency is always sharp. For example, darker shade of color in the plot indicates higher energy, which can be clearly seen in the Newhall record from 4 to 10 seconds and below 20 rad/sec (3 cps), a frequently observed feature of near-source records. The marginal Hilbert spectrum is compared with the Fourier energy spectrum obtained from the same

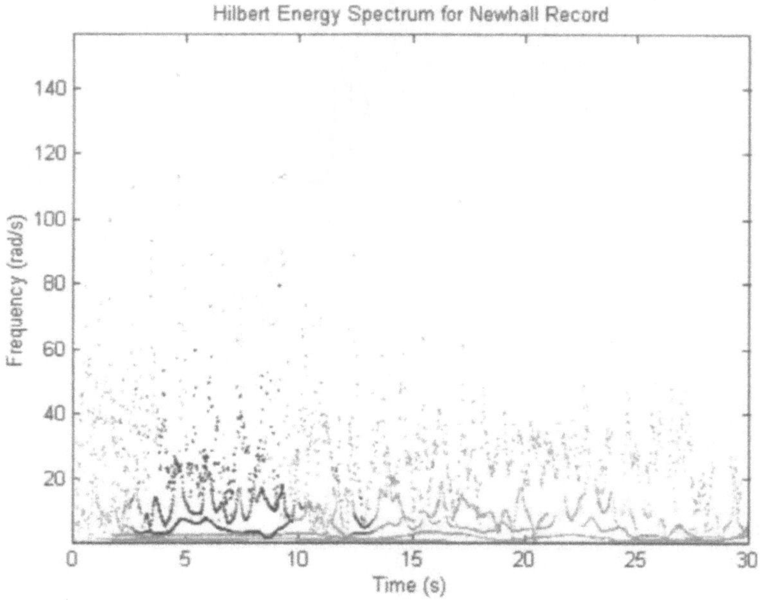

Figure 2. Hilbert Spectrum of 1994 Northridge Newhall Ground Motion

record in Figure 3. There are some obvious differences in the very low frequency range due to the two different methods of spectral description.

3. Hilbert Spectral Description and Representation of a Nonstationary process

In application of random processes to engineering problems, one is often faced with the problem of constructing a theoretical model for the real physical process that may not admit an elegant mathematical treatment. This is the case in constructing a random process model based on observational data. The observed record can be regarded as a sample of an underlying random process. The Hilbert spectrum based on the HHT+EMD method then describes the amplitude and frequency content change with time of the sample only. In theory it varies among different sample realizations. The Hilbert spectrum of the process is therefore defined as the expected value (or ensemble average) of the sample Hilbert spectra,

$$H_m(\omega, t) = E[H(\omega, t)] \tag{7}$$

which characterizes the amplitude and frequency change with time of the process. In most real physical situations, such as the case of earthquake ground

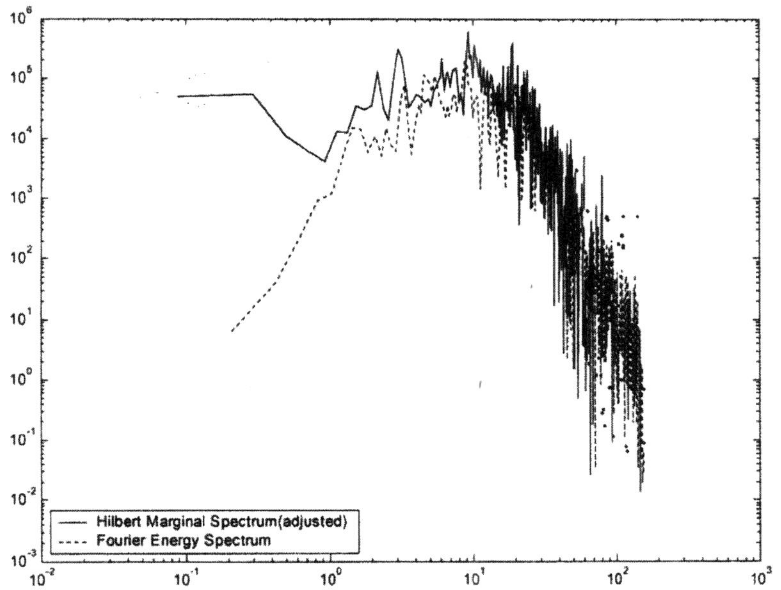

Figure 3. Comparison of Hilbert marginal spectrum (solid line) with Fourier energy spectrum (dashed line)

motions, however, only one sample can be collected. Characterization of the underlying random process is therefore a difficult problem and it is generally true in modeling a nonstationary process based on one sample observation, regardless of the method used. Under such circumstances, assumptions and approximation based on engineering judgments are necessary. For example, a parametric form of the time-dependent power spectral density or a piece-wise stationary ergodic process was often assumed in order to estimate the necessary spectral parameters of the underlying process. These assumptions could lead to serious distortion of the process. As a result, the parameter estimation has always been a weak link in modeling of nonstationary processes. In the present method, it is assumed that the sample Hilbert spectrum can be used to describe the process. It is still a compromise under the circumstance. Nevertheless, it is easier to implement, and requires no further assumptions or approximation.

The Hilbert spectral representation as given in (5) suggests that the underlying random process can be represented by introducing a random element as follows:

$$X(t) \quad = \quad \mathrm{Re}\left\{\sum_{j=1}^{n} a_j(t)e^{i[\theta_j(t)+\theta_j]}\right\} + r_n(t) \tag{8}$$

in which ϕ_j is an independent random phase angle uniformly distributed between 0 and 2π. $X(t)$ is therefore a random process. The process has the following mean, covariance, and variance functions:

$$\begin{aligned}\mu_X(t) &= E[X(t)] = \text{Re}\left\{\sum_{j=1}^{n} a_j(t)e^{i\theta_j(t)}E(e^{i\theta_j})\right\} + r_u \\ &= r_n(t) \end{aligned}\tag{9}$$

$$\kappa_{XX}(t_1,t_2) = \frac{1}{2}\sum_{j=1}^{n} a_j(t_1)a_j(t_2)\cos[\theta_j(t_1) - \theta_j(t_2)] \tag{10}$$

$$\sigma_x^2(t) = E\left\{[X(t) - \mu_X(t)]^2\right\} = \frac{1}{2}\sum_{j=1}^{n} a_j^2(t) \tag{11}$$

$X(t)$ as given in (8) has a Hilbert energy spectrum characterized by $a_j^2(t)$ with instantaneous frequency $d(\theta_j(t))/dt$, for $j = 1$ to n. Due to Central Limit Theorem, $X(t)$ approaches a Gaussian process for large n.

The above Hilbert spectral representation can be extended to a vector process $X(t)$ of m components: $[X_1(t), X_2(t), \ldots, X_m(t)]$. The k-th component has a Hilbert spectral representation given by

$$X_k(t) = \text{Re}\left\{\sum_{j=1}^{n} a_{jk}(t)e^{i[\theta_{jk}(t)+\theta_{jk}]}\right\} + r_{nk}(t). \tag{12}$$

The mean, covariance, and variance functions of time of $X_k(t)$ can be obtained as in (9)—(11) for each component. The Hilbert energy spectrum is similarly characterized by the amplitude functions and the instantaneous frequency for each component. The cross-correlation between the p-th and q-th components is described by

$$\begin{aligned}K_{x_px_q}(t_1,t_2) &= E\left\{[X_p(t_1) - r_{np}(t_1)]\cdot[X_q(t_2) - r_{nq}(t_2)]\right\} \\ &= \sum_{j=1}^{n}\sum_{k=1}^{n} a_{jp}(t_1)a_{kq}(t_2)E\left\{\cos[\theta_{jp}(t_1) + \phi_{jp}]\right. \\ &\quad \left.\cos[\theta_{kq}(t_2) + \phi_{kq}]\right\}. \end{aligned}\tag{13}$$

The correlation therefore depends on amplitude and phase functions of each component process as well as the random phase shifts. For example, if the random phase shift angles of the two components are statistically independent for all ϕ_{jp} and ϕ_{kq}, then it can be shown that the cross-correlation function vanishes. If we allow some dependence between the phase shifts of the two components, then we can build correlation between the components. One possible choice is to let the set of phase angle shifts in two components to be functionally related. For example, the IMF functions of the same order of the two components have the random phase shifts related by

$$\phi_{j1} - \phi_{jq} = \Psi_{jp} \quad \text{for } j = 1 \text{ to } n \tag{14}$$

in which Ψ_{jq} is a constant for $j = 1$, to n; and $q = 1$, to m. In other words, the random phase shifts of the IMFs of the same order of each of the components are related to each other but still independent of phase shifts of IMFs of different order. Although (14) implies that in general the random phase shifts are not all uniformly distributed between 0 and 2π, they are still uniformly distributed within a range of 2π. The mean and covariance function as given in (9)–(11) and the Hilbert energy spectrum for each component process remain unchanged. The cross-correlation between any two component- processes is equal to

$$\begin{aligned} K_{X_pX_q}(t_1, t_2) &= \frac{1}{2}\sum_{j=1}^{n} a_{jp}(t_1)a_{jq}(t_2)\cos\left[\theta_{jp}(t_1) - \theta_{jp}(t_2)\right. \\ &\quad \left. + \Psi_{jp} - \Psi_{jq}\right]. \end{aligned} \tag{15}$$

It therefore depends on the product of the amplitude functions of time of the IMFs of the components of the same order and the differences in their phase functions of time plus the difference in the random phase shifts. For example, when this total difference is very small, the cross-correlation is described by the IMFs of the two components. When the total difference is close to $\pi/2$, the cross-correlation vanishes. The constant terms Ψ_{jq} also allow some freedom in modelling cross-correlation based on observations.

4. Simulation of Nonstationary Process Based on Sample Observation

Simulation of stationary random processes by Fourier spectral representation is well developed and has been applied to many engineering problems. A summary of the development can be found in [7]. Simulation of nonstationary random processes thus far has been based on Fourier spectral analysis or some forms of its variation such as the evolutionary processes. It is a good analytical tool for response analysis of linear systems and simulation can be easily carried out if the modulation function is known [6]. Li and Kareem [2] lately used Fast Fourier Transform and digital filtering and made the method more computationally efficient. As alluded to in the foregoing, the difficulty of using this model in simulation of real physical phenomena lies in parameter estimation. In addition to estimation of the power spectral density function, one needs also estimate the modulation function of time and frequency. No completely satisfactory parameter estimation methods have been developed.

The Hilbert spectral representation model as described above suggests a simple method for simulation of a nonstationary process as follows

$$X(t) = \operatorname{Re}\left\{\sum_{j=1}^{n} a_j(t)e^{i[\theta_j(t)+\phi_j]}\right\} + r_n(t) \quad (16)$$

in which ϕ_j for $j = 1, n$ are a set of independent random variables uniformly distributed between 0 and $\pi 2$ generated on computer. The Hilbert energy distribution over time and frequency of the simulated sample processes can be shown to be the same as the target Hilbert Spectrum therefore satisfies the simulation criterion. The probability distribution of $X(t)$ at a given time will approach to Gaussian due to Central Limit theorem.

Numerical examples of simulation of the Northridge Newhall records based on the sample Hilbert energy spectrum are carried out. Sample time histories according to the proposed are shown in Figure 1. The simulated samples represent a reasonably statistical image of the intensity and frequency variation with time of the records. The recorded and simulated ground motions can be regarded as sample realizations of the underlying process characterized by the mean and covariance functions described in (9)–(11) and the time-dependent Hilbert energy spectrum by the sample spectrum.

To extend the method to generation of a vector process, in theory one needs to know the target cross-correlation function between components of the vector process in order to determine the proper correlation among the random phase shifts among the component processes in (12) in addition to $a_{jk}(t)$ and $\phi_{jk}(t)$. Again for most real situations only one sample is available; approximations and assumption based on physical or engineering arguments are again necessary. For example, an argument can be put forward that the component IMFs of the same order may represent vector wave forms due to given physical mechanism, therefore should be in phase or have the same phase shift as described by (14) with $\Phi_{jq} = 0$. Therefore in simulation, one can use the same set of random phase shifts for all components and the simulated components of the vector process will be correlated according to (15) as follows

$$K_{X_pX_q}(t_1, t_2) = \frac{1}{2}\sum_{j=1}^{n} a_{jp}(t_1)a_{jq}(t_2)\cos[\theta_{jp}(t_1) - \theta_{jp}(t_2)]. \quad (17)$$

This assumption is used to simulate the vertical and two horizontal components of the Newhall records. Result of the simulations of the three components show that the simulated ground motions represent reasonably well a statistical image of the recorded ground motions.

5. Conclusion

A new method of representation and simulation of nonstationary random processes based on the Hilbert Huang Transform (HHT) method is presented. It offers a new approach to the difficult problem of spectral analysis, parameter estimation and simulation based on sample observations for nonstationary processes. It is largely an empirical and numerical method that has some advantages over most currently available methods. It gives a sharp description of the change of spectral content of the processes with time and can be easily applied to simulation. It is specially suited for computer-aided model-based simulations. Numerical examples on earthquake ground motion spectral description and simulation show that the method has great potential in engineering applications when dealing with nonstationary processes.

Acknowledgments

This research is supported by University of Illinois Research Board. Information exchange with C. H. Loh of National Taiwan University is greatly appreciated.

References

[1] N. E. Huang, Z. Shen, S. R. Long, M. C. Wu, H. H. Shih, Q. Zheng, N.-C. Yen, C. C. Tung, and H. H. Liu, "The Empirical Mode Decomposition and the Hilbert Spectrum for Nonlinear, Nonstationary Time Series Analysis," *Proc. R. Soc. Lond. A*, vol. 454, 903-995. 213-219, 1998.

[2] Y. Li and A. Kareem, "Simulation of Multivariate Nonstationary Random Processes: Hybrid DFT and Digital Filtering Approach," *J. Engrg. Mech., ASCE*, vol. 123, no. 12, pp. 1302–1310, 1997.

[3] C. H. Loh, T. C. Wu, and N. E. Huang, "Application of EMD+HHT Method to Identify Near-fault Motion Characteristics and Structural Responses", Bullition of Seismological Society of America, Special Issue on the 1999 Chi-Chi Taiwan Earthquake, 2001.

[4] W. D. Mark, "Power Spectrum Representation for Non-Stationary Random Vibration," Random Vibration-Status and Recent Developments (Ed. I. Elishakoff and R. H. Lyon), Elsevier, pp. 211–240, 1986.

[5] M. B. Priestley, "Power Spectral Analysis of Nonstationary Random Processes", *J. Sound Vib.*, vol. 6, pp. 86–97, 1967.

[6] M. Shinozuka and C. M. Jan, "Digital Simulation of Random Processes and Its Applications," *J. Sound Vib.*, vol. 25, no. 1, pp. 111–128, 1972.

[7] M. Shinozuka and G. Deodatis, "Simulation of Stochastic Processes by Spectral Representation," *App. Mech. Reviews*, vol. 44, no. 4, pp. 191–203, 1991.

[8] P. Somerville, N. Smith, S. Punyamurthula, and J. Sun, "Development of Ground Motion Time Histories for Phase 2 of the FEMA/SAC Steel Project," Report No. SAC/BD-97/04, SAC Joint Venture, Sacramento, California, 1997.

V

PART 5: COMPUTATIONAL METHODS

Progress in the theoretical development (stochastic flow, Lyapunov exponent, multiplicative ergodic theory, stochastic bifurcation theory, large deviations, chaos theory, etc.) has led to reliable numerical algorithms for computing standard statistical measures of response and stability in random dynamical systems, system identification and stochastic control– a better understanding of nonlinear phenomena. The fifth *Lead Lecture* of the Symposium given by Walter Wedig of the University of Karlsruhe, Germany, is the lead paper of this part. The rest of the papers presented during the Symposium, on the general topic of *computational methods in stochastic and chaotic dynamics* are compiled alphabetically, based on the last name of the first author.

Numerical algorithms include the computation of both path-wise and moment responses of nonlinear systems, invariant densities of stationary solutions, and their Lyapunov exponent, exit time, etc. Walter Wedig investigated a quarter car model under base excitations by white noise or colored noise. A bi-linear damping law was assumed in the analysis, leading to non-Gaussian response of vertical vibration. Lawrence Bergman presented the concept of discrepancy sensitivity, a linear approximation for the contribution of a sample to uniformity, and quantifying the level of the provided "new" information. Christian Bucher presented an efficient importance sampling approach to the Monte Carlo solution of first-passage problems, which was capable of handling efficiently non-white and non-stationary excitations. Grant Lythe considered a stochastic partial differential equation with localized coherent structures known as kinks. The steady- state density of kinks, maintaining a balance between nucleation and annihilation of kink-antikink pairs, was determined.

One aim of the Symposium was to provide a forum to assess the current frontiers and discuss future directions in nonlinear and stochastic dynamics. Special efforts were made to promote connections between the deterministic and stochastic communities dealing with chaotic phenomena. Towards this end, Bruno Eckhardt considered two examples for the interplay between chaotic dynamics and stochastic forces in hydro-dynamical systems. Arvid Naess discussed the possibility of using the methods of stochastic differential equations to study the dynamic response of nonlinear oscillators driven by harmonic excitations. Finally, discussions moved from the analysis of a given set of nonlinear stochastic equations to the analysis of observations from a system, for the purposes of learning about the system itself. Aneta Stefanovska presented a new approach for the modeling and analysis of human cardiovascular system signals, known to be exceptionally complex, time-varying, noisy, and of necessarily limited duration.

RANDOM VIBRATIONS OF RIDING CARS WITH BILINEAR DAMPING

Walter V. Wedig
Institut für Technische Mechanik
Universität Karlsruhe (TH)
D-76128 Karlsruhe, GERMANY
wedig@itm.uni-karlsruhe.de

Abstract The paper investigates quarter car models under base or force excitations by white noise or colored noise. In case that car dashpots possess nonlinear damping characteristics, stationary densities are calculated via Fokker-Planck equations applying a nonlinear Galerkin procedure. To discuss resonant car speeds, filtered excitations are introduced leading to non-gaussian response of the vertical car vibrations. Associated root-mean-square values are calculated by means of a linear covariance analysis.

Keywords: Quarter car models, Fokker-Plank equations, nonlinear Galerkin methods, covariances of linear car vibrations.

1. Introduction into the Problem

To avoid bouncing wheel sets without any loss of comfort and safety, dashpots of cars are designed by bilinear damping characteristics [1], realized e.g., by two different hydraulic bypasses for the up and down of the wheels, respectively, or by means of two different piston valves in the hydraulic damping cylinder. Moving up the first valve is closed and the second one is opened. Moving down both valves operate vice versa. In case of different valve cross sections, the bilinear damping characteristic, shown in Figure 1, is physically evident. These characteristics are asymmetric leading to the following effects [8]:

a) The stationary response of cars driving on rough road surfaces possesses highly nonsymmetric distribution densities.

b) Their mean values are shifted to the weaker damping characteristic, meanwhile the response variances approximately remain unchanged.

N. Sri Namachchivaya and Y.K. Lin (eds.),
IUTAM Symposium on Nonlinear Stochastic Dynamics, 387-399.

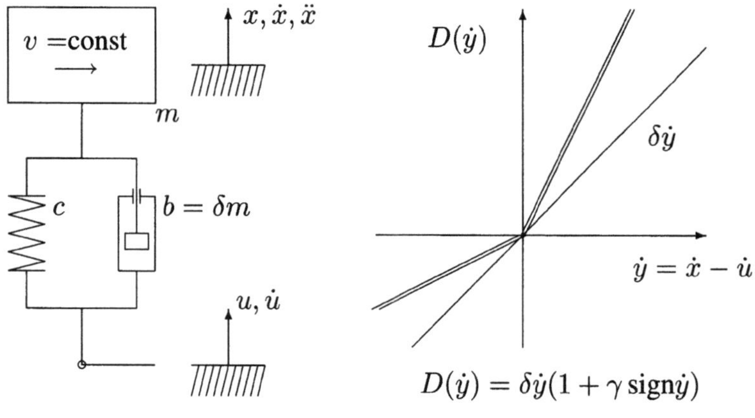

Figure 1. Quarter car model and damping characteristic $D(\dot{y})$

Both effects are not calculable by means of classical methods like equivalent linearization or stochastic averaging.

To improve such solution techniques, we apply a nonlinear Galerkin method applying the following three steps:

1 The problem is orthogonalized by appropriate coordinate transformations. This allows to derive separated densities with nonlinear distribution effects.

2 The free parameters of the separated densities are calculated to satisfy the stationary first and second order moments equations of the nonlinear system.

3 The separated densities are applied to derive orthogonal polynomials which expand the joint densities of interest by means of Galerkin methods [6].

The paper investigates the influence of the bilinear damping characteristic on the vertical vibrations of the quarter car model, shown in Figure 1. It is described by the following equation of motion:

$$\ddot{Y}_t + \delta\dot{Y}_t(1 + \gamma\,\mathrm{sign}\dot{Y}_t) + \omega_1^2 Y_t = -\ddot{U}_t, \qquad Y_t = X_t - U_t. \qquad (1)$$

Herein, Y_t denotes the relative car motion, the parameter δ determines the viscous damping of the wheel suspension, ω_1 is the natural frequency of the linear car vibrations and γ is the nonlinearity parameter. It is bounded by $|\gamma| < 1$ provided that the bilinear damping characteristic is realized by passive elements of dissipation. Bilinear dampings with $|\gamma| > 1$ are excluded in the following investigations since they can be realized only by active elements of control. The sign-function in (1) can formally be smoothed out by the arctan-function

$$D(\dot{y}) = \delta \dot{y} \left(1 + \gamma \lim_{\varepsilon \to 0} \frac{2}{\pi} \arctan \frac{\dot{y}}{\varepsilon}\right), \tag{2}$$

when the limiting procedure is restricted to $\varepsilon > 0$. In a first investigation, the acceleration $\ddot{U}_t$ of the base excitation is assumed to be white noise $\dot{W}_t$ with the intensity σ where dW_t denotes the increment of the Wiener process W_t. It is normally distributed with the mean values $E(dW_t) = 0$ and $E(dW_t^2) = dt$. Dots in (1), (2) and in all following formula are derivations with respect to the time t.

Introducing white noise by $\ddot{U}_t = -\sigma \dot{W}_t$ into the second order equation (1), it is rewritten into the first order form

$$\dot{X}_{1,t} = X_{2,t}, \qquad \dot{X}_{2,t} = -\delta X_{2,t}(1 + \gamma \text{sign} X_{2,t}) - \omega_1^2 X_{1,t} + \sigma \dot{W}_t, \tag{3}$$

where $X_{1,t}$ denotes the vertical car displacement Y_t and $X_{2,t}$ is the corresponding velocity $\dot{Y}_t$. The stationary solution of the nonlinear equations (3) is characterized by two-dimensional density distributions $p(x_1, x_2)$. Figure 2 shows

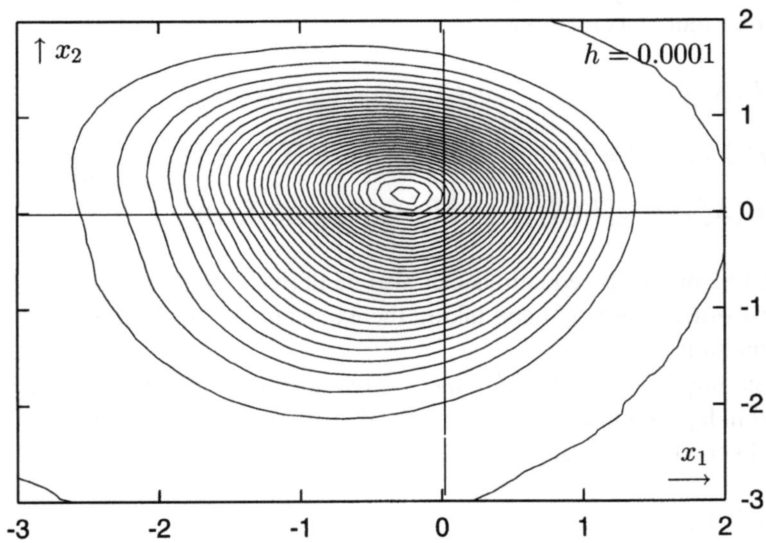

Figure 2. Contour lines obtained by Monte Carlo simulations

contour lines of $p(x_1, x_2)$ obtained by Monte Carlo simulations [7] applying Euler forward differences with the scan rate Δt and a white noise term of the form $\sqrt{\Delta t}\, R_n$, where R_n is a sequence of independent and normally distributed numbers with the mean values $E(R_n) = 0$ and $E(R_n^2) = \Delta t$. The system data, selected for the simulation, were $\omega_1 = \sigma = \delta = 1$, $\varepsilon = 0$ and $\gamma = 0.8$. The scan rate was $\Delta t = 0.01$ applied for $N = 10^8$ sample points. The contour lines in figure 2 start with level $h = 0.0001$ increased by the step size $\Delta h = 0.01$. In comparison with the perfectly symmetric densities of linear systems under white noise, figure 2 shows that the density $p(x_1, x_2)$ is highly non-symmetric with respect to x_1 and x_2 and the density maximum is shifted to the upper left side of the state space.

2. Asymptotic Solutions of the FP-Equation

The two-dimensional density $p(x_1, x_2)$, shown in Figure 2, is solution of the Fokker-Planck equation of the stationary form

$$\frac{\partial}{\partial x_1}(x_2 p) - \frac{\partial}{\partial x_2}\{[\delta x_2(1+\gamma \text{sign} x_2) + \omega_1^2 x_1]p\} = \frac{\sigma^2}{2}\frac{\partial^2 p}{\partial x_2^2}, \tag{4}$$

which follows from the two first order (3) of the quarter car model. The solution has to satisfy the normalization condition and vanishing boundary conditions at $x_1, x_2 = \pm\infty$. Integrating the Fokker-Planck equation (4) with respect to the density variables x_1 and x_2, we obtain the following first and second order moments relations:

$$E(X_{2,t}) = 0, \qquad \omega_1 E(X_{1,t}) + \delta\gamma E(X_{2,t}\text{sign}X_{2,t}) = 0, \tag{5}$$

$$E(X_{1,t}X_{2,t}) = 0, \qquad E(X_{2,t}^2) + \gamma E(X_{2,t}^2 \text{sign}X_{2,t}) = \frac{\sigma^2}{2\delta}, \tag{6}$$

$$E(X_{2,t}^2) = \omega_1^2 E(X_{1,t}^2) + \delta\gamma E(X_{1,t}X_{2,t}\text{sign}X_{2,t}). \tag{7}$$

These five moments' equations can also be derived applying Itô's calculus to the first order system (2) and taking the expectations of corresponding linear and quadratic increments.

Assuming $p(x_1, x_2) \approx p_1(x_1)p_2(x_2)$ that the system processes are approximately independent, and integrating the density $p_1(x_1)$ in the range $|x_1| < \infty$, the Fokker-Planck equation (4) yields

$$\frac{\sigma^2}{2}\frac{\partial^2 p_2}{\partial x_2^2} + \frac{\partial}{\partial x_2}\{[\delta x_2(1+\gamma\text{sign}x_2) + \alpha_2]p_2\} = 0, \quad |x_2| < \infty, \tag{8}$$

$$\int_{-\infty}^{+\infty} p_1(x_1)dx_1 = 1, \quad \int_{-\infty}^{+\infty} \frac{\partial p_1}{\partial x_1}dx_1 = 0, \quad \int_{-\infty}^{+\infty} x_1 p_1(x_1)dx_1 = \frac{\alpha_2}{\omega_1}. \tag{9}$$

Obviously, this procedure leads to an ordinary differential equation of the form (8) determined by the three integrals noted in (9). The solution of the separated Fokker-Planck equation (8) has to satisfy the boundary conditions $p_2(x_2 =$

$\pm\infty) = 0$. By integration it is calculated to

$$p_2(x_2) = C_2 \exp\{-\frac{1}{\sigma^2}[\delta x_2^2(1+\gamma \text{sign} x_2) + 2\alpha_2 x_2]\}, \quad |x_2| < \infty, \quad (10)$$

$$\alpha_2 = -\gamma\sigma(\frac{\delta}{\pi})^{1/2} + O(\gamma^3), \qquad E(X_{2,t}^2) = \frac{\sigma^2}{2\delta}[1 + \frac{3}{2}(1 - \frac{8}{3\pi})\gamma^2]. \quad (11)$$

Figure 3 shows typical numerical and analytical results of the related mean

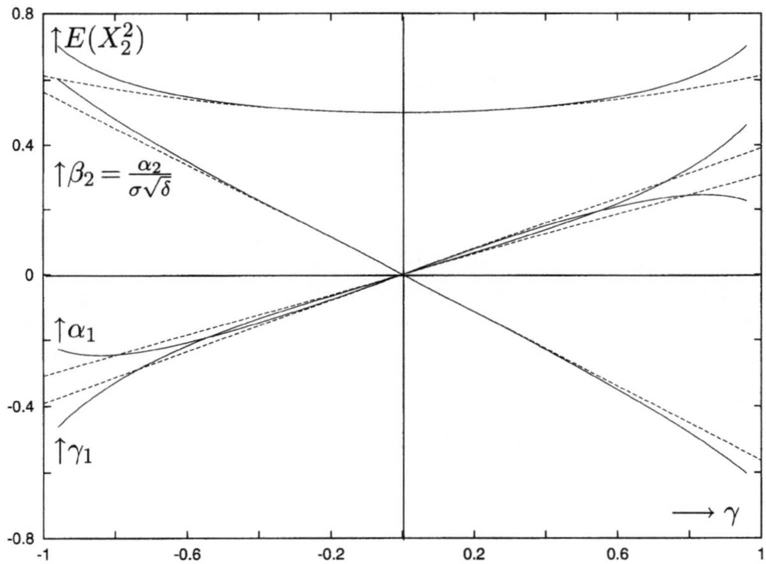

Figure 3. Analytical and numerical parameter evaluations

value $\beta_2 = \alpha_2/(\sigma\sqrt{\delta})$ and the second order moment $E(X_{2,t}^2)$ for the system data $\omega_1 = \delta = \sigma = 1$ both in dependence on the nonlinearity parameter γ in the range $|\gamma| < 1$.

The interrupted lines represent analytical approximations derived for small parameters $\gamma << 1$ by means of perturbation methods. The solid lines denote corresponding numerical evaluations of the mean value parameter α_2 determined by the zero mean condition $E(X_{2,t}) = 0$, noted in (5). The constant C_2 of integration is determined by the normalization condition $E(X_{2,t}^0) = 1$. Note that the one-dimensional solution (10) satisfies the two second order moments' equations, noted in (6), without any further adjustment.

The numerical results, given in figure 2, show that the density $p_1(x_1)$ takes a similar non-symmetric form as already discussed for $p_2(x_2)$. It is therefore reasonable to apply the same analytical density function

$$p_1(x_1) = C_1 \exp\{-\frac{\omega_1^2}{\sigma^2}[\delta x_1^2(1+\gamma_1 \text{sign} x_1) + 2\alpha_1 x_1]\}, \quad |x_1| < \infty \qquad (12)$$

but now with two free parameters α_1 and γ_1 to be calculated by the first and second order moments' equations, noted in (5) and (7). For $|\gamma| << 1$, these two moments' equations approximately represent a straight line and a twisted ellipse, respectively. Applying perturbation expansions, they lead to the following analytical solutions:

$$\gamma_1 = \gamma \frac{2}{\omega_1} \frac{\sqrt{\pi}\delta_q \omega_1 - \delta}{\sqrt{3\pi} - 2} + O(\gamma^3), \quad \delta_q^2 = \frac{3}{4} + \frac{1}{\pi}\left(\frac{\delta^2}{\omega_1^2} - 2\right), \qquad (13)$$

$$\alpha_1 = \gamma \frac{\sigma\sqrt{\delta}}{\omega_1^2} \frac{\sqrt{3}\delta - 2\delta_q \omega_1}{\sqrt{3\pi} - 2} + O(\gamma^3), \qquad \text{for } \gamma << 1. \qquad (14)$$

Figure 3 shows the analytical parameters 13) and (14) in the range $|\gamma| < 1$ by interrupted lines together with the solid lines of corresponding numerical values obtained by the moments' equations (5) and (7). The comparison of both lines shows good coincidences. Large deviations are only observable in the range of strong nonlinearities for $|\gamma| > 0.6$. Figure 4 compares

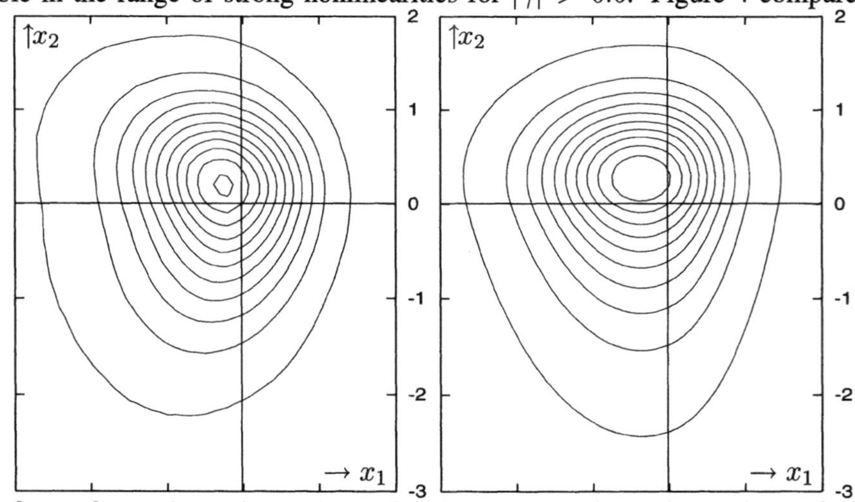

Figure 4. Comparison of MC-simulations and analytic FP-solutions

the contour lines of $p(x_1, x_2)$, already shown in Figure 2, with those obtained by the analytical FP-solutions (10) and (12) for the sytem data $\omega_1 = \sigma = \delta = 1$ and $\gamma = 0.8$.

3. Expansions by Orthogonal Polynomials

The densities $p(x)$, calculated in (10) and (12) for the stationary processes $X_{1,t}$ and $X_{2,t}$, respectively, define the orthogonal polynomials

$$P_n(x) = \sum_{k=0}^{n} a_{nk} x^k, \qquad \text{for} \qquad n = 0, 1, 2, \dots, \tag{15}$$

$$\int_{-\infty}^{+\infty} P_i(x) P_j(x) p(x) dx = \delta_{ij} \gamma_j, \qquad i, j = 0, 1, 2, \dots. \tag{16}$$

provided that they satisfy the orthogonal conditions, noted above. Herein, δ_{ij} is the Kronecker symbol and γ_j are normalization constants to be calculated for $i = j$. The evaluation of the orthogonality (16) leads to the linear inhomogeneous equation systems [4]

$$\sum_{m=0}^{n} M_{n+m} a_{nm} = \delta_{nm} \gamma_n, \qquad \text{for } n = 0, 1, 2, \dots, \tag{17}$$

$$M_j = \int_{-\infty}^{+\infty} x^j p(x) dx = E(X_t^j), \quad j = 0, 1, 2, \dots, \tag{18}$$

which have to be solved step by step for all $m \leq n$. Because of the homogeneity, one parameter of the orthogonal polynomials [5] can be fixed e.g., by $a_{nn} = 1$. Obviously, the coefficients a_{nk} of the orthogonal polynomials $P_n(x)$ are completely determined by the higher order moments $M_j = E(X_t^j)$, already needed in (5), (6) and (7).

The application of the polynomials $P_n(x)$ to the Fokker-Plank equation (4) leads to product terms like $xP_n(x)$ and differentiations of $P_n(x)p(x)$, which have to be expanded into the following finite sums:

$$x P_n(x) p(x) = \sum_{k=0}^{n+1} p_{nk} P_k(x) p(x), \qquad n = 0, 1, \dots, \tag{19}$$

$$\frac{\partial}{\partial x}[P_n(x) p(x)] = \sum_{k=0}^{n-1} d_{nk} P_k(x) p(x), \quad n = 1, 2, \dots. \tag{20}$$

The coefficients p_{nk} and d_{nk} of the product expansion (19) and the differentiation formula (20), respectively, are calculated by means of the orthogonality condition (16), as follows:

$$p_{n\ell} = \frac{1}{\gamma_\ell} \sum_{k=0}^{n} \sum_{j=0}^{\ell} a_{nk} a_{\ell j} M_{k+j+1}, \qquad n = 0, 1, \dots \tag{21}$$

$$d_{n\ell} = -\frac{1}{\gamma_\ell} \sum_{k=0}^{n} \sum_{j=0}^{\ell} j a_{nk} a_{\ell j} M_{k+j-1}, \quad \ell = 0, 1, \dots \tag{22}$$

Further terms like second order derivations and derivatives of the nonlinear damping characteristic are expanded into the infinite sum

$$\frac{\partial}{\partial x}\{[\frac{\sigma^2}{2}\frac{\partial}{\partial x}+\delta x(1+\gamma \text{sign} x)]P_m(x)p(x)\}=\sum_{k=0}^{\infty} f_{mk}P_k(x)p(x), \tag{23}$$

$$f_{mk}=\frac{1}{\gamma_k}\sum_{n=0}^{m}\sum_{\ell=0}^{k}\ell a_{mn}a_{k\ell}(\alpha_2 M_{n+\ell-1}-\frac{n}{2}\sigma^2 M_{n+\ell-2}). \tag{24}$$

In (23), the expansion coefficients f_{mk} are calculated by the formula (24) which follows from the orthogonality (16), as well.

To solve the Fokker-Planck equation (4), the orthogonal polynomials expansions are utilized by the following two-dimensional expansion [4]:

$$p(x_1,x_2)=\sum_{n=0}^{\infty}\sum_{m=0}^{\infty} c_{nm}P_n^{(1)}(x_1)P_m^{(2)}(x_2)p_1(x_1)p_2(x_2), \tag{25}$$

Herein, $P_n^{(1)}(x_1)$ and $P_m^{(2)}(x_2)$ are orthogonal polynomials of the two densities, calculated in (10) and (12), respectively, and c_{nm} are expansion coefficients to be calculated in the following. For this purpose, the two-dimensional expansion (25) is introduced into the Fokker-Planck equation (4). Subsequently, product terms, differentiations and nonlinear terms are replaced according to the corresponding relations (19), (20) and (23), respectively. This allows to cancel out the one-dimensional densities $p_1(x_1)$ and $p_2(x_2)$ and to perform a coefficient comparison of the product of the two polynomials. Therewith, the Fokker-Planck equation (4) is reduced to the inhomogeneous algebraic equations

$$\sum_{n=0}^{\infty}\sum_{m=0}^{\infty} c_{nm}\left[\delta_{nk}f_{m\ell}^{(2)}+\omega_1^2 p_{nk}^{(1)}d_{m\ell}^{(2)}-d_{nk}^{(1)}p_{m\ell}^{(2)}\right]=0, \tag{26}$$

by which the polynomial coefficients c_{nm} are determined for all indices $n, m = 0, 1, \ldots$. The equations (26) are evaluated for all indices $m+n \leq N$ assuming that the coefficients with higher indices are vanishing. Figure 5 shows contour lines of the two-dimensional density $p(x_1, x_2)$ for the parameters $\omega_1 = \delta = \sigma = 1$ and $\gamma = .8$ in the range $-3 \leq x_1, x_2 \leq 2$. The left picture of Figure 5 is the same as in Figure 3. The right one represents weak solutions obtained by the orthogonal expansion (25) for $N = 4$ of the polynomial series.

4. Riding Cars Excited by Colored Noise

To extend the above investigations to more realistic excitations of cars riding on road surfaces, stochastic road models are introduced by scalar first order

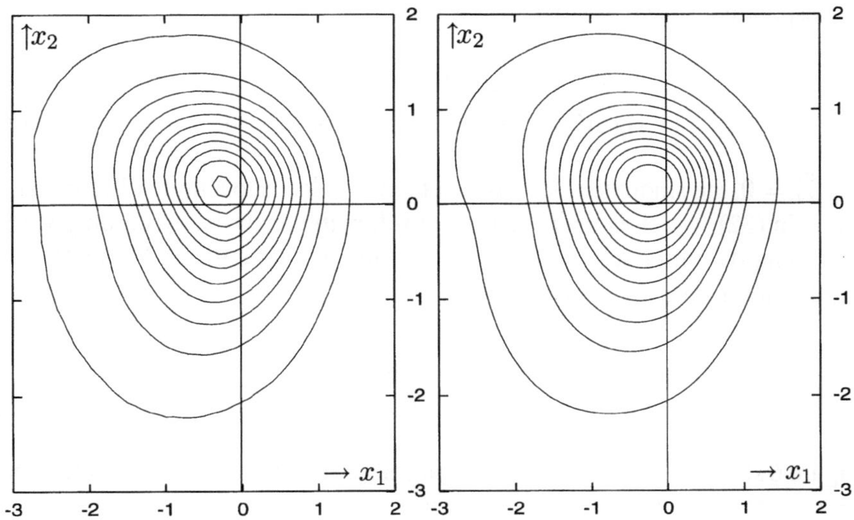

Figure 5. Comparison of MC-simulations and polynomial solutions

filter equations defined in the way domain, as follows:

$$dZ_s = -\Omega Z_s ds + \sigma dW_s, \qquad E(dW_s^2) = ds, \tag{27}$$

$$E(Z_s^2) = \frac{\sigma^2}{2\Omega}, \qquad S_z(\bar{\omega}) = \frac{\sigma^2}{\bar{\omega}^2 + \Omega^2}. \tag{28}$$

Herein, $\Omega > 0$ denotes the way corner frequency of the road excitation process Z_s and σ gives the intensity of the way noise increments dW_s which are normally distributed with zero mean and normalized mean square. The analysis of the Itô equation (27) leads to the way-invariant mean square $E(Z_s^2)$ and associated power spectrum $S_z(\bar{\omega})$, noted in (28). Applying the way increments of coordinate s and Wiener process W_s

$$ds = vdt, \qquad dW_s = \sqrt{v}dW_t, \qquad (0 < v < \infty) \tag{29}$$

the way filter (27) is transformed into the time domain, as follows:

$$dZ_t = -\Omega v Z_t dt + \sigma\sqrt{v}dW_t, \qquad Z_t = -\ddot{U}_t/\omega_1^2. \tag{30}$$

The solution Z_t of the time filter (30) is applied for the acceleration excitation of the quarter car model by $\ddot{U}_t = -\omega_1^2 Z_t$. Therewith, the road-vehicle dynamic is described by a third order system consisting of the second order equation (1) of vertical car motion and the first order equation (30) of the car excitation.

To utilize the analytical density approximations, already applied in (10) and (12) for the second order road-vehicle dynamics, the third order system has to be decoupled with respect to the linear terms. For this purpose, the filter process Z_t in (30) is eliminated by the equation (1) of motion and $\dot{Z}_t$ by the corresponding time derivative

$$\dddot{Y}_t + \delta\ddot{Y}_t(1+\gamma\text{sign}\dot{Y}_t) + \omega_1^2\dot{Y}_t = \omega_1^2\dot{Z}_t, \tag{31}$$
$$(\dot{Y}_t\text{sign}\dot{Y}_t)^{\cdot} = \ddot{Y}_t\text{sign}\dot{Y}_t, \qquad \text{for } \dot{Y}_t \neq 0. \tag{32}$$

For $\dot{Y}_t = 0$, the smoothed-out representation (2) leads to a finite differentiation value in (32), which is neglected in the following analysis. Accordingly, (32) holds and the introduction of (31) and (1) into the time filter (30) leads to the third order differential equation

$$\begin{aligned}&\dddot{Y}_t + \delta\ddot{Y}_t(1+\gamma\text{sign}\dot{Y}_t) + \omega_1^2\dot{Y}_t + \Omega v[\ddot{Y}_t + \\ &+\delta\dot{Y}_t(1+\gamma\text{sign}\dot{Y}_t) + \omega_1^2 Y_t] = \omega_1^2\sigma\sqrt{v}\dot{W}_t.\end{aligned} \tag{33}$$

Figure 6 shows contour lines of two marginal densities of the three-dimensional

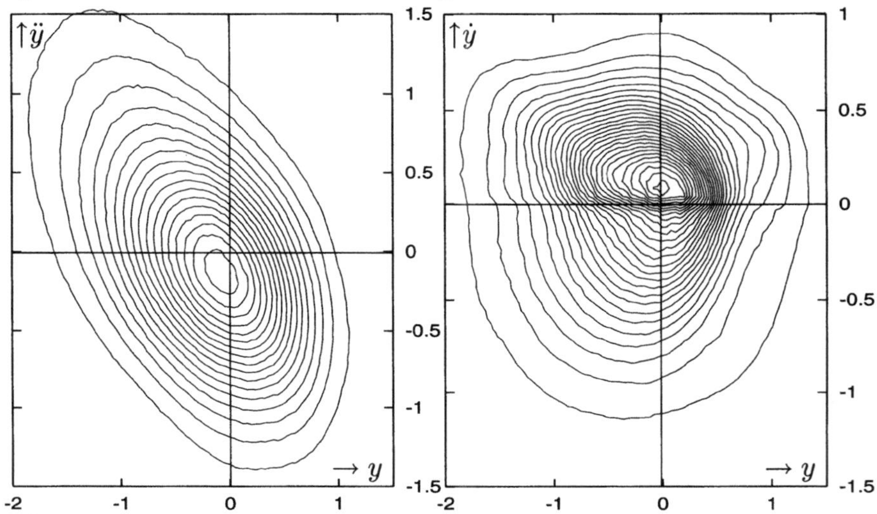

Figure 6. Contour lines obtained by Monte Carlo simulations

density $p(y, \dot{y}, \ddot{y})$ for the stationary displacement Y_t, velocity $\dot{Y}_t$ and acceleration $\ddot{Y}_t$. Both marginal densities are calculated by means of Monte Carlo simulations for the system data $\omega_1 = \delta = 1\ \text{s}^{-1}, \gamma = 0.8, \text{v} = 1\text{m}\,\text{s}^{-1}, \Omega = 1\ \text{m}^{-1}$ and $\sigma^2 = 1\ \text{m}^{-1}$. The left picture in figure 6 represents the coupled joint density $p(\ddot{y}, y)$, meanwhile the right picture shows $p(\dot{y}, y)_{\,} =$. Obviously, $p(\dot{y}, y)$ in the right picture is orthogonal in the sense that $E(\dot{Y}_t Y_t) = 0$. Because of $E(\ddot{Y}_t\dot{Y}_t) = 0$, the same orthogonality property is holding for $p(\ddot{y}, \dot{y})$.

In a second step, principle coordinates are to be determine in order to decouple $\ddot{Y}_t$ and Y_t, as well. For this purpose, the state processes of displacement Y_t, velocity $\dot{Y}_t = V_t$ and acceleration $\ddot{Y}_t = B_t$ are introduced into the third order equation (33) which gives the third order system

$$\dot{B}_t = -[(\delta + \Omega v)B_t + (\omega_1^2 + \delta\Omega v)V_t + \omega_1^2\Omega v Y_t + \delta\gamma(B_t + \\ +\Omega v V_t)\mathrm{sign}V_t] + \omega_1^2\sigma\sqrt{v}\,\dot{W}_t, \qquad \dot{V}_t = B_t, \qquad \dot{Y}_t = V_t. \tag{34}$$

For $\gamma = 0$, the system (34) is linear. Its stationary analysis leads to vanishing mean values and to the covariances

$$\begin{bmatrix} E(Y_t^2)\omega_1^2 & E(Y_t B_t) \\ E(B_t^2)/\omega_1^2 & E(V_t^2) \end{bmatrix} = \frac{\sigma_z^2\omega_1^2}{2D(1+2D\nu+\nu^2)} \begin{bmatrix} \nu + 2D & -\nu \\ \nu(1+2D\nu) & \nu \end{bmatrix}. \tag{35}$$

Herein, the parameters D, σ_z and ν denote the damping measure, respectively, the root-mean-square of the car excitation, respectively, the frequency speed of the road-vehicle system related to the natural frequency of the vertical car vibrations. These parameters are dimensionless and defined, as follows:

$$\delta = 2D\omega_1, \qquad \sigma_z = \sigma/\sqrt{2\Omega}\,, \qquad \nu = v/\omega_1\,. \tag{36}$$

Figure 7 shows numerical evaluations of the root-mean-square values of

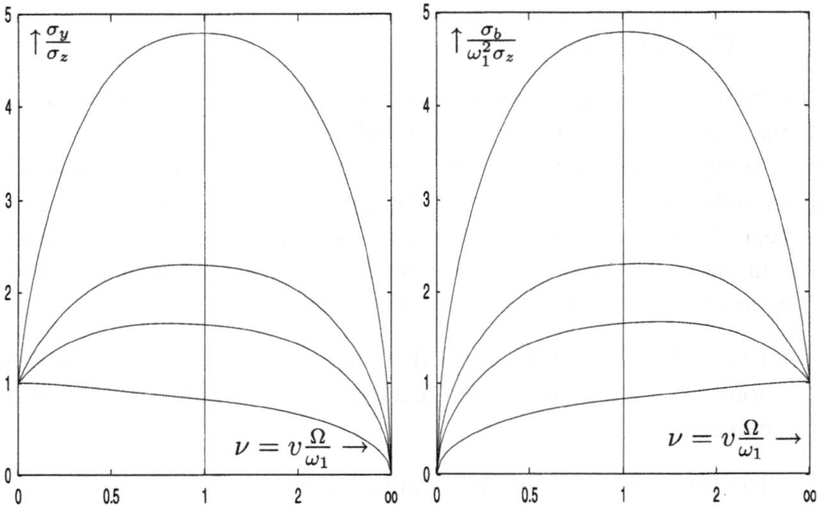

Figure 7. Root-mean-square values versus speed of riding cars

$E(Y_t^2) = \sigma_y^2$) of the vertical car vibrations and associated accelerations values $E(B_t^2) = \sigma_b^2$ in dependence on the related frequency speed $\nu = v\Omega/\omega_1$ of the car for different damping values $D = 0.011$ up to $D = 0.5$. For the displacement root-mean-square σ_y related to the excitation rms-value σ_z, the frequency amplitudes are starting with one at $\nu = 0$ and end with zero when ν tends to ∞. For the acceleration rms-values σ_b related to $\omega_1^2\sigma_z$ the frequency amplitudes start vice versa with zero and end with one. Both frequency amplitude curves become maximal for small damping measures $D <<$

1 when the frequency speeds are near one. Note, that in both pictures the scaling, applied in the first half part $0 \leq \nu \leq 1$, is linear. It is hyperbolic of the form $2 - 1/\nu$ in the second half part of the frequency axis.

By means of the stationary covariances, noted in (35), one can calculate the principal coordinates $\vec{Y}_t = (Y_{1,t}, Y_{2,t}, Y_{3,t})^T$, which completely decouple the third order system with respect to the linear terms. It allows to utilize the analytical density approximations (10) and (12) in the one-dimensional forms

$$pi(x_i) = C_i \exp\{-\frac{1}{2\sigma_i^2}[x_i^2(1 + \gamma_i \text{sign} x_i) + 2\alpha_i x_i]\}, \quad \text{for } i = 1, 2, 3. \quad (37)$$

Herein, α_i are three mean value parameters needed to satisfy the first moments' equations of the transformed system, γ_i are three non-linear coefficients determined by three covariance equations and σ_i^2 are the three linear covariances which follows by a principle axis transformation of the covariance matrix, calculated in (35). Subsequently, this analysis can be continued applying corresponding expansions by orthogonal polynomials, as previously demonstrated for second order systems.

5. Conclusion

The bilinear damping characteristic of quarter car models [3] lead to the effect, that the stationary response of cars riding on rough road surfaces [2] possesses highly nonsymmetric distribution densities. This has the consequence, that the stationary mean of the vertical car vibrations is shifted to negative values meanwhile the response variances approximately remain unchanged. The paper investigates these effects solving the associated Fokker-Planck equation by nonlinear Galerkin methods in three steps.

1 The problem is orthogonalized by appropriate coordinate transformations. This allows to derive separated densities with nonlinear distribution effects.

2 The free parameters of the separated densities are calculated to satisfy the stationary first and second order moments equations of the nonlinear system.

3 The separated densities are applied to derive orthogonal polynomials which expand the joint densities of interest by means of Galerkin methods.

For future research it is of interest to investigate the stability of the calculated vibrations for increasing nonlinearity parameter γ. For this purpose, the stationary solution is perturbed in order to determine the top Lyapunov exponent of the perturbed solution by means of the multiplicative ergodic theorem and associated invariant measure. For negative Lyapunov exponents the stationary solution is asymptotically stable. If the top Lyapunov exponent becomes positive for increasing nonlinearity parameter γ, the perturbed solution will become unstable leading to a stochastic attractor.

References

[1] D. Ammon, *Modellbildung und Systementwicklung in der Fahrzeugdynamik*, B. G. Teubner, Stuttgart, 1997.

[2] D. Cebon, *Handbook of Road-Vehicle Interaction*, Swets & Zeitlinger Publishers, Lisse, the Netherlands, 1999.

[3] K. Popp, W. Schiehlen, *Fahrdynamik*, B. G. Teubner, Stuttgart, 1993.

[4] U. von Wagner and W. V. Wedig, "On the Calculation of Stationary Solutions of Multi-Dimensional Fokker-Planck Equations by Orthogonal Functions," *Nonlinear Dynamics*, vol. 21, no. 3, pp. 289–306, 2000.

[5] W. Wedig, "Generalized Hermite Analysis of Nonlinear Stochastic Systems," in *Proc. Stochastic Structural Dynamics*, ed. by N. Sri Namachchivaya, Harry H. Hilton & Y.K. Wen, University of Illinois, Urbana-Champaign, pp. 381–389, 1989.

[6] W. Wedig, *Spectral Analysis of Non-smooth Dynamical Systems*. In: Stochastic Structural Dynamics ed. by B. F. Spencer, Jr. & E.A. Johnson, Balkema, Rotterdam, pp. 249–256, 1999.

[7] W. V. Wedig and U. von Wagner, "Oscillatory Stochastic Systems with Nonlinear Dissipations," *ZAMM*, vol. 80, no. 2, pp. 309–310, 2000.

[8] W. V. Wedig and U. von Wagner, *Stochastic Car Vibrations with Strong Nonlinearities*. ASME 2001 Design Engineering Technical Conferences, September 9-12, 2001, Pittsburgh, PA, VIB-21605, pp. 1–7, 2001.

SOLUTIONS OF THE FIRST-PASSAGE PROBLEM BY IMPORTANCE SAMPLING

Christian Bucher and Michael Macke
Institut für Strukturmechanik
Bauhaus-Universität Weimar
Marienstr. 15, D-99421 Weimar, GERMANY
christian.bucher@bauing.uni-weimar.de

Abstract The paper presents an efficient importance sampling approach to the Monte Carlo solution of first-passage problems. Based on Girsanov's theorem it is shown that an appropriate importance sampling density can be constructed utilizing "design point excitations" which are obtained by a procedure analogous to the first order reliability method. The presented approach can efficiently handle non-white and non-stationary excitations can be applied to a wide range of nonlinear problems, and yields high levels of confidence at very low numbers of samples.

Keywords: Random vibration, first-passage probability, Monte Carlo simulation, importance sampling, first order reliability.

1. Introduction

Determining the response of nonlinear dynamical systems subjected to random excitations has been a major research topic during the past decades. Nevertheless, the number of mathematically exact solutions available is still restricted to a narrow class of systems and to the knowledge of stationary probability densities or moments only [13, 14]. For assessing the reliability of a dynamical system most realistically, however, a reliability measure evolving over a time period is most instrumental. The latter can be achieved by utilizing the first-passage probability, i.e., the probability that the structural response exceeds a given failure boundary in a certain time interval for the first time. It is quite remarkable that till now a closed-form solution for the first-passage problem does not exist, not even for single-degree-of-freedom linear systems.

N. Sri Namachchivaya and Y.K. Lin (eds.),
IUTAM Symposium on Nonlinear Stochastic Dynamics, 401-413.

A generally applicable numerical solution technique for the first-passage problem is presumably Monte Carlo simulation. However, it is most often applied as a last resort only—due to numerical answers of supposedly limited accuracy or prohibitive computational costs. Whereas these drawbacks of the Monte Carlo simulation technique are an inherent characteristic of its crude form, this does not hold for variance reduction techniques [21]. Although this fact is widely recognized for time-invariant problems, for problems in random vibration the application of variance reduction techniques is still eschewed—despite different, technically sound methods proposed during the last years [8, 18, 24]. The most versatile variance reduction technique appears to be importance sampling utilizing the measure transformation method [15, 17, 23] based on Girsanov's theorem [5].

A different and at first sight much simpler approach is so called (multi-level) splitting which can be traced back at least to [10]. This method has been brought to a wider attention by [1] and [9]. A typical application to civil engineering systems can be found, for e.g., in [19]. In multi-level splitting promising sample paths are split into sub-paths at intermediate levels to increase the number of observations of rare events. As has been reported as early as in [9], however, this procedure can result in general settings in an increase of the variance of the estimator. More precarious even is an "apparent bias" in the estimator, when the levels for splitting are not chosen consistently. Indeed, multi-level splitting requires for being effective that the splitting is performed along the most likely path leading to an out-crossing [6]. However, in this case a more effective sampling method like importance sampling can be used right from start.

For the importance sampling procedure presented below, the Itô stochastic differential equations which model the structural response are changed according to a minimization criterion for the variance of the estimator. In other words, controls are introduced which allow to influence the response paths, whereby the unlikeliness of such paths is taken into account by a correction process. Although theoretically there exist controls which result in unbiased "zero-variance" estimators [15, 16], these optimal controls are in general not accessible (this would require the knowledge of the solution) and have to be replaced by sub-optimal ones.

2. Importance Sampling Method

Let us assume without any restrictions that the dynamical response of a system at time $t \in [0, T]$ is described by a q-dimensional Itô process $\mathbf{X}_t = (X_t^1, X_t^2, \ldots, X_t^q)$ in terms of the stochastic differential equations

$$\mathrm{d}X_t^j = \alpha^j(t, \mathbf{X}_t)\mathrm{d}t + \gamma^j(t, \mathbf{X}_t)\,\mathrm{d}W_t, \qquad X_0^j = x_j \tag{1}$$

with W_t as a unit Wiener process and $\mathbf{x} = (x_1, x_2, \ldots, x_r)$ as deterministic initial conditions. On the one hand, (1) can be straightforwardly extended to a vector of unit Wiener processes. On the other hand, for taking into account non-white or non-normal excitations, (1) can be augmented with filter equations of the same type (cf. [4]).

In reliability assessment of dynamical systems we are in general not interested in an exact pathwise representation of $\mathbf{X}_t$, but only in so called weak solutions of the form

$$v = \mathrm{E}[f(\mathbf{X}_T)] \tag{2}$$

whereby $f(\cdot)$ denotes some real-valued function and $\mathrm{E}[\cdot]$ is the expected value. A typical example of a first-passage problem is to determine the probability that the j-th component of $\mathbf{X}_t$ exceeds a given barrier ξ in the interval $[0, T]$, i.e.,

$$v = \mathrm{E}[\mathrm{I}(\max_{0 \leq t \leq T} X_t^j > \xi)] \tag{3}$$

with $\mathrm{I}(\cdot)$ denoting an indicator function that equals one if its argument is true, and zero otherwise.

When we utilize Monte Carlo simulation to evaluate (2), v is replaced by its sample-mean formula [21]

$$\hat{v} = \hat{\mathrm{E}}[f(\mathbf{X}_T)] = \frac{1}{n} \sum_{i=1}^{n} f(\mathbf{X}_T^{(i)}) \tag{4}$$

where $\mathbf{X}_t^{(i)}$ denotes the i-th sample path of $\mathbf{X}_t$ and n is the number of sample paths. The quantity $\hat{v}$ is an unbiased estimator of v with variance

$$\mathrm{Var}[\hat{v}] = \frac{1}{n} \mathrm{Var}[f(\mathbf{X}_T)] \tag{5}$$

The square root of this variance—the so called root mean square error—can be conceived as an indicator of the accuracy of the Monte Carlo estimator. Whereas an estimator with sufficiently small error can always be obtained by simply increasing the number n of sample paths, this is rarely a rewarding procedure. Therefore, a variance reduction technique like importance sampling becomes mandatory.

Solving the stochastic differential equations in (1) numerically, these equations are replaced by discrete time approximations. Let us subdivide the interval $[0, T]$ into m sub-intervals, such that $0 = t_0 < t_1 < \ldots < t_k = k\,\Delta_k < \ldots < t_m = T$ whereby $\Delta_k = t_{k+1} - t_k$ is the length of the sub-interval $[t_k, t_{k+1}]$. Therewith the stochastic differential equations can be written—e.g., in form of the stochastic Euler scheme—as

$$X_{k+1}^j = X_k^j + \alpha^j(t_k, \mathbf{X}_k)\Delta_k + \gamma^j(t_k, \mathbf{X}_k)\Delta W_k, \qquad X_0^j = x_j \tag{6}$$

with $\Delta W_k = W_{t_{k+1}} - W_{t_k}$. It should be noted, that for the following any other stochastic integration scheme (cf. [17]) can be utilized.

Replacing the k-th increment of the Wiener process by $\Delta W_k = \sqrt{\Delta_k}\,\zeta_k$ with ζ_k being mutually independent standard normal random variates, the functional of (2) is approximated by

$$v = \int_{-\infty}^{+\infty} (m\text{-fold}) \int_{-\infty}^{+\infty} f(\mathbf{X}_m)\,\varphi(\boldsymbol{\zeta})\,\mathrm{d}\boldsymbol{\zeta} \tag{7}$$

whereby $\varphi(\cdot)$ is the m-dimensional joint density function of the standard normal random variates $\boldsymbol{\zeta}$. Describing now the boundary of the area which will be out-crossed by the Itô process $\mathbf{X}_l$ at time t_l as

$$\bar{g}(\mathbf{X}_l) = \bar{g}(\mathbf{X}_l(\zeta_0(t_l), \ldots, \zeta_{l-1}(t_l))) = g(\zeta_0(t_l), \ldots, \zeta_{l-1}(t_l)) = 0 \tag{8}$$

then the most likely excitation leading to an out-crossing of this boundary is defined as $\boldsymbol{\zeta}^*(t_l) = (\zeta_0^*(t_l), \ldots, \zeta_{m-1}^*(t_l))$ which minimizes

$$\beta(t_l) = \left[\sum_{k=0}^{m-1} (\zeta_k^*(t_l))^2\right]^{1/2} \quad \text{subject to} \quad g(\zeta_0^*(t_l), \ldots, \zeta_{l-1}^*(t_l)) = 0 \tag{9}$$

To solve (9) standard techniques from first order reliability method can be applied (cf. [20, 22]). It should also be emphasized that an analogous formulation is given in [7, 12] for determining the mean out-crossing rate of randomly excited systems.

With the above given most likely excitation $\boldsymbol{\zeta}^*(t_l)$, (6) is modified such that

$$\tilde{X}_{k+1}^j = \tilde{X}_k^j + \alpha^j(t_k, \tilde{\mathbf{X}}_k)\Delta_k + \gamma^j(t_k, \tilde{\mathbf{X}}_k)(u^*(t_k; t_l)\Delta_k + \Delta W_k) \tag{10}$$

in which the above utilized control $u^*(t_k; t_l)$ is defined as

$$u^*(t_k; t_l) = \frac{1}{\sqrt{\Delta_k}}\,\zeta_k^*(t_l) \quad (k = 0, 1, \ldots, m-1;\ l = 1, 2, \ldots, m) \tag{11}$$

Taking into account the unlikeliness of such a modification of (10), the functional of (7) is evaluated by the importance sampling integral [21]

$$v = \int_{-\infty}^{+\infty} (m\text{-fold}) \int_{-\infty}^{+\infty} f(\tilde{\mathbf{X}}_l)\,\frac{\varphi(\tilde{\boldsymbol{\zeta}})}{h(\tilde{\boldsymbol{\zeta}})}\,h(\tilde{\boldsymbol{\zeta}})\mathrm{d}\tilde{\boldsymbol{\zeta}} \tag{12}$$

with $\tilde{\zeta}_k = \zeta_k^*(t_l) + \zeta_k$. The importance sampling density $h(\tilde{\boldsymbol{\zeta}})$ in (12) is defined as

$$h(\tilde{\boldsymbol{\zeta}}) \propto \exp\left[-\frac{1}{2}\sum_{k=0}^{m-1} (\tilde{\zeta}_k - \zeta_k^*(t_l))^2\right] \tag{13}$$

Therewith, the likelihood ratio is given by

$$\frac{\varphi(\tilde{\boldsymbol{\zeta}})}{h(\tilde{\boldsymbol{\zeta}})} = \exp\left[-\sum_{k=0}^{m-1} u^*(t_k;t_l)\Delta W_k - \frac{1}{2}\sum_{k=0}^{m-1}(u^*(t_k;t_l))^2\Delta_k\right] \tag{14}$$

It should be noted that (14) is a discrete version of the Radon-Nikodym derivative as used when applying Girsanov's theorem (cf. [5, 17]).

When determining the first-passage probability in the time interval $[0, T]$, there are—according to the time discretization—m most likely excitations leading to an out-crossing. Weighting these excitations by $\Phi(-\beta(t_l))$, i.e., their probability of occurrence, results in the m-dimensional multi-modal importance sampling density [2, 3]

$$h(\tilde{\boldsymbol{\zeta}}) \propto \sum_{l=1}^{m} \Phi(-\beta(t_l)) \exp\left[-\frac{1}{2}\sum_{k=0}^{m-1}\left(\tilde{\zeta}_k - \zeta_k^*(t_l)\right)^2\right] \tag{15}$$

Since it is quite difficult to sample from (15) directly, the sampling procedure is split. First, the out-crossing time t_j, and therewith the corresponding most likely excitation $\boldsymbol{\zeta}^*(t_j)$, is sampled with respect to the probability $\Phi(-\beta(t_j))/\sum_{l=1}^{m}\Phi(-\beta(t_l))$. This most likely excitation is then utilized to determine the response $\tilde{\mathbf{X}}_t$ and to evaluate the indicator function $\mathrm{I}(\cdot)$. Finally, the likelihood ratio $\varphi(\tilde{\boldsymbol{\zeta}})/h(\tilde{\boldsymbol{\zeta}})$ is determined by inserting the utilized excitation $\tilde{\zeta}_k = \zeta_k^*(t_j) + \zeta_k$ in the respective density functions $\varphi(\tilde{\boldsymbol{\zeta}})$ and $h(\tilde{\boldsymbol{\zeta}})$.

3. Linear Oscillator under Parametric Excitation

As a first example, let us consider a single-degree-of-freedom oscillator described by the non-dimensional equation of motion in Itô-form

$$\begin{aligned} \mathrm{d}X_t^1 &= X_t^2 \mathrm{d}t \\ \mathrm{d}X_t^2 &= (-X_t^1 - 2\eta X_t^2)\mathrm{d}t + (1-\alpha X_t^1)\sqrt{2\pi S_0}\,\mathrm{d}W_t \end{aligned} \tag{16}$$

with quiescent initial conditions, i.e., $X_0^1 = X_0^2 = 0$. In (16) $\eta = 0.05$ is a viscous damping coefficient, α is a parameter controlling the amount of parametric excitation present in the system and S_0 is the power spectral density of the Gaussian white noise excitation ($2\pi S_0 = 0.3$). In the following we determine—depending on the numerical value of α—the probability that X_t^1 up-crosses the threshold level $\xi = 10$ for the first time in the time interval $t \in [0, 50]$. Note that the stability limit for the second moments is reached at a value of $\alpha = 0.816$.

In Figures 1 to 2 the controls $u^*(t;T)$ (or "design point excitations") leading to an out-crossing at time $T = 50$ are depicted—together with the corresponding values of $\beta(T = 50)$. The destabilizing influence of the parametric excitation is clearly seen from the decrease of the $\beta(T)$-values. As is

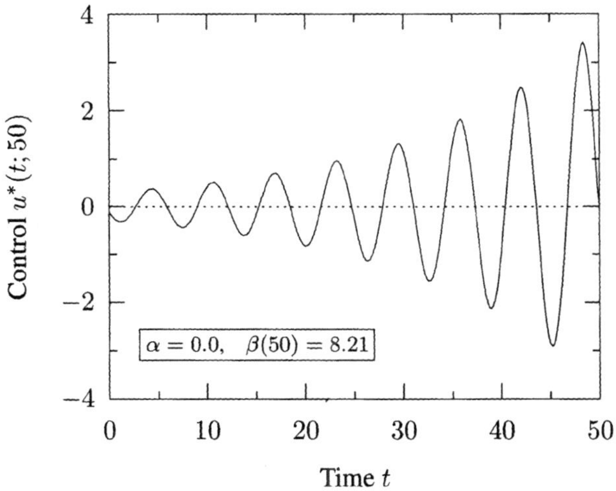

Figure 1. Control $u^*(t;T)$ of the linear oscillator (with $\alpha = 0.0$) for an up-crossing of the level $\xi = 10$ at time $T = 50$.

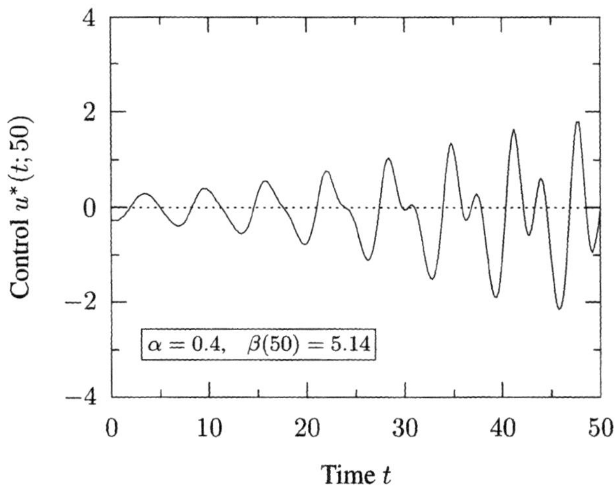

Figure 2. Control $u^*(t;T)$ of the linear oscillator (with $\alpha = 0.4$) for an up-crossing of the level $\xi = 10$ at time $T = 50$.

well known from deterministic stability analysis of parametrically excited systems, the most influential frequency region is located at twice the natural frequency of the oscillator. This is clearly seen from the controls plotted in Figures 1 to 2.

The first passage probabilities as obtained from $n = 5000$ samples for different values of α are given in Table 1 along with the root mean square errors. As can be seen, the root mean square errors vary between 5 % and 9 % of the estimated first-passage probability—with an increasing tendency when more parametric excitation is present. Nevertheless, the root mean square errors indicate a high confidence in the estimated quantities.

Table 1. Estimated first-passage probabilities vs. amount α of parametric excitation of the linear oscillator ($\xi = 10$, $T = 50$, $n = 5000$).

Parameter α	*Estimated first-passage probability*	*Root mean square error*
0.0	$2.0 \cdot 10^{-15}$	$1.2 \cdot 10^{-16}$
0.1	$8.6 \cdot 10^{-14}$	$5.3 \cdot 10^{-15}$
0.2	$2.0 \cdot 10^{-10}$	$1.7 \cdot 10^{-11}$
0.3	$9.2 \cdot 10^{-8}$	$7.6 \cdot 10^{-9}$
0.4	$5.5 \cdot 10^{-6}$	$4.5 \cdot 10^{-7}$

4. Oscillator with Nonlinear Damping

Given is an oscillator with nonlinear damping described in Itô form by the system of non-dimensional equations [25]

$$\begin{aligned} \mathrm{d}X_t^1 &= X_t^2 \mathrm{d}t \\ \mathrm{d}X_t^2 &= \left(-X_t^2 - \varepsilon |X_t^2| - X_t^1\right) \mathrm{d}t + H_t \mathrm{d}t \end{aligned} \tag{17}$$

with ε denoting a nonlinearity parameter. The excitation process H_t has the correlation function $\varrho(\tau) = \exp(-\kappa|\tau|)$ and a stationary standard normal probability density, which can be modeled by a linear stochastic differential equation with additive noise

$$\mathrm{d}H_t = -\kappa H_t \mathrm{d}t + \sqrt{2\kappa}\, \mathrm{d}W_t \tag{18}$$

where $\varepsilon = 0.8$ and $\kappa = 1.0$. We assume again quiescent initial conditions, i.e., $X_0^1 = X_0^2 = H_0 = 0$. In Figure 3 the most likely excitations h_t^* leading to an out-crossing of the levels $\xi = 2.0$, $\xi = 3.0$ and $\xi = 4.0$, respectively, at $T = 25$ are shown. In the following we want to estimate the probability that the displacement X_t^1 of the oscillator up-crosses a level ξ in the time interval $t \in [0, 50]$.

The first-passage probabilities as obtained from $n = 1000$ samples for different values of the threshold level ξ are depicted in Figure 4. As can be seen there is an excellent agreement with the results from crude Monte Carlo simulation (number of samples $n = 10^6$) for first-passage probabilities greater than $1.0 \cdot 10^{-5}$. Beyond this level, however, crude Monte Carlo simulation breaks

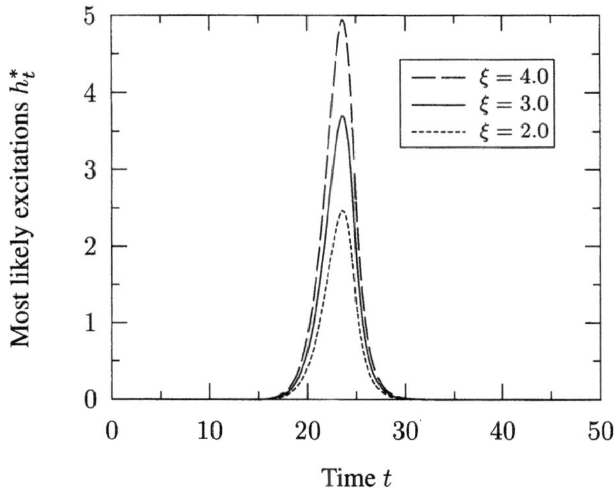

Figure 3. Most likely excitations h_t^* of the oscillator with nonlinear damping for an out-crossing of the levels $\xi = 2.0$, $\xi = 3.0$ and $\xi = 4.0$, respectively, at $T = 25$.

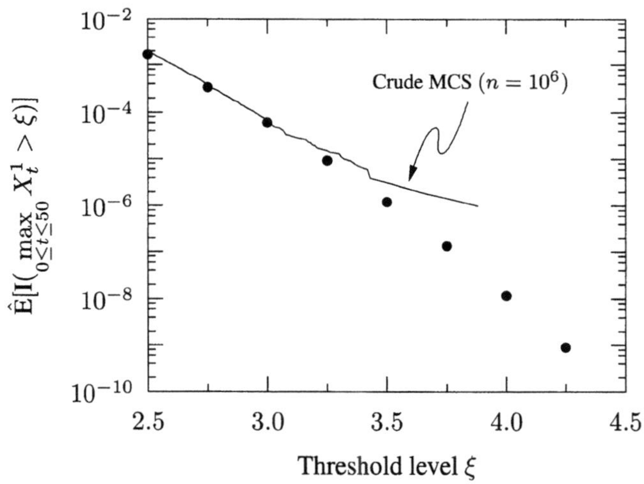

Figure 4. Importance sampling estimator (•) of first-passage probability of the oscillator with nonlinear damping for $t \in [0, 50]$ ($n = 1000$).

down—despite the enormous sample size. The importance sampling procedure, on the other hand, is capable to reach any specified level ξ with very high confidence. The root mean square errors for the first-passage probabilities shown in Figure 4 range from 7.0 % to 10.0 % of the estimated quantities.

5. Hysteretic Oscillator under Non-Gaussian Excitation

As a last example we investigate a hysteretic oscillator described by the non-dimensional Itô equation of motion

$$\begin{aligned} \mathrm{d}X_t^1 &= X_t^2 \mathrm{d}t \\ \mathrm{d}X_t^2 &= \left(-2\eta X_t^2 - \alpha X_t^1 - (1-\alpha)X_t^3\right)\mathrm{d}t + H_t \mathrm{d}t \\ \mathrm{d}X_t^3 &= \left(-\gamma |X_t^2| X_t^3 |X_t^3|^{\nu-1} - \beta X_t^2 |X_t^3|^{\nu} + A X_t^2\right)\mathrm{d}t \end{aligned} \tag{19}$$

whereby η denotes a viscous damping coefficient, α is the the post- to pre-yielding stiffness ratio and ν, γ, β as well as A are adjustable parameters to describe the hysteretic behavior. (In the following we choose $\eta = 0.05$, $\alpha = 0.05$, $\nu = 1$, $\gamma = \beta = 0.5$ and $A = 1$, respectively.) The excitation H_t is Rayleigh distributed with stationary density $f(h_t) = \lambda(\lambda h_t + 2)\exp(-\lambda h_t + 2)$, $h_t \geq -2/\lambda$ and correlation function $\varrho(\tau) = \exp(-\kappa|\tau|)$, i.e.,

$$\mathrm{d}H_t = -\kappa H_t \mathrm{d}t + \frac{1}{\lambda}\sqrt{2\kappa(\lambda H_t + 2)}\,\mathrm{d}W_t \tag{20}$$

with $\kappa = 0.5$, $\lambda = 2.5$ and W_t as a standard Wiener process (cf. [4]). The

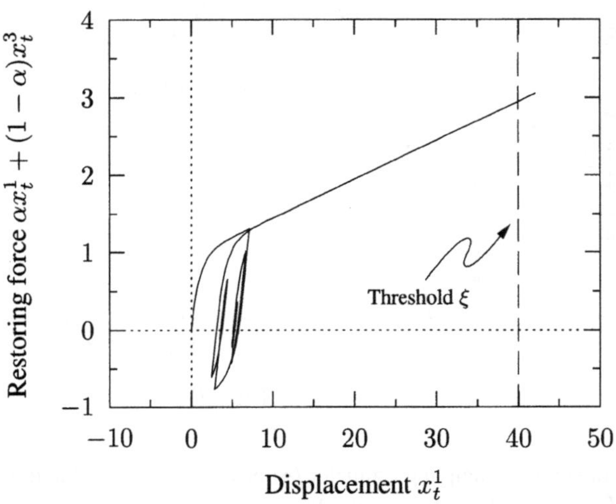

Figure 5. Sample path (restoring force vs. displacement) of the hysteretic oscillator utilizing control $u^*(t; T = 50)$ for $t \in [0, 50]$ (see Figure 6).

initial conditions of the oscillator are again quiescent. In the following we determine the probability that the displacement X_t^1 up-crosses the threshold level $\xi = 40$ in the time interval $t \in [0, 50]$.

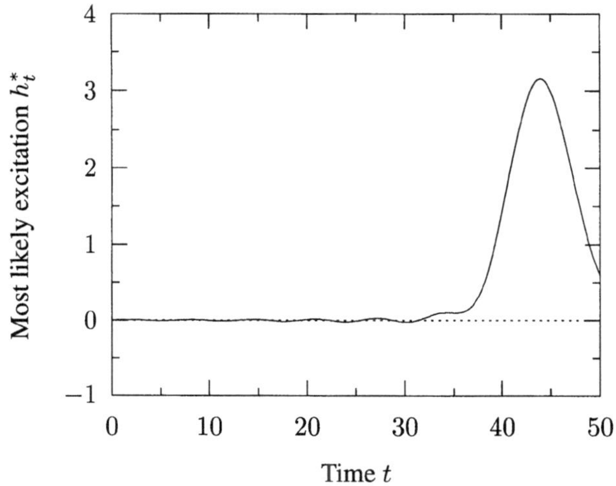

Figure 6. Most likely excitation h_t^* of the hysteretic oscillator for an out-crossing of the level $\xi = 40$ at $T = 50$.

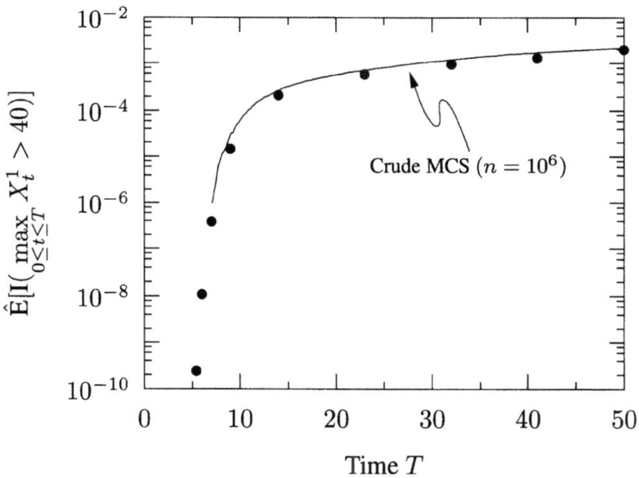

Figure 7. Importance sampling estimator (•) of first-passage probability of the hysteretic oscillator for $T \in [0, 50]$ ($\xi = 40$, $n = 1000$).

In Figure 6 the most likely excitation h_t^* leading to an up-crossing of the level $\xi = 40$ at time $T = 50$ is depicted. In Figure 5 a corresponding sample path (restoring force vs. displacement) of the oscillator utilizing this excitation h_t^* is shown. In Figure 7 the importance sampling estimators of the first-passage probability (number of samples $n = 1000$) are compared with the crude Monte Carlo simulation results (solid line, number of samples $n = 10^6$). As can be seen, the importance sampling estimators show again an excellent agreement with the crude Monte Carlo simulation results. This is also reflected in the corresponding root mean square errors which vary between 7.0 % and 10.0 % of the estimated first-passage probabilities.

6. Conclusion

A generally applicable importance sampling procedure for linear and nonlinear dynamical systems under random excitations has been presented. Utilizing sub-optimal controls constructed via the solution of an optimization problem analogous to the one from first order reliability method, the variance of the estimators can be decreased drastically as compared to crude Monte Carlo simulation. This is achieved by spending additional computational effort in the determination of a set of sub-optimal controls (or "design point excitations") which are utilized as importance sampling mean excitations. While the effort required for the computation of these controls is not negligible, it is primarily dominated by the number of random variables used for the time-discretization of the random excitation process. Consequently, when extending the herein utilized scalar excitation to a vector form, the computational effort will increase linearly with the number of additional random variables. In this form, the computational effort is acceptable for a sufficiently small number of random variables, i.e., in significantly non-stationary or transient situations. On the other hand, in the stationary situation the determination of the design point excitations turns out to be trivial for some cases. In these cases only one nonlinear dynamic analysis needs to be carried out as shown in [11]. This means that for a wide range of problems the suggested approach can be considered to be extremely efficient. The advantages become especially clear when considering the system response in the low-probability regions (distribution tails) which are not accessible to crude Monte Carlo simulation. Finally, it should be pointed out that due to the application of standard concepts from first order reliability methods it is fairly easy to incorporate randomness of system parameters (described e.g., in terms of random fields) into the analysis (see e.g., [12]).

Acknowledgments

The work reported in this paper has been partially supported by the *Deutsche Forschungsgemeinschaft (DFG)* through *Sonderforschungsbereich 524* which is gratefully acknowledged by the authors.

References

[1] A. J. Bayes, "Statistical techniques for simulation models," *Aust. Comput. J.*, vol. 2, pp. 180–184, 1970.

[2] C. Bucher, "An importance sampling technique for randomly excited systems discretized by finite elements," In J. M. Ko and Y. L. Xu, editors, *Proc. Int. Conf. on Advances in Structural Dynamics*, Hong Kong, December 13–15, 2000. Elsevier, Amsterdam, 2000.

[3] C. Bucher and M. Macke, "Time-variant reliability formulated as first passage problem," In D. M. Frangopol and H. Furuta, editors, *Proc. 1st US-Japan Workshop on Life-Cycle Cost Analysis and Design of Civil Infrastructure Systems*, Honolulu, USA, August 7–8, 2000. ASCE, Reston, 2001.

[4] G. Q. Cai and Y. K. Lin, "Generation of non-Gaussian stationary stochastic processes," *Phys. Rev. E*, vol. 54, pp. 299–303, 1996.

[5] I. V. Girsanov, "On transforming a certain class of stochastic processes by absolutely continuous substitution of measures," *Theory Probab. Appl.*, vol. 5, pp. 285–301, 1960.

[6] P. Glasserman, P., Heidelberger, P., Shahabuddin, and T. Zajic, "A large deviations perspective on the efficiency of multilevel splitting," *IEEE Trans. Automat. Contr.*, vol. 43, pp. 1666–1679, 1998.

[7] Ø. Hagen and L. Tvedt, "Vector process out-crossing as parallel system sensitivity measure," *J. Engrg. Mech., ASCE*, vol. 117, pp. 2201–2220, 1991.

[8] P. Heidelberger, "Fast simulation of rare events in queueing and reliability models," *ACM Trans. Model. Comput. Simul.*, vol. 5, pp. 43–85, 1995.

[9] A. C. M. Hopmans and J. P. C. Kleijnen, "Importance sampling in system simulation: A practical failure?" *Math. Comput. Simul.*, vol. 21, pp. 209–220, 1979.

[10] H. Kahn and T. E. Harris, "Estimation of particle transmission by random sampling," *Appl. Math. Ser.*, vol. 12, pp. 27–30, 1951.

[11] H. Koo and A. Der Kiureghian, "Design point" excitation for stationary random vibrations," In R. B. Corotis, G. I. Schuëller, and M. Shinozuka, editors, *Proc. 8th Int. Conf. on Structural Safety and Reliability*, Newport Beach, USA, June 17–21, 2001. Balkema, Rotterdam, 2001.

[12] C.-C. Li and A. Der Kiureghian, "Mean out-crossing rate of nonlinear response to stochastic input," In M. Lemaire, J.-L. Favre, and A. Mébarki, editors, *Proc. 7th Int. Conf. on Applications of Statistics and Probability*, Paris, France, July 10–13, 1995. Balkema, Rotterdam, 1995.

[13] Y. K. Lin and G. Q. Cai, "Exact stationary-response solution for second order nonlinear systems under parametric and external white-noise excitations: Part II.," *J. Appl. Mech., Trans. ASME*, vol. 55, pp. 702–705, 1988.

[14] Y. K. Lin and G. Q. Cai, G. Q., *Probabilistic Structural Dynamics: Advanced Theory and Applications*. McGraw-Hill, New York, 1995.

[15] M. Macke, "Variance reduction in Monte Carlo simulation of dynamic systems," In Melchers, R. E. and Stewart, M. G., editors, *Proc. 8th Int. Conf. on Applications of Statistics and Probability*, Sydney, Australia, December 12–15, 1999. Balkema, Rotterdam, 1999.

[16] M. Macke and N. Harnpornchai, "Importance sampling of dynamic systems—A comparative study," In Corotis, R. B., Schuëller, G. I. and Shinozuka, M., editors, *Proc. 8th*

Int. Conf. on Structural Safety and Reliability, Newport Beach, USA, June 17–21, 2001. Balkema, Rotterdam, 2001.

[17] G. N. Milstein, *Numerical Integration of Stochastic Differential Equations*, Kluwer, Dordrecht, 1995.

[18] N. J. Newton, "Variance reduction for simulated diffusions," *SIAM J. Appl. Math.*, vol. 54, pp. 1780–1805, 1994.

[19] H. J. Pradlwarter, "A selective MC simulation technique for nonlinear structural reliability assessment," In Y. K. Lin, editor, *Proc. 6th ASCE Specialty Conf. on Probabilistic Mechanics and Structural and Geotechnical Reliability*, Denver, USA, July 8–10, 1992. ASCE, New York, 1992.

[20] R. Rackwitz and B. Fiessler, "Structural reliability under combined random load sequences," *Comput. Struct.*, vol. 9, pp. 489–494, 1978.

[21] R. Y. Rubinstein, *Simulation and the Monte Carlo Method.* Wiley, New York, 1981.

[22] M. Shinozuka, "Basic analysis of structural safety," *J. Struct. Engrg., ASCE*, vol. 109, pp. 721–740, 1983.

[23] H. Tanaka, "Application of an importance sampling method to time-dependent system reliability analyses using the Girsanov transformation," In Shiraishi, N., Shinozuka, M. and Wen, Y. K., editors, *Proc. 7th Int. Conf. on Structural Safety and Reliability*, Kyoto, Japan, November 24–28, 1997. Balkema, Rotterdam, 1998.

[24] W. Wagner, "Monte Carlo evaluation of functionals of solutions of stochastic differential equations: Variance reduction and numerical examples," *Stochastic Anal. Appl.*, vol. 6, pp. 447–468, 1988.

[25] W. V. Wedig and C. Schmalfuss, "Random vibrations of driving cars with bilinear damping characteristics," *Proc. IUTAM Symposium on Nonlinear Stochastic Dynamics*, Monticello, IL, USA, August 26–30, 2002.

PASSIVE FIELDS AND PARTICLES IN CHAOTIC FLOWS

Bruno Eckhardt, Erwan Hascoët, and Wolfgang Braun
Fachbereich Physik
Philipps Universität Marburg
35032 Marburg
Germany
bruno.eckhardt@physik.uni-marburg.de

Abstract Two examples for the interplay between chaotic dynamics and stochastic forces within hydrodynamical systems are considered. The first case concerns the relaxation to equilibrium of a concentration field subject to both chaotic advection and molecular diffusion. The concentration field develops filamentary structures and the decay rate depends non-monotonically on the diffusion strength. The second example concerns polymers, modelled as particles with an internal degree of freedom, in a chaotic flow. The length distribution of the polymers turns out to follow a power law with an exponent that depends on the difference between Lyapunov exponent and internal relaxation rate.

Keywords: Chaotic advection, power law distribution, surface flows, polymers

1. Introduction

Fluid dynamics provides many prime examples for nonlinear dynamical systems. The origin of the nonlinearity is the advection of particles and fields by the underlying velocity field and takes, with $\mathbf{x}_P(t)$ the position of a particle, the form

$$\dot{\mathbf{x}}_P(t) = \mathbf{u}(\mathbf{x}_P(t), t)\,. \tag{1}$$

We assume here and in the following that the particles are neutrally buoyant so that they follow the instantaneous velocity field (for other particles, see [26]). Typically, the fields are not constant or linear, so that (1) is a system of nonlinear equations and the dynamics of the particles becomes chaotic [2, 27, 1, 3, 4]. Stochastic elements enter if in addition to the deterministic parts also random forces, such as molecular diffusion or Brownian motion, have to be taken into account. Further complications arise when the particles can react back onto the flow (e.g., by a density contrast or by modifying viscosity). They

N. Sri Namachchivaya and Y.K. Lin (eds.),
IUTAM Symposium on Nonlinear Stochastic Dynamics, 415-424.

then become dynamically active. In the situations considered below no such feedback is allowed and the particles remain dynamically passive.

Here we consider two examples of the interplay between nonlinear advection and stochastic forcing: the relaxation of scalar fields in experiments with 2-d flows [29, 33, 18] and the stretching of polymers in chaotic flows. In all cases the flow fields have a simple spatial structure and at most a periodic time dependence, i.e., we have deterministic chaos. If these deterministic flow fields are replaced by temporally fluctuating ones with prescribed spatial correlations techniques from stochastic analysis can be used to characterize the particle dynamics further [15, 23, 9]. However, both deterministic and random flows show similar behaviour.

In the first example we focus on the processes that contribute to the relaxation of a non-uniform concentration, e.g., a blob of dye, to a uniformly spread out concentration in a chaotic flow. By its chaotic nature the flow by itself will stir the blob into an object with filaments that grow longer and finer and eventually cover space. Diffusion alone will also result in a uniform spreading in space, albeit on a different time scale. When combined, these processes do not simply add up but give rise to non-trivial modulations in the relaxation rate.

In the second example we focus on the effects of the velocity gradients on particles with internal degrees of freedom, specifically long flexible polymers. The phenomenon of turbulent drag reduction [25, 32] is connected with the uncoiling of the polymers in the flow. Once uncoiled, the polymers are no longer passive but begin to change the velocity field, so that a reduced drag is observed. The passive dynamics of polymers in a prescribed flow field then is a first step towards a description of the polymer dynamics in a turbulent flow. It is also a nice example of how the presence of both additive and multiplicative fluctuating forces can give rise to a power law probability distribution [5, 6, 8, 30].

2. Relaxation in Chaotic Advection

Since the paper [2], in which the term chaotic advection was coined, many theoretical and experimental studies have addressed various aspects of the phenomenon [27, 1, 3, 33, 29]. However, in order to arrive at mixing on a molecular level, diffusion has to be added in. Already in [11] the differences between stirring and mixing are emphasized: during stirring the fluid elements are stretched and folded, but they always keep their identity. Mixing requires molecular diffusion to cause a diffusive exchange of markers between different fluid elements and to thus blur distinguishability of the origin of the fluid elements. It is obvious that stirring enhances mixing, but there will be mixing even without stirring. The persistent patterns proposed in [28] and observed experimentally in [29] and the studies in [16] highlight some of the effects that arise when chaos and diffusion interact to turn stirring into mixing.

For fast processes the effects of diffusion can be neglected. Typical times for the stirring by the flow are set by L/U_0, where U_0 and L are characteristic velocity and length scales, respectively. Diffusive spreading happens on a time scale L^2/D, where D is the diffusion constant, and is usually much slower. The experiments by Taylor on reversibility at low Reynolds number [31, 17] also indicate this vast separation of time scales: the flow is reversed before diffusion has time to destroy the memory of the initial conditions (a further discussion of such echo phenomena in chaotic flows is given in [12]).

However, this estimate has to be reconsidered when the dynamics builds up finer and finer structures and increasing gradients so that diffusion can no longer be neglected (cf. [28, 16]). Consider particle spreading in a 2-d flow designed to describe the experiments of [33, 29]. The stream function is [13]

$$\begin{aligned} \psi(x,y,t) &= \frac{f_0}{3\sqrt{9+\omega^2}} \sin(\omega t - \delta_1) \sin x \cos \sqrt{2} y \\ &\quad + \frac{5 f_0}{27\sqrt{729+\omega^2}} \sin(\omega t - \delta_2) \sin 5x \cos \sqrt{2} y \,. \end{aligned} \tag{2}$$

where f_0 is the amplitude of the periodic forcing which sets the velocity scale. The two modes respond to the driving with the phase delays $\delta_1 = \text{atan}(\omega/3)$ and $\delta_2 = \text{atan}(\omega/27)$, respectively, the difference in δ_1 and δ_2 being a prerequisite for chaotic motion. Figure 1 shows a Poincaré surface of section and Figure 2 a sequence of stroboscopic plots for many particles started in a small area. The initial distribution is stretched out along the unstable manifolds nearby and contracts along the stable ones. It soon folds back just as the manifolds do. After a few periods the particles are more or less uniformly distributed in the chaotic region.

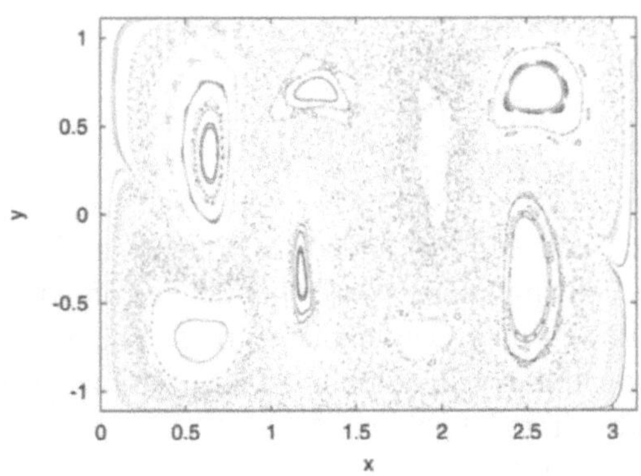

Figure 1. Stroboscopic map for the flow (2). The map was obtained by iterating 100 uniformly distributed initial conditions for 200 periods each and plotting all intermediate points. The frequency of the forcing is $\omega = 20$ and its amplitude $f_0 = 300$.

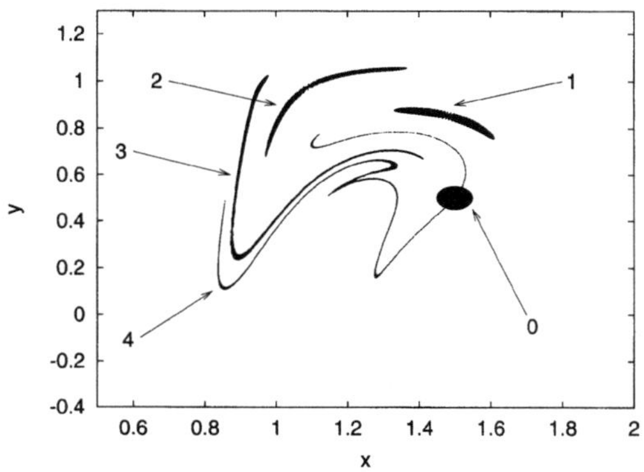

Figure 2. Evolution of a small blob of ten thousand particles after a few periods. Labels indicate multiples of a period: $t = nT$ with $n = 0$ (initial configuration) and $n = 1, 2, 3$ and 4, after which time one tendril of the stretched out blob overlaps with the initial configuration. The parameters are frequency $\omega = 40$ and the forcing amplitude $f_0 = 1500$.

The presence of islands in the flow adds a slow component due to the trapping near islands. The relaxation becomes even slower when diffusion is added since stochastic perturbations can bring particles into islands from which they can escape with difficulty only. Such localization in islands is reflected in eigenvalues and eigenmodes (O. Popovich and A. Pikovsky, in preparation).

But even if there are no islands diffusion can have a nontrivial influence on the decay rate. In order to be able to vary the diffusion constant over a wide range and still be numerically reliable we illustrate this possibility for 1-d maps on a ring, e.g.,

$$f(x) = \begin{cases} 3x & 0 < x < 1/3 \\ 2 - 3x & 1/3 < x < 2/3 \\ -2 + 3x & 2/3 < x < 1 \end{cases} . \tag{3}$$

The map is periodically continued and noise is added to simulate molecular diffusion. The calculation proceeds by taking the Frobenius-Perron equation for the first iteration step, followed by a Gaussian smearing. The eigenvalues can be calculated by numerical diagonalization in either position space or Fourier space (Hascoët and Eckhardt, to be published).

The spectrum of the unperturbed Frobenius-Perron equation can be calculated e.g., from periodic orbit theory. The eigenvalue next to 1 comes out to be 1/3. Similarly, if diffusion is large enough, the deterministic part is a small perturbation and we expect eigenvalues that depend on diffusion like $\exp(-4\pi^2 Dt)$, where t is the time interval over which diffusion acts in-between iterations of the map. The leading eigenvalue of the spectrum is shown in Figure 3.

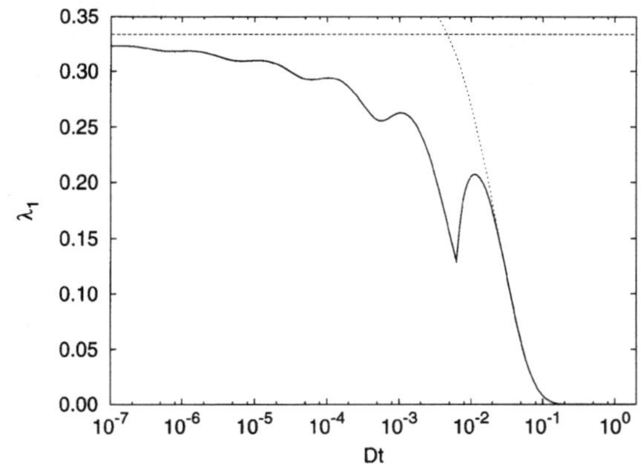

Figure 3 Variation of leading eigenvalue with diffusion for the map (??). The calculated eigenvalue is indicated by the thick line. For extremely small diffusion it approaches the eigenvalue of the deterministic map (dashed horizontal line), and for large diffusion the slowest one of the unperturbed diffusion equation (dotted line).

Instead of a simple cross over from fast chaotic relaxation for small diffusion to diffusional decay for large D we see non-monotonic variations with intriguing oscillations.

Similar non-monotonic behaviour has been observed in diffusion in chains of tent maps when the height of the tip of the tent was modified [19, 21, 22, 20]. There, the non-monotonic decay could be traced back to the non-monotonic changes in the phase space structures, the homoclinic and heteroclinic tangles and the changes in the efficiency that particles could be exchanged with between different sets of the Markov partition. We expect a similar mechanism here: the decay within the classical dynamics is connected with the efficient stretching and folding. Diffusion can help by mixing faster within cells, but it can also slow down relaxation by transporting fluid into regions with less efficient mixing.

3. Polymer Stretching in Chaotic Flows

The second example we consider is the dynamics of long flexible polymers [25, 32]. In fluids at rest these polymers are coiled up into tiny spheres. However, despite their small size, they can interact with gradients of the flow field in turbulent and also in chaotic flows. When the strains are stronger than their entropic relaxation forces they can be stretched out to form elongated structures or even thin threads. Elongated polymers influence the viscosity and change the dynamics on the smallest scales. In the case of turbulent drag reduction these microscopic modifications add up to large scale changes in the velocity field that result in noticeable reductions in drag [25, 32]. As a part of a program to understand the effects of polymers on the flow we need to understand the size distribution of the polymers.

In the simplest, and for our purposes sufficient model for the polymer, the large number of monomers and flexible elements are replaced by their inertial tensor, which can be written as a dyad formed by a vector $\mathbf{R}$, a kind of end-to-end distance vector [7]. This vector is driven by thermal fluctuations ξ, modelled as a 3-D Gaussian process. The entropic preference of the molecule for a coiled state results in an entropic restoring force, which contributes a relaxation time λ towards the equilibrium configuration. The small size of the polymers allows to linearize the flow field, so that only local gradients enter. We thus have a polymer transported along a Lagrangian path $\mathbf{x}_P(t)$, with an internal dynamics described by

$$\dot{R}_i = \left(\frac{\partial u_i}{\partial x_j}(t) - \frac{1}{2\lambda} \right) R_j + \xi_i \,. \tag{4}$$

From this vector one can form the configuration tensor, $c_{ij} = \langle R_i R_j \rangle$, where the averages are over thermal noise. The quantity used to measure extension of the polymers is $\operatorname{tr} c$. Scales can be fixed so that in equilibrium $c_{ij} = \delta_{ij}$ and $\operatorname{tr} c = 3$.

Since the polymers are advected by the flow the local gradients vary. Asymptotically in the infinite time average they approach the Lyapunov exponent of the trajectory. In order to capture some of the variations of the local stretching rate over times compatible with the internal relaxation time of the polymers, we consider distributions of finite time Lyapunov exponents,

$$\mu_T = \frac{1}{T} \log \frac{\|\delta \mathbf{x}(t_i + T)\|}{\|\delta \mathbf{x}(t_i)\|} \,. \tag{5}$$

They measure the growth rate of a separation $\delta\mathbf{x}$ over a time interval T. The equation for $\delta\mathbf{x}$ is the variational equation

$$\delta \dot{x}_i = \frac{\partial u_i}{\partial x_j}(t) \delta x_j \,. \tag{6}$$

As the time T increases the distribution narrows around the infinite time Lyapunov exponent (4).

The length distribution of the polymers results from a balance between the stretching by the gradients of the velocity field, as measured by the Lyapunov exponent μ, and the entropic relaxation. If the gradients and thus the Lyapunov exponents were constant, the thermal fluctuations would give rise to a Gaussian distribution. However, in the presence of fluctuations in the Lyapunov exponent a power law distribution of the size of the polymers follows [5, 6, 8]. A corresponding result in a more general context of multiplicative stochastic processes with a reflecting barrier is given in [30]. In

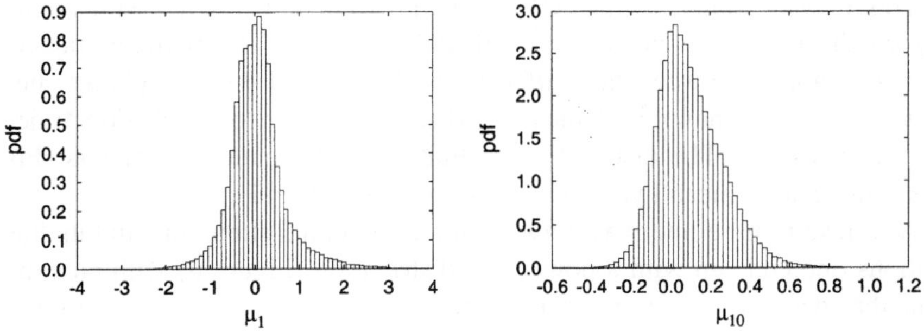

Figure 4. Finite time Lyapunov exponents for two different time intervals T and the velocity field (7). The mean in both cases is $\bar{\mu} = 0.11$, and the empirical variance decreases from 0.373 for $T = 1$ to 0.026 for $T = 10$.

fact, the noise term in 4 can be replaced by a reflecting barrier without changing the tails of the polymer length distribution. Evidently, in a power law distribution large extensions of the polymer are much more likely than in a Gaussian distribution and thus this higher probability results in a more effective and efficient back reaction of the polymers on the flow.

Since it makes no difference for the polymer dynamics if the stretching results from a spatially and temporally fluctuating turbulent flow or from a stationary flow with chaotic trajectories, we can also analyze the dynamics of polymers in a 3-d incompressible flow with chaotic trajectories, i.e.,

$$\mathbf{u} = \begin{pmatrix} (A\sin\pi y + B\sin\pi z)\sin\pi z \\ A\cos\pi x\cos\pi y + C\cos\pi z \\ B\cos\pi x\cos\pi z + C\sin\pi y \end{pmatrix} \tag{7}$$

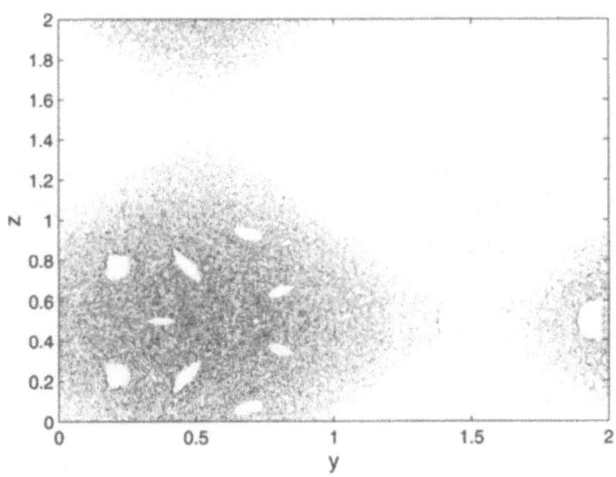

Figure 5 Poincaré surface of section for flow (7) with parameters $A = B = 1$, $C = 0.3$ at $x = 0.5$. The non-uniform distribution of points is connected with a non-uniform invariant density. The holes indicate elliptic islands.

This flow is similar to the ABC flows [10] but satisfies free slip boundary conditions in the planes $z = 0$ and $z = 1$. The particle trajectories are trapped between the surfaces, as they would be in a planar shear flow. Figure 5 shows a Poincaré surface of section of a chaotic trajectory in this flow. The distribution of finite time Lyapunov exponents (Figure 4) indicates a mean Lyapunov exponent of about 0.11.

As a result of the interplay between the exponential stretching and its fluctuations one ends up with a power law distribution (Figure 6) with an exponent that depends on the relaxation rate λ. The distribution loses its normalizability (exponent -1 or larger) when $2\mu\lambda = 1$. Numerical calculations for the polymer length support this (Figure 7).

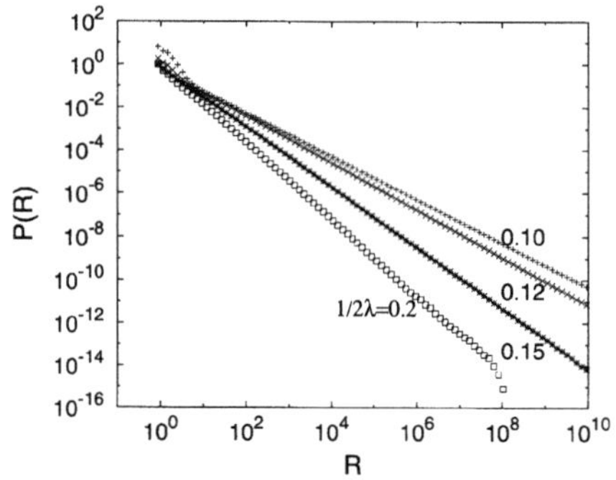

Figure 6. Polymer length distribution (not normalized) for different values of the polymeric relaxation time λ.

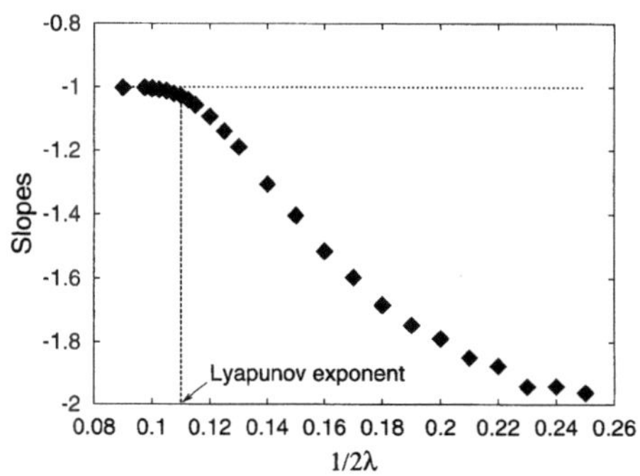

Figure 7. Exponent in the distribution of $tr\, c$ vs. the inverse of the relaxation time λ.

The power law distribution has also been identified in turbulent flows [14]. The consequences for the velocity field when the backreaction of the polymers to the flow is allowed for are under investigation.

Acknowledgments

We would like to thank O. Popovych and A. Pikovsky for stimulating exchanges on relaxation in advection diffusion equations. This work was supported in part by the EU within HPRN-CT-2000-00162 (Non-Ideal Turbulence) and by the Deutsche Forschungsgemeinschaft.

References

[1] A. Acrivos, H. Aref, and J. M. Ottino, editors (1991). *Symposium on Fluid Mechanics of Stirring and Mixing*, vol. 5 of *Phys. Fluids A*.

[2] H. Aref, "Stirring by chaotic advection," *J. Fluid. Mech.*, vol. 143, p. 1, 1984.

[3] H. Aref, editor, *Chaos Applied to Fluid Mixing*, vol. 4 of *Chaos, Solitons and Fractals*, 1994.

[4] H. Aref, "The development of chaotic advection," *Phys. Fluids*, vol. 14, p. 1315, 2002.

[5] E. Balkovsky, A. Fouxon, and V. Lebedev, "Turbulent dynamics of polymer solutions." *Phys. Rev. Lett.*, vol. 84, pp.4765–4768, 2000.

[6] E. Balkovsky, A. Fouxon, and V. Lebedev, "Turbulence of polymer solutions," *Phys. Rev. E*, vol. 64, 056301, 2001.

[7] R. B. Bird, R. C. Armstrong, and 0. Hassager, *Dynamics of polymeric liquids*, volume I. (Fluid mechanics), II. (Kinetic theory). John Wiley & Sons, New York, 1987.

[8] M. Chertkov, M. "Polymer stretching by turbulence," *Phys. Rev. Lett.*, vol. 84, p. 4761, 2000.

[9] M. Cranston and M. Scheutzow, "Dispersion rates under finite mode Kolmogorov flows," *Ann. Appl. Prob.*, vol. 12, p. 511, 2002.

[10] T. Dombre, U. Frisch, J. M. Greene, M. Henon, A. Mehr, and A. M. Soward, "Chaotic streamlines in the ABC flows," *J. Fluid. Mech.*, vol. 167, pp. 353–391, 1986.

[11] C. Eckart, "An analysis of the stirring and mixing processes in incompressible fluids," *J. Mar. Res.*, vol. 7, p. 265, 1948.

[12] B. Eckhardt, "Echoes in classical dynamical systems," *J. Phys. A*, vol. 36, p. 371, 2003.

[13] B. Eckhardt and E. Hascoët, "Chaotic advection by viscous dephasing," preprint, 2002.

[14] B. Eckhardt, J. Kronjäger, and J. Schumacher, "Stretching of polymers in a turbulent environment," *Comp. Phys. Commun.*, vol. 147, pp. 538–543, 2002.

[15] G. Falkovich, K. Gawedzki, and M. Vergassola, "Particles and fields in fluid turbulence," *Rev. Mod. Phys.*, vol. 73, p. 913, 2002.

[16] D. R. Fereday, P. H. Haynes, A. Wonhas, and J. C. Vassilicos, "Scalar variance decay in chaotic advection and Batchelor-regime turbulence," *Phys. Rev. E*, vol. 65, 35301, 2002.

[17] G. M. Homsy, H. Aref, K. S. Breuer, S. Hochgreb, J. R. Koseff, B. R. Munson, and K. G. Powell, "Multi-media fluid mechanics," *CD-ROM Cambridge University Press*, 2000.

[18] B. Jütner, D. Marteau, P. Tabeling, and A. Thess, "Numerical simulations of experiments on quasi-two-dimensional turbulence," *Phys. Rev. E*, vol. 55, 5479, 1997.

[19] R. Klages, "Transitions from deterministic to stochastic diffusion," *Europhys. Lett.*, vol. 57, p. 796, 2002.

[20] R. Klages and J. Dorfman, "Simple maps with fractal diffusion coefficients," *Phys. Rev. Lett.*, vol. 74, p. 387, 1995.

[21] R. Klages and J. Dorfman, "Simple deterministic dynamical systems with fractal diffusion coefficients," *Phys. Rev. E.*, vol. 59, 5361, 1999.

[22] R. Klages and J. R. Dorfman, "Dynamical crossover in deterministic diffusion," *Phys. Rev. E.*, vol. 55, p. 1247, 1997.

[23] V. I. Klyatskin, W. A. Woyczynski, and D. Gurarie, "Diffusing passive tracers in random imcompressible flows: statistical topography aspect," *J. Stat. Phys.*, vol. 84, p. 797, 1996.

[24] J. Kronjäger, W. Braun, and B. Eckhardt, "Polymer stretching in stationary flows," *preprint*, 2002.

[25] J. L. Lumley, "Drag reduction by additives," *Ann. Rev. Fluid Mech.*, vol. 1, p. 367, 1969.

[26] M. R. Maxey and J. J. Riley, "Equation of motion for small rigid sphere in a nonuniform flow," *Phys. Fluids*, vol. 26, p. 883, 1983.

[27] J. M. Ottino, *The kinematics of Mixing : Stretching, Chaos and Transport*, Cambridge University Press, 1989.

[28] R. Pierrehumbert, "Tracer microstructure in the large-eddy dominated regime," *Chaos, Solitons, Fractals*, vol. 4, p. 347, 1994.

[29] D. Rothstein, E. Henry, and J. P. Gollub, "Persistent patterns in transient chaotic mixing," *Nature*, vol. 401, pp. 770–772, 1999.

[30] D. Sornette and R. Cont, "Convergent multiplicative processes repelled from zero: Power laws and truncated power laws," *J. Phys. I France*, vol. 7, pp. 431–444, 1997.

[31] G. Taylor, "Low Reynolds numbers flow," *(16mm film).(Newton, Massachusetts: Educational Services Incorporated.)*, 1960.

[32] P. S. Virk, "Drag reduction fundamentals," *AIChE J.*, vol. 21, pp. 625–656, 1975.

[33] B. S. Williams, D. Marteau, and J. P. Gollub, "Mixing of a passive scalar in magnetically forced two-dimensional turbulence," *Phys. Fluids*, vol. 9, pp. 2061–2080, 1997.

SAMPLE POINT UNIFORMITY USING DISCREPANCY MEASURES

Erik A. Johnson
Department of Civil and Environmental Engineering
University of Southern California
Los Angeles, CA 90089-2531 USA
JohnsonE@usc.edu

Steven F. Wojtkiewicz
Structural Dynamics & Smart Systems, Mail Stop 0847
*Sandia National Laboratories**
Albuquerque, NM 87185 USA
sfwojtk@sandia.gov

Lawrence A. Bergman
Department of Aeronautical and Astronautical Engineering
University of Illinois at Urbana-Champaign
Urbana, IL 61801 USA
lbergman@uiuc.edu

Abstract Approximately uniform distributions of sample points are useful in certain problems in engineering and mathematics, notably optimization and uncertainty. For example, weight-controlled Monte Carlo simulation (MCS) ideally has samples approximately uniformly distributed throughout the sample space, and response surface methodologies benefit from an even distribution of sample points over input/output spaces. This paper advances the concept of discrepancy sensitivity, a linear approximation for the contribution of a sample to uniformity, quantifying the level of "new" information provided. Several examples are given of its use in choosing sample points for response surface characterization.

*Sandia is a multiprogram laboratory operated by the Sandia Corporation, a Lockheed Martin Company, for the United States Department of Energy under Contract DE-AC04-94AL85000.

N. Sri Namachchivaya and Y.K. Lin (eds.),
IUTAM Symposium on Nonlinear Stochastic Dynamics, 425-434.

Keywords: Monte Carlo simulation, variance reduction, importance sampling

1. Introduction

Problems in uncertainty and optimization are widespread in science and engineering. Their use in simulation for studying complex dynamical systems, such as real models of civil and aerospace structures, is vital to design, analysis, and reliability characterization. Great advances have occurred in computational ability in recent years, advanced by programs like the Accelerated Strategic Computing Initiative [1] and through massively-parallel capable codes such as those in development at various national laboratories (*e.g.*, DAKOTA at Sandia National Laboratories [18]). Yet, exacting study of complex dynamical systems, with all of the uncertainties that influence their behavior, taxes the computational capabilities of today's most advanced computer platforms and are likely beyond those of computers that will be available in the near future. As a result, methods are required to select and choose which simulations will provide the most useful information given the limited number of simulations that can be performed. For low-dimensional systems, it is often obvious by inspection where new sample points should be placed. In contrast, identifying empty regions of high-dimensional input/output spaces is a difficult task.

As a result, various methods have been advanced for designing computer experiments. The typical scenario is a computational model that maps a set of inputs to a set of outputs in deterministic or probabilistic fashion. Some of the variables could be design variables, and some of the outputs may be responses to be optimized by adjusting the inputs. In any case, when the computational model is very complex, it is infeasible to test all possible inputs and scenarios. Some common strategies for designing experiment inputs are Latin Hypercube sampling [9, 4], orthogonal array sampling [7], Monte Carlo and quasi-Monte Carlo methods [12], etc. These sampling methods can, then, be wrapped around existing deterministic analysis packages. Since the core analysis is typically quite computationally intensive, any overhead due to the sampling method is generally insignificant.

This paper proposes a method of quantifying the contribution of a given sample point to the uniformity of samples in the entire input or input/output domain. The concept is based on the mathematical definition of *discrepancy*. First, a discussion is given concerning some examples where sample uniformity is beneficial. Then, discrepancy is described qualitatively and mathematically. *Discrepancy sensitivity* is then derived and explained. Several numerical examples are used to demonstrate the utility of discrepancy sensitivity for choosing regions to place new sample points. The examples are chosen to represent some of the "difficult" behavior that often appears in real-world problems, such as multimodality, sharp corners and so forth.

2. Uniform Sample Distributions

An approximately uniform distribution of sample points is a useful concept in many applications. Of course, one obvious operation is pseudo-random number generators [12] that are used in science, engineering, economics and other fields. Other examples abound. The authors' primary interests in this regard are two applications: (*i*) studying the probabilistic response, statistics, and failure probabilities of complex stochastic dynamical systems, and (*ii*) the characterization of deterministic and uncertain response surfaces in simulation of complex systems.

2.1 Uniformity in stochastic dynamics

For determining probability of failure of structural dynamical systems, it is required to accurately evaluate the low-probability tails of distributions. This can be captured quite accurately with the Fokker-Planck solutions [17], but only for the simplest of problems. In most problems of practical complexity, one must resort to sampling methods. Unfortunately, accurately determining low-probability regions with standard Monte Carlo methods can be rather computationally intensive. Importance sampling and related techniques [3, 16] can improve the situation dramatically for static problems. For dynamic problems driven by white noise, however, the random variable space is of infinite dimension, rendering conventional importance sampling methods unusable.

Consider a somewhat simpler problem — the sampling from a Gaussian random variable. A typical direct Monte Carlo simulation (dMCS) might place sample points as shown in Figure 1*b*; equal weights are applied to the sam-

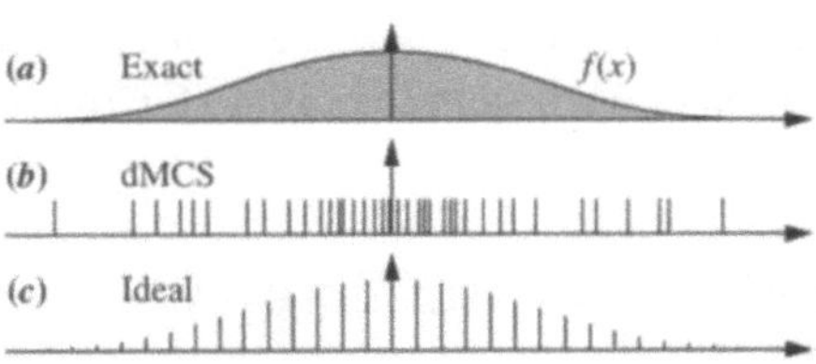

Figure 1. Ideal realization weights (adapted from [14]).

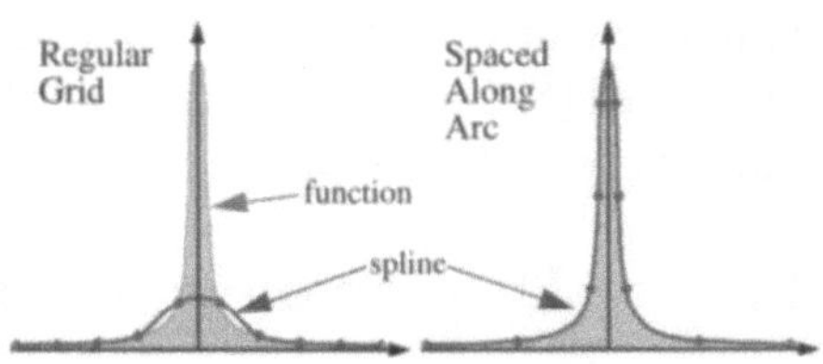

Figure 2. Ten evenly-spaced sample points to approximate an input/output map.

ples to approximate the exact density function. However, many of the samples — those in the high-probability regions — provide nearly redundant "information" whereas very few samples fall out in the tails. Here, importance sampling could be used to increase the number of samples in the tails and the weights would be adjusted so that the final PDF estimation approximates the exact solution again. However, the "ideal" set of samples, then, would be a uniform distribution of points where the weights determine the magnitude of the PDF in that region. Thus, a means of quantifying and utilizing a sense of uniformity of samples is needed.

2.2 Uniformity in response surface methodologies

Response surface methodologies [2, 11] are used to characterize input/output relationships. They often utilize some surrogate method of piecing together a surface from distinct sample points, such as splining, kriging, and so forth. The sample points themselves are generally chosen based on Latin Hypercube sampling, orthogonal array sampling, Monte Carlo and quasi-Monte Carlo methods, etc. However, these methods have two difficulties. First, given some prior distribution of sample points, these methods cannot pick a new sample point that will complement the existing set unless that method was used to choose the prior samples in the first place. Second, a spread of sample points over the entire input space does not always provide the level of uniformity desired. For example, consider the function shown in Figure 2. Sample points that are evenly spaced in the input may not capture the function behavior. However, even spacing along the arc of the function provides a more accurate estimate of the function. The limitation of the standard sampling methods is that they generally cannot directly enhance uniformity of samples in the **output** space. This is because they do not have the ability to quantify the sense of uniformity in the combined input/output space.

3. Quantitative Measure of Uniformity

The concept of *discrepancy* can be used to give a numerical value to the uniformity of samples. Based on discrepancy, the *discrepancy sensitivity* may be derived [5, 6] to quantify the contribution of each sample point to the overall uniformity of the sample set.

3.1 Discrepancy: A measure of uniformity

In the study of pseudorandom number generation and quasi-Monte Carlo methods, "discrepancy" is used to assess the difference between a distribution computed from a set of samples and the distribution used to generate the samples. Essentially, discrepancy gives an indication of the "error" in the sample distribution. Thus, discrepancy may be used to judge the uniformity of distribution of realizations of a stochastic dynamical system at a particular point in time throughout the "interesting" portion of the phase space. Furthermore, the sensitivity of the discrepancy to a realization's existence may be used to characterize the importance, or the "information content" in a realization.

The notion of discrepancy probably stems from the work of [8] on estimating the "goodness of fit" of a distribution estimated from a sequence of samples. Several different definitions of discrepancy have been advanced, and its various characteristics, such as upper and lower bounds, have been examined [10, 12]. Furthermore, several studies have shown relationships between the error in Monte Carlo methods and one or more of the discrepancy definitions [12].

The basic definition of discrepancy in the literature [10, 12, 15] may be summarized as follows: Given an infinite sequence of points $\mathbf{x}_1, \mathbf{x}_2, \ldots$ in the s-dimensional unit hypercube $[0,1]^s$, let $\wp$ be a subvolume of $[0,1]^s$ with volume $V(\wp)$. Let the N-sample-point estimate of the subvolume be given by

$$\hat{V}_N(\wp) = \frac{1}{N} \sum_{\substack{1 \le n \le N \\ \mathbf{x}_n \in \wp}} 1 = \frac{\text{number of points in } \wp}{\text{total number of points}} \tag{1}$$

Define the error between the volume and its estimate

$$R_N(\wp) = \hat{V}_N(\wp) - V(\wp) \tag{2}$$

Let $\mathcal{V}[\mathbf{a}, \mathbf{b}]$ be the s-dimensional rectangular volume given by

$$\mathcal{V}[\mathbf{a}, \mathbf{b}] = \{\mathbf{x} : x_i \in [a_i, b_i], 1 \le i \le s\} \tag{3}$$

The function generally called "the rectangle discrepancy" or sometimes the "extreme discrepancy" may be defined as

$$D_N^{(p)} = \left[\int_{[0,1]^s} \int_{a_i < b_i} |R_N(\mathcal{V}[\mathbf{a}, \mathbf{b}])|^p \, d\mathbf{a} d\mathbf{b} \right]^{(1/p)} \tag{4}$$

The p-norm "star discrepancy", which fixes one corner of the rectangular subvolumes at the origin, may be defined as

$$D_N^{*(p)} = \left[\int_{[0,1]^s} |R_N(\mathcal{V}[\mathbf{0}, \mathbf{b}])|^p \, d\mathbf{b} \right]^{(1/p)} \tag{5}$$

The ∞- and 2-norms are those generally used in the literature, since bounds on Monte Carlo integration error may be related [5, 12] to the ∞- and 2-norm discrepancies. The 2-norm discrepancies are generally written $T_N \equiv D_N^{(2)}$ or

$T_N^* \equiv D_N^{*(2)}$, whereas the ∞-norm discrepancies simplify to a *supremum*, such as

$$D_N \equiv D_N^{(\infty)} = \sup_{a_i<b_i} |R_N(\mathcal{V}[\mathbf{a},\mathbf{b}])| \tag{6}$$

Essentially, these discrepancy measures are norms of the error between a distribution and an estimate of the distribution based on a sequence of samples from the original distribution. The star discrepancy is the norm of the error in the distribution function, while the rectangle discrepancy is the norm of the density function error over all subvolumes of the random variable space. The discrepancies are bounded in $[0, 1]$; they tend to zero as the number of points tends to infinity if the sequence of points is truly uniform. For various reasons, such as computational complexity, unbiasedness, etc. [5], the T_N discrepancy is best suited for determining the contribution of individual points to uniformity.

3.2 Discrepancy sensitivity: Contribution to uniformity

Let the locations of the sample points be scaled such that the "interesting" subset of the variable space is linearly mapped onto the s-dimensional unit hypercube $[0, 1]^s$. Then, given the sequence of points $\mathbf{x}_1, \mathbf{x}_2, \ldots, \mathbf{x}_N$, the T_N discrepancy may be computed with [10]

$$\begin{aligned} T_N^2 &= \frac{1}{N^2}\left[\sum_{n=1}^{N}\sum_{m=1}^{N} p_{mn}\right] - \frac{2^{1-s}}{N}\left[\sum_{n=1}^{N} p_{nn}\right] + 12^{-s} \quad (7) \\ p_{mn} &= \prod_{i=1}^{s} \min\{x_{ni}, x_{mi}\}[1 - \max\{x_{ni}, x_{mi}\}] \quad (8) \end{aligned}$$

Introducing a fictional, integer-valued, multiplicity M_n associated with each sample point, (8) may be reformulated as

$$\begin{aligned} T_N^2 &= \frac{1}{M^2}\left[\sum_{n=1}^{N}\sum_{m=1}^{N} p_{mn}\right] - \frac{2^{1-s}}{M}\left[\sum_{n=1}^{N} p_{nn}\right] + 12^{-s}, \\ M &= \sum_{n=1}^{N} M_n \end{aligned} \tag{9}$$

Then, the discrepancy sensitivity to a point's existence, $S_{T_N n}$, may be found by taking the partial derivative with respect to the multiplicity of the n-th sample point and evaluating with the multiplicities set to unity,

$$S_{T_N n} = \frac{\partial T_N}{\partial(\text{point's existence})} = \left.\frac{\partial T_N}{\partial M_n}\right|_{\substack{M_m=1 \\ m=1,\ldots,n}}$$
$$= \frac{(B_n - A) + 2^{-s}(B - p_{nn})}{NT_N} \tag{10}$$

$$A = \frac{1}{N^2}\sum_{m=1}^{N}\sum_{n=1}^{N} p_{mn}, \quad B_m = \frac{1}{N}\sum_{n=1}^{N} p_{mn}, \quad B = \frac{1}{N}\sum_{n=1}^{N} p_{nn} \tag{11}$$

These sensitivities can be used to determine whether the discrepancy, which may now be written $T_N^2 = A - 2^{1-s}B + 12^{-s}$, will decrease if this sample is removed $(S_{T_N n} > 0)$, or if it is split or has another point added in its immediate neighborhood $(S_{T_N n} < 0)$. The magnitude can also indicate how sensitive the discrepancy is.

4. Numerical Examples

4.1 Numerical example in uncertainty

The discrepancy sensitivities of the 1000 samples in a bivariate Gaussian distribution are given in Figure 3. The sensitivities near the mean

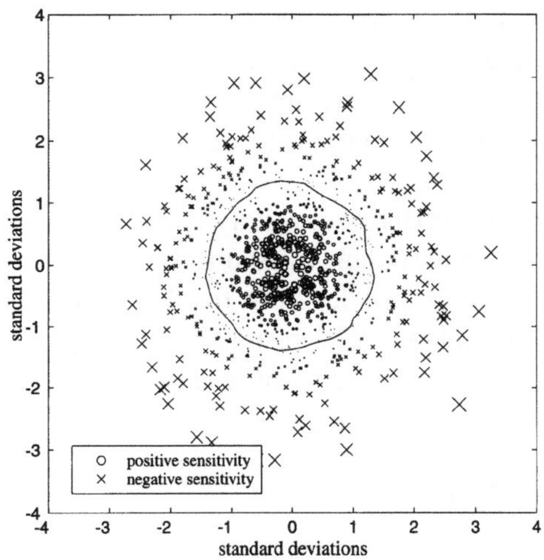

Figure 3. Discrepancy sensitivity for a bivariate Gaussian distribution.

are positive, indicating that extraneous samples are located there, and sensitivities in the "tails" are large negative values, implying that more samples should be added in the outer regions. This is exactly the type of indicator required for characterizing the contributions of the sample points to uniformity.

4.2 Discrepancy sensitivity in response surfaces

Several discrepancy sensitivity measures can be computed. Let $\mathbf{x}$ be an $n \times 1$ vector of inputs and $\mathbf{u}$ be an $m \times 1$ vector of outputs. The possible discrepancy sensitivity measures, then, include:

- $\mathbf{S_x}$ — the discrepancy sensitivity of each sample point in the n-dimensional input space
- $\mathbf{S_u}$ — ... in the m-dimensional output space
- $\mathbf{S_{xu}}$ — ... in the $(m+n)$-dimensional combined input/output space
- Sensitivity of functions of $\mathbf{x}$ and $\mathbf{u}$, such as the arc length for SISO and SIMO problems
- As well as other possibilities

Here, an example is given of a couple of these. Consider two functions: a ramp function, seen in Figure 4*c*, with sharply discontinuous slope, and a multimodal cosine function as shown in Figure 4*a*. The sample point set is started with ten points evenly distributed over the domain $[-1, 2]$, as shown in Figure 4*a*. Ninety additional points are then added based on one of two sensitivity discrepancy measures. In Figures 4*a* and 4*b*, the combined 2-dimensional input/output space discrepancy sensitivity is used. In Figures. 4*c* and 4*d*, the discrepancy sensitivity along the arc length is determined by using a piecewise linear curve through the points to determine arc length from the leftmost point.

In both cases, the discrepancy sensitivity is computed for all sample points, and then a cubic spline interpolant is used to find where the discrepancy sensitivity is most negative — the location to place a new point. This is carried out for each point added. The results are shown in Figs. 4*b*,*c*,*d*. It can be seen that both measures of sensitivity do provide coverage of the sample space in a relatively uniform manner.

5. Conclusion

An approximately uniform distribution of sample points was motivated by probabilistic and response surface methodologies. Conventional sampling methods do not directly quantify the spread of sample points, particularly over both the input and output domains. The discrepancy sensitivity is proposed as a distinct quantitative measure of the contribution of each sample point to uniformity. Several examples were given of the discrepancy sensi-

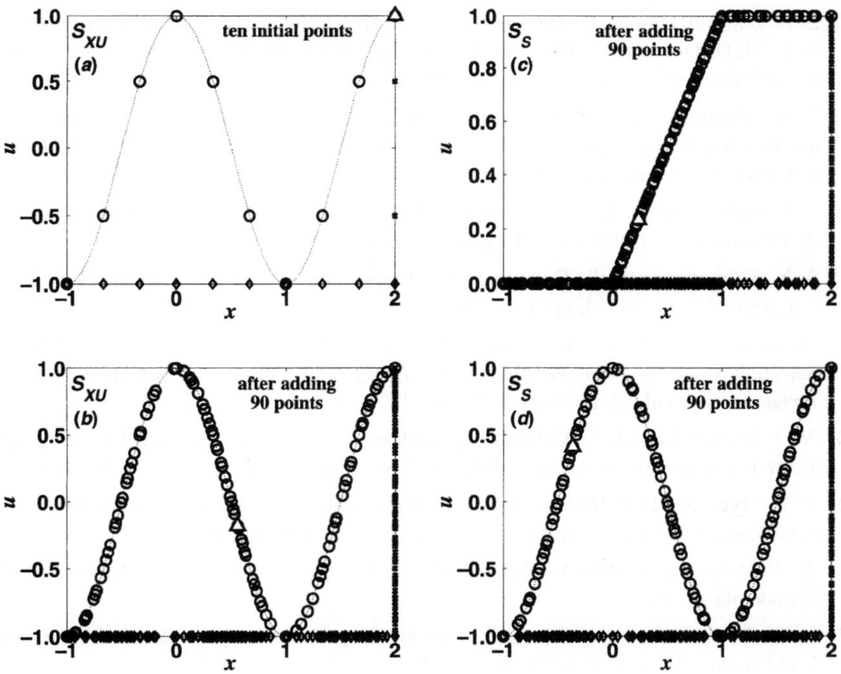

Figure 4. Sample point distribution (blue circles) and the distributions projected just in the input space (green diamonds) and just in the output space (red ×'s) for two different functions and two methods.

tivity and how it shows promise for use in uncertainty and optimization problems.

Acknowledgments

The partial support of this work by Sandia National Laboratories is gratefully acknowledged.

References

[1] *Accelerated Strategic Computing Initiative (ASCI) Program Plan.* DOE/DP-99-000010592, U.S. Department of Energy Defense Programs, 2000.

[2] G. E. P. Box and N. R. Draper, *Empirical Model-Building and Response Surfaces*, Wiley Series in Probability and Mathematical Statistics, Wiley and Sons, New York, 1987.

[3] J. M. Hammersley and D. C. Handscomb, *Monte Carlo Methods*, Methuen, London, 1964.

[4] R. L. Imam and M. J. Shortencarier, *A Fortran 77 Program and User's Guide for the Generation of Latin Hypercube Samples for Use with Computer Models*, NUREG/CR-3624, SAND83-2365, Sandia National Laboratories, Albuquerque, NM, 1984.

[5] E. A. Johnson, *On-line Monitoring, Control, and Reliability of Structural Dynamical Systems*. Ph.D. Dissertation, Dept. of Aeronautical and Astronautical Engineering, University of Illinois at Urbana-Champaign, 1997.

[6] E. A. Johnson, L. A. Bergman, and B. F. Spencer, Jr., "Intelligent Monte Carlo Simulation and Discrepancy Sensitivity." In P.D. Spanos (ed.), *Computational Stochastic Mechanics*, Balkema, Rotterdam, pp. 31–39, 1999.

[7] J. R. Koehler and A. B. Owen, *Computer Experiments*, vol. 13 of *Handbook of Statistics* (S. Ghosh and C. R. Rao, eds.), Elsevier Science, New York, 1996.

[8] A. N. Kolmogorov, "Sulla Determinazione Emprica di una Legge di Distribuz-ione." *Inst. Ital. Attuari Giorn*, vol. 4, pp. 1–11, 1933.

[9] M. D. McKay, R. J. Beckman, and W. J. Conover, "A Comparison of Three Methods for Selecting Values of Input Variables in the Analysis of Output from a Computer Code," *Technometrics*, vol. 21, no. 2, pp. 239–245, 1979.

[10] W. J. Morokoff and R. E. Caflisch, "Quasi-Random Sequences and their Discrepancies," *SIAM Journal on Scientific and Statistical Computing*, vol. 15, pp. 1251–1279, 1994.

[11] R. H. Myers and D. C. Montgomery, *Response Surface Methodology: Process and Product Optimization Using Designed Experiments*, Wiley, New York, 1995.

[12] H. Niederreiter, *Random Number Generation and Quasi-Monte Carlo Methods*, SIAM, Philadelphia, 1992.

[13] A. B. Owen, "Controlling Correlations in Latin Hypercube Samples," *Journal of the American Statistical Association*, vol. 89, no. 428, pp. 1517–1522, 1994.

[14] H. J. Pradlwarter and G. I. Schuëller, "On Advanced Monte Carlo Simulation Procedures in Stochastic Structural Dynamics," *International Journal of Non-Linear Mechanics*, vol. 32, pp. 735–744, 1997.

[15] P. D. Proinov, "Generalization of Two Results of the Theory of Uniform Distribution," *Proceedings of the American Mathematical Society*, vol. 95, pp. 527–532, 1985.

[16] R. Y. Rubinstein, *Simulation and the Monte Carlo Method*, Wiley, New York, 1981.

[17] S. F. Wojtkiewicz, L. A. Bergman, and B. F. Spencer, Jr., "High Fidelity Numerical Solutions of the Fokker-Planck Equation," *Proceedings of ICOSSAR '97, the 7th International Conference on Structural Safety and Reliability*, Kyoto, Japan, Nov. 24–28, 1997.

[18] S. F. Wojtkiewicz, M. S. Eldred, R. V. Field, Jr., A. Urbina, and J. R. Red-Horse, "Uncertainty Quantification in Large Computational Engineering Models," *Proceedings of the 42nd AIAA/ASME/ASCE/AHS/ASC Structures, Structural Dynamics, and Materials (SDM) Conference*, Seattle, WA, April 16–19, 2001. Paper AIAA-2001-1455, 2001.

[19] H. P. Wynn, "The Sequential Generation of D-Optimal Experimental Designs," *Annals of Mathematical Statistics*, vol. 41, pp. 1655–1664, 1970.

KINKS IN A STOCHASTIC PDE

Grant Lythe
Department of Applied Mathematics
University of Leeds
LS2 9JT, United Kingdom *
kinkdens@stochastic.org.uk

Salman Habib
Theoretical Division
Los Alamos National Laboratory
New Mexico 87545, USA

Abstract We consider a stochastic PDE with localized coherent structures known as kinks. The availability of exact results for the steady state and increasing computer power means that precise quantitative comparison between theory and numerics is possible. One of the quantities of interest is the steady-state density of kinks, maintained by a balance between nucleation and annihilation of kink-antikink pairs. The density as measured from numerical solutions is sufficiently accurate to resolve the difference between analytical predictions based on the exact value of the correlation length, and those based on the WKB approximation to it.

Keywords: Stochastic PDEs, kinks, convergence of algorithms.

Many extended systems have localized coherent structures that maintain their identity as they move about and are buffeted by local fluctuations [5, 4]. To study the dynamics, we must accurately solve nonlinear extended systems with noise and devise a sensible method for identifying and tracking the coherent structures. Until fairly recently, computer memory and performance restrictions were sufficiently severe that stochastic partial differential equations (stochastic PDEs) describing extended systems with noise could only be studied at low resolution. These restrictions are being overcome, at least in one space dimension [1, 3, 14], and good progress has been made towards studying, understanding, and improving the accuracy of numerical solutions.

*http://stochastic.org.uk

N. Sri Namachchivaya and Y.K. Lin (eds.),
IUTAM Symposium on Nonlinear Stochastic Dynamics, 435-443.

Numerical solutions of stochastic PDEs cannot be performed in the infinite-dimensional space in which the equation is defined [20, 13], but are carried out in a finite- (but large-) dimensional space obtained by discretizing the spatial coordinate [14]. The set of coupled stochastic differential equations (SDEs) thus obtained is evolved using a timestepping method that discretizes time. Thus a systematic study of the numerics of stochastic partial differential equations requires a study of convergence in both space and time.

In addition to questions of convergence of the numerical scheme, one is often interested in the limit of low noise intensity. This limit is convenient for analytical approximations and for unambiguous identification of coherent structures, but requires large amounts of computer time. Two reasons for the increase in computer time with decreasing temperature are as follows. Firstly, the timescales of the system typically increase exponentially as temperature decreases, so one must perform runs of increasing duration to attain the steady state from an arbitrary initial condition. Secondly, the density of coherent structures also decreases with decreasing temperature, so ever larger systems are needed to obtain good statistics.

In this paper we examine a stochastic PDE in one space dimension that can be written as follows:

$$\frac{\partial}{\partial t}\phi_t(x) - \frac{\partial^2}{\partial x^2}\phi_t(x) = \phi_t(x) - \phi_t^3(x) + (2kT)^{\frac{1}{2}}\,\xi_t(x). \tag{1}$$

The last term in (1) is space-time white noise:

$$\langle \xi_t(x)\xi_{t'}(x')\rangle = \delta(x-x')\delta(t-t'). \tag{2}$$

Angled brackets indicate mean over realizations. The amplitude of the noise is $(2kT)^{\frac{1}{2}}$ where T has the interpretation of temperature and k is Boltzmann's constant. We shall use the notation $\beta = (kT)^{-1}$. The configuration at time t is a continuous function of x denoted ϕ_t. The value at time t and position x is a random variable denoted $\phi_t(x)$. Periodic boundary conditions are used in x.

The SPDE (1), known as the ϕ^4 equation with noise, has been extensively studied as a model with nonlinear coherent structures; here we restrict ourselves to one space dimension, where they are known as kinks [4]. The one-dimensional equation, as well as serving as a general model for a chain of coupled double-well oscillators [2], has been used to model specific physical systems such as the polymer polyacetalene [18], charge-density-wave condensates [16], and Josephson-junction transmission lines [15].

1. Numerical solution

In order to solve the SPDE numerically, we first discretize in space, producing a set of coupled SDEs. We use the following notation for the SDE governing the time evolution of the stochastic process at position x [14]:

$$\mathrm{d}\phi_t(x) = \left(\phi_t(x) - \phi_t^3(x) + \tilde{\Delta}\phi_t(x)\right) \mathrm{d}t + (2kT)^{\frac{1}{2}} \mathrm{d}\mathbf{B}_t(x), \quad (3)$$

where $\langle \mathrm{d}\mathbf{B}_t(x) \mathrm{d}\mathbf{B}_{t'}(x') \rangle = \delta(x - x')\delta(t - t')\mathrm{d}t$. The discretized spatial operator is

$$\tilde{\Delta}\phi_t(x) = (\phi_t(x + \Delta x) + \phi_t(x - \Delta x) - 2\phi_t(x))\Delta x^{-2},$$

where Δx is the separation between neighboring sites. The dynamics of the system discretized in space can be thought of as that of a string of coupled overdamped oscillators subject to three competing influences: a double-well on-site potential, coupling to neighboring sites and random fluctuations.

In order to solve the system discretized in space we also discretize in time and generate a realization of the stochastic PDE, consisting of a long series of discretized configurations. Each discretized configuration consists of $\Phi(t, x)$ at a large number (typically 10^6) of values of x. In the stochastic Runge-Kutta algorithm [12, 10] used for the numerical results in this paper, each site is updated using the following set of temporary values:

$$\begin{aligned} \Psi(t, x) &= \Phi(t, x) + \left(\Phi(t, x) - \Phi^3(t, x) + \tilde{\Delta}\Phi(t, x)\right) \Delta t \\ &\quad + (2kT)^{\frac{1}{2}} (\Delta t / \Delta x)^{\frac{1}{2}} n(t, x), \end{aligned} \quad (4)$$

where $n(t, x)$ is drawn from a probability distribution, independently at each x and t. Using the temporary values (4), the update is

$$\begin{aligned} \Phi(t + \Delta t, x) &= \Phi(t, x) + \frac{1}{2} \left(\Phi(t, x) - \Phi^3(t, x) + \tilde{\Delta}\Phi(t, x)\right) \Delta t \\ &\quad + \frac{1}{2} \left(\Psi(t, x) - \Psi^3(t, x) + \tilde{\Delta}\Psi(t, x)\right) \Delta t \\ &\quad + (2kT)^{\frac{1}{2}} (\Delta t / \Delta x)^{\frac{1}{2}} n(t, x). \end{aligned} \quad (5)$$

The distribution of each $n(t, x)$ is either a Gaussian with zero mean and unit variance, or the following three-point distribution:

$$n(t, x) = \begin{cases} -\sqrt{3} & \text{with probability} \quad 1/6 \\ 0 & \text{with probability} \quad 2/3 \\ \sqrt{3} & \text{with probability} \quad 1/6. \end{cases} \quad (6)$$

Numerical runs with the $n(t, x)$ distributed according to (6) produce results that are indistinguishable, on the scale of Figure 1, from those obtained when the $n(t, x)$ are drawn from a Gaussian distribution.

In numerical realizations of the stochastic evolution described by (3), one finds that the quantity $\left\langle \phi_t^2(x) \right\rangle$, the mean square value of the field at an arbitrarily chosen point, attains a well-defined mean value after a relaxation

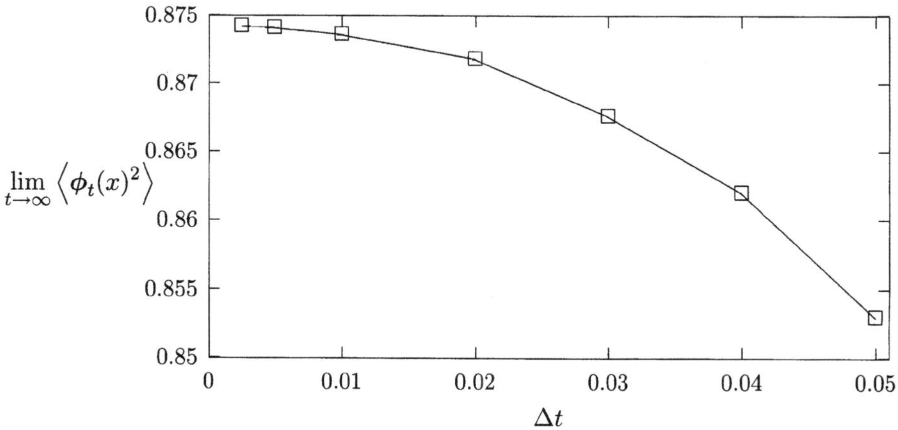

Figure 1. The measured value of the mean square versus Δt for $\Delta x = 0.4$ and $\beta = 7$.

time of order 1 [3]. This value depends both on Δx and on Δt [3]. In Figure 1 we show the dependence of $\lim_{t\to\infty}\left\langle \phi_t^2(x) \right\rangle$ on Δt when Δx is fixed.

An example of a part of a configuration (values of $\phi_t(x)$ at one time and a range of values of x) is shown in Figure 2. The configuration consists of long regions where the field is close either to 1 or to -1, occasionally interrupted by narrow interpolating regions. The latter are called kinks if they have a region where the field is negative to the left and one where the field is positive to the right; they are called antikinks if they have a region where the field is positive to the left and one where the field is negative to the right.

2. Counting kinks

Once good accuracy in space and time has been achieved, one turns to the question of identification and counting of kinks. Simply counting zero crossings of the field results in an overestimate of the number of kinks, because kinks and antikinks do not have smooth profiles, and because "phonon" fluctuations in the regions where $\phi_t(x)$ is close to ± 1 are occasionally enough to produce a zero crossing that we would not normally consider to be a kink-antikink pair. In Figure 3, we illustrate one way of dealing with this. At a chosen time in the evolution, a copy of the current configuration is sent to a subroutine that evolves the system, according to (1) but without noise ($T = 0$). The result is a very fast damping of the phonon fluctuations and smoothing of kink profiles that allows the number of kinks to be determined by counting zero crossings. Of course, if this noiseless "cool-

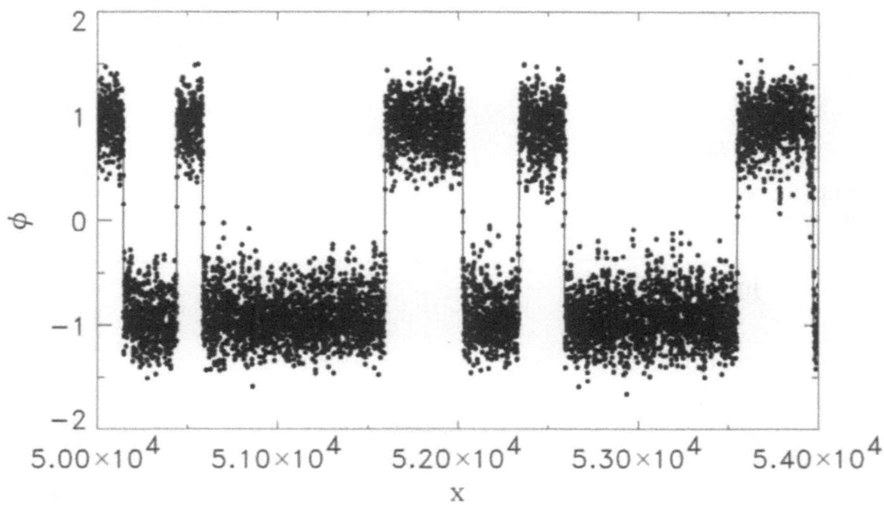

Figure 2. Part of a numerical configuration. The dots are values at grid points. Kinks and antikinks alternate along the x direction. Here $\beta = 8$.

ing" is continued indefinitely, genuine kinks and antikinks will collide and annihilate and the density of kinks will be underestimated. In practice, the separation of timescales is such that it is possible to choose a suitable intermediate value for the cooling time, as long as the equation is being solved at a temperature low enough that kinks are unambiguously defined.

3. The Steady State Density of Kinks

After a relaxation time proportional to $\exp(2E_k\beta)$, where $E_k = (8/9)^{\frac{1}{2}}$, a well-defined steady-state density of kinks (and equal density of antikinks) ρ is established [17, 7, 9]. This mean is dynamically maintained: kinks and antikinks are nucleated in pairs, follow Brownian paths and annihilate on meeting [9, 6, 11]. Under the assumption that kink-antikink pairs are nucleated at random times and positions at rate Γ with separation b, have diffusivity D and annihilate on collision, the mean density as a function of time can be calculated exactly [8]. As $b\rho \to 0$, the steady-state density of kinks is given by

$$\rho = \frac{1}{2}\left(\frac{b\Gamma}{2D}\right)^{\frac{1}{2}}. \tag{7}$$

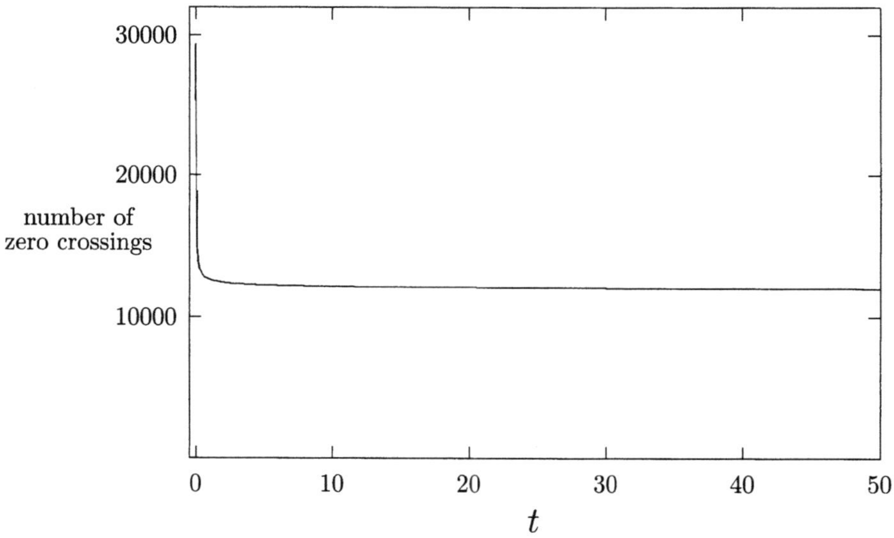

Figure 3. Counting kinks using the cooling method. The number of zero crossings is plotted as a function of cooling time. The initial rapid decrease corresponds to the damping out of small-scale fluctuations, leaving only the kinks and antikinks. The graph comes from a numerical solution at $\beta = 7$ on a spatial domain with length $L = 2097152$.

At late times the system evolving according to (1) can be thought of as sampling from a steady-state distribution of configurations, calculated as the stationary solution of a Fokker-Planck equation [17, 3]. Consequently, many late-time properties of the solutions of the stochastic PDE can be calculated analytically. For example the correlation function defined by

$$c(x) = \lim_{t \to \infty} \langle \phi_t(y) \phi_t(y+x) \rangle \tag{8}$$

is independent of y. As $x \to \infty$,

$$c(x) \to \exp(-x/\lambda). \tag{9}$$

The correlation length λ is given by

$$\lambda = (\beta(\epsilon_1 - \epsilon_0))^{-1}, \tag{10}$$

where ϵ_0 and ϵ_1 are the two smallest eigenvalues of the equation

$$\left(-\frac{1}{2\beta^2} \frac{\partial^2}{\partial u^2} - \frac{1}{2}u^2 + \frac{1}{4}u^4 \right) \psi_n(u) = \epsilon_n \psi_n(u). \tag{11}$$

The same method can be used to calculate the dependence of the correlation function on the grid spacing Δx. The lowest order corrections to the continuum in one space dimension are proportional to Δx^2 and equivalent to a corrected on-site potential [19, 3]. An algorithm has been devised, with the potential augmented by a term proportional to Δx^2, that gives improved convergence to the continuum [3, 14].

At sufficiently low temperatures, the lengthscale defined by the inverse density of kinks is very much larger than the typical separation of a kink-antikink pair at the instant of nucleation [6, 9] and the late-time distribution of kinks and antikinks is very close to a random distribution [9, 8]. In such a case, the density of kinks is related to the correlation length λ by

$$\rho = \frac{1}{4}\frac{1}{\lambda}. \tag{12}$$

In Figure 4, we compare the density of kinks as measured from very large

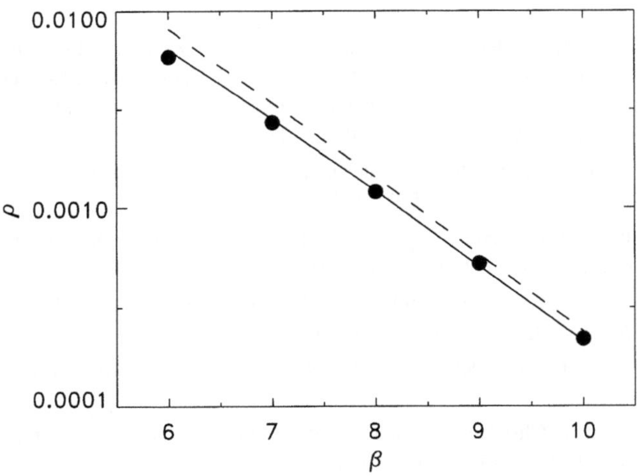

Figure 4. Density of kinks vs inverse temperature (log scale). The dots are densities measured from numerical solutions of the stochastic PDE (1). The solid line is the prediction (12), where the exact correlation length λ is obtained from (10). The dotted line uses the WKB approximation to the eigenvalues of (11).

numerical solutions with the prediction (12), where λ is given by (10). In the solid line, the eigenvalues ϵ_0 and ϵ_1 are obtained from numerical solution of (11). The dotted line uses the WKB approximation [17] to those eigenvalues. The density as measured from numerical solutions of the stochastic PDE is sufficiently accurate to resolve the difference between the exact value of λ and the WKB approximation to it. The accuracy of the latter approximation improves as the temperature T is lowered. As $T \to 0$, the number of kinks per unit length is proportional to $\exp(-E_k\beta)$.

Performing precise quantitative studies of numerical convergence of thermodynamic properties, where exact analytical results are available, makes it possible to proceed with confidence to an exploration of the fascinating dynamics of individual kinks, where few analytical results are available. Following the spacetime trajectories of individual kinks and antikinks as they are nucleated, diffuse and annihilate on collision, one finds that the mean lifetime of a kink is proportional to $\exp(E_k\beta)$. In order to understand the distribution of lifetimes, it is illuminating to classify each kink's time line according to whether it ends in annihilation with the antikink it was nucleated with, or with some other antikink. The former event is much more likely: the distance between newly nucleated pairs is much smaller than the inverse of the steady-state density of kinks, so their Brownian paths are more likely to intersect [6, 9].

References

[1] F. J. Alexander and S. Habib, "Statistical Mechanics of Kinks in (1+1)-Dimensions," *Phys. Rev. Lett.*, vol. 71, p. 955, 1993; F. J. Alexander, S. Habib, and A. Kovner, "Statistical Mechanics of Kinks in (1+1)-Dimensions: Numerical Simulations and Double Gaussian Approximation," *Phys. Rev. E*, vol. 48, p. 4284, 1993.

[2] S. Aubry, "A unified approach to the interpretation of displacive and order-disorder systems," *J. Chem. Phys.*, vol. 64, p. 3392, 1976.

[3] L. Bettencourt, S. Habib, and G. Lythe, "Controlling one-dimensional Langevin dynamics on the lattice," *Physical Review D*, vol. 60, p. 105039, 1999.

[4] A. R. Bishop, J. A. Krumhansl, and J. R. Schrieffer, "Solitons in condensed matter: a paradigm," *Physica D*, vol. 1, pp. 1–44, 1980.

[5] A. R. Bishop and T. Schneider (Eds), *Solitons and Condensed Matter Physics*, Springer, Berlin, 1978.

[6] M. Büttiker and T. Christen, "Nucleation of Weakly Driven Kinks," *Phys. Rev. Lett.* vol. 75, p. 1895, 1995; "Diffusion controlled initial recombination," *Phys. Rev. E*, vol. 58, p. 1533, 1998.

[7] M. Büttiker and R. Landauer, "Nucleation Theory of Overdamped Soliton Motion," *Phys. Rev. Lett.*, vol. 43, p. 1453, 1979; "Long-term behavior of the equilibrium Sine-Gordon chain," *J. Phys. C*, vol. 13, p. L325, 1980.

[8] S. Habib, K. Lindenberg, G. Lythe, and C. Molina-Paris, "Diffusion-limited reaction in one dimension: paired and unpaired nucleation," *J. Chem. Phys.*, vol. 115, pp. 73–89, 2001.

[9] S. Habib and G. Lythe, "Dynamics of kinks: nucleation, diffusion and annihilation," *Physical Review Letters*, vol. 84, p. 1070, 2000.

[10] K. Jansons and G. Lythe, "Stochastic Calculus: application to dynamic bifurcations and threshold crossings," *Journal of Statistical Physics*, vol. 90, pp. 227–251, 1998.

[11] D. J. Kaup, "Thermal corrections to overdamped soliton motion," *Phys. Rev. B*, vol. 27, pp. 6787–6795, 1983.

[12] P. E. Kloeden and E. Platen, *Numerical Solution of Stochastic Differential Equations*, Springer, Berlin, 1992.

[13] G. Lythe, "Defect formation in a dynamic transition," *Int. J. Theor. Phys.*, vol. 40, p. 2309, 2001.

[14] G. Lythe and S. Habib, "Stochastic PDEs: convergence to the continuum?," *Computer Physics Communications*, vol. 142, p. 29, 2001.

[15] D. W. McLaughlin and A. C. Scott, "Perturbation analysis of fluxon dynamics," *Phys. Rev. A*, vol. 18, p. 1652, 1978.

[16] M. J. Rice, A. R. Bishop, J. A. Krumhansl, and S. E. Trullinger, "Weakly Pinned Charge-Density-Wave Condensates," *Phys. Rev. Lett.*, vol. 36, p. 432, 1976.

[17] D. J. Scalapino, M. Sears, and R. A. Ferrell, "Statistical Mechanics of One-Dimensional Ginzburg-Landau Fields," *Phys. Rev. B*, vol. 6, p. 3409, 1972; J. A. Krumhansl and J. R. Schrieffer, "Dynamics and statistical mechanics of a one-dimensional model Hamiltonian for structural phase transitions," *ibid*, vol. 11, p. 3535, 1975; J. F. Currie, J. A. Krumhansl, A. R. Bishop, and S. E. Trullinger, "Statistical mechanics of one-dimensional solitary-wave-bearing scalar fields: Exact results and ideal-gas phenomenology," *ibid*, vol. 22, p. 477, 1980.

[18] W. P. Su, J. R. Schrieffer, and A. J. Heeger, "Solitons in Polyacetylene," *Phys. Rev. Lett.*, vol. 42, p. 1698, 1979.

[19] S. E. Trullinger and K. Sasaki, "Lattice-discreteness corrections in the transfer-operator method for kink-bearing chains," *Physica D*, vol. 28, p. 181, 1987.

[20] J. B. Walsh, "An introduction to stochastic partial differential equations," pp. 266–439, in *Ecole d'été de probabilités de St-Flour XIV*, ed. P. L. Hennequin, Springer, Berlin, 1986.

PATH INTEGRATION AS A TOOL FOR INVESTIGATING CHAOTIC BEHAVIOUR

Arvid Naess
Department of Mathematical Sciences
Norwegian University of Science & Technology
NO-7491 Trondheim, NORWAY
arvidn@math.ntnu.no

Christian Skaug
Consiglio Nazionale di Ricerca (CNR)
IAC "M. Picone" Sezione di Bari
I-70126 Bari, ITALY
c.skaug@area.ba.cnr.it

Abstract In the present paper the possibility of using the methods of stochastic differential equations to study the dynamic response of nonlinear oscillators driven by harmonic or perturbed harmonic excitations is discussed. The reported work continues and extends previous work done by the authors using a numerical path integration scheme for solving stochastic differential equations. It is shown that this solution method may offer an attractive means of making response predictions for nonlinear dynamic systems of the kind considered.

Keywords: Chaotic response, nonlinear oscillators, random perturbations, path integration

1. Introduction

The starting point of the work presented in this short paper is the approximation of a harmonically excited nonlinear oscillator obtained by adding white noise to the excitation. This opens up for the application of the analysis tools of stochastic differential equations (SDE). As will be be demonstrated, when applicable, such tools make it possible to actually calculate approximations to e.g., chaotic attractors of the initial system. However, the basic question of convergence immediately comes to the fore. The mathematical notions of central importance here are the invariant measures, prop-

N. Sri Namachchivaya and Y.K. Lin (eds.),
IUTAM Symposium on Nonlinear Stochastic Dynamics, 445-454.

erly defined. The study of the stability of the invariant measures of the initial system under random perturbations seems to have been initiated by Kolmogorov. For the approximation procedure indicated above to work, it must clearly be required that the invariant measures of the noise perturbed system in some specific sense converge to the corresponding invariant measures of the initial system. Recent work on these problems directly relevant to the present paper has been reported by [3]. A detailed discussion of the convergence problem is not pursued here. Some time ago the authors started to investigate the possibility of using the numerical path integration (PI) method for calculating approximations of the strange or chaotic attractors of nonlinear oscillators driven by harmonic and perturbed harmonic excitations. The primary goal of the present paper is to continue this investigation and to show that the numerical PI method holds the promise of being able to develop into a very powerful and accurate calculation tool for the study also of chaotic dynamics.

The PI method can be used to solve stochastic differential equations. However, this requires the presence of Gaussian white noise as an additional excitation source. Hence, in order to be able to use the PI method, the dynamic system is modified by introducing a stochastic perturbation in the form of additive Gaussian white noise. The state space vector of the latter system can then often be represented as a Markov diffusion process. The underlying premise of our work presented here is that the joint probability density function (PDF) of the state space vector of this Markov process, which is provided by the PI method, is closely related to the corresponding chaotic attractor of the underlying deterministic system, which is obtained as a limit of the joint PDF as the intensity of the added noise goes to zero. Previous studies have indicated that the obtained PDF for small added noise can be effectively used for prediction purposes ([4]).

2. The Nonlinear Oscillators

The nonlinear dynamic systems to be studied numerically in this work belong to the general class of systems with an equation of motion that can be expressed as follows

$$\ddot{X}_t + g(t, X_t, \dot{X}_t) \quad = \quad F_t \tag{1}$$

where g is a suitable function, and where the stochastic excitation process F_t is Gaussian white noise. That is, $F_t = b\,N_t$, where $b > 0$ and N_t denotes a standard Gaussian white noise process, that is, $E[N_t\,N_{t+\tau}] = \delta(\tau)$, where $\delta(\cdot)$ denotes a Dirac measure at $t = 0$. N_t can be considered as the formal derivative of a standard Brownian motion, cf. e.g., [9]. The mathematical implication of this, is that (1) can be rewritten as a stochastic differential equation (SDE) as shown in the next section.

The type of nonlinear dynamic system to be studied numerically in this work belongs to the general class of Duffing oscillators with an equation of motion that can be expressed as follows

$$\ddot{X}_t + c\,\dot{X}_t + k_1\,X_t + k_3\,X_t^3 = a\,\cos(\omega t) + b\,N_t \tag{2}$$

where a, b, c are non-negative constants, and k_1, k_3 are suitable constants.

Among the nonlinear oscillators with harmonic excitation ($a > 0$, $b = 0$) exhibiting chaotic behaviour, the Duffing class plays a central role ([1]). The following specific oscillator is subjected to a further study here

$$\ddot{X}_t + \frac{1}{25}\,\dot{X}_t - \frac{1}{5}\,X_t + \frac{8}{15}\,X_t^3 = a\,\cos(\omega t) + b\,N_t \tag{3}$$

Note that since $k_1 < 0$ and $k_3 > 0$, the oscillator becomes bistable, that is, it has two equilibrium points. This oscillator has been extensively studied by [8] for the case when $a = 2/5$ and $b = 0$, and he shows that the oscillator has an extremely rich bifurcation behaviour in the frequency range $0 < \omega < 0.5$. In the present paper we shall investigate the situation with $b > 0$ to see to what extent the addition of white noise will alter the character of the response.

3. The Stochastic Differential Equation and its Solution

Equation (2) can be written as an SDE. It then assumes the form

$$dY_t = \alpha(Y_t, t)\,dt + \beta\,dB_t \tag{4}$$

where $Y_t = (X_t, \dot{X}_t)^T$, $\alpha = (\alpha_1, \alpha_2)^T$, $\alpha_1(Y_t, t) = \dot{X}_t$, $\alpha_2(Y_t, t) = -c\,\dot{X}_t - k_1\,X_t - k_3\,X_t^3 + a\,\cos(\omega\,t)$, $\beta = (0, b)^T$.

B_t is a standard (unit) scalar Brownian motion process ([9, 2]).

As discussed by [6], an efficient numerical solution of the SDE (4) can be obtained by adopting a backward-forward type of Runge-Kutta-Maruyama (RKM) discretization approximation

$$Y_t = Y_{t'} + r(Y_{t'}, t')\,\Delta t + \beta\,\Delta B_{t'} \tag{5}$$

where $\Delta t = t - t'$, $\Delta B_{t'} = B_t - B_{t'}$, and $r(\cdot,\cdot)$ is a 4th order Runge-Kutta correction of $\alpha(\cdot,\cdot)$.

Since the Brownian motion process has independent increments, it follows from (5) that the sequence $\{Y_{n\Delta t}\}_{n=0}^{\infty}$ is a Markov chain. For sufficiently small Δt, this Markov chain will approximate the continuous time Markov process solution of the SDE (4). It is also observed from (5) that $p(y, t|y', t')$, which denotes the conditional PDF of Y_t given that $Y_{t'} = y'$, $t' < t$, is a Gaussian PDF since $\Delta B_{t'}$ is a Gaussian variable for every t'. For small time increments $\Delta t = t - t'$, $p(y,t|y',t')$ will be referred to as the incremental transition probability density function (TPD).

From the assumptions above we may now write the TPD corresponding to (5)

$$p(y,t|y',t') = \delta(\, y_1 - y_1' - r_1(y')\Delta t\,) \cdot \tilde{p}(y_2|y',t') \tag{6}$$

where $y = (y_1, y_2)^T$ and

$$\tilde{p}(y_2,t|y',t') = \frac{1}{\sqrt{2\pi\, b^2 \Delta t}} \cdot \exp\left\{ -\frac{(y_2 - y_2' - r_2(y')\Delta t)^2}{2\, b^2\, \Delta t} \right\} \tag{7}$$

Hence, $p(y,t|y',t')$ is a degenerate two-dimensional Gaussian PDF.

The path integration principle is based on the following basic relation

$$p(y,t) \;=\; \int_\Omega p(y,t|y',t')\, p(y',t')\, dy' \tag{8}$$

where Ω = state space. Invoking (6) and (7) it follows that for any discretized SDE of the type given by (5), the TPD can be written down explicitly and in closed form. Hence, if an initial PDF, $p_0(y) = p(y,t_0)$ say, is given, then (8) can be used repeatedly to produce the time evolution after t_0 of $p(y,t)$. The numerical implementation of the PI procedure we have developed for problems of state space dimension four or less has been described in detail by [5, 7, 6]. It is based upon using a splines interpolation scheme for continuous representation of the joint PDF $p(y,t)$ after each iteration of (8), which provides the values of the PDF at a discrete set of grid points. Experience indicates that the success of the numerical PI relies heavily on the accuracy of the continuous representation.

The PI solution provides (approximate) snap-shots of the joint PDF of the state space vector Y_t. Engineering analyses of dynamic structures typically involve estimation of the crossing rate statistics of prescribed response levels as a practical way of characterizing the structural response. We shall demonstrate how this can be achieved for nonlinear oscillators described by (4). As shown above, there is a joint PDF for the state space vector $Y_t = (X_t, \dot{X}_t)^T$, which is denoted here by $f_{X\dot{X}}(x,\dot{x};t)$. It is assumed that the mean upcrossing rate of a level ξ by X_t at time t can be calculated by using the Rice formula

$$\nu_X^+(\xi;t) \;=\; \int_0^\infty f_{X\dot{X}}(\xi,s;t)\, s\, ds \tag{9}$$

As discussed in [4], in the case of positive damping, that is $c > 0$, it can be assumed that $f_{X\dot{X}}(x,\dot{x};t) = f_{X\dot{X}}(x,\dot{x};t+t_p)$, where $t_p = 2\pi/\omega$, when

$t \geq t'$ for a suitably large t'. That is, the joint density $f_{X\dot{X}}(x, \dot{x}; t)$ eventually becomes periodic after transient effects have died down.

It follows that the mean number of upcrossings of the level ξ over one period t_p by X_t, for $t > t'$, which we denote by $n_p^+(\xi)$, is given as

$$n_p^+(\xi) = \int_{t'}^{t'+t_p} \nu_X^+(\xi; t)\, dt = t_p \int_0^{\infty} \bar{f}_{X\dot{X}}(\xi, s)\, s\, ds \tag{10}$$

where the averaged density

$$\bar{f}_{X\dot{X}}(x, \dot{x}) = \frac{1}{t_p} \int_{t'}^{t'+t_p} f_{X\dot{X}}(x, \dot{x}; t)\, dt \tag{11}$$

has been introduced.

Hence it is seen that for practical estimation of crossing rate statistics, it is the averaged density that is important and not the density at a specific point in time. This observation has practical consequences. While the instantaneous joint PDF for an underlying chaotic system perturbed by small Gaussian white noise reflects the complexity of the chaotic attractor produced by the Poincaré map, see Figures 1 and 2, the corresponding averaged density apparently has

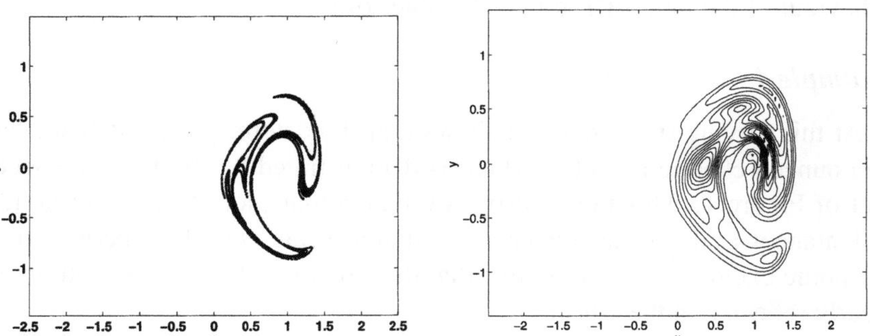

Figure 1. Left: A Poincare section plot of the deterministic oscillator for $\omega = 0.20$. Right: A contour plot of the joint PDF of the state space vector of the stochastic oscillator at the end of 30 periods of the harmonic forcing function for $\omega = 0.20$.

a somewhat simpler structure. And, significantly, the concept of stationary density makes perfect sense after averaging.

4. Numerical Examples

To explore the effect of additive white noise to the Duffing (3), that is, $b > 0$, we have calculated the stationary average joint PDF of the Markov chain associated with (5) as given by (3) for $a = 0.4$ and $b = 0.005$ for four values of the frequency of the harmonic part of the excitation. The choice of b-value is

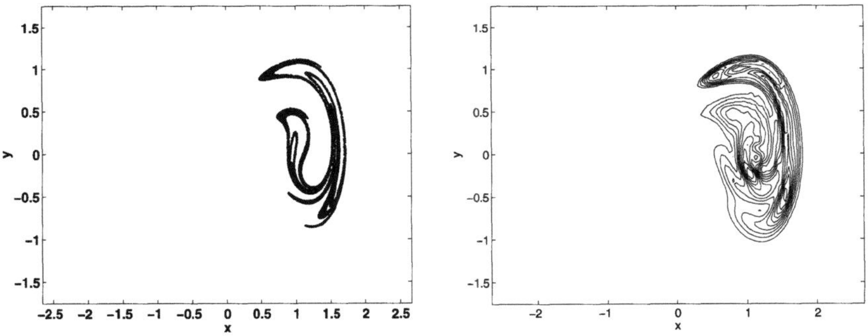

Figure 2. Left: A Poincare section plot of the deterministic oscillator for $\omega = 0.32$. Right: A contour plot of the joint PDF of the state space vector of the stochastic oscillator at the end of 30 periods of the harmonic forcing function for $\omega = 0.32$.

based on two considerations. It should be much smaller than a, but not so small as to make numerical calculations of the joint PDF of the state space vector too demanding. This is particularly relevant for the cases when the deterministic oscillator has periodic response leading to a very singular (delta-function like) behaviour of the joint PDF for small values of b.

Example 1 - $\omega = 0.20$

At this driving frequency, [8] shows that the deterministic oscillator has a pronounced chaotic attractor, which is demonstrated by the Poincaré section plot of Figure 1. This figure also shows a contour plot of the corresponding instantaneous pdf of the stochastic oscillator at the end of 30 periods of the harmonic excitation, and it is seen that the structure of the chaotic attractor is clearly reflected in the pdf.

Figure 3 shows the corresponding stationary averaged joint PDF. It is clearly seen that there is no conspicuous periodic behaviour of this oscillator.

To get an impression of the stability of the chaotic attractor to the random perturbations, calculations were carried out with increased noise intensity, viz. $b = 0.01$ and 0.05. The results are plotted in Figure 4. These results seem to indicate that the chaotic attractor is rather insensitive to the noise level, but as expected, there is an increasing smoothing of the joint PDF with increasing noise level, masking the fine structure of the chaotic attractor.

Example 2 - $\omega = 0.28$

For this driving frequency, the deterministic oscillator has a periodic stationary response, which is easily recognized from the phase space plot of the stationary response shown in Figure 5. The corresponding stationary aver-

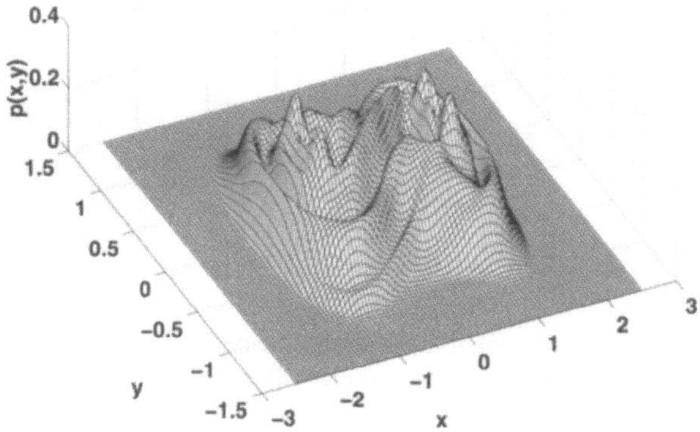

Figure 3. The averaged pdf of the stochastic oscillator for $\omega = 0.20$.

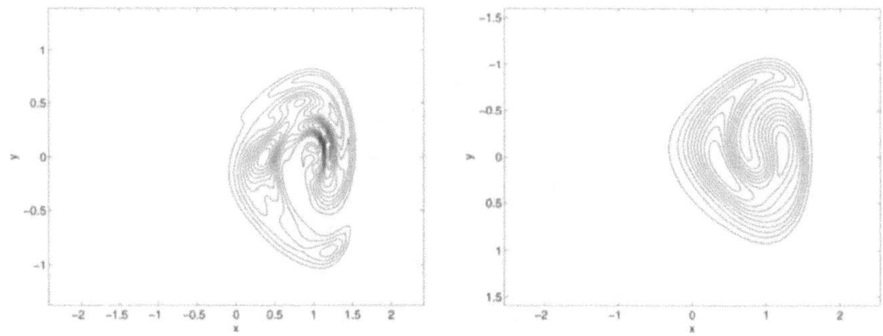

Figure 4. Contour plots of the joint PDF of the state space vector of the stochastic oscillator at the end of 30 periods of the harmonic forcing function for $\omega = 0.20$ and $b = 0.01$ (left), $b = 0.05$ (right)

aged joint PDF of the stochastic oscillator is also plotted in Figure 5, which clearly shows the periodic behaviour of the oscillator.

Example 3 - $\omega = 0.32$

In this case, the deterministic oscillator again displays chaotic behaviour. An example of a Poincaré section plot for this case is shown in Figure 2. The same figure also shows a contour plot of the corresponding instantaneous PDF of the state space vector of the stochastic oscillator at the end of 30 periods of the harmonic forcing function. It is seen that the structure of the chaotic attractor is reflected in the PDF.

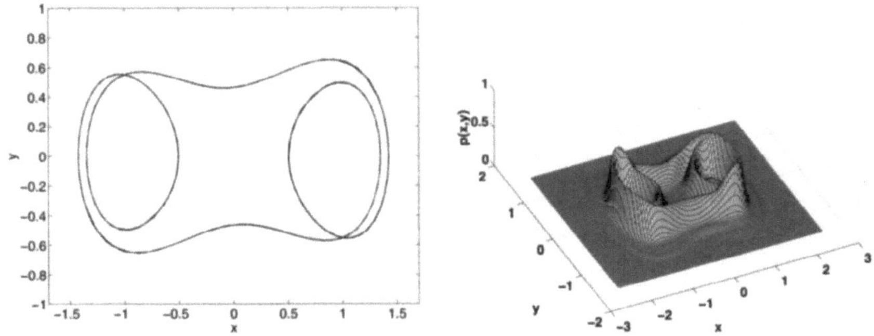

Figure 5. Left: Phase space plot of the deterministic periodic attractor for $\omega = 0.28$. Right: The averaged pdf of the stochastic oscillator for $\omega = 0.28$.

Figure 6 shows the corresponding stationary averaged joint PDF. As in Fig-

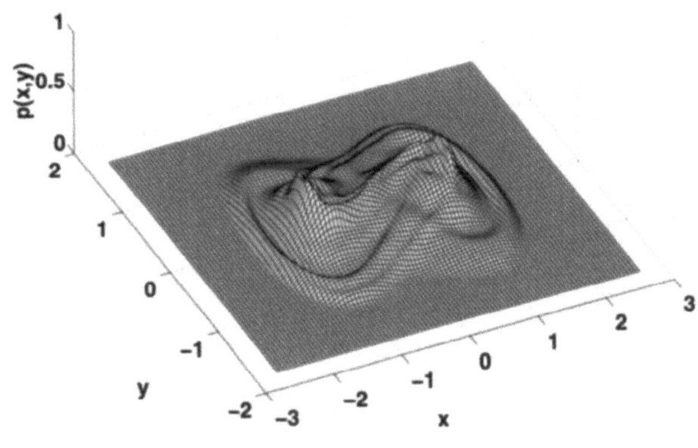

Figure 6. The averaged pdf of the stochastic oscillator for $\omega = 0.32$.

ure 3, there is no conspicuous periodic behaviour.

As in example 1, the stability of the chaotic attractor is investigated by changing the noise intensity. The results have been plotted in Figure 7. Also in this case the initial structure of the chaotic attractor seems to be rather insensitive to the noise level.

5. Conclusion

The main purpose of this short paper has been to illustrate the use of the numerical PI method as a tool for investigating the dynamics of nonlinear os-

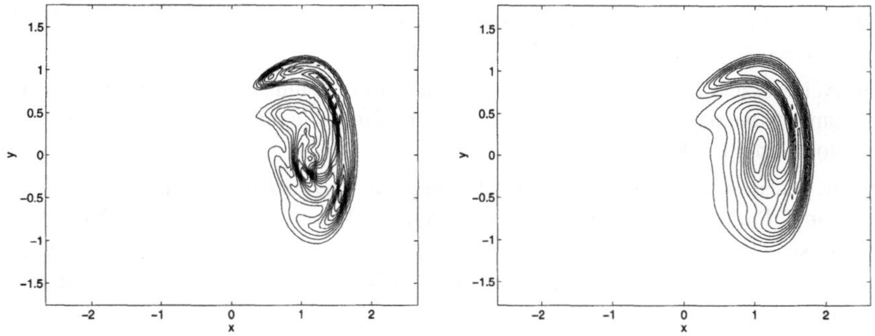

Figure 7. Contour plots of the joint PDF of the state space vector of the stochastic oscillator at the end of 30 periods of the harmonic forcing function for $\omega = 0.32$ and $b = 0.01$ (left), $b = 0.05$ (right)

cillators that may exhibit chaotic response to harmonic excitation. The underlying idea is that the response of such a nonlinear system can be considered as the limit of the response of a corresponding system obtained by adding Gaussian white noise when the intensity of this white noise approaches zero. The presence of non-zero Gaussian white noise, which is to be construed as leading to a stochastic differential equation, makes it possible to apply the numerical PI method. The results we have obtained in this paper for a harmonically driven Duffing oscillator, indicates that the suggested approach can be a very robust method. The findings in this paper support those reported by [4], which clearly indicated that the numerical PI method had the potential to be an effective tool to investigate also the dynamics of nonlinear oscillators exhibiting chaotic response.

Acknowledgments

Part of this work was carried out while the second author was a doctoral student at the Norwegian University of Science and Technology. The authors would like to acknowledge the financial support from the Norwegian Research Council (NFR), making this work possible.

References

[1] P. Holmes, "A nonlinear oscillator with a strange attractor." *Phil. Trans. of the Royal Society of London A292*, pp. 419–448, 1979.

[2] I. Karatzas and S. E. Shreve, *Brownian Motion and Stochastic Calculus*, Second Edition. New York: Springer-Verlag, 1991.

[3] Y. Kifer, "Computations in Dynamical Systems via Random Perturbations," *Discrete and Continuous Dynamical Systems*, vol. 3, no. 4, pp. 457–476, 1997.

[4] A. Naess, "Chaos and nonlinear stochastic dynamics," *Probabilistic Engineering Mechanics*, vol. 15, no. 1, pp. 37–47, 2000.

[5] A. Naess and J. M. Johnsen, "Response statistics of nonlinear, compliant offshore structures by the path integral solution method," *Probabilistic Engineering Mechanics*, vol. 8, no. 2, pp. 91–106, 1993.

[6] A. Naess and V. Moe, "Efficient path integration methods for nonlinear dynamic systems," *Probabilistic Engineering Mechanics*, vol. 15, no. 3, pp. 221–231, 2000.

[7] A. Naess and C. Skaug, "4D problems by path integration," *Proceedings of the ICASP8 Conference*, eds. R. E. Melchers and M. G. Stewart, pp. 145–150. Rotterdam: Balkema, 1999.

[8] R. Seydel, "Attractors of a Duffing Equation - Dependence on the Exciting Frequency," *Physica*, vol. 17D, pp. 308–312, 1985.

[9] E. Wong and B. Hajek, *Stochastic Processes in Engineering Systems*, New York: Springer-Verlag, 1985.

TIME-VARYING CARDIOVASCULAR OSCILLATIONS

V. N. Smelyanskiy, D. A. Timucin
NASA Ames Research Center
MS 269-2
Moffett Field, CA 94035-1000, USA

D. G. Luchinsky
Department of Physics
Lancaster University
Lancaster LA1 4YB, UK

A. Stefanovska
Department of Physics
Lancaster University
Lancaster LA1 4YB, UK
and
Group of Nonlinear Dynamics and Synergetics
Faculty of Electrical Engineering
University of Ljubljana
Tržaška 25, 1000 Ljubljana, Slovenia

A. Bandrivskyy, and P. V. E. McClintock
Department of Physics, Lancaster University
Lancaster LA1 4YB, UK
vadim@email.arc.nasa.gov, aneta@osc.fe.uni-lj.si

Abstract Signals derived from the human cardiovascular system (CVS) are exceptionally complex, being time-varying, noisy, and of necessarily limited duration. Yet an appropriate analysis of them may be expected to yield detailed information about the dynamics of the underlying physiological processes. A new approach to the analysis and modelling of CVS signals is proposed. It combines decomposition of the signals into principal modes and a novel method of pa-

N. Sri Namachchivaya and Y.K. Lin (eds.),
IUTAM Symposium on Nonlinear Stochastic Dynamics, 455-464.

rameter identification in nonlinear stochastic systems based on Bayesian inference. The scheme is tested on a noisy Van der Pol oscillator, for which it yields rapid convergence and correct inference of the known parameters. Preliminary applications to CVS data are discussed.

1. Introduction

It has long been recognized that, even for a healthy subject in repose, the human heart does not beat at a constant rate. Rather, its frequency varies in time (see [2] and references therein). This heart rate variability (HRV) reflects the instantaneous adaptation of the cardiovascular system, and the heart as its pump in particular, to the needs of the organism. In turn the HRV influences blood flow, blood pressure, and other parameters of cardiovascular regulation, which also vary in time. In fact, the influence is mutual, because of the feedback mechanisms in the system. As a result, the blood that circulates through the vessels and its pressure are regulated in a highly complex, dynamical way.

There are a number of non-invasive measurements, e.g., ECG, blood pressure, cutaneous blood flow, respiration, that can be used to characterize the state of the human cardiovascular system (CVS) (see figure 1). However, due to the extreme complexity of the system and strong intrinsic noise, it is a major challenge to find a parametric model capable of describing the numerous possible states of the system. The goal in this search for a model is to obtain further insight into the complex self-control mechanism responsible for CVS function and to improve the diagnoses of cardiovascular diseases (especially in their early stages).

One of the existing hypotheses is that the CVS can be modelled as a system of coupled oscillators (each corresponding to a certain physiological process) [14, 1, 15, 16]. Although the origin of this idea goes back to [4, 17, 8], the optimal choice of many model parameters (e.g., the type of oscillators, and the form and strength of coupling), and the attribution of physiological significance to them, still remain open problems.

In this paper we propose a new approach to model selection and to the estimation of the model parameters. It includes: (i) the simultaneous noninvasive measurement of a number of signals related to CVS dynamics; (ii) decomposition of these signals into their principal oscillatory components by empirical mode decomposition (EMD); and (iii) inference of the parameters for each of the oscillatory modes. We consider a new algorithm recently suggested by two of us, VNS and DAT, illustrate its utility in analysing a simple 1-dimensional model, the Van der Pol oscillator, and discuss the way forward.

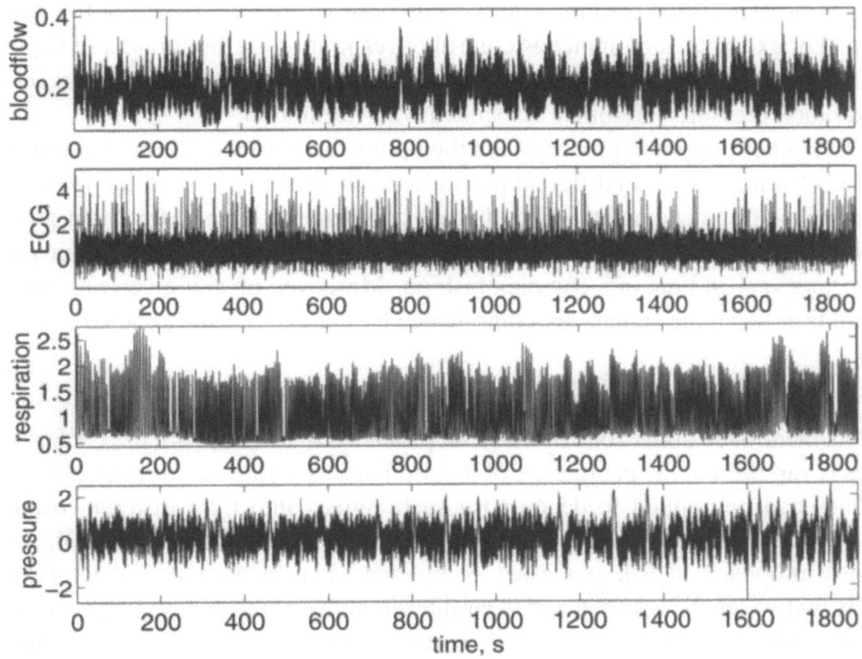

Figure 1. An example of simultaneously measured CV signals.

2. The Problem

The occurrence of oscillations in blood pressure, blood flow and heart rate has been known for many years. It was the realisation that certain of the lower frequency oscillations may be indicative of CVS control mechanisms that has stimulated particular clinical interest, because it may be possible to use measures of CVS oscillations as diagnostic tools. Analysis of the variability in heart rate, blood flow and pressure has proven to be useful in understanding cardiovascular regulation in a range of conditions including heart failure, diabetes, and hypertension [9]. It is apparent, however, that no new frequency-resolved tests have been accepted as providing diagnostic information superior to that which can be obtained by more conventional tests based on e.g., global indices of variability (see for instance [2]). One reason is that the mechanisms responsible for regulation at low frequencies are multiple, and not well established, which makes the interpretation of frequency resolved indices a difficult task. Another reason is that numerous external factors can influence the strength of these oscillations.

The difficulties related to the overlapping of the multiple control mechanisms and external factors can to some extent be overcome by simultaneous measurements of cardiovascular oscillations at different points of the system. For example, simultaneous control of respiration for both rate and depth is likely to improve the resolution of the oscillatory component in heart rate and blood pressure variability associated with respiration.

Two other important problems are the optimization of the decomposition of the CVS variability into different physiologically meaningful oscillatory components, and the characterization of these components. One of the main difficulties here is related to the fact that the oscillations are not steady; their parameters vary in time in a complex fashion. Moreover, on any given measurement time scale the signals are complicated by the presence of trends.

3. Mode Decomposition

As mentioned above, one of the main problems in dealing with the CVS signals is variation of the parameters of the oscillations in time. A number of techniques can be applied to decompose such signals into their different oscillatory components [6, 11, 5]. Empirical mode decomposition (EMD), recently proposed in [6], possesses completeness, orthogonality (at least locally, with a small leakage), and clear interpretation of the decomposed modes of the original signal $x(t)$ as a generalized Fourier expansion with varying amplitude and instantaneous frequency. We further modify this method to deal with highly noisy physiological data by imposing frequency boundaries in between which we want to obtain only one oscillatory mode. This allows one to hope for a meaningful inference of the oscillation parameters of each mode function, yielding the optimal frequency-resolved variability indexes.

The optimal decomposition of CVS variability into oscillatory components can also provide further insight into the physiological meaning of these indexes and to optimize parameters of the mathematical model of the CVS oscillations.

An example of the modified EMD applied to an HRV signal is shown in figure 2. The original HRV signal is decomposed into four oscillatory modes. Note that these modes are not the intrinsic mode functions as in the original method [6]. Additional frequency boundaries were imposed on the decomposition technique in order to obtain only one oscillatory mode corresponding to each model oscillator discussed in the following section.

4. Basic Ideas of the Modelling

An important component of our method is a mathematical model of the CVS oscillations. Such a model is not only a requirement for the optimal inference of the oscillation parameters (e.g., optimization of the frequency resolved variability indices), but is also a convenient working tool to de-

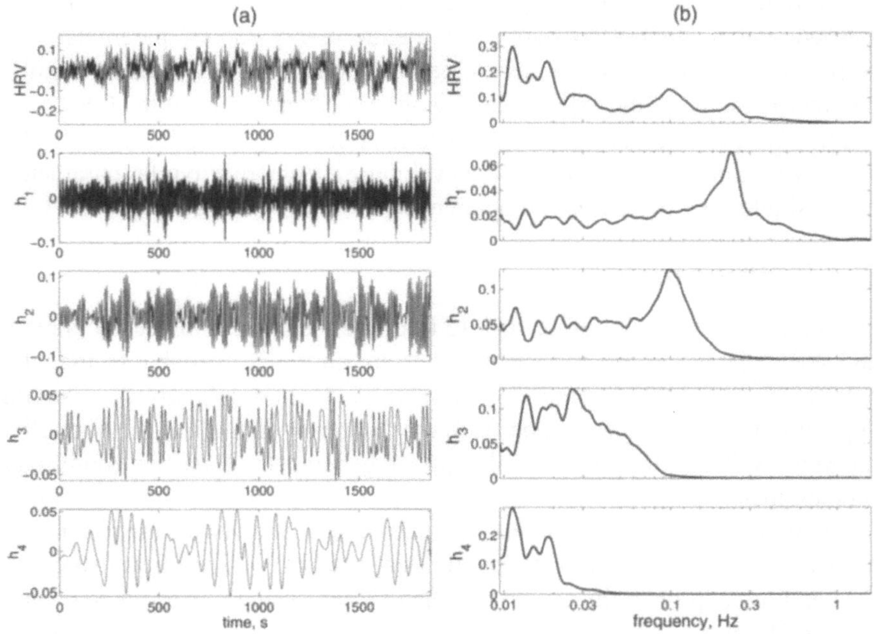

Figure 2. An example of EMD of an HRV signal. (a) Time series of the HRV signal (on the top) and four principal modes given by EMD and (b) their spectra are shown.

velop our intuition about the function and mutual interaction of the various physiological control mechanisms in CVS dynamics.

The basic assumptions of our model are well established in the literature. They include the idea that heart rate, blood flow, and blood pressure variability are composed of oscillations at a number of different frequencies and that, if such oscillations arise from the operation of control mechanisms, then their parameters may be indicative of the state of control and function of the CVS. Another assumption is that the main oscillatory components are well resolved in frequency. Evidence for the separation of the frequencies in the lower frequency range includes both direct observation [1] and measurements of the frequency response of different components of the CVS.

Most models of CVS oscillations focus on individual components of the feedback pathway responsible for the onset of oscillations in a given frequency range (see e.g., [13, 12]). In contrast, our model relies on the fact that experimentally observable oscillations in CVS variability seem to be self-sustained. It focuses on an attempt to introduce physiologically meaningful parameters to characterise these self-sustained oscillations, and to reveal possible mechanisms of interaction between the different oscillatory components. In this respect our approach is complementary to the others. Its key ingredient is an attempt to derive all the necessary model parameters directly from the observed time series, i.e. our model can be viewed as a multi-parametric model of the experimentally observed time series of the CVS.

Details of the model itself were described earlier [14, 1, 15, 16]. In this paper we are mainly concerned with developing a technique to infer parameters of the singled-out oscillatory components.

5. Inference

As a next step we therefore seek to infer the parameters of the oscillations and noise. However, no general methods are currently available to infer the parameters of a nonlinear model with internal stochastic dynamics. Existing applications of the Bayesian approach rely on heavy numerical methods (like the Markov chain Monte Carlo technique) to generate the Bayesian conditional probabilities [7].

We now introduce a new inference technique, markedly different from the earlier approach of [10]. It was recently suggested by Vadim Smelyanskiy that the conditional probability can be obtained by expressing it in terms of a white noise path integral. Taking advantage of this idea, a concise and fast iterative Bayesian inference scheme can be derived.

For the sake of simplicity we consider a one-dimensional dynamical system subject to additive noise of the form

$$\dot{x} = K(x|c) + \xi(t), \qquad (1)$$
$$\langle \xi(t) \rangle = 0, \quad \langle \xi(t)\xi(0) \rangle = D\delta(t),$$

where $K(x|c)$ is the deterministic drift force, $\xi(t)$ is the zero-mean white Gaussian noise and D is the noise intensity. We assume that the unknown deterministic force can be represented in the form

$$K(x|c) = \sum_{k=1}^{M} c_k f_k(x), \qquad (2)$$

where $f_k(x)$ are known base functions, and $c = \{c_k\}$ is a set of unknown coefficients that have to be inferred from experimental data.

Given a set of observations one can improve an estimate of the model parameters by computing a *posterior* probability density function (PDF) $p(c|x)$ taking advantage of Bayes' theorem

$$p(c|x) = \frac{L(x|c)p(c)}{\int L(x|c)p(c)dc}. \qquad (3)$$

Prior information about the model parameters is contained in a *prior* PDF $p(c)$, which is chosen to take the form of a Gaussian distribution.

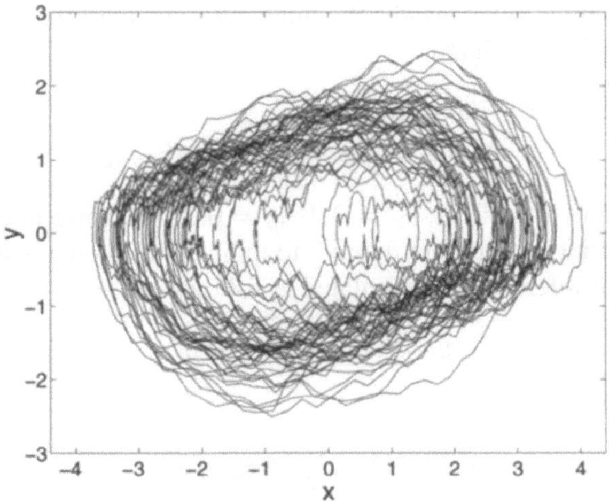

Figure 3. Phase portrait of the noisy Van der Pol oscillator whose parameters are inferred in figure 4

Writing the likelihood $L(x|c)$ as a conditional PDF (see e.g., [10, 3]) the algorithm can be written as

$$A_{p+1} = \frac{B}{D} + A_p,$$
$$c_{p+1} = -A_{p+1}^{-1}\left(\gamma - \frac{\alpha}{D} - c_p^T A\right)^T,$$

where index p counts the iterations, A is the inverse dispersion matrix in the distribution of the model coefficients, and D is the noise intensity, which is given by

$$D = \frac{2}{N}\sum_i \left(R - \alpha c_p + c_p^T \frac{B}{2} c\right).$$

Other parameters are calculated from a given realization using the following equations

$$\alpha_k = \int_{x_0}^{x_N} dx\, f_k(x), \quad x_i = x(t_i)$$
$$B_{k\,k'} = h \sum_i f_k(x_i)\, f_{k'}(x_i),$$
$$\gamma_k = \frac{h}{2}\sum_i f_k'(x_i), \quad R = \frac{1}{2h}\sum_i (x_{i+1} - x_i)^2,$$

As an example, we consider the application of the inference scheme to a noisy Van der Pol oscillator

$$\dot{x} = y,$$
$$\dot{y} = -a_1 x + a_2 y - a_3 x^2 y + \xi(t),$$

with parameters $a_1 = 0.25$, $a_2 = 0.2$, $a_3 = 0.1$, and noise intensity $D = 0.1$. Its phase portrait is plotted in figure 3 to indicate just how noisy the system is. The model for inference is formulated as follows

$$\dot{y} = K(x, y|c) + \xi(t), \tag{4}$$

where

$$K(x, y|c) = \sum_k f_k(x, y) c_k,$$

$c = \{c_1\ \ c_2\ \ c_3\}$ is a vector of unknown parameters, and $f = \{x\ \ y\ \ x^2 y\}$ is a vector of the model base functions. The prior information about the model parameters is chosen $c_0 = \{0,\ 0,\ 0\}$, $D = 0$, $A_0 = 10^{-6} I$. The time step $h = 0.1$ and number of data blocks used for inference is 60.

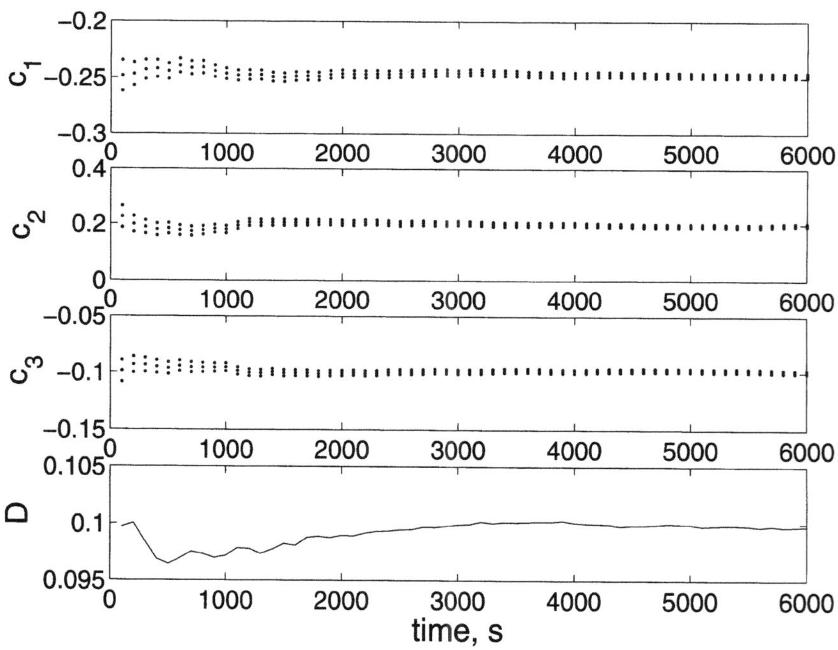

Figure 4. Inference of the parameters of the model (4). The initial point in each plot falls at zero, and is omitted.

The result of inference is shown in figure 4. All the parameters and noise intensity quickly converge to the right values. Note that our new inference technique does not just filter out the dynamical information from noisy time series. Rather, it *uses the internal noise in the system to infer dynamical information.* That is why the relatively large internal noise in the system (Figure 3) is actually an advantage for inference.

This same technique can be applied to the N-dimensional case to provide us with parameters for the oscillatory components of CVS variability in different frequency ranges. Our preliminary application to the measured CVS oscillations confirms the appropriateness of developing a stochastic nonlinear model. In particular it shows that currently available models cannot encompass the whole complexity of the experimentally observed oscillations.

Further research will include the development of the inference scheme for the case of coloured multiplicative noise and the search for more sophisticated models of the complex CVS oscillations. In this way, we believe, it will be possible to establish to what extent oscillations in the blood flow can be described in terms of nonlinear stochastic models, and to estimate physiologically significant parameters characterising a given state of the CVS from noninvasively measured data. Thus the research may pave the way to development of novel diagnostic techniques.

6. Conclusion

In conclusion, a novel approach is being developed for estimation of the frequency-resolved variability indices for cardiovascular signals. The technique has been tested on a simple model of noisy nonlinear oscillator, but it can readily be extended to tackle the twin problems inherent in all physiological investigations: time variation of the characteristic frequencies; and the inevitably limited period of observation. It will enable the inference of parameters for coupled stochastic oscillators subject to multiplicative colored noise, as will be needed for realistic cardiovascular models. The techniques of EMD and parameter identification, discussed in this paper, can be combined with analyses of the directionality and strength of inter-oscillator couplings in different frequency ranges based on bispectral analysis and on the analysis of the mutual information and predictability. The preliminary results are encouraging and will be reported in detail elsewhere.

Acknowledgments

The work was supported by EPSRC (UK), by INTAS, by the Slovenian Ministry of Education, Science and Sport, by the NASA Intelligent Systems/Intelligent Data Understanding Program (USA), by RFFI (Russia) and by the Joy Welch Trust (UK).

References

[1] M. Bračič, P. V. E. McClintock, and A. Stefanovska, "Characteristic frequencies of the human blood distribution system," in D. S. Broomhead, E. A. Luchinskaya, P. V. E. McClintock, and T. Mullin, editors, *Stochastic and Chaotic Dynamics in the Lakes*, pp. 146–153, Melville, New York. American Institute Physics, 2000.

[2] A. J. Camm, M. Malik, J. T. Bigger, and et al., "Heart rate variability – Standards of measurement, physiological interpretation, and clinical use," *Circulation*, vol. 93, pp. 1043–1065, 1996.

[3] R. Graham, "Path integral formulation of general diffusion processes," *Z. Phys. B*, vol. 26, pp. 281–290, 1977.

[4] H. Haken, "Cooperative phenomena in systems far from thermal equilibrium and in nonphysical systems," *Review of Modern Physics*, vol. 47, pp. 67–119, 1975.

[5] M. Hožič and A. Stefanovska, "Karhunen – Loève decomposition of peripheral blood flow," *Physica A*, vol. 281, pp. 587–601, 2000.

[6] N. E. Huang, Z. Shen, S. R. Long, M. C. Wu, H. H. Shih, Q. Zheng, N. Yen, C. C. Tung, and H. H. Liu, "The empirical mode decomposition and the Hilbert spectrum for nonlinear and non-stationary time series analysis," *Proc. R. Soc. Lond. A*, vol. 454, pp. 903–995, 1998.

[7] D. T. Kaplan and C. L. Bremer, "Markov chain Monte Carlo estimation of nonlinear dynamics from time series," *Physica D*, vol. 160, no. 1-2, pp. 116–126, 2001.

[8] Y. Kuramoto, *Chemical Oscillations, Waves, and Turbulence*, Springer-Verlag, Berlin, 1984.

[9] S. C. Malpas, "Neural influences on cardiovascular variability: possibilities and pitfalls," *Am. J. Physiol.: Heart. Circ. Physiol.*, vol. 282, pp. H6–H20, 2002.

[10] R. Meyer and N. Christensen, "Bayesian reconstruction of chaotic dynamical systems," *Phys. Rev. E*, vol. 62, pp. 3535–3542, 2000.

[11] M. Paluš and D. Novotna, "Detecting modes with nontrivial dynamics embedded in colored noise: enhanced Monte Carlo SSA and the case of climate oscillations," *Phys. Lett. A*, vol. 248, pp. 191–202, 1998.

[12] J. V. Ringwood and S. C. Malpas, "Slow oscillation in blood pressure via a nonlinear feedback model," *Am. J. Physiol.: Heart. Circ. Physiol.*, vol. 280, pp. R1105–R1115, 2001.

[13] H. Seidel and H. Herzel, "Bifurcations in a nonlinear model of the baroreceptor-cardiac reflex," *Physica D*, vol. 115, pp. 145–160, 1998.

[14] A. Stefanovska and M. Bračič, "Physics of the human cardiovascular system," *Contemporary Physics*, vol. 40, pp. 31–55, 1999.

[15] A. Stefanovska, M. Bračič, S. Strle, and H. Haken, "The cardiovascular system as coupled oscillators?" *Physiol. Meas.*, vol. 22, pp. 535–550, 2001.

[16] A. Stefanovska, D. G. Luchinsky, and P. V. E. McClintock, "Modelling couplings among the oscillators of the cardiovascular system," *Physiol. Meas.*, vol. 22, pp. 551–564, 2001.

[17] A. T. Winfree, *The Geometry of Biological Time*, Springer-Verlag, New York, 1980.

Index

MIX
Papier aus verantwortungsvollen Quellen
Paper from responsible sources
FSC® C105338

If you have any concerns about our products,
you can contact us on
ProductSafety@springernature.com

In case Publisher is established outside the EU,
the EU authorized representative is:
Springer Nature Customer Service Center GmbH
Europaplatz 3, 69115 Heidelberg, Germany

Printed by Libri Plureos GmbH
in Hamburg, Germany